Cyanobacteria in Symbiosis

Cyanobacteria in Symbiosis

Edited by

Amar N. Rai

North-Eastern Hill University,
Shillong, India

Birgitta Bergman

Stockholm University,
Stockholm, Sweden

and

Ulla Rasmussen

Stockholm University,
Stockholm, Sweden

SPRINGER-SCIENCE+BUSINESS MEDIA, B.V.

A C.I.P. Catalogue record for this book is available from the Library of Congress.

ISBN 978-90-481-6085-3 ISBN 978-0-306-48005-8 (eBook)
DOI 10.1007/978-0-306-48005-8

Cover page shows Nostoc *sp. isolate from stolons of* Gunnera magellanica. Nostoc *is the most common microsymbiont in cyanobacterial-plant symbioses. The filaments are composed of photosynthetic vegetative cells and nitrogen-fixing heterocysts (more transparent). Photo: Chun-Mei Wang.*

Printed on acid-free paper

Originally published by Kluwer Academic Publishers in 2002
Softcover reprint of the hardcover 1st edition 2002

TABLE OF CONTENTS

INTRODUCTION

More than 12 years ago, one of us (ANR) edited a volume on symbiotic cyanobacteria. Being the first such attempt there were several shortcomings but the book was widely appreciated and several readers gave their valuable suggestions. A great deal has happened since then. Cyanobacterial symbioses are no longer regarded as mere oddities but as important component of the biosphere, occurring both in terrestrial and aquatic habitats worldwide. It is becoming apparent that they can enter into symbiosis with a wider variety of organisms than hitherto known and there are many more such symbioses yet to be discovered, particularly in the marine environments. Large blooms of cyanobacterial-diatom symbioses in the marine environment are now regarded as major components of the global biological N_2-fixation. The *Nostoc-Geosiphon* symbiosis has opened up a new line of research. Unlike the mycobiont of lichens, *Geosiphon* is a close relative of mycorhizal fungi belonging to Glomales. The research on cyanobacterial symbioses in general has moved to unravelling molecular aspects of the symbiosis, particularly the sensing-signalling pathways. Another area that has received greater attention in the recent past is the creation of artificial N_2-fixing symbioses with crop plants. Indeed the wide host range, extant oxygen protection mechanism, and ability to colonise a variety of plant tissues and organs make cyanobacteria a very promising candidate for such artificial symbioses.

In year 2000, we had the ESF workshop on cyanobacterial symbioses in Ireland. During our discussions, we felt that it is time to bring out an up to date volume on cyanobacterial symbioses. We contacted the Kluwer Academic Publishers who readily agreed to publish such a book. The outcome is the book presented to you. It contains 16 chapters covering cyanobacterial symbioses with plants (diatoms, bryophytes, *Azolla*, cycads, *Gunnera*), cyanobacterial symbioses in marine environments, lichens, *Nostoc-Geosiphon* symbiosis and artificial associations of cyanobacteria with economically important plants. Each chapter has been written by renowned expert(s) actively involved with research on cyanobacterial symbioses. They have dealt with ecological, physiological, biochemical, molecular and applied aspects of the symbiosis. This volume on 'Cyanobacteria in Symbiosis' should complement the two earlier volumes on cyanobacteria published by Kluwer (Molecular Biology of Cyanobacteria by D.A. Bryant, and Ecology of Cyanobacteria by B.A. Whitton and M. Potts). The three volumes together should provide the most comprehensive treatment of cyanobacterial literature as a whole. The book will serve as a valuable reference work and text for teaching and research in the field of plant-microbe interactions and nitrogen fixation. We shall welcome your suggestions for improvements in future editions.

We would like to express our thanks to all the contributors for writing their chapters on time, Pernilla Lundgren for editorial assistance, and our family and friends (Anders Bergman, Urmila Rai) who have supported us in this venture directly or indirectly. We would also like to thank Claire van Heukelom from Kluwer Academic Publishers for promptly attending to our queries.

Amar N. Rai
Birgitta Bergman
Ulla Rasmussen

Plate 1. Examples of cyanobacterial symbioses. a, Rhizosolenia-Richelia *(arrow), bar 30µm; b,* Geosiphon, *bar 1mm; c,* Peltigera aphthosa, *bar 2.5cm; d,* Anthoceros *with* Nostoc *colonies (arrow), bar 3mm; e,* Azolla, *bar 1mm; f, Coralloid roots (arrow) in* Bowenia, *bar 2.5cm; g,* Nostoc *layer (arrow) in the coralloid root, bar 1mm; h,* Gunnera magellanica, *bar 1.3cm; i, section of* Gunnera *stem (bracketed in 'h') showing* Nostoc *in glands (arrow), bar 4mm.Fig. 'b' from A. Schuessler, Figs. a and c-i after Rai* et al., *2000.*

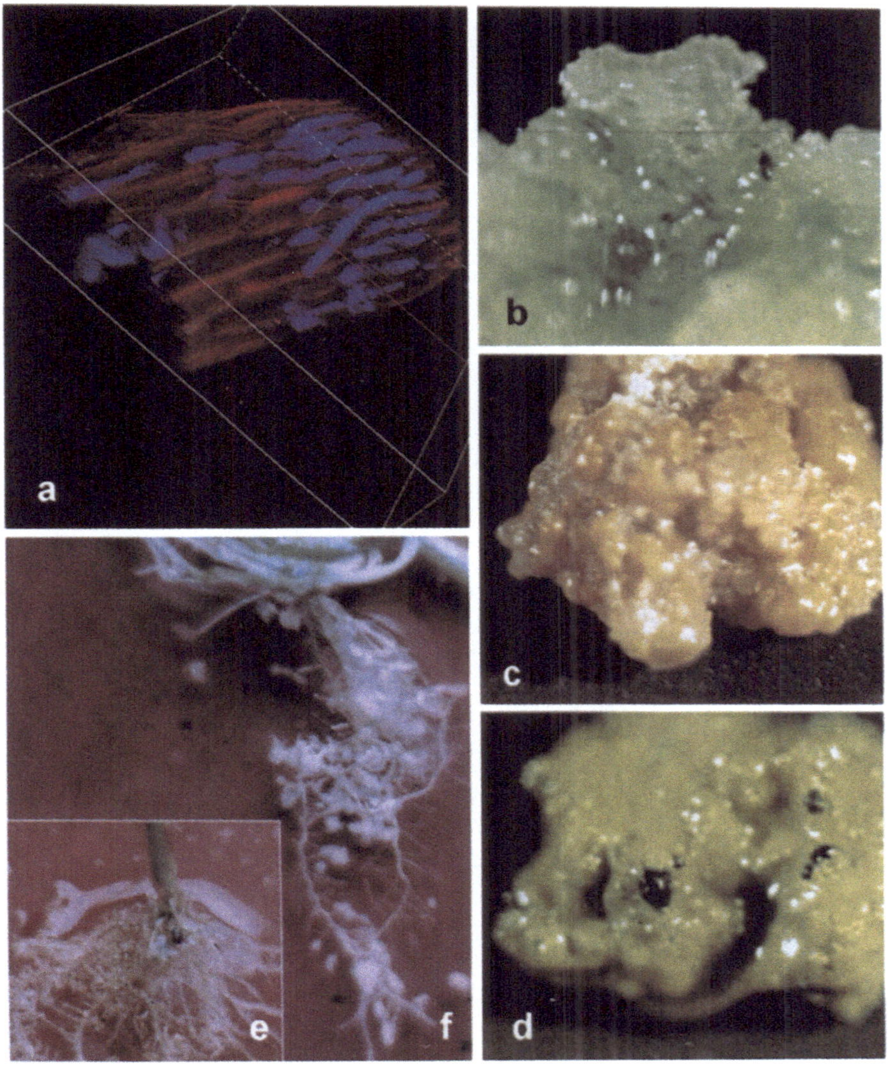

Plate 2. a, Cyanobacteria (blue) located intracellularly and epiphytically on a leaf (red) of Sanionia uncinata *(laser confocal microscopy); b, Emerald green mixed cultures of tobacco and AMC from* Azolla *after 15 days of co-culturing; c, Intense growth of satellite bacteria from* Azolla *AMC at the surface of the rice callus after 1.5 months of co-culture; d, Same as c after 2 months, note the appearance of mucoid cyanobacterial colonies; e and f, nodules formed on rice roots after co-culture of rice seedlings with AMC from* Azolla *and* Encephalartos ferox, *respectively. Fig. a from B. Solheim and M. Zielke, Figs. b-f from M.V. Gusev and co-workers.*

Chapter 1

CYANOBACTERIA IN SYMBIOSIS WITH DIATOMS

SVEN JANSON
Department of Biology and Environmental Science
Barlastgatan 1, S-391 82 Kalmar, Sweden

1. INTRODUCTION

The diatoms are unicellular algae with an armour consisting of a silica shell. They are important primary producers in aquatic systems and some very common species contain intracellular cyanobacteria as symbionts. Some diatoms serve as substratum for epiphytes. Two new symbioses involving unicellular cyanobacteria and diatoms have been described recently. One is *Climacodium frauenfeldianum* having intracellular cyanobionts, whose 16S rDNA sequence is closely related to the nitrogen-fixing *Cyanothece* sp. ATCC 51142. The other is the tripartite symbiosis involving a coccoid cyanobiont (residing extracellularly), a protist and the diatom *Leptocylindrus mediterraneus*. Genetic characterisation of heterocystous cyanobacteria living in association with several diatom species gave surprising information concerning the intra- and extracellular cyanobacterial filaments. The intracellular *Richelia intracellularis* was closely related to the extracellular *Calothrix rhizosolenia*, based on *hetR* sequences. Thus the genus *Richelia* must be revised to include also the extracellular *Calothrix rhizosolenia*. The intracellular *R. intracellularis* is most likely inherited vertically from mother cell to daughter cell because they display a high level of host specificity and re-infections seem rare.

2. WHAT ARE DIATOMS?

The diatoms belong to a group of unicellular phytoplankton chraracterised by a cell wall made of silica. When decomposing, the largely intact diatom cell wall sediment and these flakes of silica are the main component in white sandy beaches. The diatom cell contains, apart from typical eukaryotic components, two types of plastids: chloroplast and leucoplast. The chloroplast is a photosynthesising organelle and diatoms are therefore considered as one of the major contributers to the primary production in the world's aquatic systems. The micro-algae, represented by diatoms, are more common in associations with cyanobacteria than macro-algae. Apart from diatoms only a few cyanobacteria have been reported in association with algae: epiphytic *Dichothrix* on

A.N. Rai, B. Bergman and U. Rasmussen (eds.), Cyanobacteria in Symbiosis, 1-10.

Sargassum (Carpenter, 1972); *Synechococcus* sp. SF-1, isolated from a pelagic brown alga (Spiller and Shanmugan, 1987). Other loose associations most likely exist but the records of these are rather scant. In recent years, studies have been performed on microalgae in association with cyanobacteria, particularly the marine cyanobacterial–diatom symbioses. This chapter will describe the latest advances in knowledge of the cyanobacterial–diatom symbioses with special focus on the marine environment.

3. COCCOID CYANOBACTERIA IN ASSOCIATION WITH DIATOMS

Until recently there was only one well described symbiosis between unicellular cyanobacteria and diatoms; the one involving *Rhopalodia*/*Epithemia* and a coccoid cyanobacterium. However, a new cyanobacteria–diatom symbiosis has been described lately (Carpenter and Janson, 2000). The diatom *Climacodium frauenfeldianum* harbours unicellular cyanobacteria inside their cells. The host is common in open oceans in the tropics but even in modern guidebooks the intracellular cyanobacteria are not mentioned probably because they have escaped identification (Hasle and Syvertsen, 1996). With the aid of epifluorescence microscopy, Carpenter and Janson (2000) discovered the cyanobacterial nature of the intracellular inclusions.

The cells were collected on large membrane filters and the location of the cells were disclosed by epifluorescence microscopy and subsequently marked with a circle. Using a dissecting microscope the cells were located, aided by the circle, and pieces of transparent filter membranes with diatom cells on it were cut out. The pieces of filters were put directly in a PCR tube and the 16S rRNA gene from the cyanobiont was amplified. This approach to isolate autofluorescent phytoplankon is very promising and it has also been used to study the genetics of the endosymbiont *Richelia* (see section 4). The 16S rDNA sequence of the cyanobiont is related to the 16S rDNA sequence from *Cyanothece* sp. ATCC 51142 (Fig. 1). *Cyanothece* sp. ATCC 51142 is an isolate from the coast of Texas, and although it can grow in high salinity medium (Reddy *et al.*, 1993), its presence in the open oceans has not been reported. The ultrastructure of the nitrogen-fixing *Cyanothece* sp. ATCC 51142 revealed that it has unusually large starch granules (Chou *et al.*, 1994). Similar granules have been observed in the cyanobiont in *C. frauenfeldianum* (S. Janson, unpublished observations), and in some of the cyanobionts in the dinoflagellate *Histioneis* sp. (Lucas, 1991). This indicates that the cyanobiont in *C. frauenfeldianum* might have colonised other hosts as well, but this needs to be confirmed with genetic identifications. It is also possible that the cyanobiont, like *Cyanothece* sp. ATCC 51142, fixes nitrogen and that the host benefits from this in the nitrogen-poor environment of the open ocean.

As mentioned above, the first diatom–cyanobacterial association involving unicellular cyanobacteria to be described was the association betweeen cyanobacteria and members of Epithemiaceae: *Epithemia turgida* and *Rhopalodia gibba* (Drum and Pankratz, 1965). These diatoms contain intracellular coccoid cyanobacteria (Rai, 1990; DeYoe *et al.*, 1992). Two to five cyanobionts occur in the cytoplasm of each host cell. The cyanobiont is distinct from similar cytoplasmic inclusions called cyanelles in members of Glaucocystophyceae. The ultrastructure of the cyanobiont revealed that it has a thicker cell wall and the thylakoid membranes are oriented differently compared to cyanelles (Floener and Bothe, 1980). While the genetics of cyanelles has been studied in detail (Löffelhardt and Bohnert, 1994), the genetics of the cyanobionts in *Rhopalodia/Epithemia* are completely unexplored. Based on ultrastructural observations it is not likely that these symbionts are closely related to the ones in *C. frauenfeldianum,* which have large starch-granules in their cells (S. Janson, unpublished observations). The genotypes of the cyanobionts in *Rhopalodia/Epithemia* need to be investigated to confirm that they are not closely related the cyanobionts in *C. frauenfeldianum*. It has been assumed that the cyanobacteria fix nitrogen (Drum and Pankratz, 1965; Floener and Bothe, 1980), although the presence of nitrogenase gene sequences have not been shown and detection of the nitrogenase enzyme by immunolocalisation has not been demonstrated. The ability to grow without combined nitrogen and the acetylene reduction activity strongly indicates diazotrophy (Floener and Bothe, 1980). In addition, it has been shown that the number of cyanobacterial cells increases when the N/P ratio is lowered in culture experiments (DeYoe *et al.*, 1992). However, confirmations of these observations by genetic and immunological detection techniques are needed

Additional associations between coccoid cyanobacteria and diatoms have been reported occasionally. These associations were intracellular cyanobionts seen in *Streptotheca indica* and *Neostreptotheca subindica* (see Villareal, 1992). Villareal (1992) suggested that the cyanobionts were similar to the ones seen in *Epithemia/Rhopalodia.* It seems more probable that the *Streptotheca/Neostreptotheca* observed with cyanobionts are closely related to *C. frauenfeldianum* as they have similar morphology and therefore carry a similar cyanobiont. The cyanobiont of *C. frauenfeldianum* does not seem to be closely related to the cyanobiont in *Epithemia/Rhopalodia* (see above).

The only extracellular cyanobacterial–diatom association reported is the newly described tri-partite association between cyanobacteria, a protist and a diatom (Buck and Bentham, 1998). This fascinating symbiosis mainly occurs in the North Atlantic and comprises of a chain-forming diatom, *Leptocylindrus mediterraneus*, which is colonised by an aplastidic protist, *Solenicola setigera*. In the extracellular matrix surrounding *S. setigera,* cells of *Synechococcus* sp. are embedded. The maximum biomass of this consortium was estimated to be 31 **µg C** l^{-1} in the North Atlantic (Buck and Bentham, 1998). The diatom in this symbiosis appears to have mitochondria but no plastids and a very small part of the cell is occupied by the cytoplasm. In this case also, as with *Rhopalodia/Epitemia* cyanobionts, the genetic relationship with other cyanobacteria and their nitrogen-fixing potential need to be investigated.

4. HETEROCYSTOUS CYANOBACTERIA IN ASSOCIATIONS WITH DIATOMS

The heterocyst is a specialised cell devoted to nitrogen fixation and cyanobacteria are the only bacteria producing such a cell type. There are two species of heterocystous cyanobacteria reported in association with diatoms, *Calothrix rhizosolenia* and *Richelia intracellularis* Schmidt (Ostenfeld and Schmidt, 1901), occurring mainly in tropical and subtropical marine pelagic waters. The symbiosis between *R. intracellularis* and diatoms of the genera *Rhizosolenia, Hemiaulus, Chaetoceros* has been reported (Ostenfeld and Schmidt, 1901; Lemmermann, 1905; Karsten, 1907). The distributions are patchy and large blooms occur in some tropical regions, while other areas have lower abundance. This association involves two different plankton groups: cyanobacteria and diatoms, with different nutrient requirements, and it is likely that special nutrient conditions are needed for their growth. Furthermore, these conditions may only exist in certain regions at a certain time, explaining the uneven distribution. Other factors like, temperature, turbulence and light conditions might be very important as well. For example, host cells without cyanobionts were distributed further north in the North Pacific (north of 38°09'N) than host cells with cyanobionts (Venrick, 1974). The northern most observation of the *Rhizosolenia–Richelia* symbiosis is in the Baltic Sea (Pankow, 1990). It is unlikely that *Rhizosolenia–Richelia* thrives in the Baltic Sea. It may be accidentally washed in from Atlantic Ocean waters because there are no other reported observations in this area despite frequent samplings.

The intracellular cyanobiont is usually referred to as *R. intracellularis* while the epiphytic one is referred to as *C. rhizosolenia.* However, the taxonomy of the epiphytes is confusing and it has been suggested that the localisation of the cyanobiont is not important for the identification (Sundström, 1984). Sundström (1984) also acknowledged the fact that the taxonomy of the cyanobiont has not been fully reviewed. A useful genetic locus for studying species phylogenies of filamentous cyanobacteria is the *hetR* gene. The *hetR* gene is involved in heterocyst development (Buikema and Haselkorn, 1991), but probably the exact function is not restricted to this process alone because the *hetR* gene is also present in several non-heterocystous species (Janson *et al.*, 1998). The genetic characterisation, based on *hetR,* of both endo- and epiphytes of *Rhizosolenia, Hemiaulus* and *Chaetoceros* suggests that each diatom species carry its own cyanobiont species (Janson *et al.*, 1999a). The sequence of partial *hetR* genes were determined from *R. intracellularis* filaments inhabiting diatoms of the genera *Rhizosolenia* and *Hemiaulus,* as from those being attached to the outside of the diatom *Chaetoceros* sp. (corresponding to the morphospecies *C. rhizosolenia*). The epiphytes of *Chaetoceros* and endophytes of *Rhizosolenia* were clustered together and the endophytes of *Hemiaulus* belonged to a second cluster (Fig. 2). In other words, the *hetR* sequences of the epiphytes were within the same cluster, but not closely related to any of the sequences from *R. intracellularis* in different diatom hosts. This was an unexpected result because they are classified into different genera, *Calothrix* and *Richelia*, respectively. This means that the *hetR* tree is paraphyletic in respect to not only *R. intracellularis* but the genus *Richelia* as a whole. The revision of genus and species was not discussed due to the relatively low number of extant *hetR* sequences from the heterocystous cyanobacteria. However, it is clear that either *C. rhizosolenia* or the genus *Richelia* must be revised. The genus *Richelia* is closest to *Microchaete* in morphological perspectives (Ostenfeld and Schmidt, 1905) but no *hetR* sequences exist from this genus. The special habitat of *R. intracellularis* (and *C. rhizosolenia)* might be enough to retain the genus *Richelia* regardless of the outcome of any further genetic analysis. A similar situation exists for the genus *Trichodesmium,* whose 16S rDNA sequence is closely related to that of *Oscillatoria* sp. PCC 7515 (Wilmotte *et al.,* 1994). Despite this finding the authors argued that the genus *Trichodesmium* is well defined, whereas the genus *Oscillatoria* is "problematic". The situation with *Richelia* is now complicated further with the addition of a second species *R. siamensis* Hindák (Hindák,

2000). *Richelia siamensis* was isolated from a rice field in Thailand and originally designated to the genus *Anabaena*. Genetic analysis of this new species is needed to verify that it is monophyletic with the sequences derived from endosymbionts of marine diatoms. In any event, *C. rhizosolenia* shold be included within the genus *Richelia* to make it monophyletic.

The nitrogen fixation by *R. intracellularis* was first documented by Mague *et al.*, (1974). Laboratory cultures of intact symbiosis of *Rhizosolenia–Richelia* also showed nitrogenase activity (Villareal, 1990). The host could not survive in nitrogen-free medium without the cyanobiont, but could grow in medium with fixed nitrogen without the cyanobiont. The nitrogen fixation capability was confirmed by immunolablelling experiments, showing that the nitrogen-fixing enzyme nitrogenase was confined to heterocysts of *R. intracellularis* (Janson *et al.*, 1995). Due to its abundance, the nitrogen fixation associated with *R. intracellularis* aggregates is believed to be of major importance in nitrogen budgets of tropical oceans (Carpenter *et al.*, 1999).

Since both the cyanobiont and the host are potentially able to fix carbon, it is possible that the host depends on the cyanobiont for both nitrogen and carbon. The first evidence for this was provided by using microautoradiography and silver emulsion. These results indicated that most of the carbon fixation was taking place in the filaments of the cyanobiont (Weare *et al.*, 1974). The only ultrastructural study of the *Rhizosolenia–Richelia* symbiosis showed the presence of plastids in the host, located closely to the cyanobiont filament (Janson *et al.*, 1995). Thus, it might be quite impossible to distinguish a signal from the cyanobiont and a closely located plastid, or a rapid transfer of photosynthate in either direction. However, the cyanobiont has been found by immunolabelling to contain the Rubisco enzyme (Janson *et al.*, 1995), indicating that at least some of the carbon is being fixed by the cyanobiont. Immunolabelling of the glutamine synthetase indicated that the cyanobiont was poorly prepared to convert N_2-derived ammonia to glutamine. This means that N may leak out to the host in the form of ammonia. Indeed, *R. intracellularis* is capable of supporting both symbionts with fixed N when grown in N-depleted medium during laboratory conditions (Villareal, 1990).

5. EVOLUTIONARY PERSPECTIVES

It is interesting to note that the heterocyst-forming cyanobacterial epiphytes on diatoms have a morphology resembling *Calothrix*, i. e. a single heterocyst at the end of the filament, and that heterocystous epiphytes from other substrata in the aquatic environment also have this morphology. For example, the *Caltothrix*-type of epiphyte on the pelagic brown algae *Sargassum* (Carpenter, 1972), *C. epiphytica* growing on species of *Tolypothrix* (Cyanophyta), *C. parasitica* growing on *Nemalion* (Rhodophyta)(Geitler, 1932), and *C. contarenii* growing on seagrasses in mangroves (Lugomela, 2000). The relationship between these different epiphytes might be only morphological, or they could comprise a genetically closely related group. The genetic characterisation by determining the *hetR* sequences from these organisms should clarify their genetic relationships.

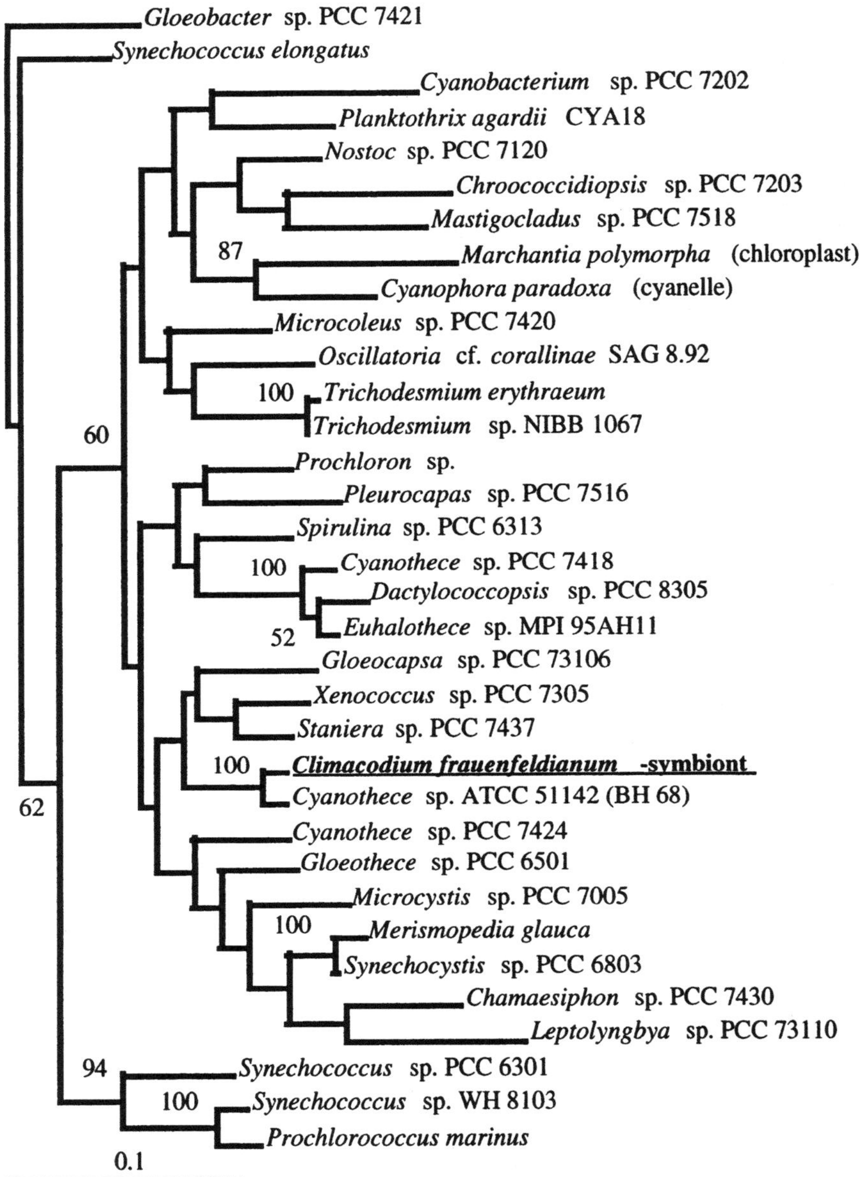

Figure 1. Maximum likelihood tree showing the relatedness of C. frauenfeldianum *cyanobiont 16S rDNA sequence to those of cyanobacteria and plastids. The numbers at the nodes are percentage of 500 bootstrap replicates, showing values over 50 only. The scale bar indicates the observed sequence divergence.*

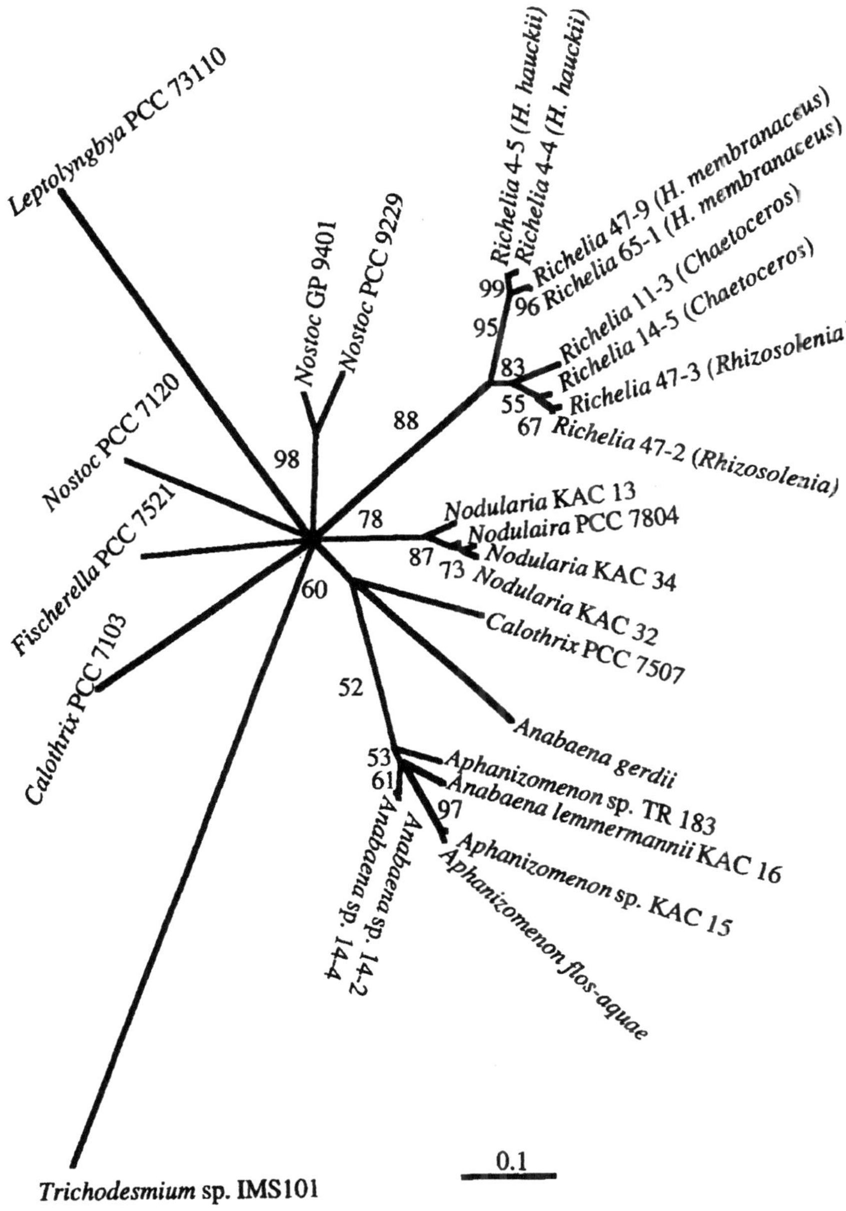

Figure 2. Phylogenetic tree inferred from hetR *sequence using quartet puzzling.* R. intracellularis *were from Atlantic and Pacific Oceans. TREE-PUZZLE 5.0 (Strimmer and Haeseler, 1996) software was used. The quartet puzzling was resampled 10,000 times; the percentage of each configuration occurring in all trees is shown at the nodes. Note the high specificity of the symbionts from several diatom species. Bar indicates the sequence divergence.*

The *hetR* gene sequences obtained from the cyanobionts of two species of *Hemiaulus* were divided over two lineages with very high statistical support (Janson *et al.,* 1999a; Fig. 1). Moreover, the sequence obtained from samples of *H. membranaceus* collected in the Caribbean Sea (Atlantic Ocean) was most closely related to the sequence obtained from *H. membranaceus* from the South Pacific Ocean. Thus the local population structure of *R. intracellularis* inside diatoms seems to be determined by which hosts are present. The genetic diversity of cyanobionts does not depend on the geographical location. The *hetR* sequences from cyanobionts of one host species never varied more than 1%. The sequence variation between cyanobionts from different host species was over 3.1%. Comparing this with *hetR* sequences from species of the non-heterocystous planktonic cyanobacterim *Trichodesmium*, where the variation within one species was 0.4% and the variation between species was over 3.3% (Janson *et al.,* 1999b). It therefore seems likely that each host species carries its own cyanobiont species. The epiphytes of *Chaetoceros* sp. were most closely related to sequences from cyanobionts of *Rh. clevei* var. *communis.* The data also indicated that the capability of infecting the diatom host has evolved twice in the cyanobiont or that the epiphytes have escaped their intracellular life and occupied a new niche by attaching on *Chatoceros* sp. cells. In order to answer which is the most likely scenario, we have to consider two things about the division cycle of the host and cyanobiont. (1) In the *Rhizosolenia–Richelia* symbiosis, the filaments of *R. intracellularis* are divided in the middle and filaments are carried over to the other end of the host cell by force of cytoplasmic streaming, before the host cell completes its division cycle (Taylor, 1982; Villareal, 1989). (2) In a laboratory culture of the *Rhizosolenia–Richelia* symbiosis, fast growing host cells that finished the formation of the cell wall septa before the cyanobiont had been transported to the daughter cells resulted in a diatom cell lacking cyanobionts. Such cells were unable to grow in nitrogen deficient deficient medium (Villareal, 1990). Thus, host cells may loose their cyanobiont and if so they are not likely to gain a new one as the re-infection is extremely unusual in laboratory culture and it has never been reported from nature. Moreover, when symbiotic host plants are continuously re-infected, e. g. in the *Gunnera–Nostoc* symbiosis (Rai *et al.,* 2000), the microbiont is usually "broad-specific" in their associations and co-evolution is only weak (Doyle, 1998). In contrast, the *Richelia*–diatom associations are highly specific and the infrequent observations of epiphytic and free-living *R. intracellularis* suggest that re-infection is a rare event. In conclusion, several facts points out that re-infection is a rare event and the common ancestor of all observed intracellular cyanobionts were living intracellularly in a common ancestor of *Hemiaulus* and *Rhizosolenia.* The heterocystous epiphytes are most likely derived from this common ancestor. Hence, the symbiosis has served as a vector for the spreading of heterocystous cyanobacteria to new environments.

The diatoms belong to a linage of algae that has probably acquired their plastids through a secondary endosymbiotic event, probably by engulfing a red algae-like organism (Medlin *et al.*, 1997). Note, however, that this was most likely a single event. This competence seem to have survived through evolution and led to the intracellular symbiosis between diatoms and several species of cyanobacteria. The dinoflagellates have a wider range of (secondary) plastid symbionts (Schnepf and Elbrächter, 1999;

Saldarriaga *et al.*, 2001), and they also have cyanobacterial symbionts of different types within the same host cell (Lucas, 1991). The process of defining extant plastids, loss of plastids, gaining new plastids and cyanobionts is an ongoing process and the exploration of this "tutti frutti" of plastids and cyanobionts will probably reveal much surprising and exciting news.

REFERENCES

Buck, K.R. and Bentham, W.N. (1998) A novel symbiosis between a cyanobacterium, *Synechococcus* sp., an aplastidic protist, *Solenicola setigera*, and a diatom, *Leptocylindrus mediterraneus*, in the open ocean, *Mar. Biol.* **132,** 349-355.

Buikema, W.J., and Haselkorn, R. (1991) Characterization of a gene controlling heterocyst differentiation in the cyanobacterium *Anabaena* 7120, *Genes Devel.* **5,** 321-330.

Carpenter, E.J. (1972) Nitrogen fixation by a blue-green epiphyte on pelagic *Sargassum.*, *Science* **178,** 1207-1208.

Carpenter, E.J., Montoya, J.P., Burns J., Mulholland M.R., Subramaniam, A., and Capone, D.G. (1999) Extensive bloom of a N_2-fixing diatom/cyanobaterial association in the tropical Atlantic Ocean, *Mar. Ecol. Prog. Ser.* **185,** 273-283.

Carpenter, E.J., and Janson, S. (2000) Intracellular cyanobacterial symbionts in the marine diatom *Climacodium frauenfeldianum.*, *J. Phycol.* **36,** 540-544.

Chou, W.-M., Chou, H.-M., Yuan, H.-F., Shaw, J.-F., and Huang, T.-C. (1994) The aerobic nitrogen-fixing *Synechococcus* RF-1 containing uncommon polyglucan granules and multiple forms of □-**Amylase,** *Curr. Microbiol.* **29,** 201-205.

DeYoe, H.R., Lowe, R.L., and Marks, J.C. (1992) Effects of nitrogen and phosphorous on the endosymbiont load of *Rhopalodia gibba* and *Epithemia turgida* (Bacillariophycaea), *J. Phycol.* **28,** 773-777.

Doyle, J.J. (1998) Phylogenetic perspectives on nodulation: evolving view of plants and symbiotic bacteria, *Trends Plant Sci* **3,** 473-478.

Drum, R.W., and Pankratz, S. (1965) Fine structure of an unusual cytoplasmic inclusion in the diatom genus *Rhopalodia*, *Protoplasma* **60,** 141-149.

Floener, L., and Bothe, H. (1980) Nitrogen fixation in *Rhopalodia gibba*, a diatom containing blue-greenish inclusions symbiotically, in: Schwemmler W. and H.E.A. Schenk (eds.), *Endo-cytobiology, Endosymbiosis and Cell Biology,* vol 1, Walter de Gruyter & Co., Berlin, pp. 541-552.

Geitler, L. (1932) Cyanophyceae, in R., Kolkwitz (ed.) *Rabenhorst's Kryptogamenflora von Deutschland, Ésterreich und der Schweiz,* vol. 14, Akademische Verlagsgesellschaft, Leipzig, pp. 597-599.

Hasle, G.R., and Syvertsen, E.E. (1996) Marine diatoms, in Tomas, C.R. (ed.), *Identifying marine diatoms and dinoflagellates,* Academic Press Inc., San Diego, pp 5-385.

Hindák, F. (2000) A contribution to the taxonomy of the nostocalean genus *Richelia* (Cyanophyta/Cyanobacteria), *Biologia* **55,** 1-6.

Janson, S., Rai, A.N., and Bergman, B. (1995) The intracellular cyanobiont *Richelia intracellularis*: Ultrastructure and immuno-localisation of phycoerythrin, nitrogenase, Rubisco and glutamine synthetase, *Mar. Biol.* **124,** 1-8.

Janson, S., Matveyev, A., and Bergman, B. (1998) The presence and expression of hetR in the non-heterocystous cyanobacterium *Symploca* PCC 8002, *FEMS Microbiol. Lett.* **168,** 173-179.

Janson, S., Wouters, J., Bergman, B., and Carpenter, E.J. (1999a) Host specificity in the *Richelia*-diatom symbiosis revealed by *hetR* gene sequence analysis, *Environ Microbiol* **1,** 431-438.

Janson, S., Bergman, B., Carpenter, E.J., Giovannoni, S.J., and Vergin, K. (1999b) Genetic analysis of natural populations of the marine diazotrophic cyanobacterium *Trichodesmium, FEMS Microbiol. Ecol.* **30,** 57-65.

Karsten, G. (1907) Das Indische Phytoplankton nach dem Material der Deutchen Tiefsee-Expedition 1898-1899. *Dtsch. Tiefsee-Exped 1898-1899,* **2,** 423-548.

Lemmermann, E. (1905) Die Algenflora der Sandwich-Inseln. Ergebnisse einer Reise nach dem Pacific, H. Schauinsland 1896/97, *Englerís Bot. Jb.* **34,** 607-663.

Löffelhardt, W., and Bohnert, H.J. (1994) Molecular biology of cyanelles, in Bryant D.A. (ed.) *The molecular biology of cyanobacteria.* Kluwer Academic Publishers, Dordrecht, pp. 65-89.

Lucas, I.A.N. (1991) Symbionts of the tropical dinophysiales (Dinophyceae), *Ophelia* **33,** 213-224.
Lugomela, C. (2000) The diversity of cyanobacteria in coastal areas of Zanzibar, Tanzania, and their role in carbon and nitrogen fixation, Licentiate thesis, Stockhom University, Sweden.
Mague, T.H., Weare, M.M., and Holm-Hansen, O. (1974) Nitrogen fixation in the north Pacific Ocean, *Mar. Biol.* **24,** 109-119.
Medlin, L.K., Kooistra, W.H.C.F., Potter, D., Saunders., G.W., and Anderssen, R.A. (1997) Phylogenetic relationship of the 'golden algae' (haptophytees, heterokont chromophytes) and their plastids, in Bhattacharya, D. (ed.),*Origins of algae and their plastids*., Springer, Berlin, pp. 187-219.
Ostenfeld, C.H., and Schmidt, J. (1901) Plankton fra det Röde hav og Adenbugten. *Vidensk. Meddel. Naturh. Forening i Kbhvn*., pp. 141-182.
Pankow, H. (1990) Ostsee-Algenflora, Gustav Fischer Verlag Jena, Leipzig, Germany, pp. 648.
Rai, A.N. (1990) Cyanobacteria in Symbioses, in Rai, A.N. (ed.), *CRC Handbook of Symbiotic Cyanobacteria*, CRC Press Inc., Boca Raton, Florida, pp. 1-7.
Rai, A.N., Söderbäck, E., and Bergman, B. (2000) Tansley Review No. 116, cyanobacterium-plant symbioses, *New Phytol* **147,** 449-481.
Reddy, K.J., Haskell, J.B., Sherman, D.M., and Sherman, L.A. (1993) Unicellular, aerobic nitrogen-fixing cyanobacteria of the genus *Cyanothece*, *J. Bacteriol.* **175,** 1284-1292.
Saldarriaga, J.F., Taylor, J.F.R., Keeling, P.J., and Cavalier-Smith, T. (2001) Dinoflagellate nuclear SSU rRNA phylogeny suggests multiple plastid losses and replacements, *J. Mol. Evol.* **53,** 204-213.
Schnepf, E., and Elbrächter, M. (1999) Dinophyte chloroplasts and phylogeny - A review, *Grana* **38,** 81-97.
Spiller, H., and Shanmugam, K.T. (1987) Physiological conditions for nitrogen fixation in a unicellular marine cyanobacterium, *Synechococcus* sp. strain SF1, *J. Bacteriol.* **169,** 5379-5384.
Strimmer, K., and von Haeseler, A. (1996) Quartet puzzling: A quartet maximum likelihood method for reconstructing tree topologies, *Mol. Biol. Evol.* **13,** 964-969.
Sundström, B.G. (1984) Observations on Rhizosolenia clevei Ostenfeld (Bacillariophyceae) and Richelia intracellularis Schmidt (Cyanophyceae), *Bot. Mar.* **27,** 345-355.
Taylor, F.J.R. (1982) Symbioses in marine microplankton, Ann. Inst. Ocanogr., Paris (Suppl) **58,** 61-90.
Venrick, E.L. (1974) The distribution and significance of *Richelia intracellularis* Schmidt in the North Pacific Central Gyre, *Limnol. Oceanogr.* **19,** 437-445.
Villareal, T.A. (1989) Division cycles in the nitrogen-fixing *Rhizosolenia* (Bacillariophyceae)-*Richelia* (Nostocaceae) Symbiosis, *Br. Phycol. J.* **24,** 357-365.
Villareal, T.A. (1990) Laboratory culture and preliminary characterization of the nitrogen-fixing *Rhizosolenia-Richelia* symbiosis, *Mar. Ecol.* **11,** 117-132.
Villareal, T.A. (1992) Marine nitrogen-fixing diatom-cyanobacteria symbioses, in Carpenter E.J., Capone D.G. and Rueter J. (eds.), *Marine Pelagic Cyanobacteria:* Trichodesmium *and Other Diazotrophs,* Kluwer Academic Publishers, Dordrecht, pp. 163-175.
Weare, N.M., Azam F., Mague T.H. and Holm-Hansen, O. (1974) Microautoradiographic studies of the marine phycobionts *Rhizosolenia and Richelia*, *J. Phycol.* **10,** 369-371.
Wilmotte A., Neefs J.-M. and DeWachter R. (1994) Evolutionary affiliation of the marine nitrogen-fixing cyanobacterium *Trichodesmium* sp. strain NIBB 1067, derived by 16S ribosomal RNA sequence analysis, *Microbiology* **140,** 2159-2164.

Chapter 2

MARINE CYANOBACTERIAL SYMBIOSES

E.J. CARPENTER AND R.A. FOSTER
Romberg Tiburon Center, San Francisco State University
3152 Paradise Drive, Tiburon CA 94920 USA

1. INTRODUCTION

Symbioses between cyanobacteria and marine organisms are abundant and widespread among marine plants and animals. Generally, they are most likely to be found in oligotrophic areas in which either N_2 fixation or dissolved organic carbon (DOC) release benefit the host organism, although a few occur in nutrient rich areas such as mudflats. Research on these symbioses is in its infancy, and generally there is very little known about the nature of many of these symbioses. Furthermore, from microscopic observations, it appears that there are many more symbiotic relationships yet to be discovered.

In the marine environment, symbioses are known to occur between cyanobacteria and sponges, Ascidians (sea squirts), and Echuroid worms in the benthos, and diatoms, dinoflagellates and a protozoan among the plankton. These symbioses can often be significant in terms of the biogeochemistry of coastal and open ocean areas. For example, the heterocystous cyanobacterium, *Richelia intracellularis* can be present in diatoms which can form blooms over 10-100 x 10^3 km^2 of oligotrophic seas, and can be significant in adding fixed nitrogen to nutrient impoverished areas. Diatom symbioses have been discussed in the previous chapter by S. Janson. Below we will briefly discuss the known marine cyanobacterial symbioses, excluding those with diatoms. Figure 1 shows some examples of these symbioses.

2. SPONGES

Associations occur between four genera of cyanobacteria and 38 genera of sponges within the sponge classes Calcarea and Desmospongia. Cyanobacteria, consisting of twelve different species, within the genera *Aphanocapsa*, *Synechocystis*, *Oscillatoria* and *Phormidium* are present in sponges, and most cyanobacterial species occur extracellularly (Adams, 2000). However, *Aphanocapsa feldmanni* occurs within specialized amoeboid sponge cells, which are called cyanocytes. Usually several hundred of these cyanobacteria occur within the same vacuole. There is limited

A.N. Rai, B. Bergman and U. Rasmussen (eds.), Cyanobacteria in Symbiosis, 11-17.

evidence of phagocytosis of cyanobacteria by sponges, and it appears that fixed carbon is transferred to sponge cells primarily as glycerol (Wilkinson, 1979). Sponges are unusual hosts, in that a variety of diverse organisms ranging from fungi, bacteria, diatoms, cyanobacteria and eukaryotic microalgae can live within sponges, apparently symbiotically.

Sponges with symbionts are either classified as phototrophs, highly dependant on the cyanobacteria for nutrition and with large populations of symbionts, or mixotrophs which have cyanobacterial symbionts but receive a portion of their nutrition by filter feeding. The phototrophic sponges typically have a flattened shape which allows maximum exposure of cyanobacteria to sunlight. These sponges are always found within the euphotic zone. Cyanobacteria are abundant in Indo-Pacific sponges, and their biomass can equal that of the host sponge. These sponges can derive over 50% of their metabolic requirements from symbiotic cyanobacteria (Wilkinson, 1983). The cyanobacteria can also contribute to the host sponge by producing secondary metabolites which function as defensive compounds (Sara *et al.*, 1998). Polybrominated secondary metabolites produced by the sponge *Dysidea herbacea* actually discourages feeding by fish (Faulkner *et al.*, 1994).

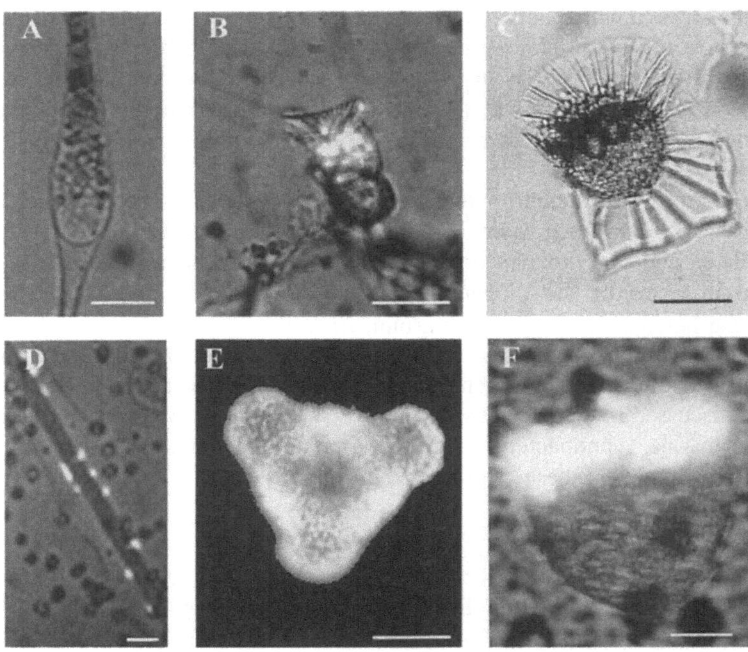

Figure 1. A, Cyanobionts within Ampisolenia ***sp.; B,*** Parahistioneis ***and. C,*** Ornithocercus magnificus ***with cyanobacteria in the girdle lists. Figures A-C are*** Dinophysis ***(Dinoflagellate) representatives frequently observed with cyanobionts. D,*** Leptocylindrus ***sp. containing fluorescing cyanobacterial cells. E, Radiolarian*** Dictyocoryne ***sp. fluorescing from the numerous cyanobacteria in its intracapsular cytoplasm; F, Unknown*** Tintinnid ***sp. containing cyanobionts in the crown of its lorica. Bars represent 10 µm.***

Two Mediterranean sponges, *Chondrilla nucula* and *Petrosia ficiformis* have cyanobacterial symbionts, but differ in their dependence on them (Arillo *et al.*, 1993). When deprived of sunlight, *C. nucula* cannot live, showing a total dependence on the photosynthetic symbionts for metabolites. However, in darkness *P. ficiformis* activates heterotrophic metabolism and is able to survive without symbionts.

Cyanobacteria may also benefit sponges through fixation of atmospheric nitrogen. The cyanobacterial species *Aphanocapsa feldmanni* located intracellularly within some sponge species has been shown to possess nitrogenase activity (Wilkinson and Fay, 1979), and low levels of N_2 fixation have also been found in sponges *Theonella swinhoei* and *Siphonochalina tabernaculata* (Rai, 1990).

3. ASCIDIANS

Ascidians, also known as sea squirts or tunicates, live either permanently attached to a solid object or buried in the sand or mud. Larvae have a notochord, thus these organisms are classified as Chordates. In one family of sea squirts, Didemnidae, there are five genera, which form associations with the cyanobacterial genera *Synechocystis* or *Prochloron*-related genera (Lambert *et al.,* 1996). There are three species, *Prochloron didemni, Prochlorococcus marinus*, and *Prochlorothrix hollandica* among the prochlorophytes living with several didemnid species. The tunicates do not appear to phagocytize and digest the cyanobacteria, and it is presumed that the hosts benefit from the release of dissolved organic carbon. Carbon fixed by the symbionts has been detected in host tissue, indicating a direct transfer (Pardy and Royce, 1992).

While it appears that most prochlorophyte cells reside outside of the host ascidian's cells, in the tropical ascidian *Lissoclinum punctatum*, the prochlorophytes can be ingested by phagocytosis and remain in the cells within a vacuole. Ingested cells appear to be healthy, and there are no morphological differences between free-living and ingested cells (Hirose *et al.*, 1996, 1998)

Compounds, known as didemnins, with antitumor and immunosuppressive activities are associated with ascidians, although it is not clearly known whether it is the host animal or cyanobacterial symbiont which is responsible for their production (Sings and Rinehart, 1996).

Recently there has been some conflicting evidence of N_2 fixation by cyanobacteria associated with ascidians. Odintsov (1991) used the acetylene reduction method to assay some ascidians for nitrogenase activity in the Seychelles and found ethylene production in encrusting forms of ascidians. However, he found no nitrogenase activity associated with isolated prochlorophytes, and there was no direct relationship between ethylene production and prochlorophyte cell numbers. Koike *et al.* (1993) studied the nitrogen budgets of two species of didemnid ascidians in Fiji and found that the estimated N requirement of the prochlorophytes was much greater than could be supplied externally. The authors discounted the possibility of N_2 fixation, as it had not been reported in this group, and they posited that N must be cycled very efficiently in this host-symbiont relationship. However, using natural abundance of $^{15}N/^{14}N$, Kline and Lewin (1999) found mean $\delta^{15}N$ values of 1.1 for isolated prochlorophyte cells from didemnid ascidians from Palau, Western Caroline Islands. Definitive proof that the prochlorophytes are responsible would await measurements on pure cultures of the prochlorophytes, but the low $\delta^{15}N$ is suggestive of nitrogenase activity in this group.

4. ECHIUROID WORMS

Cyanobacteria occur in cells of the subepidermal connective tissue of two worms, *Ikedosoma gogoshimense* and *Bonellia fuliginosa* (Rai, 1990). The former lives in muddy sand, typically found at the low tide level, and the latter on coral reefs. Virtually nothing is known about these cyanobacteria or their symbiotic relationship.

5. APLASTIDIC PROTISTS

Solenicola setigera is a heterotrophic flagellated protozoan, which lives on the frustule of the chain-forming diatom *Leptocylindrus mediterraneus* in oligotrophic open-ocean waters. Within the protozoan are unicellular cyanobacteria, which appear to be in the genus *Synechococcus* sp. (Buck and Bentham, 1998). It appears that the protozoan feeds on cell contents of the diatom, because the frustules on which they occur virtually always appear to be empty. The authors speculated that the cyanobacteria might be involved in nitrogen fixation, although virtually nothing is known about the nature of their symbiosis.

6. DINOFLAGELLATES

Dinoflagellate species exhibit a wide range of nutritional modes and may exist as free-living autotrophs, symbiotic autotrophs, mixotrophs, or heterotrophs, and as the latter, they may be phagotrophs, or even parasites. Many do not have photosynthetic pigments, and often these unpigmented dinoflagellates have bizarre shapes. It was noted by early microscopists that some unpigmented dinoflagellates had colored bodies within the girdle lists. These bodies were termed "phaeosomes" by Schütt (1895). Confirmation that they were cyanobacteria came from electron microscope observations by Lucas (1991). In studying several dinoflagellate genera, *Ornithocercus, Histoneis, Citharistes*, and *Amphisolenia*, Lucas (1991) was able to distinguish at least three (possibly four) distinct forms of symbiotic cyanobacteria. Differences were based on arrangement of thylakoids, location of carboxysomes and shape and size of the cells. Total range of cell size was from 3.5-4.8 μm for spherical cells and up to 10 μm in length for rod shaped cells. These cyanobacteria are considerably larger than the planktonic *Synechococcus* forms (< 1.0 μm), which are abundant in oceanic waters.

There are a variety of arrangements for holding the symbionts. The cyanobacteria are located externally in the girdle list in *Ornithocercus*, while in *Parahistoneis* the posterior of the cingular groove forms a pocket indentation to hold the cells. In *Histoneis* cells are within a chamber on the girdle floor. In *Citharestes* the chamber is enclosed even further and the chamber opening to the girdle is reduced to a small hole just slightly larger than the symbionts. In *Amphisolenia thrinax* cells are in a chamber in the hyposome. For all of these genera the symbionts are external to the dinoflagellate cytoplasm, however in *Amphisolenia*, the cyanobacterial cells are clearly within the host cytoplasm, as shown by TEM (Lucas, 1991).

In the Gulf of Aqaba, Red Sea, the dinoflagellate genera *Ornithocercus, Histoneis,* and *Citharistes* with cyanobacteria symbionts were typically most abundant in the autumn when nitrogenous nutrients were least available in surface waters (Gordon *et al.*, 1994). The authors hypothesized that the host dinoflagellates might provide sites of low oxygen concentration which might provide favorable conditions for N_2 fixation by the cyanobacteria.

7. LICHENS

While typically associated with terrestrial habitats, a few lichens live in marine littoral waters. It appears that there are at least seven species of truly submarine lichens, and some of these have cyanobacterial symbionts (Schenk 1992, Kohlmeyer and Kohlmeyer, 1979, Kohlmeyer and Volkman-Kohlmeyer, 1988). Schenk (1992) lists the cyanobacterium *Chroococcus* sp. in association with the ascolichen *Halographis runica* and the cyanobacterium *Hyella caespitosa* with *Arthopyrenia halodytes.*

8. SILICOFLAGELLATES

The silicoflagellate *Dictyocha speculum* Ehr. from the Indian Ocean was observed to contain coccoid cyanobacteria as symbionts within its protoplast (Norris, 1967). These cyanobacteria were described as being similar to *Synechocystis consortia*, a species which Norris (1967) has named and described as living symbiotically with the dinoflagellate *Parahistoneis* and also collected in the Indian Ocean.

9. RADIOLARIANS

The radiolarian *Dictyocoryne truncatum* is a triangular shaped spongiose skeletal species, which has been described as containing "bacterioids" by Anderson and Matsuoka (1992). The "bacteroids", are small, 0.2 x 0.5 μm and present throughout the intracapsular cytoplasm. The species is commonly found in tropical oligotrophic waters, and the "bacteroids" shown by Anderson and Matsuoka (1992) clearly have peripheral thylakoids suggestive of cyanobacteria. Using epifluorescence microscopy on a research cruise in the tropical Atlantic Ocean, these radiolarians were noted to have yellow-orange fluorescence under green excitation. This indicates the presence of fluorescing phycobiliproteins thus suggesting high densities of coccoid cyanobacteria in the radiolarians (Carpenter, unpublished results).

10. MACROALGAE

The green macroalga, *Codium*, is a siphonaceous species which has invaded European and American coastal waters. The macroalga consists of several branched shoots, each of which consist of a central medulla composed of filaments which give rise to inflated branchlets, known as utricles, on the exterior of the colony. Cyanobacteria live between the utricles and are capable of N_2 fixation (Rosenberg and Paerl, 1980). Microelectrode measurements of O_2 concentrations in the spaces between the utricles indicate low O_2 concentrations, which may be favorable for N_2 fixation by the cyanobacteria located there.

11. TINTINNIDS

Recently in the tropical oligotrophic North Atlantic, we observed cyanobionts residing in the lorica crown of an unknown Tintinnid species (Figure 1 F). Later we positively identified these as cyanobacteria since they significantly immuno-labeled phycoerythrin higher than the background (Foster, unpubl.). To the best of our knowledge no work has previously described this symbioses.

12. CONCLUSIONS

Cyanobacterial symbioses are widespread in the marine environment and, aside from diatoms include sponges, ascidians (sea squirts), protozoa (heterotrophic dinoflagellates, microflagellates, silicoflagellates, radiolaria), lichens, macroalgae, and echiuroid worms. The nature of these symbioses and cyanobacteria involved has been characterized the best in sponges and ascidians. However the remaining symbioses are virtually unstudied. Research on the nature of these symbioses should prove to be exciting.

REFERENCES

Adams, D.G. (2000) Symbiotic Interactions. in B.A. Whitton and M. Potts (eds.), *The Ecology of Cyanobacteria,* Kluwer Academic Publishers, Dordrecht. pp 523-561.

Anderson, O.R. and Matsuoka, A. (1992) Endocytoplasmic microalgae and bacteroids within the central capsule of the Radiolarian *Dictyocoryne truncatum, Symbiosis* **12**, 237-247.

Arillo, A., Bavestrello, G., Burlando, B. and Sara, M. (1993) Metabolic integration between symbiotic cyanobacteria and sponges: a possible mechanism, *Marine Biology* **117**, 159-162.

Buck, K. and Bentham, W.N. (1998) A novel symbiosis between a cyanobacterium, *Synechococcus* sp., an aplastidic protist, *Solenicola setigera*, and a diatom, *Leptocylindrus mediterraneus*, in the open ocean, *Marine Biology* **132**, 349-355.

Faulkner, J.D., Unson, M.D. and Bewley, C.A. (1994). The chemistry of some sponges and their symbionts, *Pure Appl. Chem.* **66**, 1983-1990.

Gordon, N., Angel, D.L., Neori, A., Kress, N. and Kimor, B. (1994) Heterotrophic dinoflagellates with symbiotic cyanobacteria and nitrogen limitation in the Gulf of Aqaba. *Mar. Ecol. Prog. Ser.* **107**, 83-88.

Hirose, E., Maruyama, T., Cheng, L. and Lewin, R. (1996) Intracellular symbiosis of a photosynthetic prokaryote, *Prochloron* sp., in a colonial ascidian, *Invertebrate Biology* **115**, 343-348.

Hirose, E., Maruyama, T., Cheng L. and Lewin, R.A. (1998) Intra- and extra-cellular distribution of photosynthetic prokaryotes, *Prochloron* sp., in a colonial Ascidian: Ultrastructural and quantitative studies, *Symbiosis* **25**, 301-310.

Kline, T.C. and Lewin, R. (1999). Natural $^{15}N/^{14}N$ abundance as evidence for N_2 fixation by *Prochloron* (Prochlorophyta) endosymbiotic with Didemnid Ascidians, *Symbiosis* **26**, 193-198.

Kohlmeyer, J. and Kohlmeyer, E. (1979) Submarine lichens and lichenlike associations, in J. Kohlmeyer and E. Kohlenmyer (eds.), *Marine mychology: The higher fungi*, Academic Press, New York, pp. 70-78.

Kohlmeyer, J. and Volkman-Kohlmeyer, B. (1988) *Halographis* (Opegraphales). A new endolithic lichenoid from corals and snails, *Can. J. Botany.* **66**, 1138-1141.

Koike, I., Yamamuro, M. and Pollard P.C. (1993) Carbon and nitrogen budgets of two Ascidians and their symbiont, *Prochloron*, in a tropical seagrass meadow, *Aust. J. Mar. Freshwater Res.* **44**, 173-182.

Lambert, G., Lambert, C.C. and Waaland, J.R.R. (1996) Algal symbionts in the tunics of six New Zealand ascidians (Chordata, Ascidiacea), *Invertebrate Biology* **115**,67-78.

Lucas, I.A.N. (1991) Symbionts of the tropical Dinophysiales (Dinophyceae), *Ophelia* **33**, 213-224.

Norris, R.E. (1967) Algal consortisms in marine plankton, in V. Krishnamurti (ed.), *Proceedings of the seminar on sea, salt and plants*, Central Salt and Marine Chemicals Research Institute, Bhavnagar (India), pp. 178-189.

Odintsov, V.S. (1991) Nitrogen fixation in *Prochloron* (Prochlorophyta)-Ascidian associations, Is *Prochloron* responsible?, *Endocytobiosis and Cell Research* **7**,253-258.

Pardy, R.L. and Royce, C.L. (1992) Ascidians with algal symbionts, in W. Reisser (ed.), *Algae and Symbioses, plants, Animals, Fungi, Viruses, interactions explored.* Biopress Ltd, England, pp. 215-230.

Rai, A.N., (1990) Cyanobacteria in symbiosis, in Rai. A.N. (ed.) *CRC Handbook of Symbiotic Cyanobacteria*, CRC Press, Boca Raton (Florida), pp. 1-7.

Rosenberg, G. and Paerl, H.W. (1980) Nitrogen fixation by blue-green algae associated with the siphonaceous green seaweed *Codium decorticatum*: effects of ammonium uptake, *Marine Biology* **61**, 151-158.

Sara, M., Bavestrello, G., Cattaneovietti, R. and Cerrano, C. (1998) Endosymbiosis in sponges: relevance for epigenesis and evolution, *Symbiosis* **25**, 57-70.

Schenk, H.E.A. (1992) Cyanobacterial Symbioses, in A. Balows, H.G. Trüper, M. Dworkin, W. Harder and K.H. Schleifer (eds.), *The Prokaryotes Vol. IV*, Springer-Verlog, New York, pp. 3819-3854.

Schütt F. (1895) Peridineen der Plankton-expedition, *Ergebnisse der Plankton-expedition der Humbolt Stiftung* **4**, 1-170.

Sings, H.L. and Rinehart, K. (1996) Compounds produced from potential tunicate-blue-green algal symbiosis: a review, *J. Industrial Microbiology* **17**, 385-396.

Wilkinson, C.R. (1979) Nutrient translocation from symbiotic Cyanobacteria to coral reef sponges, in C. Levi and N. Boury-Esnault (eds.), *Biologie des Spongiaries*. CNRS, Paris, pp. 373-380.

Wilkinson, C.R. (1983) Net primary productivity in coral reef sponges, *Science* **219**,410-412.

Wilkinson, C.R. and Fay, P. (1979) Nitrogen fixation in coral reef sponges with symbiotic cyanobacteria, *Nature* **279**, 527-529.

Chapter 3

THE *NOSTOC-GEOSIPHON* ENDOCYTOBIOSIS

M. KLUGE[1], D. MOLLENHAUER[2], E. WOLF[1] AND A. SCHÜßLER[1]
[1]*Institut für Botanik, Technische Universität Darmstadt, Schnittspahnstrasse 10, 64287 Darmstadt, Germany*
[2]*Forschungsinstitut Senckenberg, 60486 Frankfurt, Germany*

1. INTRODUCTION

Textbooks of lichenology list various examples of extracellular symbiotic consortia between cyanobacteria and fungi. However, up to now only one example is known of a fungus living in endocytobiotic association with a cyanobacterium, namely, the symbiotic consortium *Geosiphon pyriformis* (Kütz.) v. Wettstein / *Nostoc punctiforme*. Here we briefly describe the current knowledge about this unique consortium. For more details the other reviews on *Geosiphon* should be consulted (Mollenhauer, 1992; Kluge *et al.*, 1994; Schüßler and Kluge, 2001). In contrast to our former publications and because of the grammatical correctness, we will use the species name *G. pyriformis* (instead of *G. pyriforme*) in this review and in our future publications to denote the fungal partner (Schüßler, 2002).

Geosiphon was discovered in 1862 by Kützing and described as a siphonal alga (*Botrydium pyriforme*). Wettstein (1915) first recognised the symbiotic nature of the organism. He considered it as a symbiotic association between a heterotrophic siphonal alga (host) and the cyanobacterium *Nostoc*. Knapp (1933) realised that the macrosymbiont of the system is a fungus. The latter author was also the first who attempted to cultivate the organism, although without success. Finally Mollenhauer and Mollenhauer (1988) successfully grew *Geosiphon* in the laboratory, and due to this breakthrough sufficient samples of the organism became available for experimental work. Nevertheless, cultivation of *Geosiphon* still requires considerable experience and patience. One of the crucial factors is the phosphate content in the growth medium which has to be kept very low. High phosphate favours excessive growth of *Nostoc* so that the fungus is overpowered by the cyanobacterium and establishment of the symbiosis is prevented. In the laboratory we usually start cultures by germinating spores of *Geosiphon* on sterilised soil (obtained from the site where *Geosiphon* grows naturally) in presence of *Nostoc* filaments growing on the substrate surface (see below).

Up to now there are only five reports in the literature on location of *Geosiphon* in nature, but in the meantime, the organism has disappeared from most of the reported

A.N. Rai, B. Bergman and U. Rasmussen (eds.), Cyanobacteria in Symbiosis, 19-30.

sites. At present an arable ground near Bibergemünd (Spessart region, Germany) is the only known place where *Geosiphon* is naturally abundant and can be found occasionally.

2. GENERAL DESCRIPTION OF *GEOSIPHON*

Geosiphon pyrifomis (Kütz.) v. Wettstein represents a coenocytic fungus that spreads its mycelium in the uppermost layers of damp, oligotroph loamy soils. The fungus produces large opaque white spores, which can be easily isolated from the soil substrate (Schüßler *et al.*, 1994). In the context outlined in section 3 of this treatise it is worth mentioning that morphology and ultrastructure of the *Geosiphon* spores show typical characteristics of arbuscular mycorrhizal (AM) fungi, which form one of the most widespread symbioses on earth with terrestrical plants.

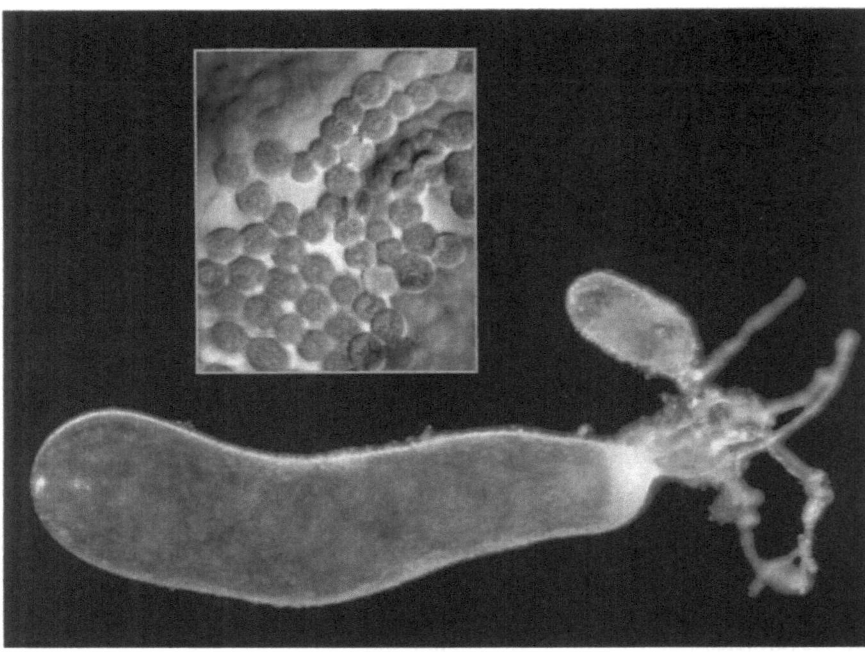

Figure 1. A mature and a young bladder of Geosiphon pyriformis. *Cells of* Nostoc punctiforme *inside the bladder (inset). Large bladder about 1.5 mm long;* Nostoc *cells 8-10µm diameter.*

Geosiphon and *Nostoc punctiforme* live together in the same habitat in and on the soil. As it will be described later in more detail, upon contact with free-living *Nostoc* cells, the tip of a fungal hypha incorporates the cyanobacterium by a mode of endocytosis not yet fully understood. Upon this incorporation the hyphal tip swells forming a pear-shaped multinucleate 'bladder', about 1-2 mm in length and 0.3 mm in diameter (Fig. 1). The bladder provides the compartment where the incorporated *Nostoc* cells reside, multiply by division, and become physiologically active. Bladders without endosymbionts have never been observed, and the endosymbionts are not located in any part other than the bladders. The dark, olive-green bladders appear at the surface of the soil and, besides the spores, are the only visible manifestation of the presence of *Geosiphon* at a given site. Since the tiny bladders are difficult to detect, it is conceivable that in nature *Geosiphon* is not so rare as it appears presently. We are now developing molecular probes to check the occurrence of *Geosiphon* in natural habitats.

3. TAXONOMIC POSITION OF THE PARTNERS

Analysis of the SSU rRNA genes of *Geosiphon* (Gehrig *et al.*, 1996) led to a comprehensive analysis of the molecular phylogeny of AM fungi (Schüßler, 1999; Schüßler *et al.*, 2001a; Schwarzott *et al.*, 2001). These studies resulted in the crection of a new fungal phylum, the Glomeromycota, comprising the AM fungi and *Geosiphon*. (Schüßler *et al.*, 2001b). *Geosiphon* unequivocally belongs to an ancestral lineage within the Glomeromycota and the description of this organism was amended recently (Schüßler, 2002). This view is consistent with the already mentioned striking similarities in the spore structures of *Geosiphon* and the AM fungi. Interestingly, this could mean that *Geosiphon* not only forms the endosymbiotic association with *Nostoc,* it also interacts with plant roots forming AM. Investigations to find out if this is true are in progress in our laboratory. Up to now there is no evidence that any member of the Glomeromycota other than *Geosiphon* can interact with photoautotrophic microorganism to form endosymbiotic systems.

The cyanobacterial endosymbiont of *Geosiphon* is *Nostoc punctiforme*. It has been observed that several strains of this organism can be incorporated by the fungus leading to the formation of functionally active bladders. *Nostoc punctiforme* strains isolated from other symbiotic systems (e.g., *Anthoceros*, *Blasia*, *Gunnera*) are particularly suitable to give rise to an active *Geosiphon* consortium. There are also strains of *N. punctiforme* whose cells are incorporated, but the formation of bladders is stopped at an early stage of development. Finally, there are strains that are never recognised and incorporated by the fungus (Mollenhauer, unpublished).

Altogether, our observations suggest that the recognition of the cyanobacterial partners by *Geosiphon* is rather specific. As discussed below, this view is supported by the fact that among the various developmental stages of *Nostoc* only the primordia are recognised and incorporated by the fungus. Better understanding of partner specificity in *Geosiphon* with respect to the endosymbiont requires progress in the taxonomy of *Nostoc* (Mollenhauer and Mollenhauer, 1996), particularly at the molecular level (Paulsrud and Lindblad, 1998; Paulsrud, 2001).

4. INITIATION AND FURTHER DEVELOPMENT OF THE CONSORTIUM

This aspect has been studied in detail by Mollenhauer *et al.* (1996) and reviewed by Schüßler and Kluge (2001). Initially *Nostoc* lives independently of the fungus at its habitat, undergoing a characteristic life cycle (Mollenhauer *et al.*, 1996, and literature quoted there). Successful interaction with the fungus requires that *Nostoc* is converted into the immobile form called primordium (Mollenhauer, 1988). The motile trichomes (hormogonia) of *Nostoc* are not recognised and incorporated by the fungus. One possible reason could be that, due to the motility of the hormogonia, the surface contacts between the partners are too short to allow proper recognition and interaction of the partners. However, there seem to be more specific mechanisms involved (see below).

The process of *Nostoc* incorporation by the fungus can be divided into two phases. In phase one, upon contact between the future partners several adjacent cells of a primordial *Nostoc* filament are progressively surrounded and enclosed by portions of fungal plasma bulging out from the tip of a hypha. The detailed mechanism of the incorporation process is not yet fully understood. In the second step no further *Nostoc* cells are incorporated; rather, the hyphal tip containing the incorporated *Nostoc* cells swells and finally forms the bladder. It is important to keep in mind that each single event of incorporation, provided there is full compatibility between the *Nostoc* strain and the fungus, gives rise to the formation of a typical *Geosiphon* bladders. Each bladder represents a polyenergid, specialised part of the fungal mycelium, coenocytic with the mycelium spreading in the soil.

Shortly after incorporation by the fungus, the *Nostoc* cells obviously suffer severe stress as indicated by bleaching of their photosynthetic pigments (Mollenhauer, unpublished). However, during the maturation of the bladder that follows, the enclosed *Nostoc* cells recover completely. They begin to multiply, with multiplication rates 2 to 3 times higher compared to the free-living filaments (Wolf *et al.*, unpublished). Moreover, the incorporated cells increase their volume up to tenfold with respect to the primordial *Nostoc* cells outside the bladder, and they considerably increase the concentration of photosynthetic pigments as compared with the free-living *Nostoc*.

It has to be mentioned that during incorporation of a *Nostoc* trichome, its heterocysts are always left outside of the fungal hypha. However, the multiplying *Nostoc* filaments inside the bladder begin to regenerate heterocysts in about the same frequency as in the free-living vegetative trichomes (Mollenhauer *et al.*, 1996).

One of the most interesting but not yet answered questions in context with *Geosiphon* concerns the mechanisms of signal transduction leading to recognition and interaction of the partners. Experimental studies in this field require synchronisation of the developmental cycle of the *Nostoc* cells used in the experiments. We achieve this synchronisation by red and green light illumination, respectively. Red light converts vegetative *Nostoc* filaments into the motile hormogonia, a stage that, under red light, lasts for two to three days. A few hours after switching illumination from red to green light, the hormogonia lose their motility and begin to convert into the pimordial stage. Shortly afterwards, first microscopically visible stages of *Nostoc* incorporation by the fungus appear.

Binding studies by Schüßler *et al.* (1997) using fluorescence labelled lectins specific for defined sugar residues revealed that the extracellular slime of the *Nostoc* primordia, susceptible to recognition and incorporation by the fungus, contains mannose. This sugar is not detected either in the cell wall of the heterocysts or in the envelope of the earlier *Nostoc* stages which do not interact with the fungus. Recently it could be shown, that the appearance of mannose containing glycoconjugates on the *Nostoc* cell surface begins three to seven hours after onset of green light illumination (Wolf, unpublished). This exactly coincides with the time point when the hormogonia convert into the immobile stage. Thus, it is tempting to assume that mannose in the *Nostoc* cell wall and slime surrounding the primordia are somehow involved in the mechanism responsible for partner recognition. However, recent experiments show that external applications of mannose or the mannose-binding lectin Concanavalin A do not affect the success of partner recognition as evident by the unchanged rates of initiation of bladder formation (Wolf, unpublished). Thus, at the present stage of knowledge there is no proof that a mannose-containing glycoconjugate is involved in the initial step of the recognition signal chain.

It is worth mentioning that the *Nostoc* cells enclosed inside *Geosiphon* bladders retain their full genomic integrity. This can be concluded from the fact that, in isotonic media, the endosymbiotic *Nostoc* can be isolated from the bladders and further cultivated without the fungal partner. This fact suggests that *Geosiphon* represents a fairly early state among the endocyanoses, in contrast to such cyanoses as *Cyanophora paradoxa* that are more advanced.

On the other hand, it is not possible till now to cultivate the *Geosiphon* fungus without its endosymbiont. This indicates that, as in the case of other AM fungi, *Geosiphon* is an obligate symbiotic organism.

5. STRUCTURE OF THE *GEOSIPHON* BLADDERS

The ultrastructure of the *Geosiphon* bladders was investigated first by Schnepf (1964). This pioneering study led to a general theory of the compartmentation of the eukaryotic cell, and it provided strong arguments supporting the endosymbiosis theory of cell evolution. Investigations by Schüßler *et al.* (1996) considerably increased our knowledge of the ultrastructure and compartmentation of the *Geosiphon* bladders. Due to this study, it is now clear that inside the bladder *Nostoc* cells are located within a single cup-shaped compartment, the symbiosome (Fig. 2). It is arranged at the periphery of the bladder. As in the case of the fungal cell wall, the symbiosome envelope bordering the *Nostoc* cells contains chitin. This finding shows that the envelope is synthesised by a fungal plasmalemma-homologue. Altogether, the symbiotic interface between *Nostoc* and the fungal cytoplasm is very similar to that found in arbuscular mycorrhiza (AM) between the fungus and the plant cell (Schüßler *et al.*, 1996; Schüßler and Kluge, 2001).

As mentioned above, the *Nostoc* cells inside the symbiosome have about 10-fold higher volume compared to free-living cells outside the bladder. Despite the increase in size, the symbiotic *Nostoc* cells show normal structures. For example, they contain high number of thylakoids and carboxysomes. Also the structure of the heterocysts is identical with that of the free-living *Nostoc*.

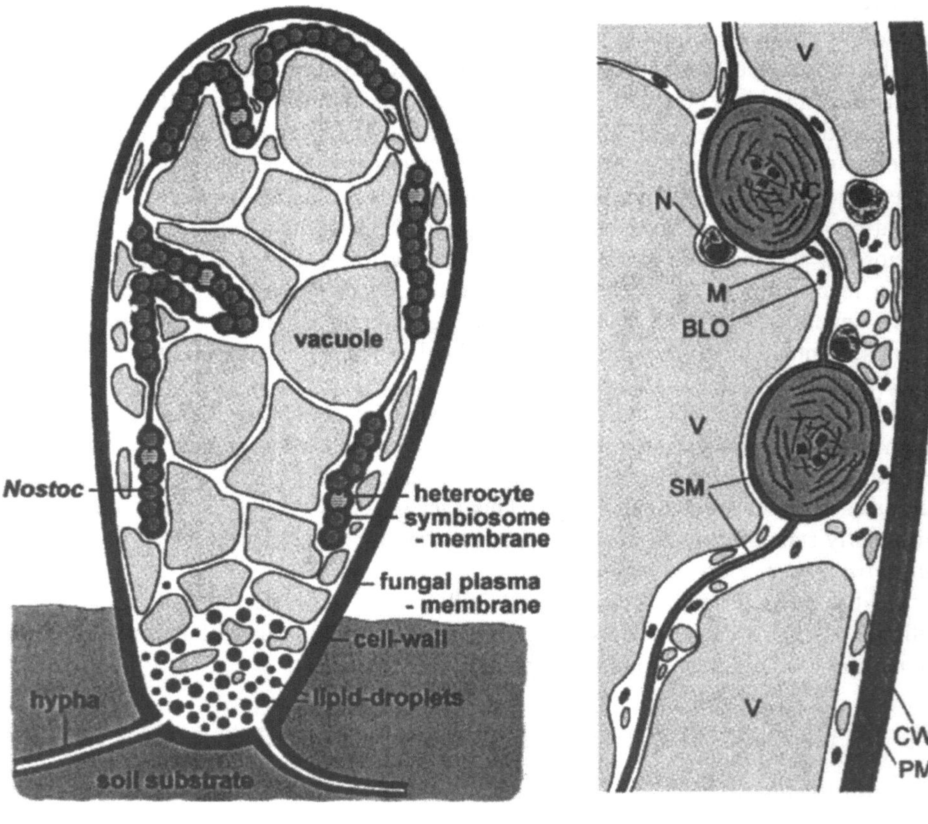

Figure 2. Schemes of the internal structure of a Geosiphon *bladder. Left, overview of the whole bladder. Right, ultrastructural details as derived from electron microscopy. BLO, bacteria like organism, CW, cell wall, M, mitochondrion, N, nucleus, PM, fungal plasma membrane, V, vacuole.*

The *Geosiphon* bladder as a whole clearly shows structural polarity. The photosynthetically active endosymbionts are located mainly in the larger apical ¾ of the bladder, which are exposed to light, while the much smaller, opaque basal part represents the storage region of the bladder. The latter part is less vacuolated and contains many lipid droplets and glycogen granules.

In particular the central cytoplasm of the bladder is highly vacuolated. It is interesting to note that in *Geosiphon* 'bacteria like organisms' (BLO) have been noted in the cytoplasm of the bladders, hyphae and spores. The said BLO represent bacteria of yet unknown phylogenetic affiliation that possess ultrastructure similar to the obligate symbiotic BLOs in AM fungi. Thus, in addition to the *Nostoc*, the *Geosiphon* bladder obviously accommodates endosymbiotic bacteria also.

6. METABOLIC ASPECTS OF THE SYMBIOSIS

The results of ^{14}C tracer studies suggest that the *Nostoc*-containing *Geosiphon* bladders fix CO_2 both in light and in darkness (Kluge *et al.*, 1991). The rate of CO_2 fixation in light is much higher than that in darkness. The ^{14}C labelling patterns are also different. In light ^{14}C labelling occurs largely in phosphate esters, polyglucans, free sugars (among them trehalose and raffinose), amino acids and some organic acids. In darkness, however, only malic acid, an fumaric acid and some amino acids are labelled. There are also differences between the symbiotic and free-living *Nostoc* in the labelling patterns resulting from $^{14}CO_2$ fixation in light. The cells of free-living *Nostoc* trap more label in phosphate esters than in sugars.

Altogether, the tracer experiments unequivocally show that *Nostoc* cells are photosynthetically active inside the bladders. The dark CO_2 fixation is also likely to be carried out by the endosymbiont using phospho-enolpyruvate carboxylase (PEPCase). This enzyme is known to be present in cyanobacteria but has not been reported in fungi. However, it cannot be ruled out that the BLOs located in the cytoplasm of *Geosiphon* contain PEPCase and contribute to the observed dark CO_2 fixation.

Photosynthetic activity of the endosymbiotic *Nostoc* cells was also shown by measurements of photosystem II chlorophyll-fluorescence kinetics (Bilger *et al.*, 1994). These authors found that, at certain quantum flux densities, the *Nostoc* cells inside the bladders achieve much higher steady-state quantum yields and much higher electron transport rates as compared to the cells of the free-living *Nostoc*.

The occurrence of heterocysts in the endosymbiotic *Nostoc* filaments suggests that they are capable of nitrogen fixation. Further evidence supporting this assumption comes from the finding that *Geosiphon* bladders exhibit considerable nitrogenase (acetylene reduction) activity (Kluge *et al.*, 1992). However, in contrast to *Nostoc* in plants symbioses, where heterocyst frequency is increased reflecting that the major role of the cyanobiont is N_2 fixation, in *Geosiphon* the heterocyst frequency does not change. This led us to conclude that here, despite its capability to fix N_2, the major role of the endosymbiotic *Nostoc* is photosynthesis.

Analyses by proton induced X-ray emission (PIXE) provided information about the composition and concentration of macro- and micro-elements in *Geosiphon* bladders (Maetz *et al.*, 1999). It was found that in *Geosiphon* grown on solution containing low amount (1μM) phosphate and 100 μM KCl, the fungal partner accumulates high amounts of P (about 1.5 % of its dry weight), Cl (2 %) and K (5 %), whilst the symbiosome (including the *Nostoc* cells) contains smaller amounts of these elements. Mo concentration was found to be much lower in the symbiosome in comparison to the other parts of the bladder. Since Mo is a constituent of nitrogenase, this finding was somewhat surprising with the assumption that in *Geosiphon* the endosymbiotic *Nostoc* cells carry out N_2-fixation. However, Mo could also be present in other enzymes or stored in the vacuole. Moreover it was shown recently, that the BLOs of AM fungi contain *nif* genes (Minerdi *et al.*, 2001), raising the question if these bacteria could fix N_2 in *Geosiphon* also.

7. LIKELY BIOLOGICAL ADVANTAGES OF THE PARTNERSHIP

Schüßler *et al.* (1995) found that the cell wall of *Geosiphon* is an impermeable barrier for molecules with a molecular radius > 0.45 nm. That is, exogenous sugar molecules such as sucrose would be excluded and cannot be taken up by the fungus. Glucose is known to permeate extremely slowly. The fungus becomes independent of an external supply of sugars by accommodating the photosynthetically active *Nostoc* cells, resulting in C and N autotrophy. Presumably *Nostoc* also derives advantage from the cooperation with the host. This is reflected by faster growth and cell division of the endosymbiotic *Nostoc* after uptake by the fungus, compared to the free-living *Nostoc*. For instance, the fungus might help to maintain homoeostasis of water relations in the immediate surrounding of the endosymbiont. It could also improve the nutrition of the endosymbiont by supplying phosphate, since free-living *Nostoc* grows poorly under the conditions used for *Geosiphon* cultures. Also the supply of other mineral nutrients and of CO_2, the substrate of photosynthesis, could be improved. We are aware that our discussion of the biological advantages of the partnership in *Geosiphon* is somewhat hypothetical, because nutrient exchange between the partners is still poorly investigated. The main reason for this is the problem to cultivate the organism in large amounts. Therefore most studies rely on micro methods and there are still many open questions that have to be answered by future work.

8. ECOLOGY OF *GEOSIPHON*

Although not yet systematically investigated, our experience with *Geosiphon* in nature and in the laboratory indicates that the organism grows only on soils poor in phosphate. For instance, at the natural stands, eutrophication by aerosols originating from excessive artificial fertilisation of land in the neighbourhood is sufficient to suppress the growth of *Geosiphon* at least for a while. Moreover, the soil where *Geosiphon* can grow has to stay constantly humid. That is, if the soil dries out because of longer lasting cessation of rain, or if excessive precipitation leads to stagnant moisture, the *Geosiphon* bladders disappear from their stands. Obviously the fungus survives in the soil such stress periods by its spores, which germinate by forming hyphae capable of establishing new bladders. It has been observed (Schüßler *et al.* 1994) that the germination of the spores is stimulated by exudations of unknown nature from mosses, which are a main component of the vegetation accompanying *Geosiphon*.

Mosses and liverworts in crop fields are the typical plant communities where *Geosiphon* can be found. In nature, *Geosiphon* grows together with the hornwort *Anthoceros*, the liverwort *Blasia*, and the moss *Dicranella*. The former two plants accommodate *Nostoc punctiforme* in cavities of their thalli. This symbiotic *Nostoc* can be isolated from the hosts and afterwards be recognised and taken up by *Geosiphon*. Moreover, it is known that *Anthoceros punctatus* at the *Geosiphon* stands forms AM like symbiosis. It was possible to establish an AM like symbiosis between *Anthoceros* and the AM fungus *Glomus claroideum*, isolated from the *Geosiphon* stand, under laboratory conditions (Schüßler, 2000). However this attempt failed with *Geosiphon*. Nevertheless, the mentioned facts and the close relationship of *Geosiphon* to the AM-fungi led us to hypothesise that, *in situ*, *Nostoc punctiforme*, *Geosiphon*, *Anthoceros*, *Blasia*, and presumably the roots of higher plants growing at the same site, are linked together in a symbiotic network (Fig. 3). This is a fascinating, although still hypothetical aspect, which is worth further investigation in the future.

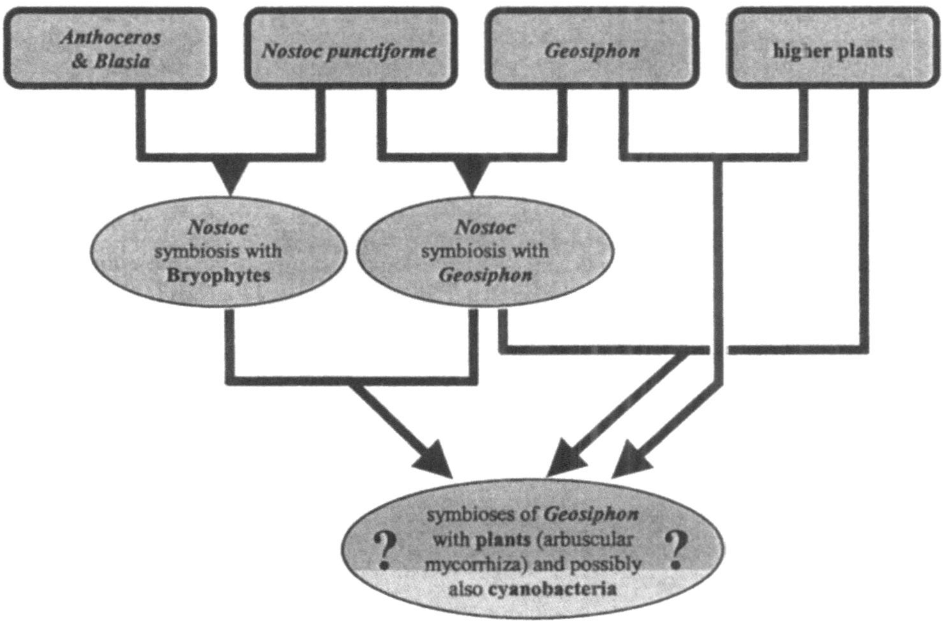

Figure 3. The position of Nostoc *in a symbiotic network at the natural stand of* Geosiphon. *The interaction between* Geosiphon *and plants to form mycorrhiza is hypothetical.*

9. IS *GEOSIPHON* A LICHEN?

The answer to the question whether or not *Geosiphon* is a lichen is first of all a matter of definition. Schwendener (1869; quoted in Hensen and Jahns, 1974) defines lichens as fungi living in symbiotic association with algae that serve as source of nutrients. According to Hensen and Jahns (1974) lichens are fungi, which for their alimentation are bound obligatorily to defined algae, forming with them morphological-physiological units.

Doubtless *Geosiphon* matches the criteria of such general definitions of a lichen. Knapp (1933) interpreted *Geosiphon* as intracellular phycomycotean lichen. Some recent lichenological textbooks also quote *Geosiphon* as an example of a lichen. However, there are definitions of lichens based on more specific criteria also. For instance, Tobler (1934) postulates that a lichen has to match four characters, namely (1) close physical contact of the partners, with the fungus clasping or braiding the alga or partially penetrating it by haustoria; (2) formation of a new morphological appearance different from that of the single partners; (3) physiological success of the symbiotic consortium; (4) special set-ups for vegetative propagation. If we apply these criteria, the classification of *Geosiphon* as a lichen has to be questioned. In contrast to the external contact between the partners in the true lichen, *Geosiphon* represents an endocytobiotic consortium, with the photobiont living inside the fungal cell. In contrast to lichens, where soredia or isidia ensure vegetative propagation by spreading both symbiotic partners together, in *Geosiphon* mechanisms of vegetative propagation of the entire symbiotic system do not exist. Moreover, true lichens and *Geosiphon* show considerable additional differences. In contrast to *Geosiphon*, no lichen is known with a non septate fungal partner and, as far as the ecophysiological behaviour is concerned, *Geosiphon* and lichens show nearly contrary properties. Whilst lichens are robust towards dehydration, the *Geosiphon* symbiosis does not survive water loss. In contrast to the temperature resistant lichens, we found that *Geosiphon* is very sensitive towards high temperatures. Finally, whilst most lichens are adapted to tolerate high light irradiance, *Geosiphon* grows only in moderate light.

Thus, we hesitate to consider *Geosiphon* as a lichen because of these differrences. On the other hand we see a strong relationships between *Geosiphon* and arbuscular mycorrhiza: these include: (1) the taxonomic position of *Geosiphon* within the Glomeromycota, comprising also the AM fungi; (2) the nearly identical structures of the symbiotic interfaces in *Geosiphon* and AM, and (3) the fact that the establishment of AM and *Geosiphon* is enhanced or induced by limitation in phosphate supply, which otherwise would be a severe stress factor for the photobiont. Thus, in our opinion *Geosiphon* is a more appropriate model for studying symbiotic interactions in AM rather than in lichens.

ACKNOWLEDGEMENT

Our work on *Geosiphon* was supported by the Deutsche Forschungsgemeinschaft (SCHU1203; SFB199; GRK340).

REFERENCES

Bilger, W., Büdel, B., Mollenhauer, R. and Mollenhauer, D. (1994) Photosynthetic activity of two developmental stages of a *Nostoc* strain (cyanobacteria) isolated from *Geosiphon pyriforme* (Mycota), *J. Phycol.* **30**, 225-230.

Gehrig, H., Schüßler, A. and Kluge, M. (1996) *Geosiphon pyriforme*, a fungus forming endocytobiosis with *Nostoc* (cyanobacteria), is an ancestral member of the Glomales: evidence by SSU rRNA analysis, *J. Mol. Evol.* **43**, 71-81.

Hensen, A. and Jahns, H.M. (1974) *Lichenes. Eine Einführung in die Flechtenkunde*, Georg Thieme Verlag, Stuttgart.

Kluge, M., Mollenhauer, D. and Mollenhauer, R. (1991) Photosynthetic carbon assimilation in *Geosiphon pyriforme* (Kützing) v. Wettstein, an endosymbiotic consortium of a fungus and a cyanobacterium, *Planta* **185**, 311-315.

Kluge, M., Mollenhauer, D., Mollenhauer, R. and Kape, R. (1992) *Geosiphon pyriforme*, an endosymbiotic consortium of a fungus and a cyanobacterium (*Nostoc*), fixes nitrogen, *Bot. Acta* **105**, 343-344.

Kluge, M., Mollenhauer, D. and Mollenhauer, R. (1994) *Geosiphon pyriforme* (Kützing) v. Wettstein, a promising system for studying endocyanoses, *Progr. Bot.* **55**, 130-141.

Kluge, M., Gehrig, H., Mollenhauer, D., Schnepf, E. and Schüßler, A. (1997) News on *Geosiphon pyriforme*, an endosymbiotic consortium of a fungus with a cyanobacterium, in H.E.A. Schenk, R. Herrmannn, K.W. Jeon, N.E. Müller, and W. Schwemmler (eds.), *Eukaryotism and Symbiosis*, Springer Verlag, Berlin Heidelberg New York, pp. 469-476.

Knapp, E. (1933) Über *Geosiphon* Fr. v. Wettstein, eine intrazelluläre Pilz-Algen-Symbiose, *Ber. Dtsch. Bot. Ges.* **51**, 210-217.

Maetz, M., Schüßler, A., Wallianos, A. and Traxel, K. (1999) Subcellular trace element distribution in *Geosiphon pyriforme, Nucl. Instrum. Meth. B* **150**, 200-207.

Minerdi, D., Fani, R., Gallo, R., Boarino, A. and Bonfante, P. (2001) Nitrogen fixation genes in an endosymbiotic *Burkholderia* strain, *Appl. Environ. Microb.* **67**, 725-732.

Mollenhauer, D. (1992) *Geosiphon pyriforme*, in W. Reisser (ed.), *Algae and Symbioses: Plants, Animals, Fungi, Viruses, Interactions Explored*, Biopress, Bristol, pp. 339-351.

Mollenhauer, D. and Mollenhauer, R. (1988) *Geosiphon* cultures ahead, *Endocyt. C. Res.* **5**, 69-73.

Mollenhauer, D. and Mollenhauer, R. (1996) *Nostoc* in symbiosis - taxonomic implications, *Arch. Hydrobiol. (Algological Studies)* **83**, 435-446.

Mollenhauer, D., Mollenhauer, R. and Kluge, M. (1996) Studies on initiation and development of the partner association in *Geosiphon pyriforme* (Kütz.) v. Wettstein, a unique endocytobiotic system of a fungus (Glomales) and the cyanobacterium *Nostoc punctiforme* (Kütz.) Hariot, *Protoplasma* **193**, 3-9.

Paulsrud, P. (2001) *The Nostoc Symbionts in Lichens. Diversity, Specificity and Cellular Modifications*, Acta Universitatis Upsaliensis, Upsala.

Paulsrud, P. and Lindblad, P. (1998) Sequence variation of the $tRNA^{Leuc}$ intron as a marker for genetic diversity and specificity of symbiotic cyanobacteria in some lichens, *Appl. Environ. Microb.* **64**, 310-315.

Schnepf, E. (1964) Zur Feinstruktur von *Geosiphon pyriforme, Arch. Mikrobiol.* **49**, 289-309.

Schüßler, A. (1999) Glomales SSU rRNA gene diversity, *New Phytol.* **144**, 205-207.

Schüßler, A. (2000) *Glomus claroideum* forms an arbuscular mycorrhiza-like symbiosis with the hornwort *Anthoceros punctatus, Mycorrhiza* **10**, 15-21.

Schüßler, A. (2002) Molecular phylogeny, taxonomy, and evolution of arbuscular mycorrhiza fungi and *Geosiphon pyriformis, Plant Soil*, in press.

Schüßler, A. and Kluge, M. (2001) *Geosiphon pyriforme,* an endosymbiosis between fungus and cyanobacteria, and its meaning as a model for arbuscular mycorrhiza research, in B. Hock (ed.), *The Mycota IX*, Springer Verlag, Berlin Heidelberg New York, pp. 151-161.

Schüßler, A., Mollenhauer, D., Schnepf, E. and Kluge, M. (1994) *Geosiphon pyriforme,* an endosymbiotic association of fungus and cyanobacteria: the spore structure resembles that of arbuscular mycorrhizal (AM) fungi, *Bot. Acta* **107**, 36-45.

Schüßler, A., Schnepf, E., Mollenhauer, D. and Kluge, M. (1995) The fungal bladders of the endocyanosis *Geosiphon pyriforme*, a *Glomus*-related fungus: cell wall permeability indicates a limiting pore radius of only 0.5 nm, *Protoplasma* **185**, 131-139.

Schüßler, A., Bonfante, P., Schnepf, E., Mollenhauer, D. and Kluge, M. (1996) Characterization of the *Geosiphon pyriforme* symbiosome by affinity techniques: confocal laser scanning microscopy (CLSM) and electron microscopy, *Protoplasma* **190**, 53-67.

Schüßler, A., Meyer, T., Gehrig, H. and Kluge, M. (1997) Variations of lectin binding sites in extracellular glycoconjugates during the life cycle of *Nostoc punctiforme*, a potentially endosymbiotic cyanobacterium, *Eur. J. Phycol.* **32**, 233-239.

Schüßler, A., Gehrig, H., Scharzott, D. and Walker, C. (2001a) Analysis of partial Glomales SSU rRNA genes: implications for primer design and phylogeny, *Mycol. Res.* **105**, 5-15.

Schüßler, A., Schwarzott, D. and Walker, C. (2001b) A new fungal phylum, the *Glomeromycota*: phylogeny and evolution, *Mycol. Res.* **103**, 1413-1421.
Schwarzott, D., Walker, C., and Schüßler, A. (2001) *Glomus*, the largest genus of the arbuscular mycorrhiza fungi (Glomales), is monophyletic, *Mol. Phylogenet. Evol.* **21**, 190-197.
Tobler, F. (1934) *Die Flechten*, Fischer Verlag, Jena.
v. Wettstein, F. (1915) *Geosiphon* Fr. v. Wettst., eine neue interessante Siphonee, *Öster. Bot. Z.* **65**, 145-156.

Chapter 4

CYANOLICHENS: AN EVOLUTIONARY OVERVIEW

JOUKO RIKKINEN
Department of Applied Biology, University of Helsinki
P.O. Box 27, FIN-00014 University of Helsinki, Finland

1. INTRODUCTION

Lichens are self-supporting and ecologically obligate associations between symbiotic fungi and green algae and/or cyanobacteria. The term 'cyanolichen' refers to all lichens with cyanobacterial symbionts, either as the sole photosynthetic component or as the second photobiont in addition to the primary photobiont (eukaryotic algae). Lichen symbioses represent a major way of life among the Fungi. Almost one-fifth of all known fungal species are lichen-forming and within the Ascomycota about two-fifths of known species are lichenized. The morphological and physiological characteristics of these associations are highly specialized and often involve intricate connections between the symbionts. As lichens include primary as well as secondary producers, and have their own carbon cycles, they resemble miniature ecosystems rather than individuals or populations. The symbiotic nature of these systems is not limited to the thallus level biology of individual lichen species. Symbiotic processes also shape the structure of lichen communities on a global scale.

2. DIVERSITY OF FUNGAL-CYANOBACTERIAL ASSOCIATIONS

Lichens are a biological phenomenon, not just a systematic group. Lichens do not have independent scientific names; all symbiotic partners have their own separate names and the name of intact 'lichen' refers to the dominating fungal partner alone. Many different types of fungi associate with cyanobacteria. These cyanophilous species, like all fungi, depend on nutrients contained in or released by other organisms. The nutritional requirements of many fungi are satisfied in the finely tuned symbioses, of which cyanolichens provide some outstanding examples. However, while cyanolichens are often quoted as premier examples of mutualism between prokaryotic and eukaryotic organisms, there is no reason to believe that anything but a continuous cline would exist between parasitic and mutualistic interactions in these symbioses.

Molecular studies have clearly shown that lichen-like symbioses have independently arisen on several occasions (Gargas *et al.*, 1995; Tehler *et al.*, 2000; Lutzoni *et al.*, 2001). This partly explains the present diversity of lichens and the mixed occurrences of

A.N. Rai, B. Bergman and U. Rasmussen (eds.), Cyanobacteria in Symbiosis, 31-72.

2001). This partly explains the present diversity of lichens and the mixed occurrences of lichenized and non-lichenized species in many fungal groups. Aptroot (1998) estimated that there would have been as many as 100 lichenization events, involving re- and de-lichenization, during the diversification of extant Fungi. Some molecular evidence suggests that the gains of lichenization have been infrequent during evolution and there may have been several independent losses of the lichen symbiosis in different ascomycete lineages (Kranner and Lutzoni, 1999; Lutzoni *et al.*, 2001). As a consequence, some lineages of exclusively non-lichen-forming fungi may have evolved form lichen-forming ancestors (Lutzoni *et al.*, 2001).

In order to demonstrate the diversity of extant fungal-cyanobacterial relationships an attempt was made to collect them all into a single list (Table 1). The list is not complete, but includes a vast majority of presently known mutualistic, commensalistic and parasitic interactions between cyanobacteria and fungi. For reasons of space the list could not be annotated. The interested reader is first referred to the Ainsworth & Bisby's Dictionary of the Fungi (Hawksworth *et al.*, 1995). For extensive lists of lichenological literature one can consult the The Bryologist's Recent Literature Lists, presently on-line at http://www.toyen.uio.no/botanisk/bot-mus/lav/sok_rll.htm

2.1. Mycobionts of Cyanolichens

Cyanobacteria form lichen symbioses almost exclusively with Ascomycota.Depending on the taxonomic classification used, 15–18 orders of ascomycetes include lichen-forming taxa (Table 1). Most of these include both lichenized and non-lichenized species and only a few groups are exclusively lichenized. Molecular data are rapidly adding to a new understanding of the phylogenetic relationships between different ascomycete groups, including those with lichenized species. However, as most ascomycete genera have yet to have any of their species sequenced, the current classification is clearly a system in transition (Hawksworth*et al.*, 1995; Alexopoulos *et al.*, 1996; Tehler *et al.*, 2000).

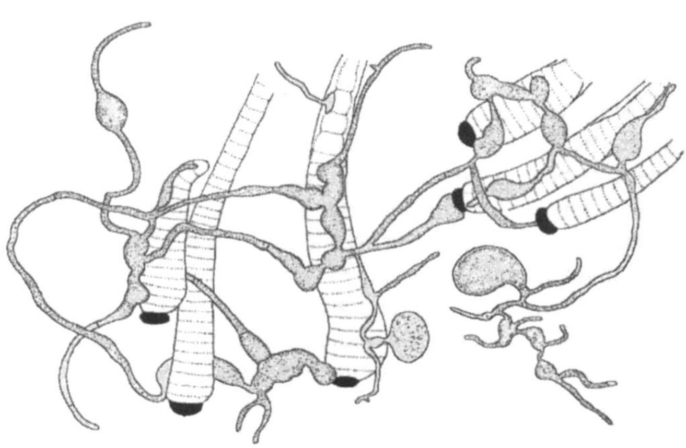

Figure 1. Polycentric chytrid feeding on filamentous cyanobacteria (after Karling, 1977).

Many zygomycetes and basidiomycetes enter into mycorrhizal symbioses, but mutualistic or parasitic associations between these types of fungi and cyanobacteria appear to be rare (Table 1). The phylum Chytridiomycota includes mainly obligate parasites. Symbiotic forms are only known from animal guts, where anaerobic rumen chytrids take part in the digestive processes of herbivores. Some aquatic chytrids are known to feed on filamentous cyanobacteria (Fig. 1). For example, different species of *Rhizophydium* have been recorded from *Anabaena*, *Aphanizomenon*, *Calothrix*, *Dichothrix*, *Lyngbya*, and *Oscillatoria* (Table 1). In addition to true fungi, some oomycetes (Oomycota) also live on free-living cyanobacteria (e.g., *Lagenidium* on *Lyngbya* and *Syzygangia* on *Tolypothrix*). However, none of these organisms have yet been recorded from lichen thalli (Sparrow, 1960; Karling, 1977).

2.1.1. Primary Mycobionts

Some 13500 species of lichen-forming ascomycetes are presently known (Hawksworth *et al.*, 1995). Cyanotrophic and lichenicolous taxa are found in many different ascomycete groups, including several pyrenocarpous orders. Lichen-forming species occur in about 50 families and 130 genera, most of which belong to two apothecial orders: Lecanorales and Lichinales. Approximately 1550 species of cyanolichens are presently known. Thus, roughly 12% of all lichen-forming fungi are associated with cyanobacteria (Table 1).

Lecanorales is one of the largest ascomycete orders. Most species in the group are lichen-forming and most lichenicolous forms appear to have evolved from lichenized ancestors (Hawksworth *et al.*, 1995, Rambold and Triebel, 1992; Lutzoni *et al.*, 2001). Overall, however, a large majority of lecanoralean lichens are green algal and only a minority house cyanobacteria as primary or accessory symbionts. The order includes foliose and fruticose macrolichens as well as many crustose species. Ascomata are typically apothecial, flat to cup-shaped, and usually with active ascospore dispersal from thick-walled asci. The anamorphs are pycnidial. In general, the multi-layered ascus wall and a more or less developed amyloid structure in the ascus apex characterize members of Lecanorales. However, molecular data also indicate large monophyletic groups within the order (Rambold and Triebel, 1992; Dellemère, 1994; Haffellner *et al.*, 1994; Hawksworth *et al.*, 1995; Tehler, 1996; Tehler *et al.*, 2000; Lutzoni *et al.*, 2001). Depending on the taxonomic classification used, the order Lecanorales includes some 40 families and well over 300 genera. The circumscription and placement of many groups still remain uncertain. Cyanophilous species occur in ca. 25 families and nearly 80 genera (Table 1).

Lecanoralean cyanolichens are found in many types of environments ranging from open tundra to rainforests. However, most species thrive in habitats that combine moderate light intensities with a relatively high atmospheric humidity. Because of their ability to utilize low photon flux densities and wavelengths of light that have filtered through a vascular plant canopy, many cyanolichens are well adopted to live in shady forest habitats (Rikkinen, 1995). Epiphytic communities rich in cyanolichen species are remarkably similar in composition over the whole circumpolar belt of boreal coniferous forests and the adjoining areas of mixed coniferous-deciduous forest.

Table 1. Fungal-cyanobacterial interactions. Most known fungi involved in lichen-like symbioses with cyanobacteria, as well as parasitic or commensalistic interactions are listed. The identity of the lichen cyanobiont or an example of the host cyanolichens is given. The last column gives a rough estimate of the number of cyanophilic species in each genus. Note that some genera also contain species that do not associate with cyanobacteria.

Fungal group	Host	Interaction type	Thallus morphology	#
CHYTRIDIOMYCOTA				
Blastocladiales				
Blastocladiaceae				
Blastocladiella	*Anabaena*	Parasitic fungus	Indeterminate	1
Inc. sed.				
Rhizosiphon	*Anabaena*	Parasitic fungus	Indeterminate	3
Chytridiales				
Chytridiaceae				
Chytridium	*Anabaena*	Parasitic fungus	Indeterminate	2
Nowakowskia	*Hormotheca*	Parasitic fungus	Indeterminate	1
Phlyctochytridium	*Stigonema*	Parasitic fungus	Indeterminate	1
Rhizophydium	*Anabaena* etc.	Parasitic fungus	Indeterminate	c. 10
Cladochytriaceae				
Coenomyces	*Calothrix* etc.	Parasitic fungus	Indeterminate	1
Spizellomycetales				
Spizellomycetaceae				
Rhizophlyctis	*Calothrix* etc.	Parasitic fungus	Indeterminate	2
ZYGOMYCOTA				
Inc. sed.				
Geosiphon	*Nostoc*	Endocyanosis	Indeterminate	1
Winfrenatia?	*Chroococcidiopsis*?	Fossil bipartite lichen	Foliose	1
BASIDIOMYCOTA				
Agaricales				
Tricholomataceae				
Fayodia	*Peltigera*	Lichenicolous fungus	Indeterminate	3
Omphalina	*Peltigera*	Lichenicolous fungus	Indeterminate	2
Rickenella	*Nostoc*?	Tripartite symbiosis?	Indeterminate	1
Aphyllophorales				
Corticiaceae				
Acantholichen	*Scytonema*	Bipartite lichen	Foliose	1
Meruliaceae				
Dictyonema	*Scytonema*	Bipartite lichen	Indeterminate	5
Tremellales				
Tremellaceae				
Tremella	*Lobaria*	Lichenicolous fungus	Indeterminate	5
ASCOMYCOTA				
Agyriales				
Agyriaceae				
Placopsis	*Nostoc* etc.	Tripartite lichen	Foliose	c. 55
Arthoniales				
Arthoniaceae				

Arthonia	*Peltigera*	Lichenicolous fungus	Indeterminate	c. 13
Roccellaceae				
Plectocarpon	*Nephroma*	Lichenicolous fungus	Indeterminate	2
Opegrapha	*Peltigera*	Lichenicolous fungus	Indeterminate	1
Diaporthales				
Valsaceae				
Obryzum	*Leptogium*	Lichenicolous fungus	Indeterminate	1
Dothideales *s. lat.*				
Chaeotothyriales				
Biciliopsis	*Leptogium*	Lichenicolous fungus	Indeterminate	1
Dacampiaceae				
Dacampia	*Peltigera*	Lichenicolous fungus	Indeterminate	2
Polycoccum	*Peltigera*	Lichenicolous fungus	Indeterminate	7
Pyrenidium	*Peltigera*	Lichenicolous fungus	Indeterminate	2
Herpotrichiellaceae				
Capronia	*Peltigera*	Lichenicolous fungus	Indeterminate	1
Hysteriaceae				
Hemigrapha	*Peltigera*	Lichenicolous fungus	Indeterminate	1
Leptosphaeriaceae				
Leptosphaeria	*Peltigera*	Lichenicolous fungus	Indeterminate	1
Microthyriaceae				
Lichenopeltella	*Peltigera*	Lichenicolous fungus	Indeterminate	7
Mycosphaerellaceae				
Sphaerulina	*Collema*	Lichenicolous fungus	Indeterminate	1
Stigmidium	*Peltigera*	Lichenicolous fungus	Indeterminate	6
Pleosporaceae				
Pleospora	*Solorina*	Lichenicolous fungus	Indeterminate	1
Leptosphaerulina	*Peltigera*	Lichenicolous fungus	Indeterminate	1
Pyrenothricaceae				
Pyrenothrix	*Scytonema*	Parasitic fungus	Indeterminate	1
Cyanoporina	Cyanobacterium	Parasitic fungus	Indeterminate	1
Pseudoperisporiaceae				
Myxophora	*Scytonema*	Parasitic fungus	Indeterminate	1
Wentiomyces	*Peltigera*	Lichenicolous fungus	Indeterminate	1
Inc. sed.				
'Arthopyrenia'	Chroococcales	Bipartite lichen?	Indeterminate	1
Cercidospora	*Peltigera*	Lichenicolous fungus	Indeterminate	4
Collemopsidium	*Xanthocapsa*	Bipartite lichen?	Indeterminate	3
'Didymella'	*Zahlbrucknerella*	Lichenicolous fungus	Indeterminate	1
Didymellopsis	*Collema*	Lichenicolous fungus	Indeterminate	4
Endococcus	*Collema*	Lichenicolous fungus	Indeterminate	3
Phycorella	*Scytonema*	Parasitic fungus	Indeterminate	1
Thelenidia	Cyanobacterium	Bipartite lichen?	Indeterminate	1
Zwackhiomyces	*Peltigera*	Lichenicolous fungus	Indeterminate	3
Erysiphales				
Erysiphaceae				
Phyllactinia	*Peltigera*	Lichenicolous fungus	Indeterminate	1
Hypocreales				
Hypocreaceae				
Nectria	*Peltigera*	Lichenicolous fungus	Indeterminate	1
Nectriopsis	*Coccocarpia*	Lichenicolous fungus	Indeterminate	3
Paranectria	*Peltigera*	Lichenicolous fungus	Indeterminate	1
Pronectria	*Peltigera*	Lichenicolous fungus	Indeterminate	4
Xenonectriella	*Collema*	Lichenicolous fungus	Indeterminate	1
Niessliaceae				
Niesslia	*Lobaria*	Lichenicolous fungus	Indeterminate	1
Lecanorales				

Acarosporaceae				
Sarcosagium	*Peltigera*	Lichenicolous lichen	Indeterminate	1
Thelocarpon	*Peltigera*	Lichenicolous lichen	Indeterminate	2
Arctomiaceae				
Arctomia	*Nostoc*	Bipartite lichen	Crustose	2
Wawea	*Nostoc*	Bipartite lichen	Fruticose	1
Bacidiaceae				
Bacidia	*Peltigera*	Lichenicolous lichen	Crustose	3
Compsocladium	*Scytonema*	Tripartite lichen	Fruticose	1
Szczawinskia	*Stigonema*	Tripartite lichen	Crustose	1
Biatoraceae				
Toninia	*Stigonema, Gloeocapsa*	Lichenicolous lichen	Crustose	c. 40
Brigantiaeaceae				
Argopsis	*Nostoc* etc.	Tripartite lichen	Fruticose	3
Candelariaceae				
Candellariella	*Stigonema* etc.	Cyanotrophic lichen	Crustose	4
Catillariaceae				
Catillaria	*Peltigera*	Lichenicolous lichen	Crustose	1
Halecania	*Stigonema* etc.	Cyanotrophic lichen	Crustose	1
Cladoniaceae				
Pilophorus	*Nostoc* etc.	Tripartite lichen	Fruticose	11
Coccocarpiaceae				
Coccocarpia	*Scytonema*	Bipartite lichen	Foliose	20
Peltularia	*Scytonema* etc.	Bipartite lichen	Foliose	3
Spilonema	*Stigonema* etc.	Bipartite lichen	Fruticose	4
Spilonemella	*Scytonema*	Bipartite lichen	Crustose	2
Steinera	*Nostoc* etc.	Bipartite lichen	Crustose	4
Collemataceae				
Collema	*Nostoc*	Bipartite lichen	Foliose	c. 75
Homothecium	*Nostoc*	Bipartite lichen	Foliose	5
Leciophysma	*Nostoc*	Bipartite lichen	Fruticose	2
Leightoniella	*Nostoc*	Bipartite lichen	Foliose	1
Leptogium	*Nostoc*	Bipartite lichen	Foliose	c. 160
Physma	*Nostoc*	Bipartite lichen	Foliose	c. 10
Ramalodium	*Nostoc*	Bipartite lichen	Foliose	4
Staurolemma	*Nostoc*	Bipartite lichen	Foliose	4
Ectolechiaceae				
Calopadia	*Scytonema*	Tripartite lichen	Crustose	1
Lasioloma	*Scytonema*	Tripartite lichen	Crustose	3
'Lopadium'	*Scytonema*	Tripartite lichen	Crustose	1
Heppiaceae				
Corynecystis	Cyanobacterium	Bipartite lichen	Foliose	1
Epiphloea	*Nostoc*	Bipartite lichen	Foliose	1
Heppia	*Scytonema*	Bipartite lichen	Foliose	6
Lecanoraceae				
Bryodina	*Stigonema*	Cyanotrophic lichen	Crustose	3
Carbonea	*Gloeocapsa*	Cyanotrophic lichen	Crustose	2
Glyphopeltis	*Peltula*	Lichenicolous lichen	Crustose	1
Lecanora	*Stigonema* etc.	Cyanotrophic lichen	Crustose	5
Miriquidica	*Stigonema* etc.	Cyanotrophic lichen	Crustose	4
Lobariaceae				
Lobaria	*Nostoc*	Bi- / tripartite lichen	Foliose	c. 60
Pseudocyphellaria	*Nostoc*	Bi- / tripartite lichen	Foliose	c. 115
Sticta	*Nostoc*	Bi- / tripartite lichen	Foliose	c. 200
'Dendriscocaulon'	*Nostoc*	Bipartite lichen	Fruticose	c. 10
Nephromataceae				

Nephroma	*Nostoc*	Bi- / tripartite lichen	Foliose	c. 35
Micareaceae				
Micarea	*Stigonema* etc.	Cyanotrophic lichen	Crustose	3
Scutula	*Peltigera*	Lichenicolous fungus	Crustose	c. 10
Pannariaceae				
Degelia	*Scytonema*	Bipartite lichen	Foliose	c. 10
Erioderma	*Scytonema*	Bipartite lichen	Foliose	16
Fuscoderma	*Nostoc*	Bipartite lichen	Foliose	3
Fuscopannaria	*Nostoc*	Bipartite lichen	Foliose	c. 30
Hosseusia	Cyanobacterium	Bipartite lichen	Foliose	1
Leioderma	*Scytonema*	Bipartite lichen	Foliose	5
Moelleropsis	*Nostoc*	Bipartite lichen	Crustose	2
Pannaria	*Nostoc*	Bipartite lichen	Foliose	12
Parmeliella	*Nostoc*	Bipartite lichen	Foliose	c. 40
Protopannaria	*Nostoc*	Bipartite lichen	Foliose	1
Psoroma	*Nostoc*	Bi- / tripartite lichen	Foliose	c. 50
Psoromidium	*Nostoc*	Tripartite lichen	Crustose	2
Santessoniella	*Nostoc*	Bipartite lichen	Crustose	9
Siphulastrum	*Scytonema*	Bipartite lichen	Foliose	4
Placynthiaceae				
Hertella	*Tolypothrix* etc.	Bipartite lichen	Fruticose	2
Koerberia	*Scytonema*	Bipartite lichen	Foliose	2
Leptochidium	*Scytonema*	Bipartite lichen	Foliose	1
Placynthiopsis	Cyanobacterium	Bipartite lichen	Crustose	1
Placynthium	*Dichothrix* etc.	Bipartite lichen	Crustose	25
Polychidium	*Nostoc* etc.	Bipartite lichen	Fruticose	5
Vestergrenopsis	*Scytonema*	Bipartite lichen	Foliose	2
Peltigeraceae				
Massalongia	*Nostoc*	Bipartite lichen	Foliose	3
Peltigera	*Nostoc*	Bi- / tripartite lichen	Foliose	c. 60
Solorina	*Nostoc*	Tripartite lichen	Foliose	10
Physciaceae				
Rinodina	*Stigonema* etc.	Cyanotrophic lichen	Crustose	6
Porpidiaceae				
Amygdalaria	*Stigonema* etc.	Tripartite lichen	Crustose	3
Psoraceae				
Psorula	*Stigonema*	Cyanotrophic lichen	Crustose	1
Sphaerophoraceae				
Sphaerophorus	*Scytonema*	Tripartite lichen	Fruticose	2
Stereocaulaceae				
Muhria	*Stigonema*	Tripartite lichen	Crustose	1
Stereocaulon	*Nostoc, Scytonema, Stigonema*	Tripartite lichen	Fruticose	c. 125
Teloschistaceae				
Caloplaca	*Stigonema*	Cyanotrophic lichen	Crustose	1
Vezdaeaceae				
Vezdaea	*Peltigera*	Lichenicolous lichen	Crustose	2
Inc. sed.				
Corticifraga	*Peltigera*	Lichenicolous fungus	Indeterminate	2
Corticiruptor	*Sticta*	Lichenicolous fungus	Indeterminate	1
Leotiales				
Dermateaceae				
Mollisia	*Collema*	Lichenicolous fungus	Indeterminate	1
Hyaloscyphaceae				
Polydesmia	*Peltigera*	Lichenicolous fungus	Indeterminate	1
Leotiaceae				

Calycina	*Peltigera*	Lichenicolous fungus	Indeterminate	2
Unguiculariopsis	*Sticta*	Lichenicolous fungus	Indeterminate	2
Orbiliaceae				
Orbilia	*Peltigera*	Lichenicolous fungus	Indeterminate	1
Inc. sed.				
Skytella	*Peltigera*	Lichenicolous fungus	Indeterminate	1
Lichinales				
Gloeoheppiaceae				
Gloeoheppia	Chroococcales	Bipartite lichen	Foliose	4
Gudelia	Chroococcales	Bipartite lichen	Foliose	1
Pseudopeltula	Chroococcales	Bipartite lichen	Foliose	3
Lichinaceae				
Anema	*Gloeocapsa*	Bipartite lichen	Foliose	7
Calotrichopsis	*Calothrix*	Bipartite lichen	Crustose	3
Cryptothele	*Gloeocapsa*	Bipartite lichen	Crustose	3
Digitothyrea	Chroococcales	Bipartite lichen	Fruticose	3
Edwardiella	*Gloeocapsa*	Bipartite lichen	Indeterminate	1
Ephebe	*Stigonema*	Bipartite lichen	Fruticose	12
Euopsis	*Gloeocapsa*	Tripartite lichen	Crustose	2
Finkia	Cyanobacterium	Bipartite lichen	Crustose	1
Gyrocollema	Cyanobacterium	Bipartite lichen	Crustose	1
Jenmania	*Gloeocapsa*	Bipartite lichen	Fruticose	2
Lecidopyrenopsis	*Gloeocapsa*	Bipartite lichen	Crustose	1
Lemmopsis	Cyanobacterium	Bipartite lichen	Crustose	3
Lempholemma	*Nostoc*	Bipartite lichen	Foliose	c. 30
Leprocollema	Cyanobacterium	Bipartite lichen	Crustose	3
Lichina	*Calothrix* etc.	Bipartite lichen	Crustose	7
Lichinella	Chroococcidiopsis	Bipartite lichen	Fruticose	11
Lichinodium	*Scytonema*	Bipartite lichen	Fruticose	4
Mawsonia	Cyanobacterium	Bipartite lichen	Crustose	1
Metamelanea	Chroococcales	Bipartite lichen	Fruticose	3
Paulia	Chroococcales	Bipartite lichen	Foliose	12
Peccania	*Chroococcidiopsis*	Bipartite lichen	Foliose	4
Phloeopeccania	*Gloeocapsa*	Bipartite lichen	Foliose	1
Phylliscidiopsis	Cyanobacterium	Bipartite lichen	Crustose	1
Phylliscidium	*Gloeocapsa*	Bipartite lichen	Foliose	1
Phyllisciella	Chroococcales	Bipartite lichen	Foliose	3
Phylliscum	Chroococcales	Bipartite lichen	Foliose	6
Porocyphus	*Calothrix*	Bipartite lichen	Foliose	7
Psorotichia	*Chroococcidiopsis*	Bipartite lichen	Crustose	c. 50
Pterygiopsis	Chroococcales	Bipartite lichen	Crustose	c. 8
Pyrenocarpon (*syn. Thelochroa)*	Chroococcales	Bipartite lichen	Crustose	1
Pyrenopsis	*Gloeocapsa* etc.	Bipartite lichen	Crustose	c. 40
Stromatella	Chroococcales	Bipartite lichen	Fruticose	1
Synalissa	*Gloeocapsa*	Bipartite lichen	Fruticose	5
Thelignya	*Calothrix*	Bipartite lichen	Crustose	1
Thermutis	*Scytonema*	Bipartite lichen	Fruticose	1
Thermutopsis	*Tolypothrix*	Bipartite lichen	Fruticose	1
Thyrea	Chroococcales	Bipartite lichen	Fruticose	c. 20
Zahlbrucknerella	*Scytonema*	Bipartite lichen	Fruticose	7
Peltulaceae				
Neoheppia	Chroococcales	Bipartite lichen	Foliose	2
Peltula	*Scytonema*	Bipartite lichen	Foliose	c. 18
Phyllopeltula	*Scytonema*	Bipartite lichen	Foliose	2
Ostropales				
Stictidaceae				

Nanostictis	*Peltigera*	Lichenicolous fungus	Indeterminate	2
Petractis	*Scytonema*	Bipartite lichen	Crustose	1
Patellariales				
Arthrorhapidiaceae				
Arthrorhapis	*Pseudocyphellaria*	Lichenicolous lichen	Crustose	1
Pertusariales				
Coccotremataceae				
Coccotrema (incl. *Lepolichen*)	*Calothrix* etc.	Tripartite lichen	Crustose	9
Phyllacorales				
Lichenochora	*Fuscopannaria*	Lichenicolous fungus	Indeterminate	1
Pyrenulales				
Pyrenulaceae				
Cyanopyrenia	Cyanobacteria	Bipartite lichen	Crustose	1
Pyrenocollema	*Gleocapsa* etc.	Bipartite lichen	Crustose	c. 20
Sordariales				
Coniochaetaceae				
Roselliniella	*Peltigera*	Lichenicolous fungus	Indeterminate	5
Roselliniopsis	*Placopsis*	Lichenicolous fungus	Indeterminate	1
Syanaptospora	*Placopsis*	Lichenicolous fungus	Indeterminate	1
Nitschkiaceae				
Acanthonitschkea	*Peltigera*	Lichenicolous fungus	Indeterminate	1
Lasiosphaeriopsis	*Stereocaulon*	Lichenicolous fungus	Indeterminate	2
Rhagadostoma	*Peltigera*	Lichenicolous fungus	Indeterminate	1
Trichosphaeriales				
Trichosphaeriaceae				
Trichosphaeria	*Peltigera*	Lichenicolous fungus	Indeterminate	1
Rhynchomeliola	*Wawea*	Lichenicolous fungus	Indeterminate	1
Verrucariales				
Verrucariaceae				
Agonimia	*Pseudocyphellaria*	Lichenicolous lichen	Indeterminate	1
Dermatocarpon	*Collema*	Lichenicolous lichen	Foliose	1
Norrlinia	*Peltigera*	Lichenicolous fungus	Indeterminate	1
Phaeospora	*Peltigera*	Lichenicolous fungus	Indeterminate	1
Polyblastia	*Peltigera*	Lichenicolous fungus	Indeterminate	1
Psoroglaena	Cyanobacterium	Bipartite lichen	Crustose	3
Ascomycota (inc. sed.)				
Aphanopsidaceae				
Steinia	*Peltigera*	Lichenicolous lichen	Crustose	1
Dactylosporaceae				
Dactylospora	*Lobaria*	Lichenicolous fungus	Crustose	3
Melaspileaceae				
Melaspilea	*Pseudocyphellaria*	Lichenicolous fungus	Indeterminate	1
Protothelenellaceae				
Protothelenella	*Peltigera*	Lichenicolous lichen	Crustose	2
Genera (inc. sed.)				
Abrothallus	*Nephroma*	Lichenicolous fungus	Indeterminate	3
Neolamya	*Peltigera*	Lichenicolous fungus	Indeterminate	1
Pleosphaeria	*Pyrenothrix*	Parasitic fungus	Indeterminate	1

Examples of mitosporic genera that include lichenicolous species on cyanolichens:
Dinemasporium, Diplodina, Epicoccum, Everniicola, Fusarium, Graphium, Hastifera, Hawksworthiana, Hobsonia, Illosporium, Karsteniomyces, Libertiella, Lichenoconium, Refractohilum, Talpapellis, Trichoconis, Phoma, Vagnia

Lichinales includes three families and over 40 genera (Table 1). All species in this order are lichenized and nearly all of them associate with cyanobacteria (Henssen *et al.*, 1987). Thalli are crustose, foliose or fruticose, and often gelatinous. Ascomata are apothecial, but may be perithecial in early stages, often immersed into the thallus. The asci are thin-walled and usually disintegrate at maturity releasing hyaline, non-septate spores. The anamorphs are pycnidial. Most species of Lichinales grow on rock or mineral soil. The group is well represented in humid coastal areas and at sites in dry regions where seepage water is periodically available. Many species are inconspicuous, but they can play important ecological roles in semi-arid ecosystems. Some species are dominant components of soil crusts in savannas, semi-deserts, deserts, and disturbed sites (Eldridge and Tozer, 1997; Schultz *et al.*, 2000).

2.1.2. Accessory Mycobionts and Cyanotrophic Fungi

Many lichenicolous fungi live on or inside lichens as parasites, commensals or saprobes (Table 1). Foliose and fruticose macrolichens can harbour a great variety of filamentous fungi and yeasts (Petrini *et al.*, 1990; Girlanda *et al.*, 1997). However, common saprophytic moulds are comparatively scarce on living lichens and many of the lichenicolous fungi are exclusively lichenicolous. For example, Hawksworth and Miadlikowska (1997) listed 87 species of lichenicolous fungi from the genus *Peltigera* alone, of which 61 were not known from any other host. Also many other large cyanolichens are rich in lichenicolous species (Alstrup and Hawksworth, 1990; Kondratyuk and Galloway, 1995; Aptroot *et al.*, 1997).

Even though lichenicolous fungi still remain a relatively undiscovered group of organisms (Hawksworth and Rossman, 1997), hundreds of species are already known. Many of them do not seem to cause major damage to their hosts. Most of them appear to exploit the photobionts of their host without direct nutritive exchange with the primary mycobiont (Fig. 2). Thus, the relationships between the fungal bionts are usually seen as commensalistic or antagonistic. However, in most cases this has not been experimentally demonstrated. Many lichenicolous fungi seem to have evolved from lichenized ancestors. The fungi continue to obtain carbohydrates from lichen symbioses without the need to find an appropriate free-living photobiont during each reproductive cycle (Rambold and Triebel, 1992; Lutzoni *et al.*, 2001).

Lichenicolous lichens grow on other lichens, either as commensals or parasites (Table 1). In these associations the two lichen mycobionts have their own photobionts, whereas no additional photobionts occur in lichenicolous fungi (Fig. 2). Numerous green algal lichens can grow on cyanolichens. Many of them, like the *Toninia* species, start their development on cyanolichens, but later become independent (Rambold and Triebel, 1992). Conversely, there are very few cases of lichenicolous cyanolichens growing on green algal host lichens. One example involves the occurrences of *Lichinodium sirosiphoideum* on parmelioid macrolichens (Henssen, 1963). Here, as in all multibiont associations, the exact relationship between cyanobacterial and green algal photobionts is not known. The mycobiont of the lichenicolous lichen appears to have a semi-antagonistic relationship with the host phycobiont, while the host mycobiont appears to have a semi-antagonistic or independent relationship with the cyanobiont of the lichenicolous lichen (Rambold and Triebel, 1992).

Some green algal lichens regularly associate with free-living cyanobacteria, usually *Stigonema* or *Gloeocapsa*, presumably in order to access an extra supply of nitrogen (Table 1). Such cyanotrophic associations may either be facultative or obligate (Poelt and Mayhofer, 1988). Rambold and Triebel (1992) viewed cyanotrophic behaviour as a compensating strategy of lichens that do not have symbiotic cyanobacteria as their primary photobionts or as additional symbionts in cephalodia. In many cyanotrophic associations the 'free-living' cyanobacteria are covered by dense mats of fungal hyphae. These compound structures have been called paracephalodia (Poelt and Mayhofer, 1988).

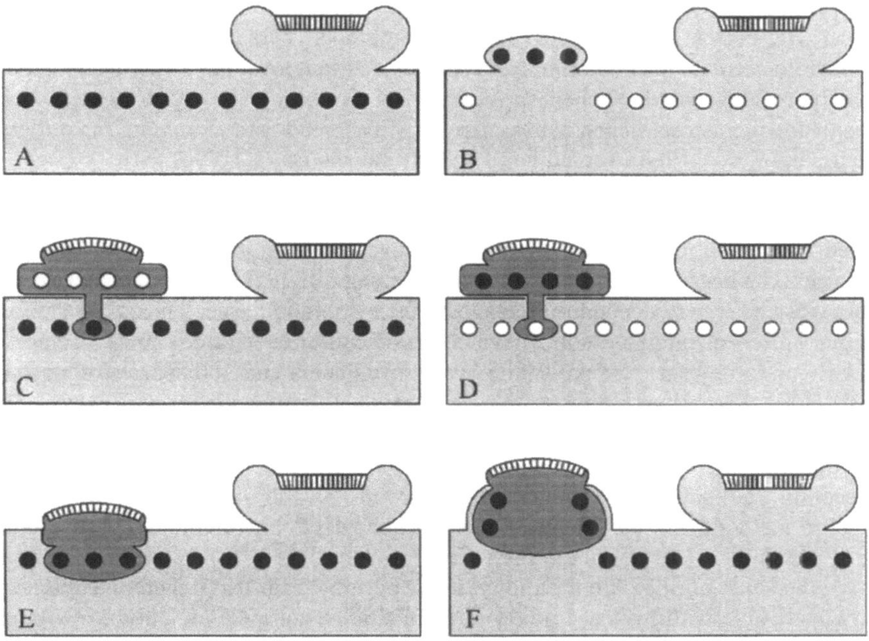

Figure 2. Examples of fungal-cyanobacterial associations: A. bipartite cyanolichen, B. tripartite (cephalodiate) cyanolichen, C. green algal lichenicolous lichen on cyanolichen, D. lichenicolous cyanolichen on green algal lichen, E. endokapylic lichenicolous fungus on cyanolichen, and F. gall-forming lichenicolous fungus on cyanolichen (after Rambold and Triebel, 1992).

In addition to the thallus-forming species, many filamentous ascomycetes obtain nutrients from free-living cyanobacteria without forming well-defined thalli (Table 1). Most of these fungi are pyrenocarpous ascomycetes, i.e., taxa with perithecium-like ascomata. The taxonomy of such fungi at family level and above remains unsettled (Hawksworth *et al.*, 1995; Aptroot, 1998). 'Non-lichenized' cyanophilous fungi are very poorly known. The nature of their interactions may be difficult to assess, but many cyanophilous fungi do not seem to seriously harm their cyanobacterial hosts in culture.

However, as almost nothing is presently known about specificity or stability in these interactions, and as many of the fungi live inside cyanobacterial colonies, they fit poorly into present definitions of a lichen. Somewhat similar symbioses are known to occur between some multicellular algae and endophytic marine fungi. Obligate associations of this type have been called mycophycobioses (Hawksworth, 1988).

2.2. Photobionts of Cyanolichens

The taxonomy of lichens is fully integrated into the classification of Fungi and most scientists working with lichens are essentially mycologists. In many lichen groups, therefore, photobiont characteristics have not been widely used in taxonomy. This partly explains why current knowledge of lichen photobionts is fragmentary at best. Serious attempts to determine the photobiont at a species or strain level have only been made for a small percentage of the lichen species. This refers not only to the photobionts of inconspicuous crustose lichens, but also to widespread and common macrolichens (Tschermak-Woess, 1988; Ahmadjian, 1993; Friedl and Büdel, 1996).

Many morphological and developmental features of lichen photobionts are not readily apparent in the symbiotic state. Often photobiont morphology is so drastically changed that even the generic position is revealed only by careful isolation and cultivation. The degree of modification varies among different photobionts and different lichen taxa, and it may depend on the age of the symbiotic tissue. Thus, the taxonomic identities of lichenized photobionts have been difficult or even impossible to determine, especially on intrageneric levels. While over thirty genera and 100 species of algae and cyanobacteria have been reported to occur as photobionts in lichens, many of the records have not been based on cultured material (Tschermak-Woess, 1988; Friedl and Büdel, 1996).

Recently the situation has greatly improved by the introduction of effective molecular methods for the identification of lichenized photobionts. These methods make it possible to identify photobionts within intact lichen thalli and offer great new tools for the study of photobiont biology. It has become clear that lichen mycobionts are strongly selective with respect to their photobionts. In many cases, only a few closely related photobiont strains serve as the appropriate symbiotic partner for individual mycobiont taxa. This indicates that the identification of photobionts will soon become a prerequisite in studies of lichen systematics (Miao *et al.*, 1997; Rambold *et al.*, 1998; Beck *et al.*, 1998; Paulsrud and Lindblad, 1998; Paulsrud *et al.*, 1998, 2000, 2001; Beck, 1999; Kroken and Taylor, 2000; Paulsrud, 2001; Lohtander, Oksanen, Paulsrud and Rikkinen, unpublished results).

2.2.1. Cyanobionts

Approximately 1700 species of fungi associate with different types of cyanobacteria (Table 1). As an adaptation to these interactions the cyanobacteria often undergo remarkable anatomical and physiological modifications (Rai *et al.,* 2000). Thus, without the help of molecular methods, isolation and cultivation is often necessary for positive identification even at the generic level. Many lichen-forming cyanobacteria can be easily isolated and brought into unialgal culture (Ahmadjian, 1993). However, as lichen thalli commonly house a rich flora of epiphytic cyanobacteria, it is quite crucial to confirm that the isolated strains actually represent the symbionts within the corresponding lichen thalli.

The genus *Nostoc* is by far the most common cyanobiont in lichens, especially in the Lecanorales (Table 1). Strains of *Nostoc* are well known for their ability to enter into different types of symbioses, sometimes serving as a source of fixed carbon and nitrogen (as in bipartite cyanolichens), or strictly as a source of nitrogen (as in other symbioses). There is still controversy over the generic delimitation of *Nostoc* and several classification schemes with slightly different taxonomic concepts are currently in use (Geitler, 1932; Rippka *et al.*, 1979; Kommarek and Anagnostidis, 1989; Castenhotz, 2001). All *Nostoc* spp. are filamentous and heterocystous, produce isopolar trichomes, lack branching, and their cells are cylindrical or spherical. Most of them posses a characteristic life cycle with distinct motile hormogonia and vegetative filaments showing different degrees of coiling. However, some lichen-forming strains do not produce hormogonia under common growth conditions and they are thus very slow to spread on culture plates. The trichomes of these strains tend to occur in pearl-like colonies eventually giving rise to grape-like clusters. While the production of hormogonia is a diagnostic feature of *Nostoc sensu* Rippka *et al.* (1979), these non-motile strains also fit well into the more classical circumscriptions of *Nostoc* (Paulsrud, 2001).

Several botanical names have been used for morphologically different strains of lichen-forming *Nostoc*. The cyanobionts of many cyanolichens, like various species of *Peltigera*, have been called *Nostoc punctiforme* (Tschermak-Woess, 1988). On the other hand, at least *Nostoc commune*, *N. microscopicum*, *N. muscorum*, *N. punctiforme*, and *N. sphaericum* have been identified as cyanobionts in different *Collema* species (Degelius, 1954). *Nostoc punctiforme* PCC73102 is a model strain for cyanobacterial symbioses, and most research on the physiology and molecular biology of symbiotic cyanobacteria have been done by using either this strain or its counterpart in the American Type Culture Collection, ATCC 29133 (Meeks *et al.*, 1999). However, in recent inoculation experiments this strain was not incorporated into the tripartite lichen *Peltigera aphthosa* (Paulsrud *et al.*, 2001). Clearly more research is needed before bacteriological or botanical species names can be used for different groups of lichen-forming *Nostoc*s (Büdel, 1992; Friedl and Büdel, 1996).

The genetic diversity of *Nostoc* cyanobionts in many cyanolichens has been studied by using nucleotide sequences of the tRNA ^{Leu} (UAA) intron as a genetic marker. Also 16S rDNA sequences have been used to resolve phylogenetic relationships (e.g., Paulsrud and Lindblad, 1998; Paulsrud, 2001; Rikkinen, Lohtander and Oksanen, unpublished results). These studies have confirmed that there is considerable genetic variation among lichen-forming *Nostoc* strains. The symbiotic strains group together with free-living *Nostoc* strains, forming a monophyletic group among the nostocalean cyanobacteria. The *Nostoc* clade is divided into several subgroups, one of which seems to only include symbiotic strains from epiphytic cyanolichens. Another group includes the cyanobionts of many terricolous cyanolichens, including *Nostoc* strains from other symbiotic systems (Lohtander, Oksanen and Rikkinen, unpublished results).

In addition to *Nostoc*, several other nostocalean genera also include lichen-forming taxa (Table 1). However, these cyanobionts have not yet been studied in detail. *Scytonema*, *Calothrix* and *Dichothrix* have been reported from a number of cyanolichens. *Scytonema* is also an important host for cyanophilous fungi. The stigonematalean cyanobacteria include some lichen-forming strains, most of which seem to belong to the genus *Stigonema*. These cyanobacteria evolve loose cyanotrophic associations with many cyanotrophic lichens (Tschermak-Woess, 1988; Friedl and Büdel, 1996).

The unicellular cyanobionts of lichens include genera that reproduce by binary fission (e.g. *Gloeocapsa*) or by both binary and multiple fission (e.g., *Chroococcidiopsis* and *Myxosarcina*). Unfortunately these cyanobionts rarely show their specific mode of reproduction when lichenized. Strains of *Gloeocapsa* and *Chroococcidiopsis* appear to be the most important photobionts in the Lichinales. Within the Lecanorales, *Gloeocapsa* forms loose associations with many green algal cyanotrophic lichens (Table 1). Many other genera of unicellular cyanobacteria have been reported from lichens, but these records have not been based on cultured material (Tschermak-Woess, 1988; Friedl and Büdel, 1996).

2.2.2. Phycobionts

The primary photobionts in most tripartite lichens are coccoid green algae. However, the most widely distributed lichen phycobiont, *Trebouxia*, is rare in cyanolichens. The phycobionts of *Stereocaulon* and *Pilophorus* belong to this group and *Dictyochloropsis*, another genus of Trebouxiophyceae, is the green algal photobiont in tripartite species of *Sticta* and *Lobaria* (Tschermak-Woess, 1988, 1995; Friedl, 1995; Friedl and Büdel, 1996; Friedl and Rokitta, 1997; Rambold *et al.*, 1998). As a whole, some 40 percent of all lichen-forming fungi associate with trebouxioid green algae. These phycobionts are particularly dominant in the in lichen floras of cool and temperate regions. *Trentepohlia* and related genera (Ulvophyceae) represent another major group of lichen phycobionts. These algae are particularly common in tropical and subtropical lichen floras. However, no associations between lichens with trentepohlioid phycobionts and cyanobacteria appear to have been reported, not even loose, cyanotrophic associations (Rikkinen, 1995; Rambold *et al.*, 1998).

Coccomyxa is the primary phycobionts of tripartite *Nephroma* and *Peltigera* species (Tschermak-Woess, 1988; Vitikainen, 1994; Friedl and Büdel, 1996; Miadlikowska and Lutzoni, 2000). This alga reproduces exclusively by autospores and its exact taxonomic position thus remains unclear. We have studied genetic variation in phycobionts of five tripartite cyanolichens. The specimens had been collected from different geographical areas, including both European and North American sites. There was almost no variation among green algal ITS sequences from *Nephroma arcticum*, *N. expallidum*, *Peltigera aphthosa*, and *P. leucophlebia*. The phycobiont of *P. britannica* was more distinct, but when analysed with other green algae, it grouped together with the phycobionts of the other tripartite lichens. Thus, the deviating phycobiont may have been a different species or subspecies of *Coccomyxa* (Lohtander, Oksanen and Rikkinen, unpublished results). Comparable patterns of genetic diversity have been reported from the photobionts of some green algal lichens (Kroken and Taylor, 2000).

2.3. Associated Organisms

In addition to the functional components of lichen symbioses, many other organisms can live closely associated with cyanolichens. For example, heterotrophic bacteria are extremely abundant in the gelatinous sheaths of cyanobacteria, including those isolated from lichens (Ahmadjian, 1989). Nothing is presently known of the possible role of these bacteria in lichen symbioses. Naturally, all lichen bionts can also be hosts for still-smaller biological entities, such as viruses.

Large cyanolichens may often house a rich flora of epiphytic algae. Both diatoms and green algae are often seen and free-living cyanobacteria are quite common. Epiphytic diatoms are most frequent in relatively dry habitats, whereas green algae and cyanobacteria are often seen in moist environments (Round, 1984; Büdel *et al.*, 1994). Small epiphytic liverworts (e.g., Lejeuneaceae) are common on large foliose cyanolichens, especially in wet tropical forests. When growing on epiphytic lichens, these organisms can be called hyperepiphytes.

While most free-living algae and cyanobacteria are clearly not acceptable symbiotic partners for lichen-forming fungi, some lichen mycobionts which are initially unable to establish associations with suitable photobionts, may preliminarily exploit free-living algae or cyanobacteria, until more suitable photobionts are encountered (Ott, 1987; Gassmann and Ott, 2000). Recently, Etges and Ott (2001) demonstrated that axenically grown and transplanted lichen mycobionts could survive for over a year in their natural habitat.

3. STRUCTURAL - FUNCTIONAL ORGANISATION OF CYANOLICHENS

Rambold and Triebel (1992) suggested that in lichens, symbiosis should be understood in a broad sense, including the phenomena of mutualism, commensalism, and possibly parasitism. They described lichens as stable two-, three- or sometimes four-biont systems. The multibiont associations include the cephalodiate cyanolichens, but also systems that have bi- or tripartite lichens as primary parts and lichenicolous fungi or lichens as accessory parts (Hawksworth, 1988; Rambold and Triebel, 1992). In many lichens the relationships between the bionts are not static, but can change depending on environmental conditions and the ontogenetic phases of the bionts or the symbiotic consortium (Richardson, 1999).

All lichen-forming fungi exploit photobionts for their own benefit. However, they also depend on the efficiency of the photobionts in capturing, transforming and translocating solar energy. Many external factors, like the intensity of ambient illumination, water availability, temperature, and the ionic environment can influence net photosynthesis (see next chapter). In this context the lichen thallus can be seen to provide as a relatively stable environment for lichen photobionts. Usually the lichen-forming fungus mediates almost all interactions between the photobionts and the outside world and, by doing so, cushions the impact of environmental extremes. For example, some lichen photobionts are relatively sensitive to strong light. Inside lichen thalli these organisms can exist in numbers that would be impossible without the protection of the fungus. This may significantly increase their ability to multiply (Rikkinen, 1995).

While lichenized algae and cyanobacteria may often show lower net productivities per unit area and lower specific growth rates than their free-living counterparts, the stature of lichen thalli often projects the photobionts high above their free-living counterparts. This may represent a significant advantage in the competition for limiting resources, such as light, water and mineral dust (Raven, 1993). A three-dimensional thallus also lifts the photobionts above the surface boundary layer of the substrate. Many foliose and fruticose lichens have high surface-to-volume ratios and low heat capacities. Thus, they provide ideal condensation points for atmospheric humidity (Rikkinen, 1995, 1997).

3.1. Types of Cyanolichens

On the basis of symbiont composition most cyanolichens can be conveniently divided into two main groups: bipartite and tripartite cyanolichens. According to their overall appearance they have traditionally been divided into three main categories: crustose, foliose and fruticose lichens. Both divisions are artificial and convergent forms have repeatedly evolved in different systematic groups. However, the categories are quite useful and widely used for descriptive purposes.

3.1.1. Bipartite and Tripartite Cyanolichens

Bipartite lichens are stable symbioses between one type of lichen-forming fungus and one type of cyanobacterial photobiont (Fig. 3A). In most bipartite cyanolichens, the cyanobionts form a more or less continuous photobiont layer below the upper cortex. Tripartite lichens contain both green algal and cyanobacterial symbionts in addition to the lichen-forming fungus (Figs. 3B and 3C). Tripartite lichens include species which house symbiotic cyanobacteria in external or internal cephalodia, species that have both types of photobionts within the same main thallus but in separate sublayers, and the species which form pairs of disparate morphs originating form the interaction of the same fungus with the two contrasting types of photobionts. In most tripartite lichens the green algal photobiont occupies much of the thallus and produces most of the photosynthate. However, there are also some species in which cyanobacteria dominate and the phycobionts are limited to restricted parts of the thallus (James and Henssen, 1976; Henssen *et al.*, 1987).

Symbiotic cyanobacteria can provide both photosynthate and fixed nitrogen to their fungal partners and the relative importance of these functions varies between bi- and tripartite symbioses. The cyanobionts of tripartite lichens tend to show higher heterocyst frequencies and higher rates of nitrogen fixation than those of bipartite species (Rai *et al.*, 2000; see also chapter 6).

Cephalodiate Lichens and Photosymbiodemes. Tripartite lichens with cephalodia are by far the best known examples of multibiont systems among lichens (Fig. 3B). Cephalodia are delimited structures containing cyanobacteria in an otherwise green algal lichen thallus. External or internal cephalodia are known from over 500 lichen species, and similar structures have clearly evolved independently in repeated instances in lichens representing different systematic groups (Table 1). The cephalodial anatomy of many tripartite species of *Lobaria*, *Nephroma*, *Peltigera*, and *Stereocaulon* have been studied in detail (Forssell, 1883; Lamb, 1951, 1968, 1976; Jordan, 1970, 1972; Jordan and Rickson, 1971; James and Henssen, 1976; Stocker-Wörgötter and Türk, 1994). In some tripartite lichens the cephalodia are located deep inside the thallus or on the lower surface. This supports a primary role of the cyanobiont in nitrogen fixation over that in photosynthesis.

Figure 3. Main types of cyanolichens: (A) bipartite Peltigera neopolydactyla, *(B) tripartite* Peltigera aphthosa, *and (C) photosymbiodeme of* Peltigera aphthosa.

The mycobionts of some cephalodiate lichens can produce different morphotypes in symbiosis with compatible green algal and cyanobacterial photobionts. Such morphotypes may either combine into a compound thallus or live separate lives. Chimeroid lichens with green algae and cyanobacteria as primary photobionts in different parts of the same thallus are called photosymbiodemes (Fig. 3C). The corresponding free-living morphotypes have been called chlorosymbiodemes or chlorotypes and cyanosymbiodemes or cyanotypes, respectively. Photosymbiodeme producing species are known from several lecanoralean genera (*Lobaria*, *Nephroma*,

Peltigera, *Pseudocyphellaria*, and *Sticta*). As a whole, however, photosymbiodemes are quite rare and most cephalodiate lichens do not seem to produce chimeroid thalli.

The biont composition and taxonomy of photosymbiodemes have evoked considerable interest from lichenologists (James and Henssen, 1976; Brodo and Richardson, 1979; Tønsberg and Holtan-Hartwig, 1983; Vitikainen, 1994; Goffinet and Bayer, 1997; Goward and Goffinet, 1998; Jørgensen, 1994, 1997, 1998; Paulsrud *et al.*, 1998, 2000; Socker-Wörgötter, 2001a,b; Tønsberg and Goward, 2001). Furthermore, photosymbiodemes have offered unique opportunities for the study of differences in the physiological performances of symbiotic green algae and cyanobacteria under similar conditions of growth, habitat, history and fungal association (Demmig-Adams *et al.*, 1990; Green *et al.*, 1993; Schelensog *et al.*, 2000; Stocker-Wörgötter, 2001a,b).

3.1.2. Growth Forms

Most cyanolichens fit well into one of the three main growth form categories: crustose, foliose and fruticose. Prominent exceptions include the species of *Dictyonema*, which do not produce thalli, but house their cyanobionts within thin, papery basidiomata. Also some mycophycobiosis-like cyanolichens are difficult to categorize. All growth form categories include both bipartite and tripartite symbioses. Many cyanolichens appear gelatinous when wet. Gelatinous species are found in all three growth form categories. Gelatinous cyanolichens tend to produce non-stratified, homoiomerous thalli in which the mycobiont hyphae run directly within the extracellular matrix of cyanobionts and there is no distinct photobiont layer. These lichens are known for their ability to absorb large quantities of water due to the abundant production of cyanobacterial mucilage.

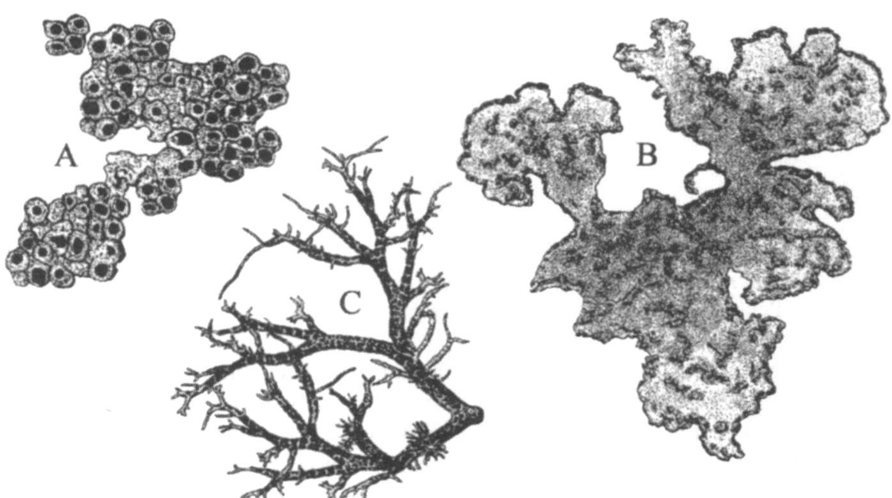

Figure 4. Growth forms of cyanolichens: (A) crustose Psorotichia diffracta, *(B) foliose* Pseudocephallaria anomala *(C) fruticose* Ephebe lanata.

Crustose Cyanolichens. Crustose lichens produce relatively undifferentiated, crust-like thalli which often grow tightly attached to their substrate. The crustose growth form is less common among cyanolichens than among green algal lichens. However, many species of Lichinales produce crustose thalli and many lichenicolous and cyanotrophic species are indeterminate or crustose (Table 1). The crustose thallus may be granular and effuse, like in *Moelleropsis*, or areolate, like in *Euopsis* and *Lemmopsis*. In areolate lichens the crustose thallus is broken up into numerous, scattered or aggregated entities. In some cases the areolae may initially develop on the surface of a fungal prothallus. Most crustose cyanolichens are episubstratic, i.e., they grow on the surface of their substrate. Some taxa, like species of *Pyrenocollema*, are endosubstratic, i.e., they mostly grow inside the substratum. Endolithic species grow in minute cracks and cavities or between mineral crystals of a rock.

Foliose Cyanolichens. Most cyanolichens produce foliose, distinctly lobate thalli (Table 1). These taxa are typically dorsiversal, flattened and their growth is predominately horizontal. This gives the lichens a more or less leafy appearance. The foliose thallus may only consist of a single lobe, which is then usually rounded, or the outward-growing thallus edge is divided into many lobes. Squamulose cyanolichens, like many species of *Heppia* and *Psoroma*, are somewhat intermediate between crustose and foliose forms. Their thalli consist of small, dorsiventral lobules or squamules, which develop individually and may remain distinct from each other. Other intermediate forms include some species of *Pannaria* (thallus partly crustose and partly lobulate) and numerous species of Lichinales (lobulate thallus margins but subfruticose central parts).

Both the form and size of thallus lobes are often quite characteristic for specific cyanolichen taxa. Some foliose cyanolichens, like *Parmeliella*, produce relatively small thalli with narrow and small thallus lobes. Others, like species of *Lobaria* and *Peltigera*, may produce very large, wide-lobed thalli. The thalli of most foliose cyanolichens are heteromerous and have a well developed cortex on the upper surface. Below the cortex there is a more or less uniform cyanobiont layer followed by a medulla of loosely intertwined hyphae. Some foliose cyanolichens, like *Collema*, produce gelatinous, homoiomerous thalli without a specialised cortex, medulla or photobiont layer. Other gelatinous forms, like *Leptogium*, have a thin cortex on the outer surface.

The lower surface of foliose cyanolichens may either be corticated, like in *Nephroma*, or ecorticate, like in *Peltigera*. Most species attach to their substrate with rhizines. In some taxa, like in species of *Coccocarpia*, entangled rhizines may project far beyond the thallus margins and form an extensive hypothallus. The thalli of some single-lobed and squamulose cyanolichens, like species of *Peltula* and *Peltularia*, attach to their substrate with single, usually central holdfasts and are thus more or less peltate.

Fruticose Cyanolichens. Fruticose, shrubby growth forms are relatively rare among cyanolichens (Table 1). Some fruticose taxa, like *Lichina* and the dendriscocauloid cyanotypes of *Sticta*, produce more or less upright, shrubby thalli that are attached to their substrate by a relatively narrow base. Others, like species of *Ephebe* and *Lichinodium*, produce mats or rosettes of minute, decumbent to semi-erect branches.

The fruticose thallus may be corticated or ecorticate, and in heteromerous forms there is a distinct photobiont layer and a central medulla in the thallus lobes.

During their development, species of *Pilophorus* and *Stereocaulon* tend to first produce a basal crust and then larger upright pseudopodetia. The pseudopodetia of *Stereocaulon* are solid, ecorticate and support innumerable corticated appendages called phyllocladia. Also wart-like cephalodia may occur in large numbers. The three-dimensional, often richly branched pseudopodetia are firmly attached to their substratum by special attachment hyphae.

3.2. The Lichen Thallus

The symbiotic lichen body is called the thallus. Most structures in lichen thalli develop through processes induced in the lichen mycobiont by the photobiont. General features of thallus morphology and anatomy have been covered in many reviews (Jahns, 1973; Jahns, 1988; Rikkinen, 1995; Büdel and Scheidegger, 1996) and more detailed accounts can be found in monographs on different taxonomic groups. General descriptions of thallus morphology are also found in many recent lichen floras (Wirth, 1980, 1995; Clauzade and Roux, 1986; Purvis *et al.*, 1992; Goward *et al.*, 1994; Goward, 1999; McCune and Geiser, 1997; Kantvilas and Jarman, 1999) and textbooks (Nash, 1996; Baron, 1999; Awasthi, 2000; Purvis, 2000). A brief account is given here.

3.2.1. Thallus Anatomy

Although lichens produce a limited variety of cell types, they show a remarkable degree of differentiation and structural complexity. This is achieved by a highly flexible developmental process and multifunctional structures (Ott, 1993). The fungal hyphae often differentiate into hygrophilic, pseudoparenchymatous zones or more loosely interwoven zones composed of aerial hyphae with water-repellent surfaces. The pseudoparenchymatous zone can form a peripheral cortex which covers or encloses a more loosely interwoven medulla. With some variations, this basic organization is seen in most lichens with a heteromerous thallus structure (Honegger, 1991; Letrouit-Galinou and Astra, 1994).

All lichenized fungi produce septate hyphae. In loosely organized lichen tissues, the hyphae are usually more or less cylindrical in form and have relatively thin walls. In modified tissues their shape may change; hyphal tips can grow, differentiate and the mutual pressure from adjoining cells influences their shape. The cell wall composition of lichen mycobionts is similar to that of other filamentous ascomycetes; the walls consist mainly of polysaccharides, with smaller amounts of proteins and lipids. While glucose has been found to be the most abundant polysaccharide monomer, mannose, galactose and glucosamine are also present. Chitin forms a dense meshwork on the inner cell wall surface of mycobiont hyphae. The ultrastructure of the hyphal walls may depend on the position and function of the hyphae; the thinnest hyphal walls, only 150–400 nm thick, are usually found in the photobiont layer. Medullary hyphae tend to be covered by layers of extracellular material. Extracellular sheaths may increase the surface area of the hyphae, hold photobiont cells in contact with the hyphae and act as channelling device for water transport and solute translocation. Within the cortex all spaces between neighbouring hyphae are often filled with amorphous substances secreted by the fungal cell walls (Jahns, 1988; Peveling, 1988; Honegger, 1991, 1997).

The thallus lobe of a typical cyanolichen consists of lengthened axial hyphae with unlimited growth which produce lateral branches with finite growth. The axial hyphae ensure the continuous elongation of the thallus lobe, while the lateral branches ensure the thickening of the thallus. The highest rates of cell division and metabolic activity of lichen bionts usually occur within a relatively narrow marginal or apical growth zone. In older parts of the thallus, growth may be limited to the slow turnover of cells in the photobiont layer and cortex and a slight volume increase of the medulla. Also the gradual accumulation of extracellular material is quite characteristic.

The morphology and anatomy of many lichens is strongly influenced by the environment. For example, the overall shape of the thallus, thickness of different thallus layers and the development of pigments and refractive structures all influence the quality of light that reaches the photobiont cells. Many lichens, particularly gelatinous ones, can exhibit major changes in dimensions during their normal wetting and drying cycles. These phenomena are generally related to the colloidal mechanism of water absorption and storage within lichen thalli (Rikkinen, 1995, 1997).

The Protective Cortex. A solid cortex protects lichen cyanobiont against physical abrasion, pathogens, excessive light and rapid desiccation. However, the cortex must be translucent enough to allow sufficient amounts of light to reach the photobiont cells. A dense cortex may also seriously hinder gas exchange and thus many structurally complex cyanolichens have special respiratory openings in their cortex. Species of *Sticta* have cyphellae, and species of *Pseudocyphellaria* pseudocyphellae, in the lower cortex. In some cyanolichens thallus aeration takes place through less specialized, multifunctional openings, such as soralia. Others, like species of *Peltigera*, lack the lower cortex altogether.

Cortex structure can vary considerably between closely related lichen species and even between different parts of a single thallus. Many examples of this can be seen in *Peltigera* (Vitikainen, 1994; Dietz *et al.*, 2000). In some cyanolichens the cortex surface is formed by dead hyphae. These can give rise to pruina or thick epinecral layers (Büdel, 1990; Büdel and Lange, 1994). Other lichens accumulate thick crusts of calcium oxalate crystals on their surface. The spectral characteristics of most lichens change quite dramatically depending on thallus water content. Minute air-cavities within the cortex scatter light effectively. When the cortex is moistened, water is absorbed into the cell walls and the interhyphal matrix. As a result, the cortical tissues swell and the number of vertical cell walls and the density of the extracellular matrix per unit area is reduced. Also cortical pigments are spread over a wider area. All these phenomena act to increase light availability within the photobiont layer (Rikkinen, 1995).

Photobiont Layer and Medulla. In heteromerous cyanolichens the cyanobiont cells are usually positioned into a relatively uniform layer in the uppermost part of the medulla. The horizontal distribution of the photobiont cells with respect to the thallus surface can show considerable variation. The exchange of metabolites requires a very close connection between lichen symbionts and several types of mycobiont-photobiont contacts occur in cyanolichens. These include wall-to-wall appositions, intracellular haustoria in some species, and intragelatinous protrusions in many heteromerous taxa. In many cyanolichens, thin-walled mycobiont hyphae penetrate the gelatinous sheaths of the cyanobionts, but there is no direct contact between the cell walls of the symbionts. Intracellular haustoria are only found in 'primitive' interactions such as those between species of *Dictyonema* and *Scytonema* (Oberwinkler, 1984). In addition to some chytrids and cyanophilous ascomycetes these lichen-forming basidiomycetes are among the few eukaryotic organisms that actually penetrate into bacterial cells in search for nutrients.

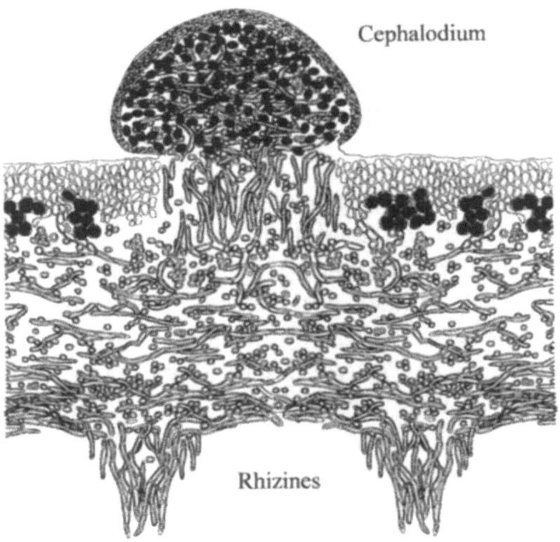

Figure 5. Diagrammatic cross-section through the thallus of the cephalodiate lichen Peltigera *species.*

Most green algal lichens with trebouxioid phycobionts produce intraparietal haustoria in which the water-repellent cell-wall surface layers of mycobiont hyphae spread to cover the algal cells. This type of haustorial complex fulfils several functions. It is the site of carbohydrate mobilisation from the photobiont to the fungus. Water, minerals, and fungal metabolites are translocated in the apoplastic continuum below the mycobiont derived surface layer of algal cells. Finally, the haustorial complex may help in the positioning of the photobiont cells within the photobiont layer (Honegger, 1997). The photobionts of most lichens are under some type of limiting control by the lichen-forming fungus. The mycobiont may control the photobiont population by draining photosynthate or by killing of surplus cells. Usually the excessive build-up of photobionts is avoided by arrests in the normal cell cycles of the photobionts (Hill,

1989, 1993). Some of these phenomena may be related to the kinetics of light-energy limited growth (Rikkinen, 1995).

Water transport within heteromerous lichens is largely a function of the mycobiont. The water-repellent surfaces of medullary hyphae help to maintain a gaseous environment within the medulla. Thus, the medulla of a tripartite *Peltigera* species, for example, remains air-filled even at water saturation (Honegger and Hugelshofer, 2000). The outer wall layers of thick medullary hyphae may hold large amounts of weakly perturbed water. This, together with the general lack of liquid water within the interhyphal spaces of fully hydrated lichens, indicates that apoplastic water transport takes place under the hydrophobic surface layers of medullary hyphae. However, the hyphae of some cyanolichens, like *Nephroma resupinatum*, have relatively thin walls and therefore, may not be very efficient in apoplastic water transport (Scheidegger, 1994).

Some lichens produce specialized strands of intertwined and conglutinated hyphae which can conduct water from the thallus surface to the photobiont cells. For example, the corticated cephalodia on the upper surface of tripartite *Peltigera* thalli may be connected to the ecorticate lower surface by strands of hydrophilic hyphae (Honegger and Hugelshofer, 2000). Water is absorbed and translocated by capillary forces along these tomental strands. As a result, the cephalodial cyanobionts have effective access to substrate moisture. The green algal photobionts, on the other hand, depend on slower apoplastic translocation of water (Honegger and Hugelshofer, 2000). While green algal photobionts can partly rehydrate from humid air, cyanobacteria tend to require liquid water for effective rehydration (Lange *et al.*, 1986, 1989; Büdel and Lange, 1991; Scheidegger, 1994). For example, in a recent study of green algal and cyanobacterial *Pseudocyphellaria* species, photosynthesis in the green algal thalli was activated at ca 20% water content, while the cyanobacterial thalli required a minimum water content of ca 50% DW (Schlensog *et al.*, 2000).

Lichen Propagules. The diversity of lichen-forming organisms makes it difficult to embrace the full range of structures and mechanisms that are involved in the reproduction of lichens. Each symbiotic partner may produce its own diaspores, and often one symbiont produces several types of propagules. For example, many lichen-forming ascomycete produce sexual ascospores as well as asexual conidia. Both types of propagules can be produced in enormous numbers. Motile hormogonia, on the otherhand, help nostocalean cyanobacteria to establish symbioses with many types of organisms, including lichen-forming fungi (Adams 2000; Rai *et al.*, 2000).

In addition to producing propagules of individual symbionts, many lichens facilitate the reproduction and simultaneous dispersal of the whole symbiotic consortium. In some lichens this is achieved simply via thallus fragmentation. Other species produce specialized symbiotic propagules, such as isidia or soredia. Isidia are small corticated outgrowths of the lichen thallus that usually break off in one piece. Soredia are minute ecorticate masses of mycobiont hyphae containing only a few photobiont cells. They are usually produced in special structures called soralia. Isidia and soredia are formed usually in specific parts of the lichen thallus. Some *Lempholemma* species produce hormocystangia at the margins or tips of their thallus lobes. These swollen structures are filled with short lichenized *Nostoc* filaments, which are liberated with the decay of the structures (Henssen, 1963).

The mycobionts of some lichens are rarely fertile and such species obviously depend on vegetative reproduction. However, as most lichen species are not known to produce specialized symbiotic diaspores, they seem to re-establish their symbiosis at each reproductive cycle. The possible importance of symbiotic dispersal in the ecology of many cephalodiate cyanolichens remains somewhat an enigma, as most of these lichens do not have symbiotic propagules. However, some *Psoroma* species do produce sorediate, isidiate or phyllidiate cephalodia (Jørgensen and Wedin, 1998).

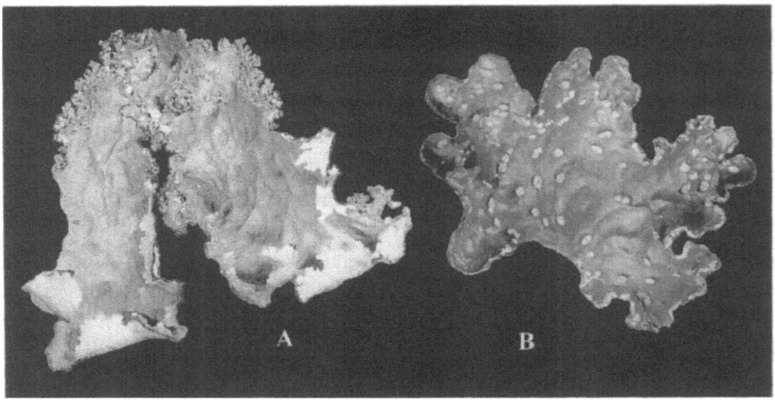

Figure 6. Symbiotic propagules of foliose cyanolichens: (A) dorsiventral, marginal phyllidia in Lobaria oregana, *(B) laminal and marginal soralia in* Pseudocyphellaria erocata.

3.2.2. Thallus Morphogenesis

Soon after the association between a compatible fungus and photobiont is established, the symbiosis starts to express morphological and chemical characteristics of a 'lichen' *e.g.*, internal stratification and the accumulation of specific secondary metabolites. As the photobionts provide energy for the whole symbiosis, the growth and subsequent differentiation of the predominately fungal thallus is probably stimulated by initial growth of the photobiont. Indeed resynthesis experiments have shown that under suitable conditions a compatible photobiont triggers the expression of the symbiotic fungal phenotype, although the mechanism remains unknown (Ahmadjian, 1993; Galun and Peleg-Zuriel, 2000).

The Symbiotic Phenotype. Aposymbiotically grown lichen mycobionts tend to produce a different hyphal arrangement than their lichenized counterparts. Apically growing and branching hyphae often form exploratory indefinite mycelia, quite similar to many non-symbiotic filamentous fungi. The colonies often have cartilaginous, conglutinate central parts with aerial hyphae at the periphery. Whereas lichenized mycobionts exhibit polarised growth, with distinct growth zones being the rule, aposymbiotic mycobionts tend to grow in a centrifugal manner (Honegger, 1991). In addition, the morphology of lichen photobionts is often greatly modified by lichenization. For example, the *Nostoc* cyanobionts may appear to reside as single cells, in honeycomb like chambers formed by the fungal partner (Kardish *et al.*, 1989; Scheidegger, 1994). Changes in the arrangement and amount of thylakoid membranes are also observed (Bergman and Hällblom, 1982). Many cyanobacteria undergo major changes in cell size and shape during the lichenization. Sometimes lichenization leads to a reduction in cell size but in most cases there is a significant increase in cell size, especially in older parts of the thallus. Presumably the drain of metabolites to the mycobiont affects the balance between reproduction and enlargement.

Until recently cyanolichens could not be resynthesized as successfully as green algal lichens (Ahmadjian, 1989, 1993). However, methods are now becoming available to perform resynthesis experiments with many types of cyanolichens, including photosymbiodemes. For more information the reader is referred to recent reviews by Stocker-Wörgötter (2001a, b). Future resynthesis experiments will undoubtedly help to elucidate many ontogenic principles in the development of cyanolichen thalli. One may also expect that new information of underlying molecular processes, such as DNA-methylations, will soon become available (Armaleo and Miao, 1998).

Resynthesis studies with lichens have generally shown that the genetic control of thallus formation is closely linked to such ecological factors as drying-wetting cycles and prevailing light intensities. For example, Stocker-Wörgötter and Türk (1994) found that relatively dry growth conditions were important for the formation of cephalodia in resynthesized *Peltigera* thalli. As long as the cultures were kept wet, *Nostoc* filaments escaped from loose precephalodial structures. During drier periods the cyanobacteria were effectively integrated into a fungal network and finally surrounded by a cortex. Similar phenomena were also reported by Yoshimura *et al.* (1994), who could not induce the formation of normal cephalodia in wet *P. aphthosa* cultures.

Recognition and Signalling. The potential partners for a lichen symbiosis are germinating fungal propagules and free-living or lichenized photobiont cells. Signalling between compatible symbionts must be mediated by chemicals produced by a symbiont (Fig. 7). Yoshimura and Yamamoto (1991) found that during the resynthesis of *Peltigera praetextata* the transformation of *Nostoc* colonies into the symbiotic state occurred even without a direct contact with the mycobiont. Thus, some sort of a diffusible soluble substance from the mycobiont controlled the transformation of *Nostoc*. Such phenomena are also known from other types of cyanobacterial symbioses (Adams, 2000). Cyanobacteria and lichen-forming fungi produce hundreds of unique secondary metabolites and novel compounds are continuously being described. Among fungal metabolites the aromatic polyketides are especially well represented. The structure and biosynthesis of these lichen compounds have been studied extensively, but their possible roles in symbiont recognition are presently unknown (Rikkinen, 1995; Huneck and Yoshimura, 1996).

Lectins attach to tissue components, notably glycoproteins, with a high degree of specificity. In fern and bryophyte symbioses, for example, the host plants produce lectins which recognize sugar residues on the cell surfaces of symbiotic *Nostocs*

(Bergman *et al.*, 1993). Kardish *et al.* (1990) studied the binding of lectin isolated from the fungal component of *Nephroma laevigatum* to *Nostoc* cells from different origins and concluded that the protein was involved in the control and regulatory processes of symbiont balance in the lichen thallus. Lectins have also been isolated from mycobiont hyphae of some *Peltigera* species (Lehr *et al.*, 1995; Rai *et al.*, 2000). In *P. aphthosa* a lectin recognizes compatible *Nostoc* cells at the initiation of cephalodium formation and this process is highly specific (Lehr *et al.*, 2000). The specificity for cyanobiont was confirmed in a recent inoculation study attempting to introduce foreign *Nostoc* strains into the cephalodia of *P. aphthosa* (Paulsrud *et al.*, 2001).

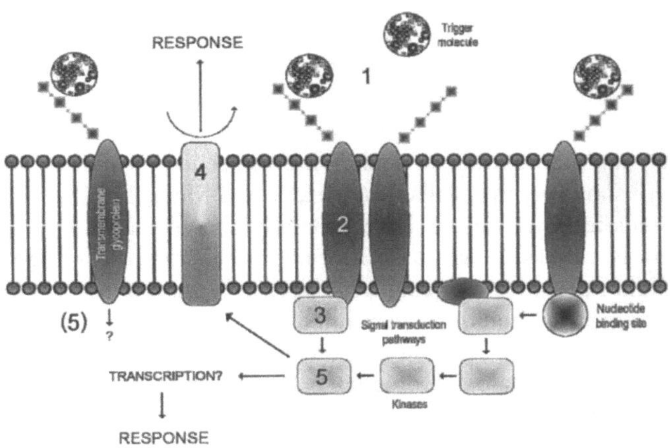

1. Trigger molecules
2. Signal recognition through transmembrane proteins.
3. The innermembrane part of the protein may function as a kinase, transmitting signals to cellular mechanisms resulting in interbiontic response.
4. Interbiontic response is transmitted out of the cell by membrane associated proteins.
5. The interbiontic response in the cell is transmitted through kinases and other factors.

Figure 7. A model for signalling between symbionts in lichen. Two types of signalling elicitors are known: general elicitors without major differences in sensitivity among responding organisms, and specific elicitors that function in specific ways. General elicitors are usually substances associated with primary metabolism, and include glucans, chitin oligomers, glycopeptides, cell wall fragments, and many phenolics. Specific elicitors may include proteins, peptides, syrincolides, and phenolic compounds (Rikkinen and Pappinen, unpublished results).

3.3. Cyanolichens as Symbiotic Processes

Taxonomic and morphological definitions do not do full justice to the biological essence of lichens. While the thalli of some species are reasonably well delimited, they never function as individuals in the conventional sense of the word. This is because phenotypic and genetic attributes do not coincide. In many lichens even a mechanistic delimitation of a single thallus is quite impossible. Many lichens rely almost exclusively on vegetative reproduction and often this leads to innumerable genetically identical thalli. This may partly explain why the morphology and ecology of many lichen species is remarkably similar throughout their range. On the other hand, some lichens start their development with the fusion of several symbiotic propagules. The genetic structure of such thalli may be quite complex.

3.3.1. Some Implications of Biont Specificity

Many studies have shown that most lichen-forming fungi are highly selective in choosing photobionts. This applies to both green algal photobionts and cyanobacterial photobionts. It thus seems that only a narrow spectrum of closely related photobiont taxa induce the transformation of a specific fungus into the symbiotic phenotype. However, the mode of cyanobiont acquisition may have a bearing on the cyanobiont diversity in lichen thalli and other cyanobacterial symbioses (Rai *et al.*, 2000).

In all bipartite cyanolichens studied so far, only one cyanobiont strain has been detected from each thallus. Also most tripartite lichens have contained the same *Nostoc* strain in all cephalodia of individual thalli (Paulsrud, 2001). The main exception is *P. venosa* which can often house different cyanobionts in different cephalodia (Paulsrud *et al.*, 2000). Ott (1988) reported that even cyonobionts resembling *Scytonema* can occur together with *Nostoc* both in the prothallus and later developmental phases of *P. venosa*. Some *Nephroma* thalli also seem to occasionally house different types of cyanobacteria in different cephalodia or possibly even in the same cephalodium (Jordan and Rickson, 1971). Some species of *Stereocaulon* are believed to evolve cephalodia with representatives of several different cyanobiont genera (Lamb, 1951). These diversity patterns have not yet been confirmed with modern methods.

3.3.2. Fungal Compatibility

Molecular studies have clearly indicated that the same fungus is responsible for both types of photobiont associations in tripartite *Peltigera* species. This information has been gained from hybridisation studies of restriction digests of total DNA and restriction site comparisons of PCR products (Armaleo and Clerc, 1991), or from comparative studies of 5.8S and ITS sequences (Goffinet and Bayer, 1997). It is still possible that more than one genetically discrete individual or genet could be involved. Subtle variations in the morphology and secondary chemistry of many *Peltigera* species could reflect different combinations of genets within specific thalli or groups of thalli. Some of them might even represent complex, three-dimensional puzzles involving a large number of genets.

Many features of sexual and vegetative compatibility in lichen mycobionts are quite identical to those of nonlichenized ascomycetes. Thus, there is no reason to believe that the genetic systems which control these phenomena would be any more dissimilar. For example, anastomoses between mycobiont hyphae are common in some lichens. In the Peltigeraceae they are often accompanied by bi- or multinucleate hyphae. In such cases, the genetic integrities of mycobiont genets could be controlled by series of bi- or multi-allelic genes at vegetative incompatibility (*het*) loci, typical of many non-lichenized ascomycetes. In addition, mating-type alleles could function as incompatibility genes, along with several other loci (Glass and Kuldau, 1992; Leslie, 1993; Coppins *et al.*, 1997). Unfortunately, many aspects of sexual and vegetative compatibility in lichen mycobionts have not yet been studied in detail (Larson and Carey, 1986; Culberson *et al.*, 1988; Goffinet and Hastings, 1995; Murthach *et al.*, 2000). Further studies on these topics are needed in to get new insights into the basic biology of lichen symbioses.

It is easy to imagine how lichens could benefits from maintaining a certain level of fungal heterogeneity within their thalli. In addition to different aspects of sexual reproduction, the mycobionts could gain through an increased resistance against viruses and other pathogens. The presence of several vegetative compatibility types within each thallus could restrict the spread of intracellular pathogenes as hyphal fusions would not develop between different compatibility groups. From this perspective, one might even hypothesise that a certain level of vc-heterogeneity is not only possible, but required for the success of slow-growing and long-lived lichen thalli.

3.3.3. Cyanolichen Guilds

Different thalli of individual cyanolichen species can often contain different strains of symbiotic cyanobacteria. Futhermore, different cyanolichen species can often share identical cyanobiont strains (Paulsrud *et al.*, 1998, 2000). Often several such lichen species co-occur in specific habitats and form characteristic communities or 'guilds'. The cyanobionts of all lichens within each guild are closely related, but not the lichen-forming fungi. Some guilds include mycobionts from many different genera or even different families. On the other hand, some closely related mycobionts associate with different types of cyanobionts and thus belong to different guilds. This implies that many cyanolichens not only share similar enviromental requirements, but also depend on a common pool of cyanobacterial symbionts. This common phenomenon influences the structure of lichen communities on all levels of community organization (Rikkinen, Oksanen and Lohtander, unpublished results).

For example, bipartite *Nephroma* species and similar epiphytic cyanolichens form a characteristic group among the epiphytes of humid boreal and temperate forests. Many of these lichens prefer old-growth forests and they have been used as biological indicators of forest antiquity. These species are usually positioned relatively low in the vertical zonation of epiphytes (McCune, 1993). They are thus well buffered from extremes of temperature, exposure to wind, and desiccation, but must survive under illumination conditions drastically different from those in more open habitats (Rikkinen, 1995). Studies of cyanobiont diversity have recently indicated that many old-growth associated cyanolichens depend on a specific group of symbiotic *Nostoc* strains that are not found in other types of cyanolichens (Fig. 8). These lichens exploit a common pool of cyanobacteria and form a horizontally linked system, the '*Nephroma* guild'. Conversely, many predominately terrestrial cyanolichens in the same forests share a different group of closely related *Nostoc* strains, thus forming the '*Peltigera* guild'. These two cyanolichen guilds meet on the basal trunks of broad-leaf trees and probably represent a fair proportion of similar guilds in boreal forests (Rikkinen, Oksanen and Lohtander, unpublished results).

Considering the dispersal of cyanolichen species it would seem unnecessary for all lichen mycobionts to spend resources in producing symbiotic propagules, when other guild members will effectively disperse the same cyanobiont. Thus, the dispersal ecology of many lichen guilds may centre around 'core species', which produce massive amounts of symbiotic diaspores. 'Fringe species', in turn, produce only fungal propagules and may largely depend on the core species in the dispersal of appropriate cyanobionts. In their natural habitat both types of cyanolichens can exist side by side without much competition for space. The competition between guild members is reduced by slightly different substrate preferences and thallus morphologies. Furthermore, the core species may actually benefit from the fringe species, as their extra cyanobionts are being deposited into neighbouring guild members rather than being completely lost. Some of these symbionts can be potentially reclaimed as, without the ability to produce symbiotic propagules, the fringe species cannot transfer the cyanobionts into new habitats. This phenomenon may explain why the existence of competition is often difficult to demonstrate in lichen communities. Further studies on guild structure are essential for a better understanding of lichen ecology and for the development of viable conservation strategies for maintaining high cyanolichen diversity in remnant old-growth forests. Special emphasis should be given to the design of experimental approaches for the study of lichens as horizontally linked symbiotic processes (Rikkinen, 1995; Rikkinen, Oksanen and Lohtander, unpublished results).

3.3.4. Continuum Between Bipartite and Cephalodiate Cyanolichens

Cephalodiate species of the Peltigerinae have often been interpreted as advanced forms of symbiosis. However, some recent findings have indicated that the cephalodiate taxa should not automatically be regarded as more 'advanced' than bipartite species. On the contrary, many bipartite lichens may have evolved from tripartite ancestors.

In a recent phylogenetic study of *Peltigera*, Miadlikowska and Lutzoni (2000) found that the cephalodiate species *P. venosa*, occupied a basal position within the genus. The other tripartite species belonged to two distinct paraphyletic groups that were nested within the bipartite species. Accordingly, the cephalodiate taxa were divided into three sections. *P. venosa* is easily distinguished form all other *Peltigera* species. In view of the above and on the basis of its unique chemistry and morphology, Miadlikowska and Lutzoni (2000) placed *P. venosa* in the monotypic section *Phlebia*. *P. venosa* has *Nostoc* cyanobionts in external cephalodia on the lower surface of the thallus. In addition, it regularly produces free-living cyanomorphs with homoiomerous, *Leptogium*-like thalli (Ott, 1988; Paulsrud *et al.*, 2000). All other tripartite *Peltigera* species produce external cephalodia on their upper surface. Two such species, *P. aphthosa* and *P. britannica*, together with some bipartite species (*e.g.*, *P. malacea*), belong to sect. *Peltidea*. The tripartite species are able to form photosymbiodemes, but these are not nearly as common as in *P. venosa*. When produced, the cyanomorphs are more or less identical with bipartite species of the section, but generally show poor fitness in natural environments. Finally, the third group of *Peltigera* with tripartite species, sect. *Chloropeltigera*, only includes cephalodiate taxa. These species are not known to produce photosymbiodemes in nature (Miadlikowska and Lutzoni, 2000).

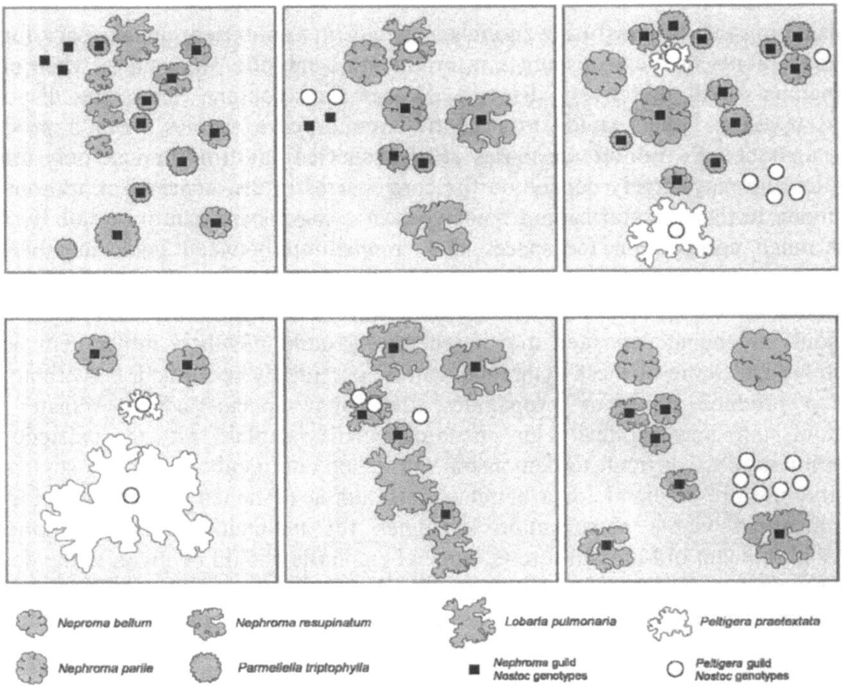

Figure 8. Distribution of symbiotic Nostoc *strains on 6 tree trunk sample plots in two old-growth forests of central Finland. The cyanobionts of* Nephroma bellum, N. parile, N. resupinatum, *and* Parmeliella triptophylla *belong to four closely related tRNA*^{Leu} *(UAA) intron genotypes (solid symbols). These epiphytes exploit a common pool of cyanobionts and thus form a horizontally linked '*Nephroma *guild'. The cyanobionts of* Peltigera praetextata *and most free-living* Nostoc *strains constitute a genetically more variable '*Peltigera *guild' (open symbols). From Rikkinen, Oksanen and Lohtander, (unpublished results).*

Within *Nephroma*, a sister group of *Peltigera*, the tripartite species house typical '*Peltigera* guild' *Nostoc* strains, while all bipartite species have typical '*Nephroma* guild' cyanobionts. This segregation is quite significant as, like in *Peltigera*, the tripartite taxa do not form a monophyletic group. Conversely, some cephalodiate taxa are more closely related to bipartite species than to each other. This implies that within *Nephroma*, evolutionary transitions between bi- and tri-partite symbioses could not have been occurred simply via the acquisition or loss of the green algal photobiont; they would have also required concurrent switches in cyanobiont composition (Lohtander, Oksanen and Rikkinen, unpublished results). This may explain the poor fitness of many *Nephroma* cyanomorphs: the bipartite thalli appear to have lost their green algal symbionts, but may still continue to house 'wrong' cyanobionts.

The cyanobionts of all *Peltigera* species belong to the same main group of symbiotic *Nostoc* strains (Lohtander, Oksanen and Rikkinen, unpublished results). However, this group is quite diverse and there are major differences in the cyanobiont spectra of individual *Peltigera* species. For example, *P. venosa* has exhibited a higher level of cyanobiont diversity than any other lichen species studied so far. Conversely, only two different *Nostoc* strains have been found from both *P. aphthosa* and *P. britannica* (Paulsrud and Lindblad, 1998; Paulsrud *et al.*, 1998, 2000, 2001). This indicates that there may be a stepwise increase in cyanobiont specificity from sect. *Phlebia* to *Peltidea*, and finally to *Chloropeltigera*. This could, in turn, explain some group specific differences in the establishment of cephalodia and in the tendency to produce photosymbiodemes.

Stöcker-Wörgötter and Türk (1994) resynthesized *P. leucophlebia* in culture. This tripartite lichen belongs to sect. *Chloropeltigera* and thus does not produce photosymbiodemes in nature. However, four different types of thallus primordia developed in culture: primordia containing *Nostoc*, primordia containing *Coccomyxa*, primordia containing both photobionts, and cyanobacterial lobules with green algal outgrowths. The primordia with both photobionts soon died and the primordia with *Nostoc* did not differentiate into heteromerous thalli. Eventually the cultures were dominated by small *P. leucophlebia* chlorotypes arising from a layer of *Nostoc*. Quite remarkably, the final establishment of cephalodia did not occur through the capture of free-living cyanobacteria by the chlorotypes: only previously lichenized cyanobacterial primordia were attached to the green algal thalli. Colonies of free-living *Nostoc* were even purposely inoculated on the thallus surface, but these were not incorporated into cephalodia (Stöcker-Wörgötter and Türk, 1994). These results were in clear contrast with many previous and later reports of cephalodia being generated through the capture of free-living *Nostoc*s by cortical hyphae or by rhizines (Jordan, 1970; Jordan and Rickson, 1971; Jahns, 1988; Lehr *et al.*, 2000). However, the later group of observations had been made from *P. aphthosa* or other lichens that are not closely related to *P. leucophlebia*. Clearly, several mechanisms of cephalodial establishment might exist among different groups of tripartite lichens.

It is tempting to speculate that *P. venosa* has many features that once characterized the ancestor of modern *Peltigera* species and probably also the common ancestor of *Peltigera* and *Nephroma*. These include the ability to form associations with many types of *Nostoc* and the ability to produce different morphotypes, including both tripartite and bipartite thalli. Furthermore, it seems quite possible that some extant continuums between bipartite and cephalodiate cyanolichens could reflect genetic polymorphism within the fungal components of these symbioses. Nobody has shown that the cyanobacterial and green algal morphotypes would, in fact, represent different genets of the same mycobiont species, showing different photobiont preferences. However, as the mating of many lichen-forming fungi is likely to require the fusion of propagules with previously established thalli, and as the cephalodial establishment of some tripartite lichens seems to require previously lichenized propagules, there might be a direct link between sexual processes and the early evolution of cephalodia. These processes, that are readily visible in tripartite cyanolichens, might also occur in many bipartite lichens. In these types of symbioses the fusion of sexual or symbiotic propagules with established thalli could play its original role in fungal reproduction. In some cephalodiate lichens this process may have acquired a secondary role in keeping a compatible cyanobiont within the symbiotic consortium.

4. EVOLUTION OF CYANOLICHENS

Cyanobacteria, plants and fungi have affected each other profoundly during the course of evolution. The ultimate examples of this are the plastids of eukaryotic algae and plants which once evolved from cyanobacterial ancestors. Furthermore, it seems quite likely that the initial diversification of terrestrial biota was closely linked to the appearance of lichen-like and mycorrhizal symbioses, the two major types of mutualistic interactions between photosynthetic organisms and fungi. The oldest fossil accounts of both occur in the Early Devonian over 400 million years ago (Taylor and Taylor, 1993; Taylor *et al.*, 1997). However, both types of symbioses may have evolved much earlier, deep within the Precambrian. Cyanobacteria are an ancient group of organisms and recent molecular clock estimates have indicated that all major lineages of extant Fungi were already present at least 1000 million years ago. These estimates have also indicated that land plants appeared by 700 Ma (Heckman *et al.*, 2001).

4.1. Evidence of Antiquity

The degree of mutual dependence of symbiotic partners may often correlate with the evolutionary age of the association. Almost all lichen mycobionts are obligately lichen-forming and many of them seem to be unable to complete their normal life-cycles in the aposymbiotic state. This feature is particularly pronounced among the mycobionts of cyanolichens. For example, in a recent study Crittenden *et al.* (1995) attempted to isolate and bring into pure culture over 1000 species of lichen-forming and lichenicolous fungi from diverse ecosystems and systematic groups. Almost 500 species were successfully isolated from spores or from thallus macerates. However, only 22% of cyanobacterial lichens yielded fungal isolates compared with 46% and 43% of those containing chlorococcoid and trentepohlioid green algae, respectively. Success with bipartite cyanolichens was particularly low; less than 10% of studied species in the Pannariaceae and Collemataceae, for example, yielded fungal isolates (Crittenden *et al.*, 1995).

Most lichen-forming fungi depend on trebouxioid green algae. Species of *Trebouxia* are more or less confined to lichens and they seem to have given up much of their individual lives. Some of them may depend on their fungal partners to such an extent that they are poorly equipped for independent existence in natural environments (Rikkinen, 1995). Conversely, many cyanobacteria, including symbiotic forms, seem to have remained more or less unchanged since the Precambrian. Upon isolation and culture they easily revert to a typical free-living form. However, the fact that some lichen-forming *Nostoc* strains do not readily produce hormogonia in culture might reflect an adaptation to symbiotic dispersal (Paulsrud *et al.*, 2001).

Despite the lack of obvious morphological adaptations to a symbiotic lifestyle, cyanobacteria most probably were already involved in the earliest lichen-like symbioses. Green and Lange (1994) suggested that such lichens could have been early colonizers of terrestrial habitats, their prime adaptation being the capacity to tolerate desiccation. Also the adaptive significance of the lichen thallus as a shield against UV radiation may have been important in promoting the early colonization of terrestrial habitats (Rikkinen, 1995). The dual ability to produce photosynthate and to fix atmospheric nitrogen was, of course, the central factor in promoting the early recruitment of cyanobacteria into lichen-like symbioses.

4.2. Fossil Lichens

Lichens are unlikely candidates for fossilization and thus few well preserved lichen fossils have ever been found. As pointed out by Green and Lange (1994) problems of fossilization are probably the most important reason for the scarcity of lichens in the fossil record. Fossil assemblages are indeed notorious for their many biases; most notably for the unequal preservation of hard and soft structures.

Taylor *et al.* (1997) described an exquisitely preserved cyanolichen from the 400 million-year-old Rhynie chert. The same paleoecosystem has also provided important information on many other types of fungal organisms (Taylor and Taylor, 1993, 1997; Taylor, 1994; Hass *et al.*, 1994; Taylor *et al.*, 1992, 1999). Specimens of the fossilized lichen consist of a thallus of superimposed layers of aseptate hyphae and, on the upper surface, numerous depressions. Extending into the bases of the depressions are fungal hyphae that form a three-dimensional netlike structure. Enclosed within the net are coccoid cyanobacteria. The cyanobacterial cells have thick sheaths and some of them have divided in three planes, resulting in colonial clusters. Old cells are parasitized by the fungus in the base of the hyphal net, while new cyanobacterial cells are formed distally. This results in the production of soredia-like symbiotic propagules (Taylor *et al.*, 1997).

The Early Devonian lichen was placed into the new genus *Winfrenatia* and the authors suggested that its fungal component could have been an early zygomycete (Taylor *et al.*, 1997). The cyanobionts are quite similar to modern pleurocapsalean forms, like *Chroococcidiopsis*. While the non-septate hyphae and chlamydospores of the fossilized fungus support a zygomycetous affinity, the fossilized thallus also has many characteristics of some extant gelatinous ascomycetes. For example, species of *Anema*, *Paulia* and *Phylliscum* all house unicellular cyanobionts in loose networks of hyphae, quite similar to the hyphal nets of *Winfrenatia*. Also the habitat requirements of these lichens correspond with the environment postulated for the Rhynie chert paleoecosystem. Schultz *et al.* (1999) suggested that modern *Paulia* species would have already evolved on the continental center of Pangaea. Thus, their present widely disjunct distributions would mainly reflect the effects of continental drift.

Perfectly preserved amber fossils have shown that many modern lichen genera, and possibly even species, were already present in the Tertiary (Poinar and Poinar, 1999; Peterson, 2000; Poinar *et al.*, 2000; Rikkinen and Poinar, unpublished results). For example, two fossil species of *Parmelia s. lat.* were recently described from Dominican amber and several other fossils are known from Mexican and Baltic ambers (Poinar *et al.*, 2000; Rikkinen, unpublished results). For example, two well preserved specimens of the foliose green algal lichen *Anzia* were recently found from Baltic amber (Rikkinen and Poinar, unpublished results). The fossils show that all characteristics in the thallus morphology of *Anzia* sect. *Anzia* have remained unchanged for at least 40 million years. As there is no reason to believe that the fossilisation would have been immediately preceded by a period of more rapid evolution, the initial divergence of anzioid lichens must have happened in the distant past, probably in the Cretaceous. No amber fossils of cyanolichens have yet been described. However, a 12–24 million year old impression of a foliose species belonging to Lobariaceae was recently reported by Peterson (2000).

Even a single fossil has the potential to give a minimum age estimate for the origins of several evolutionary lineages. Eventually, after more detailed phylogenetic hypotheses have been generated for different groups of lichens, the few available fossils will become invaluable for timing branching events and calibrating molecular clocks (Parks and Wendel, 1990; Hibbett *et al.*, 1997; Xiang *et al.*, 1998; Taylor *et al.*, 1999). Printzen and Lumbsch (2000) used vegetation history and paleoclimatic data to calibrate a molecular clock based on fungal ITS sequences from two genera of epiphytic green algal lichens. Their results indicated that diversification within *Biatora* started already in the Late Cretaceous and took place during periods of climate cooling, when many new forest vegetation types evolved and spread in the Northern Hemisphere.

Because lichen fossils are rare, most of what is presently known of lichen evolution is based on comparative studies of extant taxa. The proposed antiquity of many modern cyanolichens, like *Paulia* species, is supported by their present range. Some lichens show classic disjunct ranges involving East Asia and eastern North America. In the case of *Anzia* this distribution, together with the European amber fossils, clearly indicated that these lichens once had a circum-Laurasian range (Rikkinen and Poinar, 2000). Later they became extinct from Europe, but were preserved in East Asia and eastern North America. Both regions have acted as centres of survival for many groups of organisms that previously had a semi-continuous range across the Holarctic, but suffered major constrictions in range as a consequence of climatic deterioration during the Pleistocene. Thus, the present distributions of some lichens and fungi correspond with the relict ranges of gymnosperms, like *Metasequoia* and *Ginkgo*, and of angiosperms, like *Liriodendron* and *Magnolia* (Tiffney, 1985; Redhead, 1989; Galloway, 1994; Jørgensen, 1994, 2000; Wu and Mueller, 1997; Xiang *et al.*, 1998; Rikkinen and Poinar, 2000; Wu *et al.*, 2000).

4.3. Possible Relations with Mycorrhizal Symbioses

Cyanobacteria initially evolved oxygenic photosynthesis and so changed the Earth's atmosphere from anoxic to oxic. As a consequence, most nitrogen-fixing bacteria became confined to anoxic environmental niches. This is mainly because nitrogenase, the enzyme complex responsible for nitrogen fixation, is highly sensitive to oxygen. In the cyanobacteria several strategies evolved to protect nitrogenase from oxygen, including a temporal separation of oxygenic photosynthesis and nitrogen fixation or, in some filamentous groups, the differentiation of a specialized cell, the heterocyst, to protect the functioning of nitrogenase. However, this sensitivity may also have enhanced the evolution of cyanobacterial symbioses, as some symbiotic structures help to protect cyanobacterial cells from atmospheric oxygen.

The evolution of some cyanobacterial symbioses could even correlate temporally with geological periods of increased oxygen in the atmosphere. Especially during the Carboniferous high atmospheric oxygen concentrations (possibly over 30 %) might have seriously interfered with nitrogen-fixation. This would have given adaptive value to all structures and mechanisms that help to isolate nitrogen-fixing cells from the atmosphere. Hence the birth of some symbiotic structures, like the coralloid roots of cycads, might represent a parallel case to the development of arthropod and amphibian gigantism, which appear to have been directly facilitated by the hyperoxic atmosphere (Dudley-Robert, 1998). While the later phenomena, like many others, were subsequently lost during the late Permian transition to hypoxia, most types of cyanobacterial symbioses were not. Lower O_2 concentrations did not directly threaten the symbiotic associations which continued to function under hypoxic condition as well. However, the ecological significance of cyanobacterial nitrogen fixation may have been greatly reduced when rhizosphere symbioses became a viable option for nitrogen starved plants.

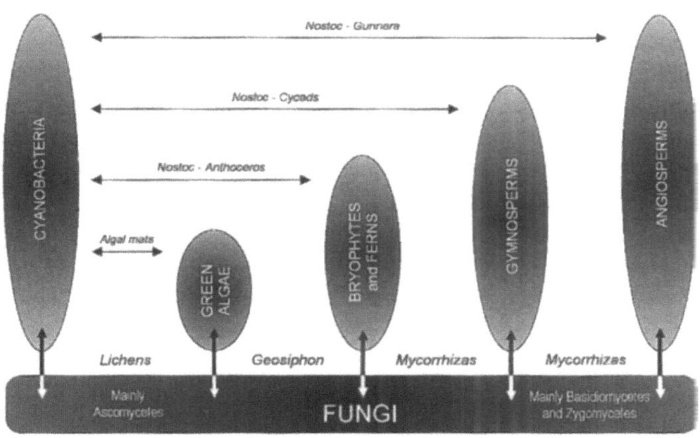

Figure 9. Mutualistic interactions between cyanobacteria, green plants and fungi. Symbiotic associations between cyanobacteria and fungi evolved before the initial evolution of mycorrhizal symbioses, the rise of vascular plants, and the subsequent diversification of parasitic and saprophytic fungi. Both lichens and mycorrhizal symbioses are polyphyletic and have evolved repeatedly. The diversification of mycorrhizal symbioses, in close associations with nitrogen-fixing soil bacteria, has reduced the relative importance of cyanobacterial symbioses during evolution. However, isolated representatives in all major groups of vascular plants can form stable symbioses with nitrogen-fixing cyanobacteria.

The mycelia of mycorrhizal fungi are known to associate with soil bacteria capable of using organic nitrogen compounds and/or fixing atmospheric nitrogen (Perez-Moreno and Read, 2000; Sen, 2000). The early evolution of these associations is not known, but there may have been a real triumph of rhizosphere interactions in the late Permian, when free-living soil bacteria, following the atmospheric transition to hypoxia, became more efficient in nitrogen-fixation. This may have given a boost to the evolution of mycorrhizal symbioses and the subsequent diversification of vascular plants. Similar expansions soon followed in parasitic and saprophytic interactions between plants and fungi. Also the radiation of fungal endophytes and modern lichens must have been closely linked to the diversification of vascular plants. For example, most families of extant lichens include many epiphytic forms.

The earliest fungal symbioses may well have been primitive lichens, most probably those with cyanobacterial photobionts. Littoral habitats along ancient shore-lines brought a wide range of free-living cyanobacteria, green algae and fungi into close contact under conditions where there were good opportunities for the evolution of new symbiotic interactions (Rikkinen, 1995). The earliest lichens may have developed long before the initial evolution of mycorrhizal symbioses, the subsequent rise of vascular plants and the later diversification of parasitic and saprophytic fungi. Accordingly, some modern cyanolichens may preserve biological features from very early stages of terrestrial evolution.

Mutualistic and parasitic interactions between cyanobacteria, plants and fungi are clearly polyphyletic and have repeatedly evolved from each other during the course of evolution. The trend towards more versatile and efficient fungal and bacterial symbioses may have reduced the relative importance of cyanobacterial symbioses during fungal and vascular plant evolution. However, isolated representatives of all major groups of extant plants and fungi form associations with nitrogen-fixing cyanobacteria (Fig. 9). The closely intertwined evolutionary history of all these organisms indicates that there are many basic similarities in the molecular recognition systems of symbiotic cyanobacteria, green plants and fungi.

REFERENCES

Adams, D.G. (2000) Symbiotic interactions in Whitton, B.A. and Potts, M. (eds.), *The Ecology of Cyanobacteria*, Kluwer Academic Publishers, Dordrecht, pp. 523–561.

Ahmadjian, V. (1989) Studies on the isolation and synthesis of bionts of the cyanolichen *Peltigera canina* (Peltigeraceae), *Plant Systemat. Evol.* **165**, 29–38.

Ahmadjian V. (1993) *The Lichen Symbiosis*, John Wiley and Sons, New York.

Alexopoulos, C.J., Mims, C.W., and Blackwell, M. (1996) *Introductory Mycology*, John Wiley and Sons, USA.

Alstrup, V. and Hawksworth, D.L. (1990) The lichenicolous fungi of Greenland. Meddelelser om Grønland, *Bioscience* **31**, 1–90.

Aptroot, A. (1998) Aspects of the integration of the taxonomy of lichenized and non-lichenized pyrenocarpous ascomycetes, *Lichenologist* **30**, 501–514.

Aptroot, A., Diederich, P., Sérusiaux, E. and Sipman, H.J.M. (1997) Lichens and lichenicolous fungi from New Guinea, *Bibliotheca Lichenologica* **64**, 1–220.

Armaleo, D. and Clerc, P. (1991) Lichen chimeras: DNA analysis suggests that one fungus forms two morphotypes, *Exp. Mycol.* **15**, 1–10.

Armaleo, D. and Miao, W. (1998) Symbiosis and DNA methylation in the *Cladonia* lichen fungus, *Symbiosis* **26**, 143–163.

Awasthi, D.D. (2000) *A Handbook of Lichens*, Bishen Singh Mahendrapal Singh, Dehradun, India.

Baron, G. (1999) *Understanding Lichens*, The Richmond Publishing Co. Ltd., Slough,UK.

Beck, A. (1999) Photobiont inventory of a lichen community growing on heavy metal-rich rock, *Lichenologist* **31**, 501-510.

Beck, A., Friedl, T., and Rambold, G. (1998) Selectivity of photobiont choice in a defined lichen community: inference from cultural and molecular studies, *New Phytol.* **13**, 709-720.

Bergman, B. and Hällbom, L. (1982) *Nostoc* of *Peltigera canina* when lichenized and isolated, *Can. J. Bot.* **60**, 2092–2098.

Bergman, B., Rai, A.N., Johansson, C. and Söderbäck, E. (1993) Cyanobacterial-plant symb oses, *Symbiosis* **14**, 61–81.

Brodo, I.M. and Richardson, D.H.S. (1979) Chimeroid associations in the genus *Peltigera*, *Lichenologist* **10**, 157–170.

Büdel, B. (1990) Anatomical adaptations to the semiarid/arid environment in the lichen genus *Peltula*, *Bibliotheca Lichenologica* **38**, 47–61.

Büdel, B. (1992) Taxonomy of lichenized prokaryotic blue-green algae, in Reisser, W. (ed.), *Alage and Symbioses*, Biopress, Bristol, pp. 301-324.

Büdel, B. and Lange, O.L. (1991) Water status of green and blue-green phycobionts in lichen thalli after hydration by water vapour uptake: do they become turgid? *Acta Botanica* **104**, 361–366.

Büdel, B. and Lange, O.L. (1994) The role of cortical and epinecral layers in the lichen genus *Peltula*, *Cryptogam. Bot.* **4**, 262–269.

Büdel, B., Lüttge, U., Stelzer, R., Huber, O., and Medina, E. (1994) Cyanobacteria of rocks and soils of the Orinoco Lowlands and the Guayana Uplands, Venezuela, *Acta Botanica* **107**, 422–431.

Büdel, B. and Scheidegger, C. (1996) Thallus morphology and anatomy, in Nash III, T.H. (ed.), *Lichen Biology*. Cambridge University Press, Cambridge, pp. 37-64.

Castenholz, R.W. (2001) Phylum BX. Cyanobacteria, in Garrity, G.M. (ed.), *Bergey's Manual of Systematic Bacteorology Vol I*, Springer-Verlag, Berlin, pp. 473–599.

Clauzade, G. and Roux, C. (1986) Likenoj de okcidenta Europa. Ilustrita determinlibro, *Bull. Soc. Bot. Centre-Quest Novelle serie* **7**, 1-893.

Coppins, E., Debuchy, R., Arnaise, S., and Picard, M. (1997) Mating types and secual development in filamentous ascomycetes, *Microbiol. Mol. Biol. Rev.* **61**, 411–428.

Crittenden, P.D., David, J.C., Hawksworth, D.L., and Campbell, F.S. (1995) Attempted isolation and success in the culturing of a broad spectrum of lichen-forming and lichenicolous fungi, *New Phytol.* **130**, 267–297.

Culberson, C.F., Culberson, W.L. and Johnson, A. (1988) Gene flow in lichens, *Amer. J. Bot.* **75**, 1135-1139.

Degelius, G. (1954). The lichen genus *Collema* in Europe, *Symbolae Botanicae Upsalienses* **13**, 1–499.

Dellemère, A. (1994) Asci and ascospores in ascomycete systematics, in Hawksworth, D.L. (ed.), *Ascomycete Systematics: Problems and Perspectives in the Nineties*, Plenum Press, New York, pp. 111–126.

Demmig-Adams, B., Adams III, W.W., Green, T.G.A., Czygan, F.C. and Lange, O.L. (1990) Differences in the susceptibility to light stress in two lichens forming a phycosymbiodeme, one partner possessing and one lacking the xanthophyll cycle, *Oecologia* **84**, 451–456.

Dietz, S., Büdel, B., Lange, O.L., and Bilger, W. (2000) Transmittance of light through the cortex of lichens from contrasting habitats, *Bibliotheca Lichenologica* **75**, 171–182.

Dudley-Robert (1998) Atmospheric oxygen, giant palaeozoic insects and the evolution of aerial locomotor performance, *J. Exp. Biol.* **201**, 1043–1050.

Eldridge, D. and Tozer, M.E. (1997) *A Practical Guide to Soil Lichens and Bryophytes of Australia's Dry Country*, New South Wales Department of Land and Water Conservation, Australia.

Etges, S. and Ott, S. (2001) Lichen mycobionts transplanted into the natural habitat, *Symbiosis* **30**, 191–206.

Forssel, K.B.J. (1883) Studier öfver cephalodierna. Bidrag till kännedomen om lafvarnes anatomi och utvecklingshistoria, *Bihang till Kongliga Svenska Vetensaps. Akaemiens Handlingar* **8**, 1–112.

Friedl, T. (1995) Inferring taxonomic positions and testing genus level assignments in coccoid green algae; a phylogenetic analysis of 18S ribosomal RNA sequences from *Dictyochloropsis reticulata* and form members of the genus *Myrmecia* (Chlorophyta, Trebouxiophyceae cl. nov.), *J. Phycol.* **31**, 632–639.

Friedl, T. and Büdel, B. (1996) Photobionts, in Nash III, T. H. (ed.), *Lichen Biology*, Cambridge University Press, Cambridge, pp. 8–23.

Friedl, T. and Rokitta, C. (1997) Species relationships in the lichen alga *Trebouxia* (Chlorophyta, Trebouxiophyceae): Molecular phylogenetic analyses of nuclear-encoded large subunit rRNA gene sequences, *Symbiosis* **23**, 125–148.

Galloway, D.J. (1994) Biogeography and ancestry of lichens and other ascomycetes, in Hawksworth, D.L. (ed.), *Ascomycete Systematics: Problems and Perspectives in the Nineties*, Plenum Press, New York, pp. 175–184.

Galun, M. and Peleg-Zuriel, L. (2000) Growth enhancement of lichen-forming fungi, *Bibliotheca Lichenologica* **75**, 105–111.

Gargas, A., DePriest, P.T., Grube, M. and Tehler, A. (1995) Multiple origins of lichen symbioses in Fungi suggested by SSU rDNA phylogeny, *Science* **268**, 1492–1495.

Gassmann A. and Ott, S. (2000) Growth-strategy and the gradual symbiotic interactions of the lichen

Ochrolechia frigida, *Plant Biology* **2**, 368–378.
Geitler, L. (1932) Cyanophyceae, in Kolkwitz (ed.), *Kryptogamenflora von Deutschland, Österreich und der Schweiz Vol. 14*, Akademische Verlagsgesellschaft, Leipzig, pp. 1–1196.
Girlanda, M., Isocrono, D., Bianco, C., and Luppi-Mosca, A.M. (1997) Two foliose lichens as microfungal ecological niches, *Mycologia* **89**, 531–536.
Glass, N.L. and Kuldau, G.A. (1992) Mating type and vegetative incompatibility in filamentous ascomycetes., *Annu. Rev. Phytopathol.* **30**, 201–224.
Goffinet, B. and Bayer R.J. (1997) Characterization of mycobionts of phytomorph pairs in the Peltigerineae (lichenized ascomycetes) based on internal transcribed spacer sequences of the nuclear ribosomal DNA, *Fungal Genetic. Biol.* **21**, 228–237.
Goffinet, B. and Hastings, R.I. (1995) Two new sorediate taxa of *Peltigera*. *Lichenologist* **27**, 43-58.
Goward, T. (1999) *The Lichens of British Columbia, Illustrated Keys Part 2, Fruticose Species*, Ministry of Forests (Research Branch), Victoria, Canada.
Goward, T. and Goffinet, B. (1998) Is *Nephroma silvae-veteris* the cyanomorphs of *Lobaria oregana*? Insights from molecular, chemical and morphological characters, in Glenn, M.G., Harris, R.C., Dirig, R. and Cole, M.S. (eds.), *Lichenographia Thomsoniana: North American Lichenology in Honour of John W. Thompson*, Mycotaxon Ltd, Ithaca, pp. 41–52.
Goward, T., McCune, B. and Meidinger, D. (1994) *The Lichens of British Columbia, Illustrated Keys Part 1, Foliose and Squamulose Species*, Ministry of Forests (Research Branch), Victoria, Canada.
Green, T.G.A., Büdel, B., Heber, U., Meyer, A., Zellner, H., and Lange O.L. (1993) Differences in photosynthetic performance between cyanobacterial and green algal components of lichen photosymbiodemes measured in the field, *New Phytol* **125**, 723–731.
Green, T.G.A. and Lange, O.L. (1994) Photosynthesis in poikilohydric plants: a comparison of lichens and bryophytes. in Schulze, E.D. and Caldwell, M.C. (eds.), *Ecophysiology of Photosynthesis*, Springer-Verlag, Berlin., pp. 319–341.
Haffellner, J., Hertel, H., Rambold, G. and Timdal, E. (1994) Discussion 4: Lecanorales, in Hawksworth, D.L. (ed.), *Ascomycete Systematics: Problems and Perspectives in the Nineties*, Plenum Press, New York, pp. 379-397.
Hass, H., Taylor, T.N. and Remy, W. (1994) Fungi from the Lower Devonian Rhynie chert: mycoparatism, *Amer. J. Bot.* **81**, 29–37.
Hawksworth, D. L. (1988) The variety of fungal-algal symbioses, their evolutionary significance, and the nature of lichens, *Bot. J. Linn. Soc.* **96**, 3–20.
Hawksworth, D.L. and Miadlikowska, J. (1997) New species of lichenicolous fungi occurring on *Peltigera* in Ecuador and Europe, *Mycol. Res.* **101**, 1127–1134.
Hawksworth, D.L., Kirk, P.M., Sutton, B.C. and Pegler, D.N. (1995) *Ainsworth and Bisby's Dictionary of the Fungi*, International Mycological Institute, Egham, UK.
Hawksworth, D.L. and Rossman, A.Y. (1997) Where are all the undescribed fungi? *Phytopathology* **87**, 888–891.
Heckman, D.S., Geiser, D.M., Eidell, B.R., Stauffer, R.L., Kardos, N.L., and Hedges, S.B. (2001) Molecular evidence fot the early colonization of land by fungi and plants, *Science* **293**, 1129–1133.
Henssen, A. (1963) Eine Revision der Flechtenfamilien Lichinaceae und Ephebaceae, *Symbolae Botanicae Upsalienses* **18**, 1–123.
Henssen, A, Büdel, B., and Titze, A. (1987) *Euopsis* and *Harpidium*, genera of the Lichinaceae (Lichenes) with rostrate asci, *Botanica Acta* **1**, 49–55.
Hibbett, D.S., Grimaldi, D. and Donoghue, M.J. (1997) Fossil mushrooms from Miocene and Cretaceous ambers and the evolution of homobasidiomycetes, *Amer. J. Bot.* **84**, 981–991.
Hill, D.J. (1989) The control of cell cycle in microbial symbionts, *New Phytol.* **112**, 175–184.
Hill, D.J. (1993) The co-ordination of development of symbionts in mutualistic symbiosis with reference to the cell cycle of the photobiont in lichens, *Symbiosis* **14**, 325–333.
Honegger, R. (1991) Fungal evolution: symbiosis and morphogenesis, in Margulis, L. and Fester, R. (eds.), *Symbiosis as a Source of Evolutionary Innovation*, The MIT Press, Cambridge, pp. 319–340.
Honegger, R. (1997) Metabolic interactions at the mycobiont-photobiont interface in lichens, in Carroll, G.C. and Tudzynski, P. (eds.), *The Mycota V Part A. Plant Relationships*, Springer-Verlag, Berlin, pp. 209–221.
Honegger, R. and Hugelshofer, G. (2000) Water relations in the *Peltigera aphthosa* group visualised with LTSEM techniques. *Bibliotheca Lichenologica* **75**, 113–126.
Huneck, S. and Yoshimura, I. (1996) *Identification of Lichen Substances*, Springer-Verlag, Heidelberg.

Jahns, H.M. (1973) Anatomy, morphology and development, in Ahmadjian, V. and Hale, Jr. M.E. (eds.), *The Lichens*, Academic Press, New York, pp. 3-58.
Jahns, H.M. (1988) The lichen thallus, in Galun, M. (ed), *CRC Handbook of Lichenology*, CRC Press Inc, Boca Raton, pp. 3–16.
Jahns, H.M. and Ott, S. (1997) Life strategies in lichens – some general considerations, *Bibliotheca Lichenologica* **67**, 49–67.
James, P.W. and Henssen, A. (1976) The morphological and taxonomic significance of cephalodia, in Brown, D.H., Hawksworth, D.L. and Bailey, R.H. (eds.), *Lichenology, Progress and Problems*, Academic Press, London, pp. 27–77.
Jordan, W.P. (1970) The internal cephalodia of the genus *Lobaria*, *Bryologist* **73**, 669–681.
Jordan, W.P. (1972) Erumpent cephalodia, an apparent case of phycobial influence on lichen morphology, *J. Phycol.* **8**, 112–117.
Jordan, W.P. and Rickson, R. (1971) Cyanophyte cephalodia in the lichen genus *Nephroma*, *Amer. J. Bot.* **58**, 562–568.
Jørgensen, P.M. (1994) On the nomenclature of lichen phototypes, *Taxon* **45**, 663–664.
Jørgensen, P.M. (1997) Lichen phototypes, nature's unmanageable misprints? *Taxon* **46**, 721–722.
Jørgensen, P.M. (1998) What shall we do with the blue-green counterparts? *Lichenologist* **30**, 351–356.
Jørgensen, P.M. (2000) Survey of the lichen family Pannariaceae on the American continent, north of Mexico, *The Bryologist* **103**, 670–704.
Jørgensen, P.M. and Wedin, M. (1998) On *Psoroma* species from the Southern Hemisphere with cephalodia producing vegetative dispersal units, *Lichenologist* **31**, 341–347.
Kantivilas, G. and Jarman, S.J. (1999) *Lichens of Rainforest in Tasmania and South-Eastern Australia*. The Australian Biological Resource Study, Canberra, Australia.
Kardish, N., Kessel, M. and Galun, M. (1989) Characterization of symbiotic and cultured *Nostoc* of the lichen *Nephroma laevigatum* Ach, *Symbiosis* **7**, 257–266.
Kardish, N., Rotem-Abarbanell, D., Ziberstein, A. and Galun, M. (1990) Comparison between the symbiotic *Nostoc* of the lichen *Nephroma laevigatum* Ach. and its cultured, isolated *Nostoc* by recombination DNA, *Symbiosis* **8**, 135–145.
Karling, J.S. (1977) *Chytridiomycetarium Iconographia*, J. Cramer, Vaduz.
Komárek, J., Anagnostidis, K. (1989) Modern approach to the classification system of cyanophytes 4, Nostocales, *Arch. Hydrobiol. Suppl.* **82**, 247–345.
Kondratyuk, S. and Galloway, D.J. (1995) Lichenicolous fungi and chemical patterns in *Pseudocyphellaria*, *Bibliotheca Lichenologica* **57**, 327–345.
Kranner, I. and Lutzoni, F. (1999) Evolutionary consequences of transition to a lichen symbiotic state and physiological adaptation to oxidative damage associated with poikilohydry, in Lerner, H.R. (ed.), *Plant Responces to Environmental Stresses: From Phytohormones to Genome Reorganization*. Marcel Deccer, New York, pp. 591–628.
Kroken, S. and Taylor, J.W. (2000) Phylogenetic species, reproductive mode, and specificity of the green alga *Trebouxia* forming lichens with the fungal genus *Letharia*, *The Bryologist* **103**, 645–660.
Lamb, I.M. (1951) On the morphology, phylogeny, and taxonomy of the lichen genus *Stereocaulon*, *Can. J. Bot.* **29**, 522–584.
Lamb, I.M. (1968) The species of *Stereocaulon* with protosacculate cephalodia, *Japan. J. Bot.* **43**, 291–301.
Lamb, I.M. (1976) Structurally unusual types of cephalodia in the lichen genus *Stereocaulon* (subgen. *Holostedium*), *Japan. J. Bot* **51**, 353–359.
Lange, O.L., Kilian, E. and Ziegler, H. (1986) Water vapor uptake and photosynthesis in lichens: performance differences in species with green and blue-green algae as phycobionts, *Oecologia* **71**, 104–110.
Lange, O.L., Bilger, W. And Schreiber, U. (1989) Chlorophyll fluorescence of lichens containing green and blue-green algae during hydration by water vapour uptake and by addition of liquid water, *Botanica. Acta* **102**, 306-313.
Larson, D.W. and Carey, C.K. (1986) Phenotype variation within 'individual' lichen thalli, *Amer. J. Bot.* **73**, 214-223.
Lehr, H., Fleminger, G. and Galun, M. (1995) Lectin from the lichen *Peltigera membranacea* (Ach.) Nyl.: characterization and function, *Symbiosis* **18**, 1–13.
Lehr, H., Galun, M., Ott, S., Jahns, H.M. and Fleminger, G. (2000) Cephalodia of the lichen *Peltigera aphthosa* (L.) Willd. Specific recognition of the compatible photobiont, *Symbiosis* **29**, 357–365.
Leslie, J.F. (1993) Fungal vegetative incompatibility, *Annu. Rev. Phytopathol.* **31**, 127–150.

Letrouit-Galinou, M.A. and Asta, J. (1994) Thallus morphogenesis in some lichens, *Cryptogam. Bot.* **4**, 274–282.

Lutzoni, F., Pagel, M. and Reeb, V. (2001) Major fungal lineages are derived from lichen symbiotic ancestors, *Nature* **411**, 937–940.

McCune, B. (1993) Gradients in the epiphyte biomass in three *Pseudotsuga-Tsuga* forests of different ages in Western Oregon and Washington, *Bryologist* **96**, 405-411.

McCune, B. and Geiser, L. (1997) *Macrolichens of the Pacific Northwest*, Oregon State University Press, Corvallis.

Meeks, J.C., Campbell, E., Hagen K., Hanson, T., Hitzeman, N. and Wong, F. (1999) Developmental alternatives of symbiotic *Nostoc punctiforme* in response to its plant partner *Anthoceros punctatus*, in Peschek, G.A., Loffelhardt, W. and Schmetterer, G. (eds.), *The Photosynthetic Prokaryotes*, Kluwer Academic Publishers, pp. 665-678.

Miadlikowska; J. and Lutzoni, F. (2000) Phylogenetic revision of the genus *Peltigera* (lichen-forming Ascomycota) based on morphological, chemical, and large subunit nuclear ribosomal DNA data, *Internat. J. Plant Sci.* **16**, 925–958.

Miao, V.P.W., Rabenau, A. and Lee, A. (1997) Cultural and molecular characterization of photobionts of *Peltigera membranacea*, *Lichenologist* **29**, 571–586.

Murthach, G.J., Dyer, P.S. and Crittenden, P. D. (2000) Reproductive systems: Sex and the single lichen, *Nature* **404**, 564.

Nash III, T.H. (1996) *Lichen Biology*, Cambridge University Press, Cambridge.

Oberwinkler, F. (1984) Fungus-alga interactions in basidiolichens, *Nova Hedwigia* **79**, 739–774.

Ott, S. (1993) Experimental research in regulation of developmental processes in lichens, in *Abstracts of the XV International Botanical Congress*, Yokohama, Japan.

Ott, S. (1988) Photosymbiodemes and their development in *Peltigera venosa*, *Lichenologist* **20**, 361–368.

Ott, S. (1987) Sexual reproduction and developmental adaptations in *Xanthoria parietina*, *Nordic J. Bot.* **7**, 219–228.

Parks, C.R. and Wendel, J.F. (1990) Molecular divergence between Asian and North American species of *Liriodendron* (*Magnoliaceae*) with implications for interpretation of fossil floras, *Amer. J. Bot.* **77**, 1243–1256.

Paulsrud, P. (2001) The *Nostoc* symbiont of lichens. Diversity, specificity and cellular modifications, *Acta Universitatis Upsaliensis Comprehensive Summaries of Uppsala Dissertations from the Faculty of Science and Technology* **662**, 1–55.

Paulsrud, P. and Lindblad, P. (1998) Sequence variation of the tRNALeu intron as a marker for genetic diversity and specificity of symbiotic cyanobacteria in some lichens, *Appl. Environ. Microbiol.* **64**, 310–315.

Paulsrud, P. Rikkinen, J. and Lindblad, P. (1998) Cyanobiont specificity in some *Nostoc*-containing lichens and in a *Peltigera aphthosa* photosymbiodeme, *New Phytol.* **139**, 517–524.

Paulsrud, P., Rikkinen, J. and Lindblad, P. (2000) Spatial patterns of photobiont diversity in some *Nostoc*-containing lichens, *New Phytol.* **146**, 291–299.

Paulsrud, P., Rikkinen, J. and Lindblad, P. (2001) Field experiments on cyanobacterial specificity in *Peltigera aphthosa*, New Phytol. **152**, 117–123.

Perez-Moreno, J. and Read, D.J. (2000) Mobilization and transfer of nutrients from litter to tree seedlings via the vegetative mycelium of ectomycorrhizaö plants, *New Phytol.* **145**, 301-309.

Peterson, E.B. (2000) An overlooked fossil lichen (Lobariaceae), *Lichenologist* **32**, 298–300.

Petrini, O., Hake, U. and Dreyfuss, M. (1990) Analysis of fungal communities isolated from fruticose lichens, *Mycologia* **82**, 444-451.

Peveling, E. (1988) Beziehungen zwischen den Symbiosepartnern in Flecten, *Naturwissenschaften* **75**, 77–86.

Poelt, J. and Mayhofer, H. (1987) Über Cyanotrophie bei Flechten, *Plant Systemat. Evol.* **158**, 265–281.

Poinar, Jr., G. and Poinar, R. (1999) *The Amber Forest*, Princeton University Press, Princeton.

Poinar, Jr., G., Peterson, E.B. and Platt, J.L. (2000) Fossil *Parmelia* in New World amber, *Lichenologist* **32**, 263–269.

Printzen, C. and Lumbsch, T. (2000) Molecular evidence for the diversification of extant lichens in the Late Cretaceous and Tertiary, *Mol. Phylogenet. Evol.* **17**, 379–387.

Purvis, W. (2000) *Lichens*, Smithsonian Institute Press and Natural History Museum, London.

Rai, A.N., Söderback, E. and Bergman, B. (2000) Cyanobacterium-plant symbioses, *New Phytol.* **147**, 449-481.

Rambold, G., Friedl, T., and Beck, A. (1998) Photobionts in lichens: possible indicators of phylogenetic

relationships? *The Bryologist* **101**, 392–397.
Rambold, G. and Triebel, D. (1992) The inter-lecanoralean associations, *Bibliotheca Lichenologica* **48**,1–201.
Raven, J.A. (1993) Energy and nutrient acquisition by autotrophic symbioses and their asymbiotic ancestors, *Symbiosis* **14**, 33–60.
Redhead, S.A. (1989) A biogeographical overview of the Canadian mushroom flora, *Can. J. Bot.* **67**, 3003–3062.
Richardson, D.H.S. (1999) War in the world of lichens: parasitism and symbiosis as exemplified by lichens and lichenicolous fungi, *Mycol. Res.* **103**, 641-650.
Rikkinen, J. (1995) What's behind the pretty colours? A study on the photobiology of lichens, *Bryobrothera* **4**, 1–239.
Rikkinen, J. (1997) Habitat shifts and morphological variation of *Pseudevernia furfuracea* along a topographic gradient, *Symbolae Botanicae Upsalienses* **32**, 223–245.
Rikkinen, J. and Poinar, Jr., G. (2000) A new species of resinicolous *Chaenothecopsis* (*Mycocaliciaceae, Ascomycota*) from 20 million year old Bitterfeld amber, with remarks on the biology of resinicolous fungi, *Mycol. Res.* **104**, 7–15.
Rippka, R., Deruelles, J., Waterbury, J.B., Herdman, M. and Stanier, R.Y. (1979) Generic assignements, strain histories and properties of pure cultures of cyanobacteria, *J. Gen. Microbiol.* **111**, 1–61.
Round, F.E. (1984) *The Ecology of Algae*, Cambridge University Press, Cambridge.
Scheidegger, C. (1994) Low-temperature scanning electron microscopy: the localization of free and preturbed water and its role in the morpholgy of the lichen symbionts, *Cryptogam. Bot.* **4**, 290–299
Schelensog, M. Schroeter, B., and Green, T.G.A. (2000) Water dependent photosynthetic activity of lichens from New Zealand: differences in the green algal and the cyanobacterial thallus parts of photosymbiodemes, *Bibliotheca Lichenologica* **75**, 149–160.
Schultz, M., Brown, G., and Büdel, B. (2000) Cyanophilous lichens from Kuwait, *Nova Hedwigia* **70**,193–216.
Schultz, M., Mies, B. and Al-Gifri, A.N. (1999) New localities of some *Paulia* species (Lichinaceae, lichenized Ascomycota) from Socotra (Indian Ocean), *Bryologist* **102**, 61–66.
Sen, R. (2000) Budgeting for the wood-wide web, *New Phytol.* **145**, 161-165.
Sparrow, Jr., F.K. (1960) *Aquatic Phycomycetes*, The University of Michigan Press, Ann Arbor.
Socker-Wörgötter, E. (2001a) Experimental studies of the lichen symbiosis: DNA-analyses, differentiation and secondary chemistry of selected mycobionts, artificial resynthesis of two- and tripartite symbioses, *Symbiosis* **30**, 207–227.
Stocker-Wörgötter, E. (2001b) Experimental lichenology and microbiology of lichens: culture experiments, secondary chemistry of cultured mycobionts, resynthesis, and thallus morphogenesis, *Bryologist* **104**, 576–581.
Stocker-Wörgötter, E. and Türk, R. (1994) Artificial resynthesis of the photosymbiodeme *Peltigera leucophlebia* under laboratory conditions, *Cryptogam. Bot.* **4**, 300–308.
Taylor, T.N. (1994) The fossil history of ascomycetes, in Hawksworth, D.L . (ed.), *Ascomycete Systematics: Problems and Perspectives in the Nineties*, Plenum Press, New York, pp. 167–174.
Taylor, T.N. and Taylor, E.L. (1993) *The Biology and Evolution of Fossil Plants*, Prentice Hall, Englewood Cliffs, New Jersey.
Taylor, T.N. and Taylor, E.L. (1997) The distributions and interactions of some Paleozoic fungi, *Rev. Palaeobot. Palynol.* **95**, 83–94.
Taylor, T.N., Hass, H. and Kerp, H. (1997) A cyanolichen from the Lower Devonian Rhynie chert, *Amer. J. Bot.* **84**, 992–1004.
Taylor, T.N., Hass, H. and Kerp, H. (1999) The oldest fossil ascomycetes, *Nature* **399**, 648.
Taylor, T.N., Remy, W. and Hass, H. (1992) Fungi from the Lower Devonian Rhynie chert: chytridiomycetes, *Amer. J. Bot.* **79**, 1233–1241.
Tehler, A. (1996) Systematics, phylogeny and classification, in Nash III, T. H. (ed.), *Lichen Biology*, Cambridge University Press, Cambridge, pp. 217–239.
Tehler, A., Farris, J.S., Lipscomb, D.L. and Källersjö, M. (2000) Phylogenetic analyses of the fungi based on large rDNA data sets, *Mycologia* **92**, 459–474.
Tiffney, B.H. (1985) Perspectives of the origin of the floristic similarity between eastern Asia and eastern North America, *J. Arnold Arboretum* **66**, 73–94.
Tønsberg, T. and Goward, T. (2001) *Sticta oroborealis* sp.nov. and other Pacific North American lichens forming dendriscocauloid cyanotypes, *Bryologist* **104**, 12–23.

Tønsberg, T. and Holtan-Hartwig, J. (1983) Phycotype pairs in *Nephroma*, *Peltigera* and *Lobaria* in Norway, *Nordic J. Bot.* **3**, 681–688.

Tschermak-Woess E. (1988) The algal partner, in Galun, M. (ed.), *CRC Handbook of Lichenology Vol. I,* CRC Press, Boca Raton, pp. 39–92.

Tschermak-Woess E. (1995) The taxonomic position of the green phycobiont of *Sticta canariensis* (Ach.) Bory ex Delise and its extraordinary modification in the lichenized state, *Bibliotheca Lichenologica* **58**, 433–438.

Vitikainen, O. (1994) Taxonomic revision of *Peltigera* (lichenized Ascomycotina) in Europe, *Acta Botanica Fennica* **152**, 1–96.

Wirth, V. (1980) *Flectenflora*, Verlag Eugen Ulmer, Stuttgart.

Wirth, V. (1995) *Die Flecten Baden-Wurttenbergs*, Verlag Eugen Ulmer, Stuttgart.

Wu, Q.X. and Mueller, G.M. (1997) Biogeographic relationships between macrofungi of temperate eastern Asia and eastern North America, *Can. J. Bot.* **75**, 2108–2116.

Wu, Q.X., Mueller, G.M., Lutzoni, F.M., Huang, Y.Q. and Guo, S.Y. (2000) Phylogenetic and biogeographic relationships of eastern Asian and eastern North American disjunct *Suillus* species (Fungi) as inferred from nuclear ribosomal RNA ITS sequences, *Mol. Phylogenet. Evol.* **17**, 37–47.

Xiang, Q.Y., Soltis, D.E. and Soltis, P.S. (1998) The eastern Asian and eastern and western North American floristic disjunction: congruent phylogenetic patterns in seven diverse genera, *Mol. Phylogenet. Evol.* **10**, 178–190.

Yoshimura, I. and Yamamoto, Y. (1991) Development of *Peltigera praetextata* lichen thalli in culture, *Symbiosis* **11**, 109–117.

Yoshimura, I., Kurokawa, T., Yamamoto, Y., and Kinoshita, Y. (1994) In vitro development of the lichen thallus of some species of *Peltigera*, *Cryptogam. Bot.* **4**, 314–319.

Chapter 5

CYANOLICHENS: CARBON METABOLISM

K. PALMQVIST
Department of Ecology and Environmental Science
Umeå University, SE-901 87 Umeå, Sweden

1. INTRODUCTION

Lichens are symbiotic associations between a fungus (mycobiont), and a photobiont, which can be an alga, and/or a cyanobacterium. Estimates of the number of lichenized fungi range from 13 000 to 17 000 of which 10-15% have a cyanobacterial symbiont (Friedl and Büdel, 1996; Richardsson, 1999). Among these, the filamentous and heterocyst containing cyanobacterial genus *Nostoc* is the most common. In addition, about 500 lichenized fungi can form tripartite associations with a green alga as primary carbon fixing symbiont, and a cyanobacterium as a secondary N_2-fixing partner (Tschermak-Woess, 1988). In these, the cyanobacterium is located in either external or internal cephalodia (see the previous chapter by Rikkinen and the next chapter by Rai).

Depending on the species involved, the structural organization and complexity of lichen thalli vary (Hawksworth, 1988). In the homoiomerous genus *Collema* the fungus does not alter the organization of the *Nostoc* filaments, and the thallus resembles a *Nostoc*-colony penetrated by fungal hyphae. In this genus and other homoiomerous associations such as some *Leptogium* spp. the photobiont contributes significantly more to the total thallus biomass than in heteromerous thalli (Hawksworth & Hill, 1984). In the genus *Peltigera*, on the other hand, the thallus is much more complex resembling the organization of a higher plant leaf (Honegger, 1991). In some *Peltigera* spp., *Nostoc* is found evenly distributed in a distinct zone below the upper cortex of the thallus, a location that increases their light absorption efficiency.

Recent data have shown that the identity of the *Nostoc* symbiont of bipartite and tripartite lichens is more dependent on genetic affiliation of the mycobiont than collection site (Paulsrud *et al.*, 2000). Moreover, the same *Nostoc* can be modified by the fungal host to participate either as primary photobiont in a bipartite association, or as N_2-fixing partner in a tripartite association (Paulsrud *et al.*, 2001). Even though the latter example reflects the particular case of lichen-forming fungi being able to form a so called photosymbiodeme, the findings of Paulsrud and co-workers emphasize that the host-symbiont specificity might be very high for *Nostoc*-containing lichens, and that the fungus can regulate the specific function of the cyanobacterium to optimize its own fitness (Hyvärinen *et al.*, 2002). Lichens should then be viewed as a nutritional strategy

A.N. Rai, B. Bergman and U. Rasmussen (eds.), Cyanobacteria in Symbiosis, 73-96.

of the fungus rather than uniquely new organisms, where the primary function of the photobiont is to provide the fungus with photo-assimilated carbon (Honegger, 1996; Richardsson, 1999). From the fungal perspective it then appears to be particularly advantageous to choose a cyanobacterial photobiont being capable of both CO_2 and N_2 fixation. This chapter will mainly be focused on such bi-partite cyanolichens were the symbiont apparently has this dual nature. Among these, lichens containing *Nostoc* have been most extensively investigated ranging from ecological to biochemical studies.

2. CARBON METABOLISM AND THE POIKILOHYDRIC LIFE-STYLE

Cyanolichens as well as their free-living cyanobacterial relatives can be found in most terrestrial ecosystems of the world, ranging from arctic to tropical regions (Nash, 1996; Paulsrud, 2001). Many species are epiphytic on trees in boreal, temperate and tropical forests, while others are terricolous and can be found during early successions on bare soil. Some of the more prominent *Peltigera* species can even be found among mosses, apparently being able to compete with these for space, light and nutrients. The biomass contribution of lichens to an ecosystem varies from insignificant to major, depending on habitat and species. Lichens with cyanobacteria may even play an important role of mineral cycling in their ecosystem (Kappen, 1988). This is because they can grow relatively fast (their biomass increases by 20-50% per year) and fix N_2 (Nash, 1996).

The ecological success of lichens, including cyanolichens, can in part be explained by their poikilohydrous nature, and their ability to resist desiccation and low temperatures (Kappen, 1988). The extent to which lichens can tolerate drought stress is partly related to the moisture conditions to which they are adapted in their natural habitat. For example, xeric species recover more quickly and from longer periods of drying than mesic species (Bewley, 1979). Desiccation tolerance of a lichen involves desiccation tolerance of both mycobiont and photobiont. This is a trait that the lichen symbionts also share with their free-living relatives (Raven, 1992; Qui and Gao, 2001).

For poikilohydric organisms such as lichens, both the uptake and the loss of water are physical processes without metabolic control (Blum, 1973). Rates of water movement between lichen thalli and their environment therefore varies between species, as being dependent on the amount of water the thallus can hold at water saturation, as well as thallus morphology, anatomy and color (Rundel, 1982). For instance, due to their larger surface area to volume ratios, filamentous and fruticose species take up and lose water more rapidly than flat, foliose species. For the same reason, a thick foliose lichen will equilibrate slower with the surrounding air than a thinner lichen (Gauslaa and Solhaug, 1998). Typically, lichens contain 1 to 3 g water per g dry weight at maximal hydration (Blum, 1973). However, homoiomerous lichens may hold 20-30 g water per g dry weight (Lange *et al.*, 1993). Depending on environmental conditions, lichen water contents (WC) can fluctuate from complete dryness to full hydration and *vice versa* within a few minutes (Lange *et al.*, 1993; Palmqvist and Sundberg, 2000).

When lichens desiccate, both photosynthesis and respiration gradually decline (Cowan *et al.*, 1979a, 1979b; Rundel, 1982) until cell turgor is lost (Beckett, 1997). Although some rainforest species (Green *et al.*, 1991) and aquatic lichens are irreversibly damaged by drying (Bewley, 1979), the majority of lichens can survive prolonged periods in a metabolically inactive state when their thallus WCs are at or below 10 % of their dry weight (Bewley, 1979). In many lichens, the photosynthetic apparatus can be preserved during desiccation without damaging photosynthetic pigments (see section 4). In cyanolichens, photooxidative damage seems to be avoided by detachment of the light harvesting phycobilisome antenna (PBS) from photosystem II when the lichens desiccate (Bilger *et al.*, 1989).

Metabolism recovers very quickly when dry lichens are re-hydrated, and respiratory CO_2 efflux is detectable within two to four minutes (Smith and Molesworth, 1973) (Fig. 1A). Photosynthesis in lichens with cyanobacterial photobionts always requires addition of liquid water to recover from desiccation (Lange *et al.*, 1986). The reason for this has not been elucidated, but the PBS antenna is not functionally attached to PSII again until liquid water is added (Bilger *et al.*, 1989; Lange *et al.*, 1989). Time required to fully induce metabolism after re-hydration varies depending on the particular lichen species involved. Fast recovery of net CO_2 fixation upon re-hydration has been shown for the *Nostoc*-lichen *Nephroma resipunatum* (Lange *et al.*, 1986), and the basidiomycete cyanolichen *Dictyonema glabratum* (Lange *et al.*, 1994). In contrast, *Peltigera leucophlebia* and *Collema auriculatum* required 30-40 min for complete photosynthetic induction (Lange *et al.*, 1986). A similarly long activation period was recorded for *Peltigera canina* (Fig. 1) in an experiment where net CO_2 fixation (Fig. 1A), respiratory CO_2 efflux (Fig. 1A) and variable chlorophyll *a* fluorescence yield (Fig. 1B) were measured simultaneously. In this particular experiment, photosynthesis started after a lag-period of 10-15 minutes (Fig. 1B), while a high rate of respiration was observed immediately upon addition of the water (Fig. 1A). This initially high, so-called 'resaturation respiration' (Brown *et al.*, 1983), decreased thereafter to a lower steady-state rate during the first 15-20 minutes, while photosynthetic induction continued for at least 1 h when a positive net CO_2 fixation rate was reached (Fig. 1A).

The underlying biochemical mechanisms for resaturation respiration are not known, although there are several suggestions. These include an increased energy demand for repair of damaged membranes (Smith and Molesworth, 1973) and a burst in respirable substrates related to the drought damage of membranes (Farrar and Smith, 1976). The amplitude of the burst as well as the time required to reach steady state may depend on the time the thallus was active during its previous active period and the rate of drying. Faster drying results in a larger burst (Brown *et al.*, 1983). It also appears that the burst is less prominent when CO_2 fluxes are followed *in situ* (cf. Lange *et al.*, 1994; Zotz *et al.*, 1998).

Most lichens undergo frequent cycles of drying and wetting in their natural habitat, and the length and frequency of these cycles will have a large impact on their carbon budget. For example, too short and infrequent periods of metabolic activity might lead to high rates of resaturation respiration, incomplete recovery of photosynthesis, and high carbon losses because of leakage from the thallus (Dudley and Lechowicz, 1987). In addition, and as will be discussed later, both respiration and photosynthesis increase with temperature (Figs. 3A, C), and photosynthesis strongly depends on prevailing light conditions (Fig. 3B). Taken together, this emphasizes the significant impact of environmental conditions (water, light and temperature) on the carbon budget of lichens (Kappen, 1988; Nash, 1996).

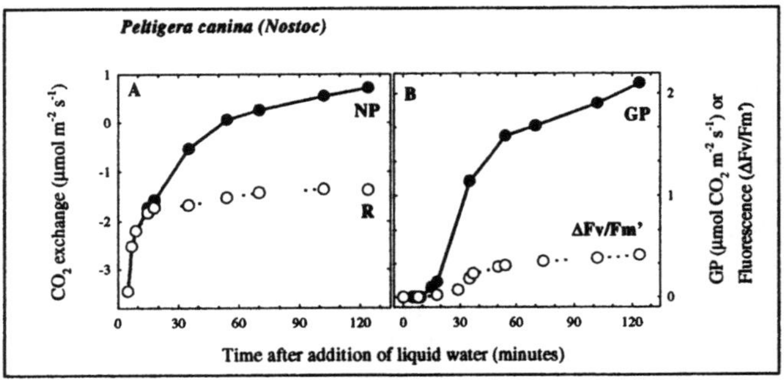

Figure 1. Metabolic induction in Peltigera canina *after addition of liquid water at time 0. CO_2 gas exchange (A) and chlorophyll a fluorescence yield (B) were measured simultaneously (Palmqvist and Sundberg, 2000). The desiccated sample was mounted on a wire tray, placed in a CO_2 gas-exchange cuvette, and equilibrated to 10°C for a few minutes before adding the water and closing the cuvette. The light (50 μmol m^{-2} s^{-1}) was switched on and off at 2-5 min intervals to record both fixation and efflux of CO_2. Gross CO_2 fixation (GP) was calculated by adding the mean respiration rate (R) before and after a light period to the net CO_2 fixation rate (NP) in the middle of the intervening light period.*

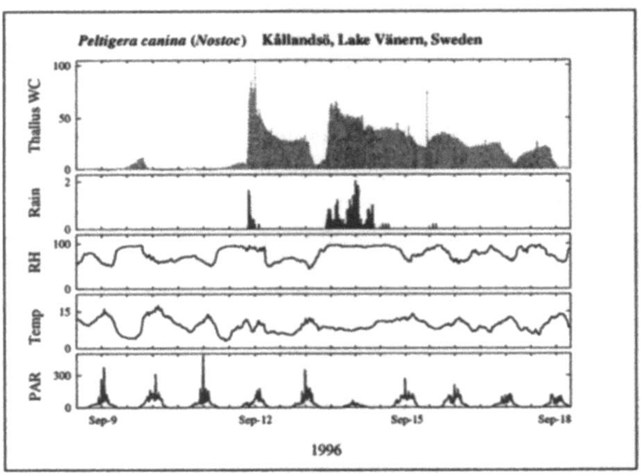

Figure 2. Water content (WC) and climatic conditions of a Peltigera canina *thallus during 10 days in Sept 1996. WC was measured with an impedance technique. PAR and thallus Temp were also measured with specially designed sensors mounted on the thallus, as described in Palmqvist and Sundberg (2000).*

Numerous studies have followed CO_2 gas exchange, thallus water content, and microclimatic conditions of cyanolichens for extended periods in the field (*e.g.*, Lange *et al.*, 1993; Lange *et al.*, 1994; Leisner *et al.*, 1996; Zotz *et al.*, 1998; Palmqvist and Sundberg, 2000). These studies have shown that the respiratory CO_2 losses during night can be significant, and that photosynthesis can be depressed during desiccation caused by high irradiance levels and occasionally lead to a negative daily carbon balance (Lange *et al.*, 1994; Zotz *et al.*, 1998). The negative effects of high ambient temperatures at night may explain the absence of macrolichens in such habitats as lowland rainforests (Lange *et al.*, 1994). Nevertheless, in habitats where environmental conditions are favorable, cyanolichens can achieve high net carbon gain. For example, *P. canina* showed an annual biomass increase of 50-60% (Palmqvist and Sundberg, 2000). In this study, the lichens were wetted and activated by rainfall. They remained wet and presumably metabolically active for several days after the rainfall (Fig. 2). On average, this species got wet 39 times between May and Sept, each wet period lasting 30-40 h, and achieved a 30-40% net weight gain. A model using environmental and laboratory CO_2 gas exchange data predicted very accurately the growth in this species, so carbon losses associated with re-hydration were probably minor. In addition, temperatures were higher during daytime than at night (Fig. 2), and the lichens were wet in the light during c. 70% of the total wet time. These factors also favored the high positive net carbon gain (Palmqvist and Sundberg, 2000).

However, supraoptimal water contents in the thallus can be detrimental to lichen carbon budgets. This is mainly related to the 10 000 times lower CO_2 diffusion rate in water compared to air, and a subsequent decrease in photosynthesis due to lack of substrate (Cowan *et al.*, 1992). The total water holding capacity and the relative water content required for optimal photosynthesis vary widely depending on lichen species. At least four different types of responses can be distinguished (Lange *et al.*, 1993). Cyanolichens have a photosynthetic CO_2 concentrating mechanism (CCM) (Badger *et al.*, 1993) that might partly compensate for the reduced CO_2 diffusion rate at supraoptimal WC. Despite this, photosynthesis in some cyanolichens, as well as in free-living terrestrial *Nostoc flagelliforme*, may still be significantly inhibited by high water contents (Lange *et al.*, 1993; Qui and Gao, 2001). This emphasizes that the CCM cannot fully compensate for the CO_2 diffusion limitation. However, the beneficial effects of being able to remain metabolically active for longer periods must also be taken into account when quantifying the adverse effect of depressed CO_2 acquisition at high thallus WCs. Hence, even if a supraoptimal WC may limit CO_2 fixation rates during parts of active periods, being able to maintain a high WC for prolonged periods may still be of competitive advantage. To study such trade-offs, the recently proposed parameter Lichen Water Use Efficiency (LWUE), measuring carbon gain versus water loss might be useful (Máguas *et al.*, 1997).

3. CARBON REQUIREMENTS

In addition to the above environmental constraints on lichen carbon budgets, recent data suggest that there must also be some internal regulation of carbon gain and expenditure reactions in lichens (Palmqvist *et al.*, 2002). In plants, it has long been established that the magnitudes of carbon gain and expenditure reactions vary together. The general view is that photosynthesis is feedback inhibited by carbohydrate demands and not *vice versa* (cf. Lambers *et al.*, 1998). Assuming that the mycobiont is the largest carbon sink in lichens we may then speculate that the fungus is somehow able to control its carbohydrate availability by regulating the size of its photobiont population. This idea goes back to Schwendener (1873), and much evidence tends to support this viewpoint (cf. Honegger, 1991; Hyvärinen *et al.*, 2002). So, it makes sense to first analyze the carbon requirements of cyanolichens (see Table 1) before going into the details of their carbon acquisition capacities (see Section 4).

3.1. Growth and Thallus Composition

The major part of lichen biomass (~95-98% w/w) is made of carbohydrate (CH_2O) or its derivatives, and the weight gain is primarily dependent on photosynthetic carbon assimilation minus respiratory CO_2 losses (Palmqvist and Sundberg, 2000). In addition, the N_2-fixing cyanolichens contain nitrogen which is 2-5% of their dry weight (Rai, 1988; Palmqvist *et al.*, 1998, 2002). As already mentioned, the terricolous species *P. canina* could increase its biomass by 50-60% annually. Two epiphytic *Nostoc* lichens, *Lobaria oregana* and *Pseudocyphellaria rainierensis*, also display a relatively high growth rate, with an annual increase of 20-30% in their biomass (Sillett and McCune, 1998). We can assume that such significant increases in lichen biomass also result in formation of new tissue, because area and biomass increases are apparently tightly coupled in un-stressed lichens (Sundberg *et al.*, 2001; Dahlman *et al.*, 2002). This is probably in part related to the fact that growth of both biotrophic partners must be synchronized and sharply coordinated to avoid disintegration of the thallus (Smith *et al.*, 1969).

A significant amount of reduced carbon compounds (c. 0.1 g g^{-1} DW for each 10% increase) was required for making the new tissue, and for energization of the growth process, *i.e.* growth respiration (cf. Lambers *et al.*, 1998). As in plants, cell walls may constitute a large fraction of the lichen tissue, particularly in green algal lichens where 60-70% (w/w) of the thallus weight may be attributed to cell walls. In the *Nostoc*-lichen *P. canina*, 36% of the dry weight could be attributed to cell wall compounds (Boissière, 1987). Additional carbohydrate skeletons are used for amino acids and proteins, which may constitute c. 75% of the total nitrogen content of a *P. canina* thallus (Rai, 1988). Considering that *P. canina* has a nitrogen content of 4% and that the nitrogen content of compounds such as proteins and amino acids may be 10-15%, up to 30% of the thallus dry weight can be attributed to proteins and amino acids (Table 1). In addition to the requirements for hyphal construction, carbohydrates and nitrogen are also needed for the growth and maintenance of the photobiont. For example, *Nostoc*-filaments are enclosed within a gelatinous sheath composed of fibrillar polysaccharides (presumably glucans) (Honegger, 1991; Bogner *et al.*, 1993). This sheath can constitute 5-17% of the dry mass in isolated *Nostoc* (Bogner *et al.*, 1993).

Table 1. Summary of the major carbohydrate sinks in P. canina. *Requirements are tentative and based on an annual growth rate of 50%. It is assumed that N_2-fixation matches carbohydrate input so that the lichen can maintain balanced growth (Palmqvist and Sundberg, 2000; Sundberg et al., 2001; Dahlman et al., 2002).*

Sink	Pool size (% DW)	Annual Requirement ($g.g^{-1}$ DW)	Comments
Cell walls	36	0.18	Low maintenance costs once formed. 13% of the cell wall consists of chitin having 6% N (Boissiere, 1987).
Proteins and amino acids	30	0.15	Assuming that 4% of the thallus dry wt is N, and that 75% of this is in the form of structural proteins, enzymes and amino acids with N content of 10% (Rai, 1988).
Soluble C (mannitol)	5-10	>>0.05	Drought resistance requires a high C level (Farrar, 1988). The C pool is also used as respiratory substrate. Major losses can occur during re-wetting (Dudley and Lechowicz, 1987). Therefore, the pool must be replenished constantly.
Nostoc	?	?	Photobionts constitute 36% of the thallus proteins (Rai, 1988) but probably lesser of the thallus dry mass. The surrounding polysaccharide sheath constitutes 5-17% of the dry mass in symbiotic *Nostoc* grown in culture (Bogner *et al.*, 1993).
Secondary lichen acids	?	?	Generally absent in cyanolichens (Galun and Shomer-Ilan, 1988).
Missing pools	25-30	<0.12	
Respiration	35-50% of net photo-synthesis	35-50% of net photosynthesis	Data are scarce on the the relative energy demands of various processes such as maintenance vs growth respiration (Lambers *et al.*, 1998), resaturation respiration (Brown *et al.*, 1983), or carbon costs for N_2-fixation and additional nutrient assimilation

3.2. Soluble Carbohydrates and Secondary Metabolites

Sugar alcohols (polyols) are the dominant soluble carbohydrates in lichens. They constitute 2-10% of the thallus dry weight, depending on species and season. The highest polyol concentration has been recorded during late summer in the *Nostoc* lichen *Peltigera polydactyla* (Lewis and Smith, 1967). Cyanolichens contain mannitol and arabitol; the green algal products erythritol and ribitol are lacking in cyanobionts, because cyanobacteria release glucose to the mycobiont (cf. Honegger, 1991). Arabitol is depleted more rapidly than mannitol under conditions of stress (Farrar, 1976), suggesting that this polyol may function as a short-term carbohydrate reserve, but mannitol has also been proposed to serve as a substrate for respiration during prolonged periods of darkness (Drew, 1966). These compounds, together with non-reducing sugars such as trehalose and sucrose, may also be involved in desiccation tolerance, since they can substitute for water and stabilize proteins and membranes under dry conditions (Farrar, 1988; Leprince *et al.*, 1993; Jennings and Lysek, 1996). It has therefore been speculated that the polyol pool might be the largest carbon sink in lichens, particularly in slow growing crustose species in harsh environments where drought periods can be particularly long and severe, and where brief re-hydration events might result in significant C losses (Smith, 1975). As already mentioned, lichen growth is also a three-dimensional process involving cell division and expansion of the photobiont cells, and growth of the hyphae. The driving force for hyphal expansion is probably the turgor pressure, with major structural wall components being manufactured directly on the extending apical plasma membrane (Wessels, 1993). Osmotically active carbon compounds such as mannitol are then required, together with water, to create the turgor pressure needed for hyphal extension (Jennings and Lysek, 1996).

In general, most of the organic compounds found in lichens are secondary metabolites of the fungus, which are deposited on the surface of the hyphae. These products may amount to between 0.1 and 10% (sometimes up to 30%) of the thallus dry weight (Galun and Shomer-Ilan, 1988). These low molecular weight metabolites, often referred to as lichen substances or lichen acids, are one of the more intensively investigated aspects of lichenology. However, these are usually absent in lichens with cyanobacterial photobionts (Galun & Shomer-Ilan, 1988), probably due to the cyanobiont's ability to fix N_2 and thereby have a larger access to nitrogen for biosynthesis (cf. Palmqvist, 2000). Synthesis of complex secondary carbon compounds may then simply be a way to make use of excessive carbon when nitrogen is a limiting resource, which evidently does not seem to be the case in many cyanolichens. However, some secondary carbon products can serve a defensive role against parasitizing fungi or bacteria, or against browsing animals, a biological function that should be particularly beneficial for lichens with cyanobacterial photobionts due to their high tissue nitrogen concentrations (cf. Chapin *et al.*, 1987).

3.3. Respiration

In plants, a significant portion of the photo-assimilated carbohydrates becomes the main substrate for respiration (Amthor, 1995). This is apparently also the case for lichens where up to 50% of the photo-assimilated carbon might be "lost" in respiration (Table 1). However, respiration should not be viewed as a futile carbon-loss process, because the function of respiration is to convert photo-assimilate into substances used for growth and maintenance, to energize this, as well as transport and nutrient assimilation processes (Amthor, 1995; Lambers *et al.*, 1998). During respiration, and growth, CO_2 is then released as a byproduct. The fraction of carbon and assimilate in the photo-assimilate that is "lost" during respiratory metabolism is dependent on the pathways of respiration, the mitochondrial ADP:O ratio, and substrate composition (Amthor, 1995). Plant and fungal respiration appear to be fundamentally similar (Fahselt, 1994; Lambers *et al.*, 1998), so only a few specific characteristics of cyanobacterial lichen respiration are given here.

Lichens with low nitrogen concentrations, and hence high C:N ratios, have the lowest rates of respiration both when related to area and dry weight (Sundberg *et al.*, 1999). This implies that species with low respiration may be relatively rich in carbon compounds that contribute to biomass but with low metabolic turnover and low maintenance costs. Indeed, and as for plants, lichen respiration increases with thallus nitrogen concentration, being an effect of increased energy demand related to protein turnover (Lambers, 1985). For lichens, however, this relation is less tight, due to the nitrogen content of the cell wall compound chitin that does not turn over as rapidly as proteins (Sundberg *et al.*, 1999). A recent study of CO_2 assimilation capacity and steady-state respiration rates in 75 lichen species (Palmqvist *et al.*, 2002) showed that at 15°C, 38% of the variation in respiration across species could be attributed to variation in photosynthetic capacity and chlorophyll *a* concentration in such a way that respiration increased when these two parameters increased. However, among cyanobacterial lichens, 30% of the variation in respiration across samples could be attributed to a variation in photosynthetic capacity alone (Fig. 6). The implications of this finding will be discussed later (Section 5).

Since mycobiont biomass dominates in most lichens, one may easily jump to the conclusion that mycobiont respiration should dominate over photobiont respiration. This, however, has not been thoroughly investigated and does not necessarily have to be true. For instance, in *Peltigera canina* as much as 36% of the thallus protein is located in *Nostoc* (Rai, 1988), although the photobiont contributes much less to the thallus dry weight (Hawksworth and Hill, 1984). Because protein-rich tissues have higher rates of maintenance respiration compared to the carbon-rich tissues (Lambers, 1985), photobiont respiration is likely to be relatively high in this cyanolichen. Support for this can be found in another cyanolichen *Peltigera polydactyla*, where highest respiration was found in the photobiont region (Smith, 1960). On the other hand, high respiration in this region might also be related to high metabolic rates in the hyphal tips invading the cyanobacterial gelatinous sheath (see Section 5).

Several environmental factors affect respiration rates in lichens (Green and Lange, 1995; Kershaw, 1985; Nash, 1996). As in plants (Lambers, 1985), respiration in lichens increases significantly with increasing temperature. A 10°C increase in temperature may result in a 2- to 3-fold increase in respiration (Fig. 3C). This is mainly related to an overall increased metabolism at higher temperatures, including increased maintenance costs (Lambers *et al.*, 1998). However, as in plants, respiration can acclimate to increased temperatures so that individuals adapted to a higher temperature display relatively lower increases in respiration with increasing temperature compared to a low temperature adapted population (Sancho *et al.*, 2000). Respiration also increases with increased water content, probably reflecting increased metabolism in the fungus (MacFarlane and Kershaw, 1982). As already discussed, respiration may increase further when dry lichens are re-hydrated, due to 'resaturation respiration'.

In addition to the higher respiratory demands of maintaining a tissue with high nitrogen concentrations, the acquisition of nitrogen presents a further cost, which can be predominant in higher plants. This is because plants require this compound in great quantity, and it frequently limits growth (Chapin *et al.*, 1987). Carbon costs for nitrogen acquisition includes the construction costs of roots, tissue for translocation, and energetic requirements of N-assimilation. In lichens, which do not construct specific absorption or translocation tissues, carbon requirements for nitrogen acquisition are mainly restricted to assimilation costs. However, various methods of nitrogen assimilation differ greatly in their energy requirements. For example, the energy cost for N_2 fixation is higher than that for nitrate assimilation, which in turn is higher than that for ammonia assimilation (Chapin *et al.*, 1987). This emphasizes that the N_2 fixation process of cyanolichens might also be a significant carbon sink.

To conclude, even though we have too little information to construct a complete and quantitative carbon expenditure budget for cyanolichens (Table 1), it appears that these lichens may have a relatively expensive life-style compared to slower growing and nitrogen limited green algal lichens. Their relatively high growth rates require significant amounts of carbon for new tissues and energy, and their thallus nitrogen concentrations are high, thus demanding higher rates of maintenance respiration (Lambers *et al.*, 1998). High rates of hyphal expansion also require a stable supply of osmotically active carbohydrates for creating the driving force for turgor pressure, and finally the N_2 fixation process is more expensive than passive acquisition of ammonium or nitrate.

4. CARBON ACQUISITION

Since lichen growth is well correlated with net CO_2 fixation (Nash, 1996; Palmqvist and Sundberg, 2000), we may assume that the carbohydrate expenditure, as discussed above and summarized in Table 1, can be met from photobiont photosynthesis. It then makes sense to analyze the photosynthetic characteristics of cyanobacteria. However, the following presentation will not go into all the details of the photosynthetic performance of cyanobacterial lichens and cyanobacteria, and a more comprehensive information can be found elsewhere (cf. Kershaw, 1985; Kappen, 1988; Reuter and Müller, 1993; Campbell *et al.*, 1998; Palmqvist, 2000; Rai *et al.*, 2000).

Maximal net CO_2 assimilation capacity (P_{max}) of cyanobacterial lichens, measured at light saturation, 15 °C and ambient CO_2, can vary between -2 and 43 nmol g^{-1} DW s^{-1} (Fig. 6), or between -0.2 and 8.5 μmol m^{-2} s^{-1} (Palmqvist *et al.*, 2002; Palmqvist, unpublished). These rates are in agreement with other studies on cyanolichens but are well below those for higher plant leaves (Green and Lange, 1995). On the other hand, when related to chlorophyll, P_{max} of lichens and their photobionts is more similar to the rates in higher plants (Palmqvist, 2000). However, variation in P_{max} can be attributed to differences in carboxylation capacity rather than to chlorophyll (Björkman, 1981), more specifically to the variation in amount and activity of Rubisco (ribulose 1,5-bisphosphate carboxylase/oxygenase). For lichens, it has been found that variation in P_{max} across species is well correlated to their Rubisco and chlorophyll *a* concentrations. The cyanobacterial lichen *Peltigera canina* has a somewhat higher photosynthetic efficiency in relation to both chlorophyll *a* and Rubisco than lichens with green algal photobionts (Palmqvist *et al.*, 1998). This may be explained by a lower chlorophyll *a* concentration per photosynthetic unit (PSU) of cyanobacteria (cf. Campbell *et al.*, 1998) and an inherently higher maximal rate (V_{max}) of the cyanobacterial Rubisco (Badger and Andrews, 1987).

Thus, at the cellular level, P_{max} seems to be determined by the same factors in lichen photobionts, including cyanobacteria, as in plants. The relatively lower photosynthetic capacities per thallus area, or dry weight, of lichens can then be explained by their relatively low PSU concentrations (cf. Palmqvist, 2000). The following may explain the relatively low PSU concentrations in lichens: First, in contrast to plants and bryophytes, lichens do not construct two-dimensional surfaces entirely made of photosynthetic tissue, and all photobiont cells are surrounded by fungal tissue that, even though maintaining structural integrity, probably limits photobiont expansion within the thallus (Honegger, 1991). Second, photobiont development may be constrained by lack of sufficient nitrogen required for the proteins of the photosynthetic apparatus The latter is supported by a strong correlation between chlorophyll *a* and thallus nitrogen concentration, both in green algal and cyanobacterial lichens (Palmqvist *et al.*, 2002). Third, even if cyanobacterial lichens are characterized by high thallus N concentrations these species still have low PSU densities in comparison with plant leaves with similar nitrogen status (Palmqvist *et al.*, 1998). This is apparently caused by the requirement of nitrogen for fungal growth and biosynthesis, a trade-off that is not shared by plants and bryophytes. Fourth, the poikilohydric nature of lichens restricts photosynthetic activity to occasions when the thallus is able to maintain sufficient hydration. Since exposure to high irradiances enhances the evaporative losses of water, lichen photosynthesis is most often restricted to periods when irradiances are relatively low, such as during rainfall photobiont density may then be limited by increased self-shading.

4.1. Environmental Limitations of Photosynthesis

Photosynthesis always displays a characteristic response to variations in irradiance (Fig. 3B). Typically, the relation between photosynthesis and irradiance has three different phases: the light-limited part where photosynthesis is limited by irradiance, the light- and early morning hours. The beneficial effects on lichen productivity by a high saturated part where carboxylation efficiency is the limiting factor; and the transition zone between these two phases (denoted convexity). The efficiency of photosynthesis will be highest if the organism operates at an environmental irradiance level below their light saturation value, and if the transition zone from light-limitation to light-saturation is narrow (see Palmqvist, 2000 and references therein). It is therefore useful for the organism to be able to acclimate to the prevailing light conditions so that nitrogen investments are directed to light harvesting proteins when irradiances are low, and to Rubisco when irradiances are high. This is indeed the characteristic of most photosynthetic organisms including cyanobacteria (Björkman, 1981; Bryant, 1987). As a result, photosynthesis becomes saturated at lower irradiances in low-light acclimated than in high-light acclimated cells, and as already discussed, due to the higher Rubisco levels of high-light acclimated cells, these also display higher P_{max} rates (Björkman, 1981). Moreover, because highlight acclimated cells usually have higher nitrogen concentrations, as a consequence of their higher Rubisco levels, dark respiration rates are increased in these cells. Irradiance required to reach light compensation is therefore also increased (Björkman, 1981), implying that high Rubisco concentrations might be too expensive to maintain in a low light environment. This again emphasizes the necessity for a tight acclimation of the photosynthetic apparatus to varying light conditions, for instance between the seasons. Few attempts have been made to study the cellular acclimation of photosynthesis in lichens (cf. Kershaw, 1985) but recent advances using molecular techniques to quantify *de novo* Rubisco synthesis, in combination with fluorescence analysis of electron transport capacity, have proved that such studies are possible for lichens (MacKenzie *et al.*, 2001). Indeed lichen data have shown intra-specific variation in photosynthetic performance depending on habitat characteristics, indicating that these organisms are able to acclimate to changing irradiance levels (cf. Kershaw, 1985; Palmqvist, 2000 and references therein).

Apart from the strong influence of irradiance, and water (Section 2) on lichen photosynthesis, CO_2 assimilation also increases with increasing temperature (Fig. 3A) due to increased activities of the Calvin cycle enzymes. However, in contrast to respiration that increases linearly (Fig. 3C), net photosynthesis usually shows an optimum, and generally does not increase as much as respiration with increasing temperature. This is in part related to the increased oxygenase activity of Rubisco at higher temperatures (Björkman, 1981).

4.2. Light Capture and Electron Transport in Cyanobacteria

The function and molecular structure of the cyanobacterial photosynthetic apparatus is similar to that of eucaryotic algae and higher plants (Bryant, 1987). However, in contrast to the chloroplasts of eucaryotic cells, the thylakoids of cyanobacteria are invaginations of the cytoplasmic membrane and mostly localized at the periphery of the cells, forming concentric circles parallel to the cytoplasmic membrane (Reuter & Müller, 1993). In addition, Rubisco is localized in specific sub-cellular structures called carboxysomes, whose function is discussed later. As in chloroplasts, four of the five multiprotein complexes of the photosynthetic electron transport apparatus, PS II, PS I, plastoquinone-plastocyanin oxidoreductase, and the ATP synthase, are localized within the thylakoid membrane. However, the principal light-harvesting complexes of cyanobacteria, the phycobilisomes, are located peripheral to the thylakoid membranes, in contrast to the integral chlorophyll a/b binding proteins which capture light in green algal and plant chloroplasts (cf. Campbell *et al.*, 1998). This localization allows the phycobilisome to diffuse along the thylakoid surface from PSII to PSI within 100 ms (Mullineaux *et al.*, 1997). As a result, the cell can rapidly re-direct electron flow to where it is best needed, and protect PSII from over-excitation. The detachment of phycobilisomes from PSII also occurs during desiccation (Bilger *et al.*, 1989). Further, both photosynthetic and respiratory electron flow occur in the cyanobacterial thylakoid membrane (Jones and Myers, 1963), sometimes simultaneously, and they share several electron transport intermediates. This results in a highly flexible system that can respond rapidly to environmental changes as well as to changes in metabolic demands. Additional mechanisms for protecting PSII from over-excitation is to re-direct electrons to oxygen (Mehler-reactions), or replacement of the PSII reaction center D1 protein having a high energy capture efficiency with a lower efficiency D1 (see Campbell *et al.*,

1998 and references therein). Taken together, these various strategies for rapid adjustment and regulation of electron transport, in combination with their desiccation tolerance, provide an explanation to why cyanobacteria are so successful in exploiting highly contrasting environments (cf. Lüttge *et al.*, 1995).

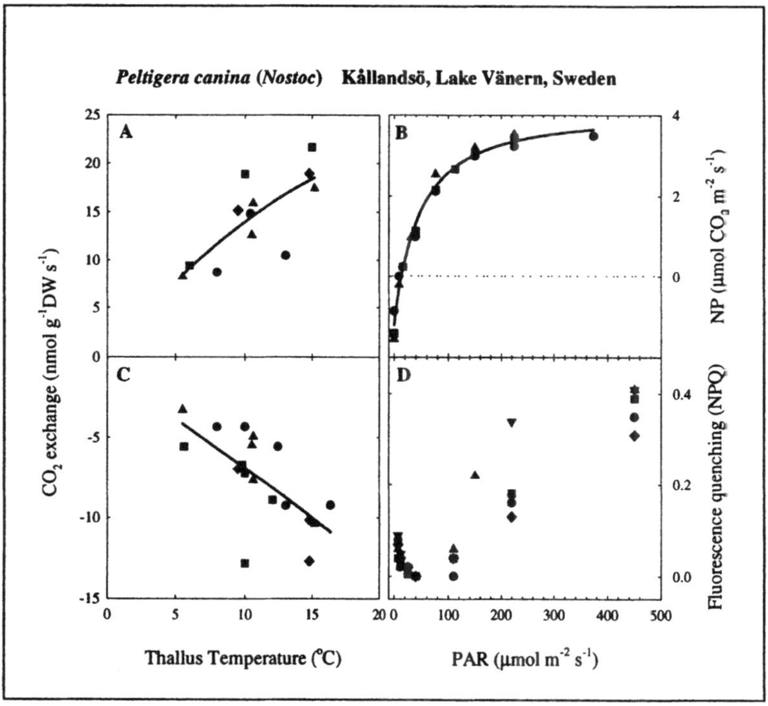

Figure 3. Light saturated net CO_2 fixation (NP) and respiratory CO_2 efflux (R) as a function of thallus temperature (A, C), and NP as a function of irradiance (B) for 3-4 thalli cf Peltigera canina. *Chlorophyll a fluorescence quenching (NPQ) was obtained for 5 thalli with a PAM-2000 and calculated as in (Campbell* et al., *1998). The light response data are for 2 thalli at 10 °C.*

Even though most of the above insights are based on studies with non-symbiotic and unicellular cyanobacteria such as *Synechococcus* and *Synechocystis* (cf. Reuter & Müller, 1993; Campbell *et al.*, 1998), these characteristics appear to be valid for lichenized cyanobacteria also. This can be inferred from similar chlorophyll *a* fluorescence characteristics of cyanolichens and free-living cyanobacteria (Bilger *et al.*, 1989; Leisner *et al.*, 1996; Sundberg *et al.*, 1997). Chlorophyll fluorescence analysis is a fast non-invasive tool for measuring key aspects of photosynthetic electron transport and can be used to detect some of the characteristics unique to cyanobacteria described above (Campbell *et al.*, 1998). For instance, non-photochemical quenching of variable

PSII fluorescence reflects the state-transition mechanism for distribution of excitation energy between the photosystems. Cyanobacteria display a characteristic decline in non-photochemical quenching during a shift from darkness to increasing irradiance, where the minimum can be used to estimate the light level to which a cyanobacterial population is photosynthetically acclimated (Campbell *et al.*, 1998). The potential use of this fluorescence parameter also for lichenized cyanobacteria has been demonstrated in a laboratory study (Sundberg *et al.*, 1997). Unfortunately, the light acclimation level of the investigated cyanolichens were not known in that particular study, so it was not possible to test the validity of the proposed method. However, the light acclimation level was better known in the previously mentioned field study of *P. canina*, where the lichens were exposed to a mean irradiance of 80µmol m^{-2} s^{-1} during photosynthetically active periods (Palmqvist and Sundberg, 2000). When a fluorescence quenching analysis of those thalli was made (Palmqvist, unpublished), a minimum in non-photochemical quenching was indeed evident between 40-100 µmol photons m^{-2} s^{-1} (Fig. 3D). However, more studies combining mechanistic approaches of lichen performance with field microclimatic data will be needed to evaluate whether the non-photochemical quenching parameter is useful for describing environmental light conditions of cyanolichens.

4.3. The Cyanobacterial CO_2 Concentrating Mechanism

CO_2 is the substrate for the primary carboxylating enzyme Rubisco, and as in terrestrial plants, lichens predominantly obtain CO_2 directly from the atmosphere, while aquatic photosynthetic organisms obtain their CO_2 from the surrounding water. As already mentioned, the acquisition of CO_2 from an aquatic environment presents problems because its diffusion rate is 10 000 times slower in water than in air. So the organisms in aquatic environments have evolved strategies to overcome the problem of CO_2 limitation of photosynthesis. One major strategy, that has evolved at all systematic levels, is a mechanism for the active transport and accumulation of CO_2 and/or HCO_3^- within the cell (Raven, 1991; Badger and Price, 1994; Price *et al.*, 1998). The need of a CO_2 concentrating mechanism (CCM) is also closely related to the kinetic properties of Rubisco. This enzyme is bifunctional and can both carboxylate and oxygenate ribulose-1,5-bisphosphate. O_2 inhibits CO_2 fixation competitively and leads to photorespiration (Björkman, 1981). As the CCM functions to increase CO_2 around the active site of Rubisco, photorespiration will be depressed and carboxylation increased (Fig. 5A).

A simplified model of the cyanobacterial CCM as it appears in fresh-water and marine strains of *Synechocystis* and *Synechococcus* is presented in Figure 4 (Badger and Price, 1994; Price *et al.*, 1998; Kaplan and Reinhold, 1999; Tchernov *et al.*, 2001). The inorganic carbon transport system is of central importance for the functioning of the CCM. So far it has not been possible to isolate any pump(s) but the uptake is active and requires photosynthetically transduced energy. Some evidences support a single transporter model in which a plasma membrane located pump is able to use either CO_2 or HCO_3^- as a substrate. (Badger and Price, 1994). However, CO_2 uptake might occur passively via water channels in the plasma membrane (aquaporins) with subsequent energy-dependent conversion to HCO_3^- within the cell using a $\Delta\mu H^+$ locally generated by PSI electron transport (Tchernov *et al.*, 2001).

The carboxysome is another prerequisite for the cyanobacterial CCM. These are small polyhedral protein bodies present in the cytosol of cyanobacteria. The possibility of carboxysomes being the actual site of increase in CO_2 concentration (an important part of the CCM) emerged gradually, and was formalized in a model by Reinhold *et al.* (1989). This model has been experimentally tested and confirmed. It is now clear that the accumulated HCO_3^- is indeed dehydrated to CO_2 within the carboxysomes where much of the Rubisco is located, and that this dehydration is facilitated by a low level of carbonic anhydrase (CA) (cf. Badger and Price, 1994). The model also postulated that CA should be absent from the cytosol, so that the slow uncatalyzed conversion between HCO_3^- and CO_2 would minimize wasteful leakage of CO_2 out of the cell. This prediction has also been experimentally tested and confirmed (Badger and Price, 1994).

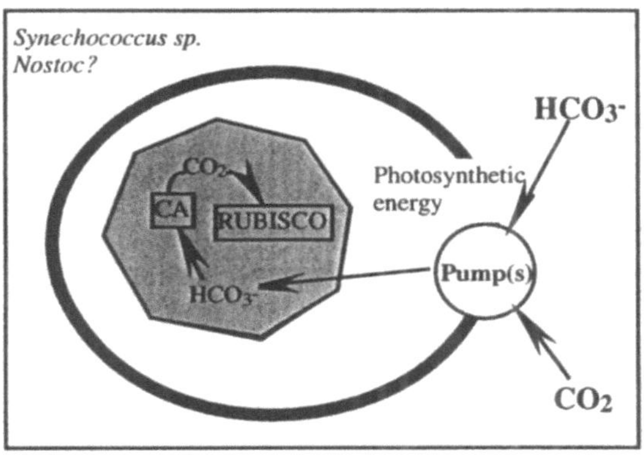

Figure 4. A simplified model of the cyanobacterial CCM. CA, carbonic anhydrase.

Several reasons made us test the hypothesis whether a CCM might be operating in cyanolichens also (Badger *et al.*, 1993; Máguas *et al.*, 1993; Palmqvist, 1993). First, a CCM is present in most free-living cyanobacteria; second, lichens might benefit from a CCM to partly overcome CO_2 diffusion limitation of photosynthesis when the thallus contains supraoptimal water contents (Cowan *et al.*, 1992); and third, earlier data from lichens suggested that a CCM was present (Coxson *et al.*, 1982; Raven *et al.*, 1990). Indeed, the presence of a CCM was demonstrated for both cyanobacterial lichens and for green-algal *Trebouxia* lichens. For cyanolichens, there is data from *Nostoc* and *Calothrix* only, and most studies have been focused on intact *Nostoc*-lichens. These accumulate a large pool of inorganic carbon (Fig. 5B) that is larger than that in *Trebouxia*-lichens (Badger *et al.*, 1993), and have high affinity for CO_2, *i.e.* a low $K_{0.5}$ (CO_2) of photosynthesis (Fig. 5A). Rubisco has not been isolated from any lichenized cyanobacterium but based on indirect evidence (Palmqvist, 2000), K_m (CO_2) of *Nostoc* Rubisco is probably as high as for free-living cyanobacteria (Badger and Andrews, 1987; Tabita, 1999). Furthermore, as in free-living cyanobacteria with a well-developed

CCM, Rubisco of *Nostoc* is located in carboxysomes (Bergman and Rai, 1989) where the accumulated inorganic carbon may be released as CO_2, causing the local elevation required for efficient photosynthesis (Badger and Price, 1994). There are no indications of photorespiration in cyanolichens (Palmqvist, 1993). The CO_2 accumulation and Rubisco's affinity for CO_2 were both sensitive to ethoxyzolamide, a potential inhibitor of carbonic anhydrase (Badger *et al.*, 1993). The size of the CO_2 pool increases with increasing irradiance emphasizing its energy dependence (Sundberg *et al.* 1997), and CO_2 continues to accumulate in light even after the inhibition of photosynthetic CO_2 fixation (Badger *et al.*, 1993). Moreover, the lower discrimination of the carbon isotope ^{13}C in a range of cyanobacterial lichens (Lange *et al.*, 1988) can, at least in part, be explained by the presence of a CCM (Máguas *et al.*, 1993).

As initially mentioned, the requirement of a CCM is closely related to the kinetic properties of Rubisco in a particular organism. Indeed, the cyanobacterial Rubisco has a lower affinity for CO_2 than the Rubisco of higher plants and green algae (Tabita, 1999), but its V_{max} rates are significantly higher (Badger and Andrews, 1987). An additional significance of the cyanobacterial CCM might be the enhanced nitrogen use efficiency in photosynthesis (Raven, 1991), because the higher V_{max} rates might allow relatively lower nitrogen investments in Rubisco. Support for this is provided in the earlier cited study of Rubisco concentrations and P_{max} rates of *P. canina* (Palmqvist *et al.*, 1998). However, more extensive studies of nitrogen investments and Rubisco concentrations in cyanobacterial lichens are required to clarify this.

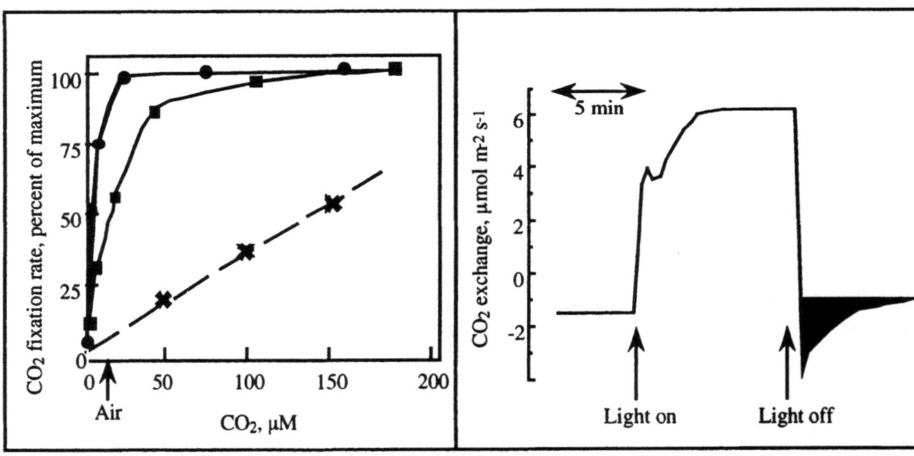

Figure 5. A. Photosynthetic CO_2 response of the Nostoc *lichen* Peltigera neopolydactyla *(squares), and cultured cells of* Synechococcus *(circles); CO_2 affinity of a cyanobacterial Rubisco (crosses). Data from Palmqvist (1993). B. Measurements of CO_2-uptake and release pools (shaded area) in a* Nostoc *lichen. The size of the pools can be calculated by integrating their area as in Badger* et al. *(1993).*

5. CARBON TRANSLOCATION AND REGULATION OF THE CARBON BUDGET

Lichens lack specific cells or tissues for the translocation of metabolites, water, and nutrients, between their symbionts. Translocation patterns thereby vary among species, and depend on factors such as the chemical composition of the symbiont cell walls and their integration (Honegger, 1991). There are also biochemical differences: the green algae release a polyhydric sugar alcohol (polyol) to the mycobiont whilst cyanobacteria release glucose (Richardson and Smith, 1966, 1968; Hill and Smith, 1972). Once taken up by the mycobiont, the carbohydrate is rapidly and irreversibly metabolized into mannitol, via the pentose phosphate pathway (Lines *et al.*, 1989), and thereby made unavailable to the photobiont (Galun, 1988). The mechanisms behind the induction of carbohydrate export and mass transfer from photobionts to mycobiont is still a matter of debate, and to date no specific polyol or glucose transporter has been isolated from lichens, even though such a carrier has been postulated (Collins and Farrar, 1978).

Data on carbohydrate movement from photobiont to mycobiont are scarse, but it appears that the rate of transfer and the extent of carbohydrate that eventually ends up in the mycobiont are faster and larger in cyanobacterial lichens than in green algal species (Honegger, 1991; Sundberg, 1999). In *Nostoc* containing *Peltigera* species photosynthetically fixed carbohydrates can be found in the mycobiont within 60s, with 40% of fixed $^{14}CO_2$ being transferred within 4h (Richardsson and Smith, 1966). However, this was probably an underestimation because the ^{14}C labeled assimilates may have been diluted with already existing, and unlabelled, soluble carbohydrates in the photobiont (Richardson *et al.*, 1968). So, depending on the sizes and turnover rates of carbohydrate sinks in the photobiont, ^{14}C label will be more or less diluted and have a longer or shorter retention time in the photobiont. To quantify carbon fluxes through lichens we would hence need to adopt compartmental analysis in conjunction with isotopic labeling.

As already discussed (Table 1), the *Nostoc* filaments of heteromerous thalli are enclosed within a gelatinous sheath composed of fibrillar polysaccharides (Honegger, 1991). Thin-walled fungal protrusions invade this gelatinous sheath, presumably by means of hydrolytic enzymes. These hyphal tips are found close to, but never inside, the cyanobacterial cells. The intragelatinous hyphae are lateral outgrowths of the aerial hyphae of the uppermost part of the lichen thallus (Honegger, 1991), so carbohydrates assimilated by the hyphal tips may then be translocated to other hyphal parts located more distantly from the photobiont. Both symbiotic and non-symbiotic *Nostoc* invests a high proportion of assimilated carbohydrates in the synthesis of the polysaccharide sheath (cf. Fay, 1983), which in the symbiotic system is readily hydrolyzed by enzymes such as glucanases. Such enzymes are released by the intragelatinous fungal protrusions (Hill, 1972), producing the glucose as the transfer metabolite in cyanobacterial lichens (Honegger, 1991). However, polysaccharides produced in *Collema* are largely unavailable to the fungus; the latter receives newly synthesized glucose (Henriksson, 1964). This suggests that the lower lichenization of this homoiomerous genus might be related to a lack of hydrolyzing enzymes in the fungal partner.

Relations between environmental conditions and carbohydrate allocation patterns have been studied even less (Armstrong and Smith, 1994), although thallus hydration status and temperature seem important (Feige, 1978). For instance, in the *Nostoc* lichen *Peltigera polydactyla* mannitol formation was significantly enhanced when water contents were increased (MacFarlane and Kershaw, 1982). It has therefore been assumed that alternating wetting and drying cycles may function to control adequate carbon supply to each one of the symbionts.

5.1. Regulation of Carbon Expenditure and CO_2 Assimilation

From the above analysis it may be concluded that lichen metabolism, with respect to carbohydrate expenditure and CO_2 assimilation, is strongly regulated by environmental conditions. For instance, water availability determines length and frequency of metabolically active periods, temperature increases may enhance respiratory CO_2 losses, and high irradiance levels may cause desiccation, even though an ample supply of light is necessary for photosynthetic activity. However, despite this passive dependence on the environment for metabolic activity, lichens are still able to grow in a controlled and coordinated way in a wide range of habitats. This emphasizes that they are indeed able to maintain a positive energy (carbon) balance in a regulated manner, having some capacity to respond to variations in environmental resource supply.

In addition to carbon, the lichen must also acquire mineral nutrients such as nitrogen and phosphorous for the synthesis of new proteins, membranes and DNA. To maintain a balanced growth, the acquisition of carbon must therefore be balanced in relation to mineral availability and *vice versa* (Chapin, 1991). Further, acquired resources must be allocated to different cells and organs in a regulated manner to secure future development (Grace, 1997). Presently, there is no mechanistic explanation for how this might be achieved in plants, but according to functional equilibrium models (Brouwer, 1962) it is assumed that acquired resources are allocated so that pool sizes of key elements remain constant within and between organs and that environmental limitations or excesses that reduce resource use will also reduce uptake (Chapin, 1991). This model has been summarized as: if the C:N ratio of your tissue is too high – grow root, otherwise grow shoot, assuming that each species has a specific C:N ratio optimum (Grace, 1997). It was recently suggested that a similar model might be applicable also to lichens by replacing shoot with photobiont and root with fungal hyphae (Palmqvist, 2000). This hypothesis was tested by gathering data from 75 lichen associations with various photobionts and habitat preferences (Palmqvist *et al.*, 2002). It was then found that investments in photobiont *versus* mycobiont tissue were surprisingly equally balanced in relation to each other, across the investigated lichen species. This was evident from a similar chlorophyll *a* to ergosterol ratio across species (ergosterol was used as a marker for metabolic activity of the fungus). Maximal photosynthetic capacity was well balanced in relation to steady-state respiration rates, particularly for lichens from Antarctica and for cyanolichens (Fig. 6). Taken together, this indeed suggests that lichens are somehow able to optimize their resource investments between carbohydrate input and expenditure tissue, again emphasizing some mode of internal regulation.

Despite their ability to optimize resource investments in the thallus, and the ability of cyanobacterial cells to rapidly adjust their photosynthetic apparatus to variations in resource supply (Reuter and Müller, 1993; Campbell *et al.*, 1998), cyanolichens can still be vulnerable to drastic changes in light, nitrogen or water supply. For example, *Collema curtisporum*, *C. furfuraceum*, *Lobaria oregana* and *Pseudocyphellaria rainierensis* displayed reduced growth rates on trees remaining after forestry actions (Hedenås, 2002; Sillett and McCune, 1998). These lichens are confined to old natural forests with a long continuity, and other studies have shown that such lichens are highly susceptible to sudden increases in the light level, causing significant photoinhibition and subsequently decreased net CO_2 uptake (Gauslaa and Solhaug, 1996). However, in species from more exposed habitats, photoinhibition and bleaching of chlorophyll pigments seem to be avoided as long as the lichen is dry and its photosynthetic apparatus has been adjusted to desiccating conditions (Gauslaa and Solhaug, 1996). Production of photoprotective and antioxidative agents such as specific carotenoids, can be further induced in high-light acclimated cyanolichens (Leisner *et al.*, 1993, 1994).

Finally, fertilization of cyanolichens with combined nitrogen such as NH_4^+ may also disturb the resource equilibrium between photobiont and mycobiont (cf. Dahlman *et al.*, 2002). This is because NH_4^+ may inhibit their N_2 fixation activity (Rai, 1988), possibly resulting in reduced nitrogen flow to the mycobiont and disturbing relative strengths of the various carbohydrate sinks in the thalli. So, depending on the nature of the regulatory mechanism to maintain a balanced C:N ratio in the thallus, the lichen will be more or less disturbed by such fertilization.

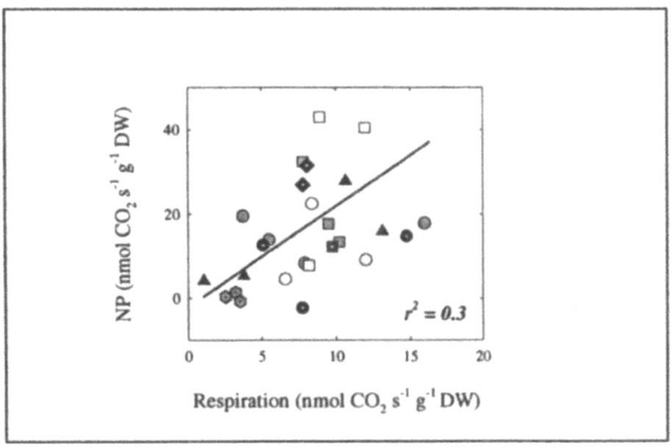

Figure 6. Photosynthetic net CO_2 fixation (NP) as a function of steady-state respiration rate, measured at 15 °C and a WC optimal for photosynthesis, in 10 lichens with cyanobacterial photobionts. Each symbol represents one type of association. From Palmqvist et al. *(2002).*

6. CONCLUDING REMARKS

So, even though we have quite extensive data on the ecology and physiology of photosynthesis in cyanolichens, we have little knowledge about the mechanisms that regulate resource flow between photobiont and mycobiont tissues in these symbiotic associations. More efforts must therefore be devoted to quantify major carbohydrate sinks, and we need to study how these sinks vary in sizes and turnover rates in relation to environmental constraints. Such information will help us understand how the lichen partners regulate and integrate their respective cellular activities. We also need to separate fungal respiration from photobiont respiration, because even though the fungus might be dominating in terms of biomass, energy requirements might be higher in the photobiont. This idea finds support from the notion that fungal hyphae are generally most active at their growing tips, whereas the rest of the hyphae are merely composed of cell walls and vacuoles with low metabolic turnover (Jennings and Lysek, 1996). In contrast, cyanobacterial photobionts are rich in proteins and membranes involved in photosynthesis or N_2 fixation. The photobionts may then have relatively higher rates of maintenance respiration, per unit cell volume, due to their higher effective nitrogen concentrations.

ACKNOWLEDGEMENTS

Time for writing was provided by a grant from FORMAS, Stockholm, Sweden. Henrik Hedenås (Umeå, Sweden) gave valuable comments for improvements. The 900 odd lichen reprints that Professor Kerstin Huss-Danell (Umeå, Sweden) handed over when I started as a lichenologist, have again proven to be invaluable. Unfortunately, however, it has not been possible to cite all those studies.

REFERENCES

Amthor, J.S. (1995) Higher plant respiration and its relationships to photosynthesis, in E-D. Schulze and M.M. Caldwell MM (eds.), *Ecophysiology of Photosynthesis*, Springer, Berlin, pp. 71-101.

Armstrong, R.A and Smith, S.N. (1994) The levels of ribitol, arabitol and mannitol in individual lobes of the lichen *Parmelia conspersa* (Ehrh. Ex Ach.) Ach, *Environ. Exp. Botany* **34**, 253-260.

Badger, M.R and Andrews, T.J. (1987) Co-evolution of Rubisco and CO_2 concentrating mechanisms, in J. Higgins (ed.), *Progress in Photosynthesis Research. Vol III*, Martinus Nijhoff, Dordrecht, pp. 601-609.

Badger, M.R. and Price, G.D. (1994) The role of carbonic anhydrase in photosynthesis, *Annu. Rev. Plant Physiol. Plant Mol. Biol.* **45**, 369-392.

Badger, M.R., Pfanz, H., Büdel, B., Heber, U. and Lange, O.L. (1993) Evidence for the functioning of photosynthetic CO_2 concentrating mechanisms in lichens containing green algal and cyanobacterial photobionts, *Planta* **191**, 57-70.

Beckett, R.P. (1997) Pressure-volume analysis of a range of poikilohydric plants implies the exisyance of negative turgor in vegetative cells, *Ann. Botany* **79**, 145-152.

Bergman, B. and Rai, A. (1989) The *Nostoc-Nephroma* symbiosis: localization, distribution pattern and levels of key proteins involved in nitrogen and carbon metabolism of the cyanobiont. *Physiologia Plantarum* **77**, 216-224.

Bewley, J.D. (1979) Physiological aspects of desiccation tolerance, *Annu. Rev. Plant Physiol.* **30**, 195-238.

Bilger, W., Rimke, S., Schreiber, U. and Lange, O.L. (1989) Inhibition of energy-transfer to Photosystem II in lichens by dehydration: different properties of reversibility with green and blue-green phycobionts, *J. Plant Physiol.* **13**, 261-268.

Björkman, O. (1981) Responses to different quantum flux densities, in O.L. Lange, P.S. Nobel, C.B. Osmond and H. Ziegler (eds.), *Physiological Plant Ecology I. Responses to the Physical Environment, Encyclopedia Plant Physiol. 12 A*, Springer, Berlin, pp. 57-108.

Blum, O.B. (1973) Water relations, in V. Ahmadjian and M.E. Hale (eds.), *The Lichens*, Academic Press, New York, pp. 381-400.

Bogner, E., Wastlhuber, R., Schlegl, I. and Loos, E. (1993) Glycogen, amylase and α-glucosidase as possible components in the glucose release system of the Cyanobiont in *Peltigera horizontalis*. Partial purification and characterization, *Symbiosis* **14**, 485-494.

Boissière, J.C. (1987) Ultrastructural relationship between the composition and the structure of the cell wall of the mycobiont of two lichens, *Bibliotheca Lichenologica* **25**, 117-123.

Brouwer, R. (1962) Distribution of dry matter in the plant, *Netherland J. Agricultural Sciences* **10**, 399-408.

Brown, D., MacFarlane, J.D. and Kershaw, K.A. (1983) Physiological-environmental interactions in lichens. XVI. A re-examination of resaturation respiration phenomena, *New Phytol.* **9**, 237-246.

Bryant, D.A. (1987) The cyanobacterial photosynthetic apparatus: comparison of those of higher plants and photosynthetic bacteria, *Can. Bull. Fish. Aquat. Sci.* **214**, 423-500.

Campbell, D., Hurry, V., Clarke, A.K., Gustafsson, P. and Öquist, G. (1998) Chlorophyll fluorescence analysis of cyanobacterial photosynthesis and acclimation, *Microbiol. Mol. Bbiol. Rev.* **62**, 667-683.

Chapin III, F.S. (1991) Integrated responses of plants to stress, *BioScience* **41**, 29-36.

Chapin III, F.S., Bloom, A.J., Field, C.B. and Waring, R.H. (1987) Plant responses to multiple environmental factors, *BioScience* **37**, 49-57.

Collins, C.R. and Farrar, J.F. (1978) Structural resistances to mass transfer in the lichen *Xanthoria parietina*, *New Phytol.* **31**, 71-78.

Cowan, D.A., Green, T.G.A. and Wilson, A.T. (1979a) Lichen metabolism 1. The use of tritium labeled water in studies of anhydrobiotic metabolism in *Ramalina celastri* and *Peltigera polydactyla*, *New Phytol.* **82**, 489-503.

Cowan, D.A., Green, T.G.A. and Wilson, A.T. (1979b) Lichen metabolism 2. Aspects of light and dark physiology, *New Phytol.* **83**, 761-769.

Cowan, I.R., Lange, O.L. and Green, T.G.A. (1992) Carbon-dioxide exchange in lichens: determination of transport and carboxylation characteristics, *Planta* **187**, 282-294.

Coxson, D.S., Harris, G.P. and Kershaw, K.A. (1982) Physiological-environmental interactions in lichens. XV. Contrasting gas exchange patterns between a lichenized and non-lichenized terrestrial *Nostoc* cyanophyte, *New Phytol.* **92**, 561-572.

Dahlman, L., Näsholm, T. and Palmqvist, K. (2002) Growth, nitrogen uptake, and resource allocation in the two tripartite lichens *Nephoma arcticum* and *Peltigera aphthosa* during nitrogen stress, *New Phytol.* **153**, 000-000.

Drew, E.A. (1966) *Some Aspects of the Carbohydrate Metabolism of Lichens*, PhD Thesis, University of Oxford, U.K.

Dudley, S.A. and Lechowicz, M.J. (1987) Losses of polyol through leaching in subarctic lichens, *Plant Physiol.* **83**, 813-815.

Fahselt, D. (1994) Carbon metabolism in lichens, *Symbiosis* **17**, 127-182.

Farrar, J.F. (1976) Ecological physiology of the lichen *Hypogymnia physodes*. II. Effects of wetting and drying cycles and the concept of physiological buffering, *New Phytol.* **77**, 105-113.

Farrar, J.F. (1988) Physiological buffering, in M. Galun (ed.), *CRC Handbook of Lichenology, vol 2*, CRC Press, Boca Raton, pp. 101-105.

Farrar, J.F. and Smith, D.C. (1976) Ecological physiology of the lichen *Hypogymnia physodes*. III. The importance of the rewetting phase, *New Phytol.* **77**, 115-125.

Fay, P. (1983) *The Blue-Greens*, Edward Arnold, London.

Feige, G.B. (1978) Probleme der flechtenphysiologie, *Nova Hedwigia* **30**, 725-774.

Friedl, T. and Büdel, B. (1996) Photobionts, in T.H. Nash (ed.), *Lichen Biology*, Cambridge University Press, Cambridge, pp. 8-23.

Galun, M. (1988) Lichenization, in M. Galun (ed.), *CRC Handbook ofLlichenology, vol 2*, CRC Press, Boca Raton, pp. 153-169.

Galun, M. and Shomer-Ilan, A. (1988) Secondary metabolic products, in M. Galun (ed.), *CRC Handbook of Lichenology, vol 3*, CRC Press, Boca Raton, pp. 3-8.

Gauslaa, Y. and Solhaug, K.A. (1996) Differences in the susceptibility to light stress between epiphytic lichens of ancient and young boreal forest stands, *Functional Ecol.* **10**, 344-354.

Gauslaa, Y. and Solhaug, K.A. (1998) The significance of thallus size for the water economy of the cyanobacterial old-forest lichen *Degelia plumbea*, *Oecologia* **116**, 76-84.

Grace, J. (1997) Toward models of resource allocation by plants, in F.A. Bazzaz and J. Grace (eds.), *Plant Resource Allocation*, Academic Press, San Diego, pp. 279-291.

Green, T.G.A. and Lange, O.L. (1995) Photosynthesis in poikilohydric plants: a comparison of lichens and bryophytes, in E-D. Schulze and M.M. Caldwell (eds.), *Ecophysiology of Photosynthesis*, Springer, Berlin, pp. 71-101.

Green, T.G.A., Kilian, E. and Lange, O.L. (1991) *Pseudocyphellaria dissimilis*: a desiccation-sensitive, highly shade-adapted lichen from New Zealand, *Oecologia.* **85**, 498-503.

Hawksworth, D.L and Hill, D.J. (1984) *The Lichen Forming Fungi*, Blackie, Glasgow.

Hawksworth, D.L. (1988) The variety of fungal-algal symbiosis, their evolutionary significance and the nature of lichens, *Bot. J. Linnean Soc.* **96**, 3-20.

Hedenås, H. (2002) *Epiphytic Lichens on* Populus tremula*: Implications for Conservation*, PhD Thesis. Umea University, Sweden

Henriksson, E. (1964). *Studies in the physiology of the lichen Collema*, PhD Thesis, Uppsala University, Sweden, *Acta Universitatis Upsaliensis* **38**, 1-13.

Hill, D.J. (1972) The movement of carbohydrate from the alga to the fungus in the lichen *Peltigera polydactyla*, *New Phytol.* **71**, 31-39.

Hill, D.J. and Smith, D.C. (1972) Lichen physiology XII. The "inhibition technique", *New Phytol.* **71**, 15-30.

Honegger, R. (1991) Functional aspects of the lichen symbiosis, *Annu. Rev. Plant Physiol. Plant Mol. Biol.* **42**,553-578.

Honegger, R. (1996) Mycobionts, in T.H. Nash (ed.), *Lichen Biology*, Cambridge University Press, Cambridge, pp. 24-36.

Hyvärinen, M., Härdling, R. and Tuomi, J. 2002. Cyanobacterial lichen symbiosis: the fungal partner as an optimal harvester, *Oikos* (in press).

Jennings, D.H. and Lysek, G. (1996) *Fungal Biology: Understanding the Fungal Lifestyle*, BIOS Scientific Publishers, Guildford.

Jones, L.W. and Myers, J. (1963) A common link between photosynthesis and respiration in a blue-green alga, *Nature* **199**, 670-672.

Kaplan, A. and Reinhold, L. (1999) CO_2 concentrating mechanisms in photosynthetic microorganisms, *Annu. Rev. Plant Physiol. Plant Mol. Biol.* **50**, 539-570.

Kappen, L. (1988) Ecophysiological relationships in different climatic regions, in M. Galun (ed.), *CRC Handbook of Lichenology, vol*, CRC Press, Boca Raton, pp. 37-100.

Kershaw, K.A. (1985) *Physiological Ecology of Lichens*, Cambridge University Press, Cambridge.

Lambers, H. (1985) Respiration in intact plants and tissues: its regulation and dependence on environmental factors, metabolism and invaded organisms, in R. Douce and D.A. Day (eds.), *Higher Plant Respiration*, Springer, Berlin, pp. 418-465.

Lambers, H, Chapin III, F.S. and Pons, T.L. (1998) Photosynthesis, respiration, and long distance transport, in H. Lambers, F.S. Chopin and T.L. Pons (eds.), *Plant Physiological Ecology*, Springer, Berlin, pp. 10-95.

Lange, O.L, Büdel, B., Heber, U., Meyer, A., Zellner, H. and Green, T.G.A. (1993) Temperate rainforest lichens in New Zealand: high thallus water content can severely limit photosynthetic CO_2 exchange, *Oecologia* **95**, 303-313.

Lange, O.L., Büdel, B., Zellner, H., Zotz, G. and Meyer, A. (1994) Field measurements of water relations and CO_2 exchange of the tropical, cyanobacterial basidiolichen *Dictyonema glabratum* in a Panamian Rainforest, *Bot. Acta* **107**, 279-290.

Lange, O.L., Green, T.G.A. and Ziegler, H. (1988) Water status related photosynthesis and cabon isotope discrimination in species of the lichen genus *Pseudocyphellaria* with green or blue-green photobionts and in photosynthesis, *Oecologia* **75**, 494-501.

Lange, O.L., Kilian, E. and Ziegler, H. (1986) Water vapour uptake and photosynthesis of lichens: performance differences in species with green and blue-green algae as phycobionts, *Oecologia* **71**, 104-110.

Leisner, J.M.R, Bilger, W., Czygan, F-C. and Lange, O.L. (1993) Lipophilous carotenoids of cyanobacterial lichens from different habitats, including an extreme desert site, *Cryptogamic Bot.* **4**, 74-82.

Leisner, J.M.R., Bilger, W., Czygan, F-C. and Lange, O.L. (1994) Light exposure and the composition of lipophilous carotenoids in cyanobacterial lichens, *J. Plant Physiol.* **143**, 514-519.
Leisner, J.M.R., Bilger, W. and Lange, O.L. (1996) Chlorophyll fluorescence characteristics of the cyanobacterial lichen *Peltigera rufescens* under field conditions. I. Seasonal patterns of photochemical activity and the occurrence of photosystem II inhibition, *Flora* **191**, 261-273.
Leprince, O., Hendry, G.A.F. and McKersie B.D. (1993) The mechanisms of desiccation tolerance in developing seeds, *Seed Sci. Res.* **3**, 231-246.
Lewis, D.H. and Smith, D.C. (1967) Sugar alcohols (polyols) in fungi and green plants. I. Distribution, Physiology and Metabolism, *New Phytol.* **66**, 143-184.
Lines, C.E.M., Ratcliffe, R.G., Rees, T.A.V. and Southon, T.E. (1989) A ^{13}C NMR study of photosynthate transport and metabolism in the lichen Xanthoria calcicola Oxner, *New Phytol.* **111**, 447-456.
Lüttge, U., Büdel, B., Ball, E, Strube, F. and Weber, P. (1995) Photosynthesis of terrestrial cyanobacteria under light and desiccation stress as expressed by chlorophyll fluorescence and gas-exchange. *J. Exp. Bot.* **46**, 309-319.
MacFarlane, J.D. and Kershaw, K.A. (1982) Physiological-environmental interactions in lichens. XIV. The environmental control of glucose movement from alga to fungus in *Peltigera polydactyla*, *P. rufescens*, and *Collema furfuraceum*, *New Phytol.* **91**, 93-101.
MacKenzie, T.D.B., MacDonald, T.M., Dubois, L.A. and Campbell, D.A. (2001) Seasonal changes in temperature and light drive acclimation of photosynthetic physiology and macromolecular content in *Lobaria pulmonaria*, *Planta* **214**, 57-66.
Máguas, C., Griffiths, H., Ehleringer, J. and Serôdio, J. (1993) Characterization of photobiont associations in lichens using carbon isotope discrimination techniques, in J. Ehleringer, A. Hall and G. Farquhar (eds.), *Stable Isotopes and Plant Carbon-Water Relations*, Academic Press, New York, pp. 423-458.
Maguás, C., Valladares, F. and Brugnoli, E. (1997) Effects of thallus size on morphology and physiology of foliose lichens: new findings with a new approach, *Symbiosis* **23**, 149-164.
Mullineaux, C.W., Tobin, M.J. and Jones, G.R. (1997) Mobility of photosynthetic complexes in thylakoid membranes, *Nature* **390**, 421-424.
Nash, T.H. (1996) Photosynthesis, respiration, productivity and growth, in T.H. Nash (ed.) *Lichen Biology*. Cambridge University Press, Cambridge, pp. 88-120.
Palmqvist, K. (1993) Photosynthetic CO_2 use efficiency in lichens and their isolated photobionts: The possible role of a CO_2 concentrating mechanism in cyanobacterial lichens, *Planta* **191**, 48-56.
Palmqvist, K. (2000) Tansley Review No. 117: Carbon economy in lichens, *New Phytol.* **148**, 11-36.
Palmqvist, K. and Sundberg, B. (2000) Light use efficiency of dry matter gain in five macro-lichens: relative impact of microclimate and species-specific traits, *Plant Cell Environ.* **23**, 1-14.
Palmqvist, K., Campbell, D., Ekblad, A. and Johansson, H. (1998) Photosynthetic capacity in relation to nitrogen content and its partitioning in lichens with different photobionts, *Plant Cell Environ.* **21**, 361-372.
Palmqvist, K., Dahlman, L., Valladares, F., Tehler, A., Sancho, L.G. and Mattsson, J-E. (2002) A broad scale comparison of CO_2 exchange processes and thallus nitrogen statuses across 75 lichen associations of contrasting photosynthetic partners, habitats and morphologies, *Oecologia* (in press)
Paulsrud, P. (2001) *The Nostoc symbiont of lichens. Diversity, Specificity and cellular modifications*, PhD Thesis, Uppsala University, Sweden.
Paulsrud, P., Rikkinen, J. and Lindblad, P. (2001) Field investigations on cyanobacterial specificity in *Peltigera aphthosa*, *New Phytol.* **152**, 117-123.
Paulsrud, P., Rikkinen. J. and Lindblad, P. (2000) Spatial patterns of photobiont diversity in some *Nostoc*-containing lichens, *New Phytol.* **146**, 291-299.
Price, G.D., Sültemeyer, D., Klughammer, B., Ludwig, M. and Badger, M.R. (1998) The functioning of the CO_2 concentrating mechanism in several cyanobacterial strains: a review of general physiological characteristics, genes, proteins, and recent advances, *Can. J. Bot.* **76**, 973-1002.
Qui, B.S. and Gao, K.S. (2001) Photosynthetic characteristics of the terrestrial blue-green alga *Nostoc flagelliforme*, *Eur. J. Phycol.* **36**, 147-156.
Rai, A.N. (1988) Nitrogen metabolism, in M. Galun (ed.), *Lichenology vol. 1*, CRC Press, Boca Raton, pp. 1-237.
Rai, A.N., Söderback, E. and Bergman, B. (2000) Tansley Review No. 116: Cyanobacterium-plant symbioses, *New Phytol.* **147**, 449-481.
Raven, J.A. (1991) Implications of inorganic carbon utilization: ecology, evolution and geochemistry, *Can. J. Bot.* **69**, 908-924.

Raven, J.A. (1992) Energy and nutrient acquisition by autotrophic symbioses and their asymbiotic ancestors, *Symbiosis* **14**, 33-60.

Raven, J.A., Johnston, A.M., Handley, L.L. and McInroy, S.G. (1990) Transport and assimilation of inorganic carbon by *Lichina pygmea* under emersed and submersed conditions, *New Phytol.* **114**, 407-417.

Reinhold, L., Zviman, M. and Kaplan, A. (1989) A quantitative model for inorganic carbon fluxes and photosynthesis in cyanobacteria, *Plant Physiol. Biochem.* **27**, 945-954.

Reuter, W. and Müller, C. (1993) Adaptation of the photosynthetic apparatus to light and CO_2, *J. Photochem. Photobiol. B. Biology* **21**, 3-27.

Richardson, D.H.S. and Smith, D.C. (1966) The physiology of the symbiosis in *Xanthoria aureola* (Ach.) Erichs, *Lichenologist* **3**, 202-206.

Richardsson, D.H. (1999) War in the world of lichens: parasitism and symbiosis as exemplified by lichens and lichenicolous fungi, *Mycol. Res.* **6**, 641-650.

Richardson, D.H.S. and Smith, D.C. (1968) Lichen physiology. IX. Carbohydrate movement from the Trebouxia symbiont of *Xanthoria aureola*, *New Phytol.* **67**, 61-68.

Richardson, D.H.S., Hill, D.J. and Smith, D.C. (1968) Lichen physiology XI. The role of the alga in determining the pattern of carbohydrate movement between lichen symbionts, *New Phytol.* **67**, 469-486.

Rundel, P.W. (1982) Water uptake by organs other than roots, in O.L. Lange, P.S. Nobel, C.B. Osmond, and H. Ziegler (eds.), *Physiological Plant Ecology II. Water Relations and Carbon Assimilation, Encyclopedia Plant Physiol. 12*, Springer, Berlin, pp. 111-134.

Sancho, L.G., Valladares, F., Schroeter, B. and Kappen, L. (2000) Ecophysiology of Antarctic versus temperate populations of a bipolar lichen: The key role of the photosynthetic partner, in W. Davison, C.H. Williams and P. Broady (eds.), *Antarctic Ecosystems: Models for Wider Ecological Understanding*, New Zealand Natural Sciences Publications, Christchurch, pp 190-194.

Schwendener, S. (1873) Die Flechten als Parasiten der Algen, Scweighauser, Basel (Reprinted from Verh. Naturf. Ges. Basel 1873).

Sillett, C.S. and McCune, B. (1998) Survival and growth of cyanolichen transplants in Douglas-fir forest canopies, *The Bryologist* **101**, 20-31.

Smith, D.C. (1960) Studies in the physiology of lichens 3. Experiments with dissected discs in *Peltigera polydactyla*, *Ann. Bot.* **24**, 186-199.

Smith, D.C. (1975) *Symbiosis and the Biology of Lichenized Fungi*, The University Press, Cambridge.

Smith, D.C. and Molesworth, S. (1973) Lichen physiology. XIII. Effects of rewetting of dry lichens, *New Phytol.* **72**, 525-533.

Smith, D.C., Muscatine, L. and Lewis, D.H. (1969) Carbohydrate movement from autotrophs to heterotrophs in parasitic and mutualistic symbiosis, *Biol. Rev.* **44**, 17-90.

Sundberg, B. (1999) *Physiological Ecology of Lichen Growth*, PhD Thesis, Umeå University, Sweden.

Sundberg, B., Campbell, D. and Palmqvist, K. (1997) Predicting CO_2 gain and photosynthetic light acclimation from fluorescence yield and quenching in cyanolichens, *Planta* **201**,138-145.

Sundberg, B., Ekblad, A., Näsholm, T. and Palmqvist, K. (1999) Lichen respiration in relation to active time, temperature, nitrogen and ergosterol concentrations, *Functional Ecology* **13**, 119-125.

Sundberg, B., Näsholm, T. and Palmqvist, K. (2001) The effect of nirogen on growth and key thallus components in the two tripartite lichens, *Nephroma arcticum* and *Peltigera aphthosa*, *Plant Cell Environ.* **24**, 517-527.

Tabita, F.R. (1999) Microbial ribulose 1,5-bisphosphate carboxylase/oxygenase: A different perspective, *Photosynth. Res.* **6**, 1-28.

Tchernov, D., Helman, Y., Keren, N., Luz, B., Ohad, I., Reinhold, L., Ogawa, T. and Kaplan A. (2001) Passive entry of CO_2 and its energy-dependent intracellular conversion to HCO_3^- in cyanobacteria are driven by a photosystem I-generated $\Delta\mu H^+$, *J. Biol. Chem.* **276**, 23450-23455.

Tschermak-Woess, E. (1988) The algal partner, in M. Galun (ed.) *CRC Handbook of Lichenology, vol 1*, CRC Press, Boca Raton, pp. 39-94.

Wessels, J.G.H. (1993) Tansley Review No 45: Wall growth, protein excretion and morphogenesis in fungi, *New Phytol.* **12**, 397-413.

Zotz, G., Büdel, B., Meyer, A., Zellner, H. and Lange O.L. (1998) In situ studies of water relations and CO_2 exchange of the tropical macrolichen, *Sticta tomentosa*. *New Phytol.* **139**, 525-535.

Chapter 6

CYANOLICHENS: NITROGEN METABOLISM

A.N. RAI
Department of Biochemistry, North-Eastern Hill University,
Shillong – 793022, India.

1. INTRODUCTION

This chapter deals with nitrogen metabolism in cyanolichens, in particular the aspects of N_2 fixation, nitrogen assimilation and transfer of fixed nitrogen. Other aspects of cyanolichens are covered in the previous two chapters. Although several cyanobacterial genera occur in lichen symbioses, most studies on aspects of nitrogen metabolism relate to *Nostoc*-containing bipartite and tripartite cyanolichens (see Rai, 1990; Rai *et al.*, 2000).

Most cyanobionts in lichen symbioses are heterocystous forms, fix N_2, and provide fixed-nitrogen to the mycobiont. Some unicellular cyanobacteria also occur as cyanobionts in lichens and probably fix nitrogen as well, but this has not been experimentally verified.

Unlike the host plants in other cyanobacterial symbioses, the mycobiont is nonphotosynthetic. The cyanobiont as the sole photobiont, is responsible for provision of fixed-carbon as well as fixed-nitrogen to the mycobiont in bipartite cyanolichens. In tripartite cyanolichens however, fixed-carbon to the mycobiont is provided by the phycobiont (the green algal partner); the cyanobiont provides little or no fixed-carbon, except meeting its own requirements. In this review, the cyanobacterial partner is referred to as cyanobiont, the green algal partner as phycobiont, and fungal partner as mycobiont. The term photobiont refers to both the cyanobacterial and the green algal partners.

In lichen thalli, the cyanobiont undergoes structural-functional changes that enable close interaction and nutrient transfer between the partners (Fig. 1; Honegger, 1991). The extent of the structural-functional changes varies progressively from younger to older parts of the thallus. The changes develop in a coordinated manner optimising nutrient transfer and ensuring a consistent overall nutrient availability. Whether and to what extent these structural-functional changes are caused directly by the mycobiont and/or the cyanobiont themselves (as a result of the special environmental conditions in the lichen thallus), remains to be worked out. Some of the changes, which relate to nitrogen fixation and nitrogen transfer, are discussed in this article.

A.N. Rai, B. Bergman and U. Rasmussen (eds.), Cyanobacteria in Symbiosis, 97-115.

2. NITROGEN CONTENT, SOLUBLE NITROGEN POOLS AND UTILIZATION OF EXOGENOUS NITROGEN

Cyanolichens have higher nitrogen content than the lichens with only green algae as photobionts (see Rai, 1988, 1990). Hitch and Stewart (1973) reported nitrogen content of 2.2% in cyanolichens and 0.85% in other lichens. Green *et al* (1980) found these values to be 3.4% and 0.5%, respectively. In tripartite lichens, the cephalodia contain higher nitrogen content than the rest of the thallus (Hitch and Stewart, 1973; Englund, 1977; Rai, 1980), probably due to the fact that the N_2-fixing cyanobiont is located in the cephalodia. Sampling away from cephalodia on the *Peltigera aphthosa* thalli, a progressive decrease in nitrogen content is observed (Englund, 1977). Of the total thallus nitrogen in *Peltigera polydactyla*, 75% is insoluble nitrogen and 25% soluble nitrogen. Ammonia, amino acids and amide-N constitute nearly 50% of the soluble nitrogen pool (Smith, 1960a).

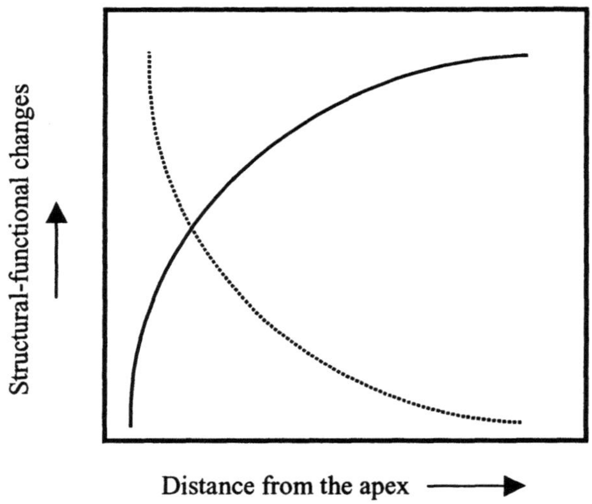

Figure 1. General pattern of structural-functional changes in symbionts along a lichen thallus (from apex towards the centre). Solid line: cell size, heterocyst frequency, N_2 fixation and CO_2 fixation by the cyanobiont, and glutamate dehydrogenase of the mycobiont. Dotted line: growth rate, cell division and glutamine synthetase of the cyanobiont.

Nostoc, *Coccomyxa* and the mycobiont constitute 6%, 14.4% and 79.6%, respectively, of the total thallus protein in *P. aphthosa* (Rai, 1980). However, *Nostoc* constitutes 80% of the total soluble protein in cephalodia (Sampaio *et al.*, 1979). In *Peltigera canina*, *Nostoc* and mycobiont constitute 36% and 64% of the total thallus protein, respectively (Sampaio *et al.*, 1979). A comparision of the free-living and symbiotic *Coccomyxa* from the *P. aphthosa* thallus, showed that both contain similar nitrogen content and therefore the *Coccomyxa* in the lichen thallus does not seem to be nitrogen-limited (Rai, 1988).

Major soluble amino acids in lichen thalli are glutamate, glutamine, alanine and aspartate. However, their proportions in the mycobiont and photobiont(s) may vary. For example, in *P. canina* the cyanobiont contains about a third of the total thallus protein as well as a third of the ammonia, glutamate and glutamine pools. However, pools of aspartate and alanine are proportionately much lower in the cyanobiont. Probably, much of the alanine and aspartate is located in the mycobiont (see Rai, 1988 and the references therein).

Amines seem to be common in most cyanolichens and may be involved in some of the physiological processes (Bernard and Goas, 1968; Bernard and Larher, 1971; Rai, 1988, 1990). Rowell *et al* (1985) found that *P. canina* can metabolise methylamine using the enzyme glutamine synthetase of the cyanobiont and convert it to γ-methylglutamine. In plants, polyamines may provide protection against thermal and enzymic destruction of nucleic acids, membranes and ribosomes by binding and stabilizing them under extreme environmental conditions (Cohen, 1971; Smith, 1971, 1975). The polyamines may have a similar role in lichens too.

Sarcosine was found to be a major component of free amino acid pools of *Peltigera praetextata* and it may have a role in regulation of glutamine synthetase (GS) and nitrogenase (Hallbom, 1984). Nitrogenase activity increased, GS activity decreased and ammonia liberation occurred when free-living *Nostoc* from *P. canina* was subjected to sarcosine treatment (Hallbom, 1984). Rowell *et al* (1985) did not find any significant level of sarcosine in *P. canina*. Although they found a peak eluting close to sarcosine, it did not co-chromatograph with sarcosine. In other respects, the amino acid pool composition of the two lichens was similar.

Cyanolichens show a very slow rate of nitrate uptake as compared to green algal lichens, and they do not seem to assimilate it. However, cyanolichens such as *P. aphthosa, P. canina* and *P. polydactyla* do take up and assimilate ammonia, although much of it is by the mycobiont and the phycobiont. The cyanobiont in the lichen thalli assimilates ammonia at much slower rate than its free-living counter part (Smith, 1960a; Rai *et al.*, 1980; Rai, 1988) apparently because of the repression of the GS. Ammonia absorption by *P. polydactyla* thalli is enhanced by the addition of glucose (Smith, 1960a). Protein synthesis is slow, in small amounts, and increases slightly by the addition of glucose. Ammonia absorption leads to an increase in ammonia and amino-N of the thallus, but not the amide-N. It has been suggested that the amides do not play an important quantitative role in N-metabolism of *P. polydactyla* (Smith, 1960a). These observations are consistent with the repression of cyanobiont GS and lack of GS in the mycobiont. Smith (1960b,c) studied the uptake and utilization of amino acids glutamine, glutamate, aspartate and asparagines in *P. polydactyla*. Asparagine uptake was faster than the rest of the amino acids. All amino acids were taken up and led to increases in ammonia and amino-N of the thallus. The cyanobiont region absorbed more asparagine and ammonia than the medullary region, indicating higher metabolic activity in the cyanobiont region of the thallus. In contrast to ammonia absorption, absorption of asparagine was adversely affected by addition of glucose. The utilization of absorbed asparagine was slow and involved deamidation to ammonia. High absorption capacity and slow utilization could be a useful adaptation enabling accumulation of nutrients during periods of plenty and their use during periods of scarcity (Smith, 1960b,c).

3. HETEROCYSTS

Under N-limited conditions, 5-10% of the total cells in filaments of free-living cyanobacteria differentiate into heterocysts. These are the sites of aerobic N_2-fixation (Stewart, 1980; Bergman *et al.*, 1986) and provide oxygen protection to nitrogenase (Gallon, 1992). Millbank (1972) suggested that in *P. canina* nitrogen fixation may occur even in vegetative cells of the cyanobiont, considering the microaerobic environmnt created by the surrounding fungal hyphae. However, nitrogenase has been found only in heterocysts of the cyanobionts in lichen thalli (Bergman *et al.*, 1986; Bergman and Rai, 1989; Janson *et al.*, 1993). Heterocyst differentiation in free-living cyanobacteria may be adversely affected by microaerobiosis (Madan and Nierzwicki-Bauer, 1993). However, in lichen thalli, the cyanobiont continues to develop heterocysts despite the presence of fixed-N and microaerobic conditions (see Bergman *et al.*, 1992; Rai *et al.*, 2000). There are two reports of bipartite lichen cyanobionts losing their filamentous character (becoming unicellular) and heterocysts altogether: *Scytonema* in *Heppia echinulata* (Marton and Galun, 1976) and *Calothrix/Dichothrix* in *Placynthium nigrum* (Geitler, 1934).

Higher heterocyst frequency and altered heterocyst pattern have been noted in all the tripartite cyanolichens but not in bipartite ones. In bipartite cyanolichens, heterocyst frequency of the cyanobiont ranges between 2.1 to 7.8%, which is similar to that in free-living forms. In contrast, heterocyst frequency of 15-36% has been reported among cyanobionts of tripartite lichens (Griffiths *et al.*, 1972; Hitch and Millbank, 1975a,b; Millbank, 1976; Kershaw, 1985). The increase in heterocyst frequency correlates with the age of the thallus. Heterocyst frequency increases from apical to central parts of the lichen thalli (see Hill, 1989; Rai, 1990). Englund (1977) found heterocyst frequencies of 14% in apical parts of *P. aphthosa* thalli, but the central and basal parts showed heterocyst frequencies of 21%.

It is not clear how heterocyst differentiation is altered in tripartite lichens. There does not seem to be any correlation to the decrease in levels of glutamine synthetase (GS) since the repression of GS occurs in cyanobionts of bipartite as well as tripartite lichens but heterocyst frequency increases only in cyanobionts of tripartite lichens. Heterocyst frequency in lichen cyanobionts correlates well with their carbon nutrition. In bipartite lichens, the cyanobionts, in addition to meeting their own requirements, provide both fixed-N and fixed-C to the mycobiont. In tripartite lichens, the cyanobiont provides only fixed-N but little or no fixed-C to the partners; the fixed-C requirement of the mycobiont is met by the phycobiont. Addition of exogenous sugars leads to an increase in heterocyst frequency of *Nostoc* in laboratory cultures (Bergman *et al.*, 1992; Rai *et al.*, 1996). In other cyanobacterial-plant symbioses, the cyanobiont receives fixed-C from the host plant and its heterocyst frequency can reach upto 80% (see Bergman *et al.*, 1992; Rai *et al.*, 2000). It is possible that in tripartite lichens increased heterocyst formation occurs due to photosynthate movement from phycobiont to the cyanobiont (Hitch and Millbank, 1975b). This possibility remains to be explored but it seems almost a certainty in tripartite lichens like *Peltigera venosa* where cephalodia develop on the under surface of the thallus. Owing to the limitations of light, the cyanobiont in them can not photosynthesise and the Rubisco levels may be drastically reduced. The cyanobiont in internal cephalodia of *Nephroma arcticum* (occurring below the fungal cortex and the green algal layer) contains 75% fewer carboxyomes than the cyanobiont in *P. canina* (bipartite lichen where the cyanobiont layer is just below the fungal cortex) (Bergman and Rai, 1989). Regulation of heterocyst differentiation in relation to C-nutrition and altered C:N ratios in lichen thalli merits further research. Other culture conditions reported to cause increased heterocyst differentiation in cyanobacteria include immobilization (slow growth), green light, fructose addition, and Ca or P limitation (see Bergman *et al.*, 1992). Some of these conditions are analogous to those found in lichen thalli.

An understanding of the altered heterocyst frequency and spacing pattern in lichen cyanobionts needs study of the regulation of genes involved in heterocyst differentiation, spacing pattern and overall control of nitrogen metabolism under symbiotic conditions. Several of the genes involved and their regulatory mechanisms have been identified in free-living cyanobacteria (Frias *et al.*, 1994; Haselkorn, 1998; Yoon and Golden, 1998; Adams and Duggan, 1999; Flores *et al.*, 1999; Lee *et al.*, 1999; Meeks *et al.*, 1999). It is possible that endogenous regulation by the cyanobiont in response to the environmental conditions in lichen thalli, alters or disrupts the regulatory cascade for heterocyst differentiation and spacing pattern. Alternatively, an effector from the mycobiont may be responsible, but this seems unlikely since the same mycobiont seems not to do so in bipartite lichens (see Rai *et al.*, 2000).

4. GLUTAMINE SYNTHETASE

Glutamine synthetase (GS) is the primary ammonia-assimilating enzyme in N_2-fixing heterocystous cyanobacteria (Wolk *et al.*, 1976; Thomas *et al.*, 1977; Stewart, 1980). GS is present both in heterocysts and vegetative cells but the intracellular concentrations, as well as the activity, are two-fold higher in heterocysts than in vegetative cells (Bergman *et al.*, 1985). The enhanced level of GS in heterocysts and the expression of nitrogenase in them, are correlated (Renstrom-Kellner *et al.*, 1990). Repression of nitrogenase in heterocysts leads to a repression of GS also, lowering it to the level similar to that in vegetative cells. The enhanced GS seems to be essential for assmilation of the ammonia being generated by N_2-fixation in the heterocysts.

Table 1. GS protein levels in heterocysts and vegetative cells of cyanobionts in lichen thalli or in free-living cultures. Values are expressed as number of gold particles per μm^2 (from immunogold labelling experiments). After Hallbom et al *(1986), Bergman and Rai (1989); Janson* et al *(1993)*

Cyanobiont	GS in Heterocysts	GS in Vegetative Cells
In *Lichina confines*	1.2	0.9
Free-living	20.3	8.1
In *Nephroma arcticum*	3.2	4.0
Free-living	39	19
In *Peltigera* spp.	1-1.5	0.8
Free-living	35	17

Early studies on the lichens *P. aphthosa* and *P. canina* reported a decrease of >90% in GS activities of the cyanobiont (Stewart and Rowell, 1977; Rai *et al.*, 1980). This decrease varies from young to more mature symbiotic tissues (i.e., from apical to central parts of a lichen thallus) and the maximum decrease in GS coincides with high nitrogenase activity and ammonia release in the thallus (Rowell *et al.*, 1985; Rai, 1988, 1990; Fig. 2). The decrease in GS activity of the cyanobiont occurred due to a repression of the GS synthesis (Stewart *et al.*, 1983). Since whole lichen thalli were used in these experiments, the differences in younger and older parts of the thalli were obscured. Furthermore, the levels of residual GS in heterocysts (sites of N_2-fixation and assimilation of the resulting ammonia) and vegetative cells could not be ascertained. Later studies using immunogold localization on *Lichina confinis* (Janson *et al.*, 1993), *Nephroma arcticum* (Bergman and Rai, 1989), *P. aphthosa* and *P. canina* (Hallbom *et al.*, 1986) answered these questions. These studies confirmed the repression of cyanobiont GS synthesis in lichen thalli and showed that the repression occured in heterocysts as well as in vegetative cells. The residual GS protein represents <10% of that present in free-living cyanobionts, is uniformely distributed within the cells, and there is no difference in the GS levels of heterocysts and vegetative cell. This contrasts with free-living cyanobacteria where heterocysts have a 2-fold higher level of GS than the vegetative cells (Table 1). Repression of GS has been noted in heterocysts of cyanobionts in all cyanobacterial-plant symbioses, except cycads, and may be the reason why cyanobionts release fixed-N as N_2-derived ammonia to the host plants (see Rai *et al.*, 2000).

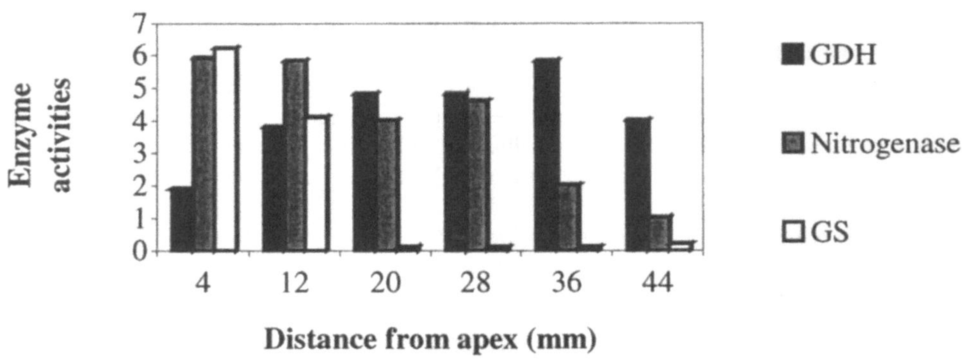

Figure 2. Localisation of nitrogenase, glutamine synthetase and NADP$^+$-glutamate dehydrogenase activities in thalli of Peltigera canina. *Activities are expressed as nmol product formed min^{-1} mg protein^{-1}. Activities were determined in a series sections cut from lichen thalli from the apex to the base. GDH (x10^{-2}), Nitrogenase (x10). After Rowell* et al. *(1985).*

In addition to GS, glutamate synthase (GOGAT) activities of the cyanobiont are also affected in the lichen thalli of *P. aphthosa* and *P. canina* (Rai *et al.*, 1981a). The phycobiont (*Coccomyxa*) showed normal levels of GS and GOGAT activities, and the mycobiont lacked both these activities (Rai *et al.,* 1980, 1981a; Rai, 1988, 1990). However, the mycobiont shows very high levels of glutamate dehydrogenase (GDH) activity, particularly in the hyphae that are in direct contact with the cyanobiont (Stewart and Rowell, 1977; Bernard and Goas, 1979; Rai, 1980; Rai *et al.*, 1980). The mycobiont GDH activity also varied along the lichen thallus. Higher GDH activities occurred in the part of the thallus where cyanobiont had high nitrogenase activity and low GS activity. Thus, the higher mycobiont GDH coincided with the regions of high ammonia release (Fig. 2; Rowell *et al.*, 1985).

Several other enzymes, directly or indirectly involved in nitrogen metabolism, have been analysed in different lichens, particularly *P. canina* (bipartite) and *P. aphthosa* (tripartite). These include, alanine dehydrogenase, glutamate oxaloacetate transaminase, glutamate pyruvate transaminase, aspartate pyruvate transaminase, carbamoyl phosphate synthetase, glutamate glyoxylate transaminase, hydrogenase, and serine transhydroxymethylase (Bernard and Goas, 1969; Stewart and Rowell, 1977; Rai *et al.*, 1980, 1981b,c, 1983a, 1992). For more details see Rai (1988, 1990).

The mechanism by which GS is repressed in lichen thalli is still unclear. Does an effector from the mycobiont selectively affect *glnA* expression in the cyanobiont? If so, this effector must be specific for the prokaryotic GS only since the GS in the phycobiont (*Coccomyxa*) is unaffected. In addition, the effector must also be effective against GOGAT. It seems more likely that the decreased expression of GS in the cyanobiont is a result of the endogenous regulation by the cyanobiont itself due to the altered growth conditions in the lichen thalli. For example, Duan *et al* (1994) noted a decrease in *glnA* expression and GS levels after immobilization of *Anabaena*. Another interesting observation is the enhanced *nif* transcription and repression of GS and ammonium transport due to microaerobic conditions in the *Rhizobium*-legume symbiosis (Quispel, 1992). Lichen thalli too are microaerobic with bulk of the thallus being the heterotrophic mycobiont. Could this be the answer? Another suggestion has been made by Hallbom (1984). He noted high levels of sarcosine in *P. praetextata*, and found that treatment of the free-living *Nostoc* cyanobiont with sarcosine causes increase in nitrogenase activity, decrease in GS activity, and release of ammonia. However, Rowell *et al* (1985) could not find sarcosine in another lichen, *P. canina*.

5. NITROGEN FIXATION

All cyanolichens with heterocystous cyanobacteria fix N_2 and the cyanobiont is responsible for it (see Rai, 1988, 1990). As in free-living heterocystous forms, nitrogenase in cyanobionts of lichen thalli is localized in heterocysts only (Bergman *et al.,* 1986; Bergman and Rai, 1989; Janson *et al.,* 1993). The subcellular distribution pattern for nitrogenase is similar in free-living and symbiotic forms except that cyanobiont heterocysts in older parts of the lichen thalli have a somewhat lower level of nitrogenase protein (Bergman and Rai, 1989).

Lichens with unicellular, non-heterocystous, cyanobionts remain to be critically assessed whether or not they fix N_2. Several non-heterocystous cyanobacteria do fix N_2, although some of them do so microaerobically or during darkness (Bergman *et al.*, 1997).

Nitrogenase activity of the cyanobiont in bipartite lichens resembles that in free-living cyanobacteria. This is keeping with the similar heterocyst frequency in both cases. However, nitrogenase activity of the cyanobiont in tripartite cyanolichens is considerably higher and it matches the 2-3-fold higher heterocyst frequency in cyanobionts of tripartite lichens compared to their free-living counterparts (Hitch and Stewart, 1973; Stewart and Rowell, 1977; Kershaw, 1985; Rai, 1988). Nitrogenase activity also varies along the lichen thallus (Figs. 1, 2; Englund, 1977; Hitch and Stewart, 1973; Rowell *et al.,* 1985). In the bipartite lichen *P. canina*, nitrogenase activities are highest in the apical to middle parts of the lichen thallus and decline in the older parts towards the base of the thallus (Hitch and Stewart, 1973; Rowell *et al.,* 1985). In the tripartite lichen thalli of *P. aphthosa*, the apical regions show low nitrogenase activity and central parts the highest. This correlates with heterocyst frequency, which is lower in the apical parts. However, nitrogenase activity declines in basal parts of the thallus despite high heterocyst frequency (Englund, 1977). It is believed that in older parts of lichen thalli inactive heterocyst may occur. Bergman and Rai (1989) did find that in *Nephroma arcticum* older heterocysts contained lesser nitrogenase.

When removed from *Stereocaulon paschale* thalli, rates of N_2 fixation by cephalodia were adversely affected (Huss-Danell, 1979). It is possible that in this lichen the cyanobiont receives some photosynthate from other partners (phycobiont and/or the mycobiont). However, such a decline in N_2 fixation was not found when cephalodia were detached from the *P. aphthosa* thalli. In *P. aphthosa*, the cyanobiont is self sufficient in its requirements of fixed carbon (Feige, 1976; Rai *et al.,* 1980, 1981b,c, 1983b, Rai, 1990).

In some lichen habitats, molybdenum may be scarce. Being a constituent of nitrogenase, molybdenum is an important nutritional requirement for diazotrophs. Horstmann *et al* (1982) reported that addition of 1 ppm molybdenum enhances nitrogenase activity by 180% in *Lobaria pulmonaria* and by 50% in *Lobaria oregana*. These lichens were collected from a habitat where the molybdenum levels were only 0.01 ppm. Addition of high concentrations of molybdenum (10 ppm) however, had an inhibitory effect on nitrogenase activity.

Several factors affect N_2 fixation by lichens in the natural environment. Kershaw (1985) has reviewed the ecophysiological aspects of the environmental control of nitrogen fixation in lichens. Therefore, these are not covered here. However, two regulatory aspects of the N_2 fixation by cyanobionts differ in lichen thalli. These are the effect of combined-N and the N_2 fixation under darkness.

Nitrogenase activity in the cultured *Nostoc* isolate is fully inhibited by ammonium but only partially (45%) in the bipartite *P. canina* thalli. Similarly, while nitrate inhibits 50-60% of the nitrogenase activity in the cultured isolate, there is no inhibition of nitrogenase activity in the *P. canina* thalli (Stewart and Rowell, 1977). This is apparently due to low ammonia assimilation in the cyanobiont since the primary ammonia-assimilating enzyme GS is repressed. Probably, the nitrate was not assimilated at all. Indeed, the bipartite lichen *P. polydactyla* shows a very slow nitrate uptake and does not assimilate it (Smith, 1960a).

In the tripartite lichen *P. aphthosa*, ammonia does not affect nitrogenase activity in excised cephalodia, while 100% inhibition occurs in the cultured *Nostoc* isolate (Rai *et al.*, 1980). However, when whole thalli are used, nitrogenase inhibition by ammonia does occur (Fig. 3). This inhibition does not occur in presence of MSX (methionine sulphoximine, an inhibitor of GS). This is surprising since the cyanobiont had little GS any way. Rai *et al* (1980) concluded that addition of ammonium to thalli of *P. aphthosa* leads to ammonia assimilation by GS of the phycobiont (*Coccomyxa*) and production of high levels of glutamine. The latter accumulates throughout the thallus, including cephalodia where it inhibits nitrogenase of the cyanobiont. Unlike the cyanobiont where GS is repressed, the phycobiont GS is fully active, and when provided with exogenous ammonia the freshly isolated phycobionts release glutamine (Rai *et al.*, 1980; Rai, 1988). Addition of MSX inhibits phycobiont GS preventing ammonia assimilation and glutamine production. Hence, MSX relieves inhibition of nitrogenase activity. Ammonia assimilation does occur in the mycobiont but this is through the glutamate dehydrogenase and not GS, which is absent in the mycobiont. These observations also point to the fact that the phycobiont is able to influence the cyanobiont even though the two are not in direct contact but separated by the mycobiont.

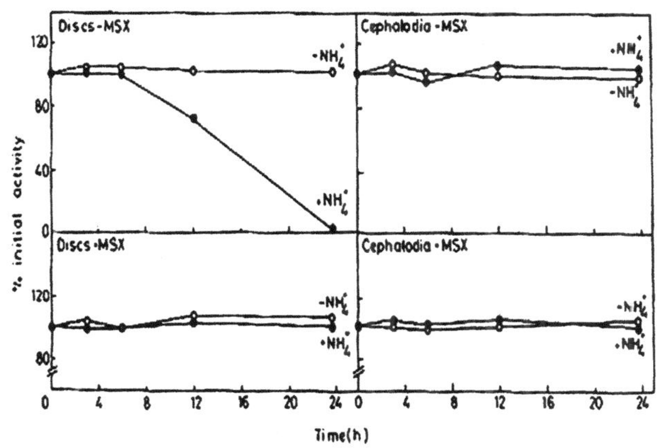

Figure 3. Effect of NH_4^+ on nitrogenase activity of P. aphthosa *thallus discs and cephalodia) in presence or absence of MSX. Initial nitrogenase activities of thallus discs and cephalodia were 0.4 and 6.35 nmol acetylene reduced h^{-1} mg fresh wt^{-1}, respectively. After Rai* et al. *(1980).*

There is evidence to suggest that ammonium *per se* can inhibit nitrogenase in cyanobacteria (Singh *et al.*, 1983; Mackerras and Smith, 1986). Rai *et al* (1983b) showed that in *P. aphthosa* also, ammonia *per se* inhibits nitrogenase of the cyanobiont but only in absence of CO_2 under darkness when ammonia accumulated at high concentrations due to lack of its assimilation (carbon skeletons in the mycobiont become limiting). In other situations, inhibition of nitrogenase by ammonia *per se* does not occur because much of the ammonia is assimilated by the mycobiont via GDH preventing its accumulation and the cyanobiont assimilates little ammonia since its GS is highly repressed. Mycobiont hyphae are in very close contact with the cyanobiont (Honegger, 1991) and these have very high levels of GDH. Thus, levels of both ammonium and glutamine remain below inhibitory concentrations. Even in free-living cyanobacteria, ammonia causes little or no inhibition of nitrogenase when GS activity is inhibited (Stewart and Rowell, 1975; Ownby, 1977; Ladha *et al.*, 1978). It has been suggested that in symbiosis, ammonium transport system may be lacking in the cyanobionts and hence ammonium cannot enter the cells (Kleiner *et al.*, 1981). However, this seems unlikely, at least in some symbioses (Rai *et al.*, 1984).

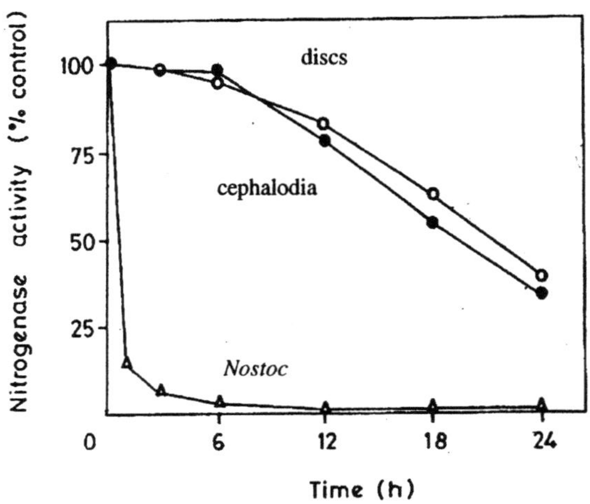

Figure 4. Nitrogenase activity of discs, excised cephalodia and free-living Nostoc *from* P. aphthosa, *placed in dark at zero time and incubated under air throughout. Control rates were those determined under light. After Rai* et al. *(1981c).*

Free-living heterocystous cyanobacteria fix N_2 mainly during the light. N_2 fixation also occurs in the darkness but at a much slower rate and for short period only, unless there is an exogenous supply of utilizable fixed carbon (see Stewart, 1980). In *P. aphthosa* N_2 fixation occurs in the darkness at higher rates and for longer period than that in the cultured *Nostoc* isolate (Fig. 4). Excised cephalodia and digitonin-treated cephalodia (digitonin destroys the mycobiont) still carry out dark N_2 fixation at rates similar to that when they are intact and attached to the thallus (Rai *et al.*, 1983b). Thus, the mycobiont or the phycobiont are not essential as sources of energy or reductant for dark N_2 fixation by the cyanobiont. The dark N_2 fixation was apparently supported by catabolism of the cyanobiont's own glycogen reserves accumulated during the light, which supplied reductant and ATP (Rai *et al.*, 1981c). Indeed the cyanobiont in *P. aphthosa* stores large amounts of carbon reserves as glycogen granules, and there is a linear decrease in glycogen content of the cyanobiont during dark N_2 fixation. The cyanobiont fixes CO_2 in light and provides little or no fixed carbon to the mycobiont (Richardson *et al.*,1968; Fiege, 1976).

Accumulation of large amounts of glycogen by the cyanobiont of *P. aphthosa*, to support prolonged N_2 fixation seems surprising considering the fact that a larger proportion of cells become non-photosynthetic heterocysts. However, several factors make it possible. These include: i) a low demand of fixed carbon due to very slow cyanobiont growth; ii) a low demand of carbon skeletons for ammonia assimilation because most of the ammonia (>90%) is released to the mycobiont that assimilates it; iii) little or no movement of fixed carbon from cyanobiont to mycobiont, the phycobiont meets mycobiont's fixed carbon requirements (Smith *et al.*, 1969; Rai *et al.*, 1981b,c, 1983b). Apart from the glycogen reserves, CO_2 fixation during darkness is essential for prolonged N_2 fixation during darkness (Rai *et al.*, 1981c, 1983b). Dark CO_2 fixation occurs via PEP case in all the partners. In the mycobiont, dark CO_2 fixation replenishes TCA cycle and provides carbon skeletons for assimilation of ammonia being released by the cyanobiont during dark N_2 fixation. In absence of dark CO_2 fixation, N_2-derived ammonia accumulates due to lack of its assimilation, and causes inhibition of nitrogenase. Indeed, treatment of cephalodia with digitonin to disrupt the mycobiont and thereby facilitate release of ammonia into the medium (*i.e.*, prevention of ammonia accumulation in the thallus) during dark N_2 fixation, allowed dark nitrogenase activity to continue even in the absence of dark CO_2 fixation (Rai *et al.*, 1981c, 1983b). That the ammonia *per se* inhibited the dark nitrogenase activity in absence of dark CO_2 fixation, was concluded by Rai *et al* (1983b) based on the following facts: i) internal pools of ammonia in the cephalodia increased significantly and correlated inversely to dark nitrogenase activity; ii) inhibition of GS by MSX, preventing any increase in glutamine, did not prevent inhibition of dark nitrogenase activity.

The ability of the cyanobiont to fix N_2 and transfer fixed nitrogen to its partners even during darkness may be of importance in meeting the increased demand for fixed nitrogen during thallus growth when other natural conditions are favourable.

6. TRANSFER OF FIXED NITROGEN AND PARTITIONING AMONG THE PARTNERS

In heterocystous cyanobacteria, N_2 fixation and the primary assimilation of the resulting ammonia occur in the heterocysts. Fixed nitrogen moves from heterocysts to vegetative cells and fixed carbon from vegetative cells to heterocysts (Wolk *et al.*, 1994). During N_2 fixation, free-living cyanobacteria release little or no fixed nitrogen to the outside. However, early studies during 1950s suggested that transfer of fixed nitrogen from cyanobiont to other partners occurs in lichens (see Rai, 1988, 1990). Detailed studies, including the use of ^{15}N tracer, have shown that 55% of the total N_2 fixed by the cyanobiont in *P. canina*, and up to 95% in *P. aphthosa*, is liberated (Table 2) and taken up by the mycobiont (Stewart and Rowell, 1977; Rai *et al.*, 1980, 1981b, 1983a). In *P. aphthosa* such nitrogen transfer may also occur during darkness (Rai *et al.*, 1981c, 1983b). ^{15}N tracer studies on *P. aphthosa* also showed that the partitioning of fixed nitrogen between cephalodia and the main thallus was in the ratio 15:85. All the partners received fixed nitrogen resulting from N_2 fixation in the cyanobiont, roughly in proportion to their biomass in the thallus (Fig. 5). Kershaw and Millbank (1970) had earlier suggested that the phycobiont gets only 3% of its expected share of the fixed-nitrogen from the cyanobiont but studies by Rai *et al* (1981b) clearly established that this is not so. In fact *Coccomyxa* receives >60% of its expected share of the fixed nitrogen from the cyanobiont (5-6% of the total N_2 fixed).

Table 2. Fate of fixed nitrogen resulting from N_2 fixation in P. canina *and* P. aphthosa. *Values are given as % of the total N_2 fixed. After Rai* et al. *(1980, 1983a)*

Lichen	Retained by the cyanobiont	Liberated as Ammonia	Liberated as Other Compounds
Peltigera canina	44.5	53.7	1.8
Peltigera aphthosa	5	95	0

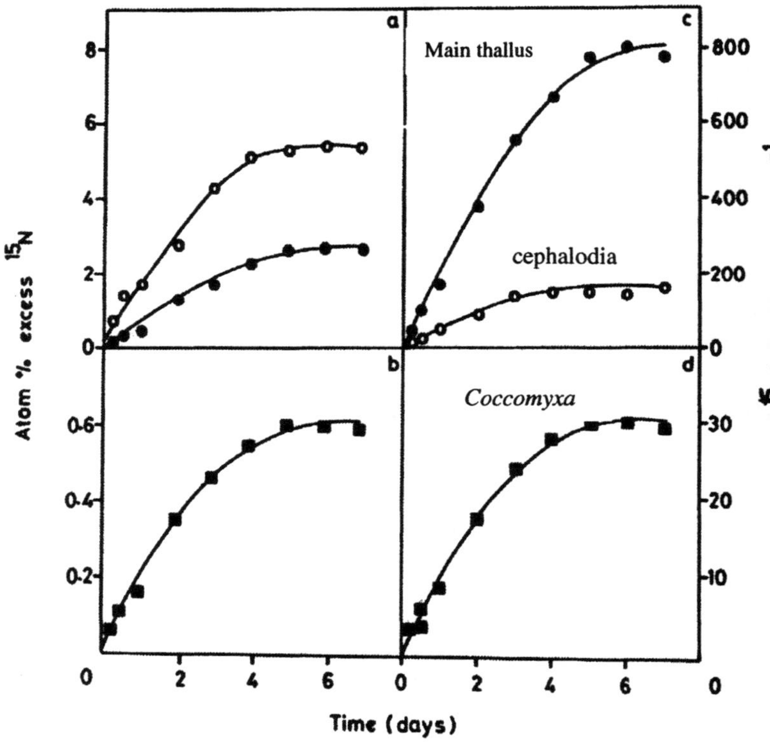

Figure 5. Partitioning of fixed nitrogen after $^{15}N_2$ fixation in P. aphthosa. *a and b, ^{15}N labelling of each component (atom % excess ^{15}N); c and d, total ^{15}N label in each component (µg ^{15}N per gram cephalodial dry weight). After Rai* et al. *(1981b).*

Early studies using whole lichen thalli or intact cephalodia detected a mixture of nitrogenous compounds being liberated by *Peltigera* spp (Millbank, 1974). However, these could not indicate the actual form of fixed nitrogen liberated by the cyanobiont. What Millbank (1974) detected was the forms of fixed nitrogen after the mycobiont had assimilated the transferred nitrogen. It was pointed out that to detect the form of fixed nitrogen being transferred by the cyanobiont in lichen symbioses, the mycobiont's metabolism has to be disrupted (see Rai, 1990). From a number of studies where the mycobiont was disrupted by using digitonin, and using ^{15}N tracer, it is now clearly established that the cyanobiont releases fixed nitrogen as ammonia, resulting directly from N_2 fixation (Stewart and Rowell, 1977; Rai *et al.*, 1980, 1981b,c, 1983a,b). Similar conclusions have been drawn in the cases of *N. arcticum* (Bergman and Rai, 1989) and *L. confinis* (Janson *et al.*, 1993).

In free-living heterocystous cyanobacteria, ammonia assimilation occurs via GS-GOGAT pathway (Wolk *et al.*, 1976). GS is present in both heterocysts and vegetative cells but GOGAT is present only in vegetative cells (absent in heterocysts) (Rai *et al.*, 1982; Bergman *et al.*, 1985). Assimilation of ammonia produced during N_2 fixation occurs in heterocysts via GS followed by movement of fixed nitrogen to the vegetative cell. In lichens, GS and GOGAT activities are repressed in the cyanobiont (see section 4), including in the heterocysts. Thus, heterocysts do not have sufficient GS to assimilate all the ammonia produced during N_2 fixation, and this leads to release of ammonia to the outside of the cell (Fig. 6). Indeed, even in free-living cyanobacteria, inhibition of GS leads to ammonia release into the culture media (Stewart and Rowell, 1975; Ladha *et al.*, 1978). The fact that in *P. apthosa* and *P. canina*, the ammonium released by the cyanobiont results directly from N_2 fixation, supports this view, as do the ^{15}N kinetics studies (Rai *et al.*, 1981b, 1983a). More recent studies on *N. arcticum* (Bergman and Rai, 1989) and *L. confinis* (Janson *et al.*, 1993) also show severe repression of GS in their cyanobionts and suggest liberation of fixed nitrogen as ammonia to the mycobiont.

The mechanism of ammonia release across the cyanobiont cell membrane (heteroctst or vegetative cell) and its uptake across the mycobiont membrane remain unknown. In the cell, ammonia occurs in two forms: NH_3 and NH_4^+. The former can diffuse across the cell membrane. An ammonium transport system (ATS) may recycle the diffusing NH_3 (Kleiner, 1985). The ATS in bacteroids of legume nodules is reported to be repressed, probably due to the microaerobic conditions (Pargent and Kleiner, 1985). This together with the elevated levels of cellular ammonia (due to repression of GS) and near zero level of ammonia outside the bacteroids leads to a continuous diffusion of ammonia out of bacteroids (Quispel, 1992). A similar mechanism may operate in lichens where the mycobiont occurring in close contact with the cyanobiont cells and possessing high levels of GDH, efficiently mops up ammonia and ensures a continuous diffusion out of the cyanobiont.

Cyanobacteria also possess an ATS driven by transmembrane electrical potential (Rai *et al.*, 1984). There has been a suggestion that mycobiont may produce compounds which may specifically lower transmembrane electrical potential in cyanobionts, and thereby inhibit ATS activity (Kleiner *et al.*, 1981). However, this seems unlikely because transmembrane electrical potential is also required for nitrogenase to function

(Hawkesford *et al.*, 1981). In addition, ATS is present in the *Azolla* cyanobiont, and yet it liberates ammonia produced during N_2 fixation, the only thing being common is the repression of GS in heterocysts (see Rai *et al.*, 2000).

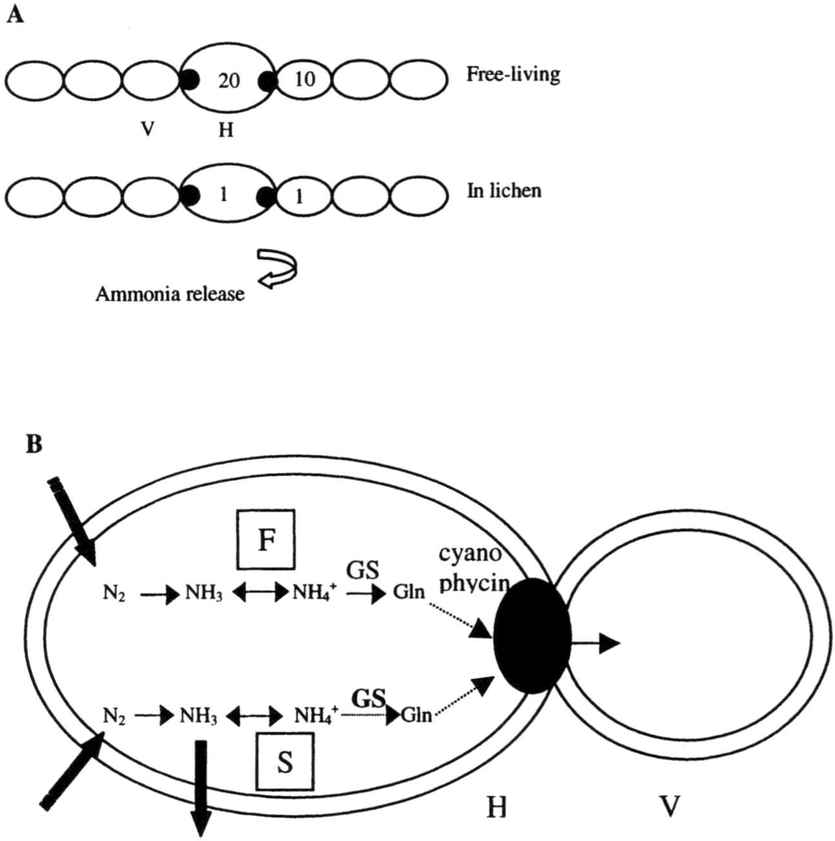

Figure 6. (A) Distribution pattern of GS in heterocysts and vegetative cells of cyanobionts. Note that in the free-living isolate intracellular concentration of GS in heterocysts is twice that in vegetative cells (2:1). In lichens, the synthesis of GS is repressed by >90% and the residual GS is distributed uniformly in the heterocysts and vegetative cells. (B) Release of ammonia from the cyanobiont. In free-living cyanobacteria (F), ammonia originating from N_2 fixation is assimilated in heterocysts and then transferred to vegetative cells, probably via the cyanophycin pool. In lichen symbioses (S), repression of GS synthesis curtails such ammonia assimilation in heterocysts, leading to ammonia release. H, heterocyst; V, vegetative cell. After Rai et al. *(2000).*

7. PATHWAYS OF NITROGEN METABOLISM

Mycobiont hyphae occur in close contact with the cyanobiont cells in lichen thalli and these hyphae possess very high levels of the ammonia-assimilating enzyme GDH. The mycobiont very effectively mops up the ammonia released by the cyanobiont, with the result that hardly any ammonia escapes outside the thallus. Most probably the ammonium uptake by the mycobiont is an active transport. The pathways of ammonia assimilation and metabolism in *P. aphthosa* and *P. canina* have been worked out in detail using $^{15}N_2$ tracer (Rai, 1980; Rai *et al.*, 1981b, 1983a). In *P. canina*, ammonia retained by the cyanobiont is assimilated via GS-GOGAT pathway by the residual activities of these enzymes still present in the cyanobiont. The ammonia liberated by the cyanobiont and taken in by the mycobiont, is assimilated by GDH followed by aminotransferases. In *P. aphthosa*, it is the same except that the ammonia released by the cyanobiont is taken up and assimilated by the cephalodial mycobiont and much of the assimilated nitrogen ends up as alanine. It has been suggested that alanine moves from the cephalodial mycobiont to the rest of the thallus including the phycobiont *Coccomyxa* (Rai, 1980; Rai *et al.*, 1981b; see Fig. 7).

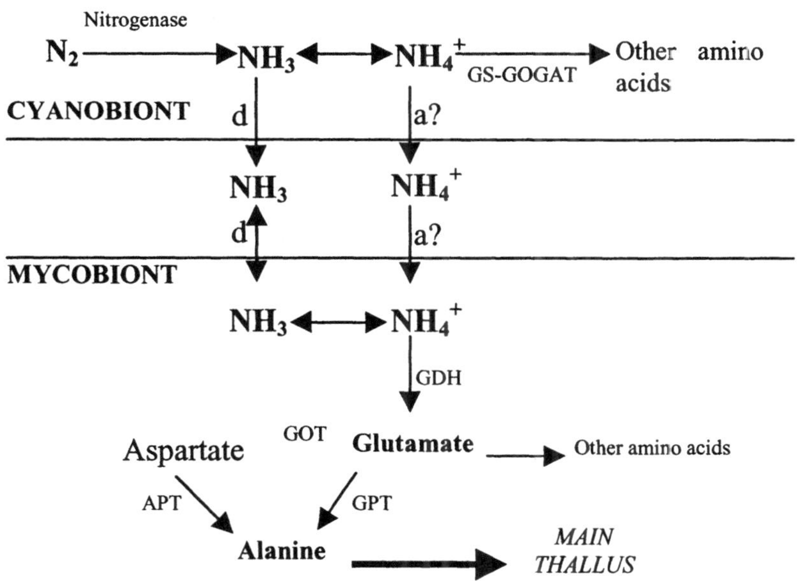

Figure 7. Routes of nitrogen metabolism in Peltigera aphthosa *cephalodia. a, active transport; d, diffusion; GOGAT, glutamate synthase; GOT, glutamate-oxaloacetate transaminase; GS, glutamine synthetase; GPT, glutamate-pyruvate transaminase; APT, aspartate-pyruvate transaminase*

REFERENCES

Adams, D.G. and Duggan, P.S. (1999) Heterocyst and akinete differentiation in cyanobacteria, *New Phytol.* **144**, 3-33.

Bergman, B. and Rai, A.N. (1989) The *Nostoc-Nephroma* symbiosis: localization, distribution pattern and levels of key proteins involved in nitrogen and carbon metabolism of the cyanobiont, *Physiol. Plant.* **77**, 216-224.

Bergman, B., Pettersson, A., Renstrom, E. and Tiberg, E. (1985) Immuno-gold localization of glutamine synthetase in a nitrogen-fixing cyanobacterium (*Anabaena cylindrica*), *Planta* **166**, 329-334.

Bergman, B., Lindblad, P. and Rai, A.N. (1986) Nitrogenase in free-living and symbiotic cyanobacteria: immunoelectronmicroscopic localization, *FEMS Microbiol. Letts.* **35**, 75-78.

Bergman, B., Rai, A.N., Johansson, C. and Soderback, E. (1992) Cyanobacterial-plant symbioses, *Symbiosis* **14**, 61-81.

Bergman, B., Gallon, J.R., Rai, A.N. and Stal, L. (1997) N_2 fixation by non-heterocystous cyanobacteria, *FEMS Microbiol. Rev.* **19**, 139-185.

Bernard, T. and Goas, G. (1968) Contribution a l'etude du metabolisme azote des lichens. Characterisation et dosages des methylamines de quelques especies de la famille des Stictacees, *Comptes Rendus* **267**, 622-624.

Bernard, T. and Goas, G. (1969) Contribution a l'etude du metabolisme azote des lichens. Mise en evidence de quelques transaminases, activite de la glutamate oxaloacetate transaminase dans cinq especes de la famille des Stictacees, *CR Acad. Sci. Paris Ser. D* **269**, 1657.

Bernard, T. and Goas, G. (1979) Contribution a l'etude du metabolisme azote des lichens. Glutamate dehydrogenases du lichen *Lobaria laetevirens* (Lightf.) Zahlbr. Caracteristiques de l'enzyme du Champignon, *Physiol. Veg.* **17**, 535-546.

Bernard, T. and Larher, F. (1971) Contribution a l'etude du metabolisme azote des lichens. Role de la glycine ^{14}C-2 dans la formation des methylamines cher *Lobaria laetevirens*, *Comptes Rendus* **272**, 568-571.

Cohen, S.S. (1971). *Introduction to the Polyamines*, Prentice-Hall, Englewood Cliffs, N.J.

Duan, X.Y., Shi, D.J., Wu, X.Q., Wu, Y.H. and Qiu, Z.S. (1994) The effects of immobilization on the activity of glutamine synthetase and the expression of the *glnA* gene in *Anabaena* sp. Strain 2B, *Chinese J. Bot.* **6**, 102-106.

Englund, B. (1977) The physiology of the lichen *Peltigera aphthosa* with special reference to the blue-green phycobiont (*Nostoc* sp.), *Physiol. Plant.* **41**, 298-304.

Feige, G.B. (1976) Untersuchungen zur Physiologie der Cephalodien der Flechte *Peltigera aphthosa* (L.) Willd. II. Das phtosynthetische ^{14}C-Mearkierungsmuster und der Kohlenhydrattransfer zwischen Phycobiont und Mycobiont, *Z. Pflanzenphysiol.* **80**, 386-394.

Frias, J.E., Flores, E. and Herrero, A. (1994) Requirements of the regulatory protein NtcA for the expression of nitrogen assimilation and heterocyst development genes in the cyanobacterium *Anabaena* sp. PCC 7120, *Mol. Microbiol.* **14**, 823-832.

Flores, E., Muro-Paster, A.M. and Herrero, A. (1999) Cyanobacterial nitrogen assimilation genes and NtcA-dependent control of gene expression, in G.A. Peschek, W. Loffelhardt, and G. Schmetterer (eds.), *The Photosynthetic Prokaryotes*, Kluwer Academic Publishers, Dordrecht, pp. 463-477.

Gallon, J.R. (1992) Reconciling the incompatible: N_2-fixation and O_2, *New Phytol.* **122**, 571-609.

Geitler, L. (1934) Beitrage zur kenntnis der Flechtensymbiose IV-V, *Arch. Protistenk.* **82**, 51-85.

Green, T.G.A., Horstmann, J., Bonnett, H., Wilins, A. and Silvester, W.B. (1980) Nitrogen fixation by members of Stictaceae (Lichenes) of New Zealand, *New Phytol.* **84**, 339-348.

Griffiths, H.B., Greenwood, A.D. and Millbank, J.W. (1972) The frequency of heterocysts in the *Nostoc* phycobiont of the lichen *Peltigera canina*, *New Phytol.* **71**, 11-13.

Hallbom, L. (1984) Sarcosine: a possible regulatory compound in the *Peltigera praetextata-Nostoc* symbiosis, *FEMS Microbiol. Letts* **22**, 119-121.

Hallbom, L., Bergman, B. and Rai, A.N. (1986) Immunogold localization of glutamine synthetase in the cyanobionts of the lichen *Peltigera aphthosa* and *Peltigera canina*, *Lichen Physiol. Biochem.* **1**, 27-34.

Haselkorn, R. (1998) How cyanobacteria count to 10, *Science* **282**, 891-892.

Hawkesford, M.J., Reed, R.H., Rowell, P. and Stewart, W.D.P. (1981) Nitrogenase activity and membrane electrogenesis in the cyanobacterium *Anabaena variabilis* Kutz, *Eur. J. Biochem.* **115**, 519-523.

Hill, D.J. (1989) The control of cell cycle in microbial symbionts, *New Phytol.* **112**, 175-184.

Hitch, C.J.B. and Millbank, J.W. (1975a) Nitrogen metabolism in lichens VI. The blue-green phycobiont content, heterocyst frequency and nitrogenase activity in *Peltigera* species, *New Phytol.* **74**, 473-476.

Hitch, C.J.B. and Millbank, J.W. (1975b) Nitrogen metabolism in lichens VII. Nitrogenase activity and heterocyst frequency in lichens with blue-green phycobionts, *New Phytol.* **75**, 239-244.

Hitch, C.J.B. and Stewart, W.D.P. (1973) Nitrogen fixation by lichens in Scotland, *New Phytol.* **72**, 509-524.

Horstmann, J.L., Denison, W.C. and Silvester, W.B. (1982) $^{15}N_2$-fixation and molybdenum enhancement of acetylene reduction by *Lobaria* spp., *New Phytol.* **92**, 235-241.

Huss-danell, K. (1979) The cephalodia and their nitrogenase activity in the lichen *Stereocaulon paschale*, *Z. Pflazenphysiol.* **95**, 431-440.

Honegger, R. (1991) Functional aspects of the lichen symbioses. *Annu. Rev. Plant Physiol. Plant Mol. Biol.* **42**, 553-578.

Janson, S., Bergman, B. and Rai, A.N. (1993) The marine lichen *Lichina confinis* (O.F. Mull.) C. Ag.: ultrastructure and localization of nitrogenase, glutamine synthetase, phycoerythrin and ribulose 1,5-bisphosphate carboxylase/oxygenase in the cyanobiont, *New Phytol.* **124**, 149-160.

Kershaw, K.A. (1985) *Physiological Ecology of Lichens*, Cambridge University Press, Cambridge.

Kershaw, K.A. and Millbank, J.W. (1970) Nitrogen metabolism in lichens II. The partition of cephalodial fixed nitrogen between the mycobiont and phycobiont of *Peltigera aphthosa*, *New Phytol.* **69**, 75-79.

Kleiner, D. (1985) Bacterial ammonium transport, *FEMS Microbiol. Rev.* **32**, 87-100.

Kleiner, D., Phillips, S.C. and Fitzke, E. (1981) Pathways and regulatory aspects of N_2 and NH_4^+ assimilation in N_2-fixing bacteria, in H. Bothe and A. Trebst (eds.), *Biology of Inorganic Nitrogen and Sulphur*, Springer-Verlag, Berlin, pp. 131-140.

Ladha, J.K., Rowell, P. and Stewart, W.D.P. (1978) Effects of 5-hydroxylysine on acetylene reduction and NH_4^+ assimilation in the cyanobacterium *Anabaena cylindrica*, *Biochem. Biophys. Res. Commun.* **83**, 688-696.

Lee, H.M., Vazques-Bermudez, M.F. and Tandeau de Marsac, N. (1999) The global nitrogen regulator NtcA regulates transcription of the signal transducer P_{II} (GlnB) and influences its phosphorylation level in response to nitrogen and carbon supplies in the cyanobacterium *Synechococcus* sp. Strain PCC 7942, *J. Bacteriol.* **181**, 2692-2702.

Madan, A.P. and Nierzwicki-Bauer, S.A. (1993) *In-situ* detection of transcripts for ribulose 1,5-bisphosphate carboxylase in cyanobacterial heterocysts, *J Bacteriol.* **175**, 7301-7306.

Mackerras, A.H. and Smith, G.D. (1986) Evidence for direct repression of nitrogenase by ammonia in the cyanobacterium *Anabaena cylindrica*, *Biochem. Biophys. Res. Commun.* **134**, 835.

Marton, K. and Galun, M. (1976) *In vitro* dissociation and reassociation of the symbionts in the lichen *Heppia echinulata*, *Protoplasma* **87**, 135-143.

Meeks, J.C., Campbell, E., Hagen, K., Hitzeman, N. and Wong, F. (1999) Developmental alternatives of symbiotic *Nostoc puntiforme* in response to its plant partner *Anthoceros punctatus*, in G.A. Peschek, W. Loffelhardt, and G. Schmetterer (eds.), *The Photosynthetic Prokaryotes*, Kluwer Academic Publishers, Dordrecht, pp. 665-678.

Millbank, J.W. (1972) Nitrogen metabolism in lichens IV. The nitrogenase activity of the *Nostoc* phycobiont in *Peltigera canina*, *New Phytol.* **71**, 1-10.

Millbank, J.W. (1974) Nitrogen metabolism in lichens V. The forms of nitrogen released by the blue-green phycobiont in *Peltigera* spp., *New Phytol.* **73**, 1171-1181.

Millbank, J.W. (1976) Aspects of nitrogen metabolism in lichens, in D.H. Brown, D.L. Hawkesworth and R.H. Bailey (eds.), *Lichenology: Progress and Problems*, Academic press, London, pp. 441-455.

Ownby, J.D. (1977) Effects of amino acids on methionine-sulphoximine-induced heterocyst formation in *Anabaena*, *Planta* **136**, 277-279.

Pargent, W. and Kleiner, D (1985) Characterization and regulation of ammonium (methylammonium) transport in *Rhizobium meliloti*, *FEMS Microbiol. Letts.* **30**, 257-259.

Quispel, A. (1992) A search for signals in endophytic microorganisms, in D.P.S. Verma (ed.), *Molecular signals in Plant-Microbe Communications*, CRC Press, Boca Raton, FL, pp. 471-491.

Rai, A.N. (1980) *Studies on the Nitrogen-Fixing Lichen* Peltigera aphthosa *Willd*, PhD Thesis, Dundee University, Scotland.

Rai, A.N. (1988) Nitrogen metabolism, in M. Galun (ed.), *Lichenology Vol.I*, CRC Press, Boca Raton, FL, pp. 201-237.

Rai, A.N. (1990) Cyanobacterial-fungal symbioses: the cyanolichens, in A.N. Rai (ed.), *Handbook of Symbiotic Cyanobacteria*, CRC Press, Boca Raton, FL, pp. 9-41.

Rai, A.N., Rowell, P. and Stewart, W.D.P. (1980) NH_4^+ assimilation and nitrogenase regulation in the lichen *Peltigera aphthosa* Willd, *New Phytol.* **85**, 545-555.

Rai, A.N., Rowell, P. and Stewart, W.D.P. (1981a) Glutamate synthase activity in symbiotic cyanobacteria, *J. Gen. Microbiol.* **126**, 515-518.

Rai, A.N., Rowell, P. and Stewart, W.D.P. (1981b) $^{15}N_2$ incorporation and metabolism in the lichen *Peltigera aphthosa* Willd, *Planta* **152**, 544-552.

Rai, A.N., Rowell, P. and Stewart, W.D.P. (1981c) Nitrogenase activity and dark CO_2 fixation in the lichen *Peltigera aphthosa* Willd, *Planta* **151**, 256-264.

Rai, A.N., Rowell, P. and Stewart, W.D.P. (1982) Glutamate synthase activity of heterocysts and vegetative cells of the cyanobacterium *Anabaena variabilis* Kutz, *J. Gen. Microbiol.* **128**, 2203-2205.

Rai, A.N., Rowell, P. and Stewart, W.D.P. (1983a) Interactions between cyanobacterium and fungus during $^{15}N_2$-incorporation and metabolism in the lichen *Peltigera canina*, *Arch. Microbiol.* **134**, 136-142.

Rai, A.N., Rowell, P. and Stewart, W.D.P. (1983b) Mycobiont-cyanobiont interactions during dark nitrogen fixation by the lichen *Peltigera aphthosa*, *Physiol. Plant.* **57**, 285-290.

Rai, A.N., Rowell, P. and Stewart, W.D.P. (1984) Evidence for an ammonium transport system in free-living and symbiotic cyanobacteria, *Arch. Microbiol.* **137**, 241-246.

Rai, A.N., Borthakur, M., Soderback, E. and Bergman, B. (1992) Immunogold localization of hydrogenase in the cyanobacterial-plant symbioses *Peltigera canina*, *Anthoceros punctatus* and *Gunnera magellanica*, *Symbiosis* **12**, 131-144.

Rai, A.N., Borthakur, M. and Paul, D. (1996). Symbiotic cyanobacteria: biotechnological applications, *J. Sci. Indust. Res.* **55**, 742-752.

Rai, A.N., Soderback, E. and Bergman, B. (2000) Cyanobacterium-plant symbioses, *New Phytol.* **147**, 449-481.

Renstrom-Kellner, E., Rai, A.N. and Bergman, B. (1990) Correlation between nitrogenase and glutamine synthetase expression in the cyanobacterium *Anabaena cylindrica*, *Physiol. Plant.* **80**, 12-19.

Richardson, D.H.S., Hill, D.J. and Smith, D.C. (1968) Lichen Physiology XI. The role of the alga in determining the pattern of carbohydrate movement between lichen symbionts, *New Phytol.* **67**, 469-486.

Rowell, P., Rai, A.N. and Stewart, W.D.P. (1985) Studies on the nitrogen metabolism of the lichens *Peltigera aphthosa* and *Peltigera canina*, in D.H. Brown (ed.), *Lichen Physiology and Cell Biology*, Plenum Press, London, pp. 145-160.

Sampaio, M.J.A.M., Rai, A.N., Rowell, P. and Stewart, W.D.P. (1979) Occurrence, synthesis and activity of glutamine synthetase in N_2-fixing lichens, *FEMS Microbiol. Letts.* **6**, 107-110.

Singh, H.N., Rai, U.N., Rao, V.V. and Bagchi, S.N. (1983) Evidence for ammonia as an inhibitor of heterocyst and nitrogenase formation in the cyanobacterium *Anabaena cycadae*, *Biochem. Biophys. Res. Commun.* **111**, 180-187.

Smith, D.C. (1960a) Srudies in the physiology of lichens I. The effects of starvation and of ammonia absorption upon the nitrogen content of *Peltigera polydactyla*, *Ann. Bot.* **24**, 52-62.

Smith, D.C. (1960b) Studies in the physiology of lichens 2. Absorption and utilization of some simple organic nitrogen compounds by *Peltigera polydactyla*, Ann. Bot. **24**, 172-185.

Smith, D.C. (1960c) Studies in the physiology of lichens 3. Experiments with dissected discs of *Peltigera polydactyla*, *Ann. Bot.* **24**, 186-199.

Smith, D.C., Muscatine, L. and Lewis, D.H. (1969) Carbohydrate movement from autotrophs to heterotrophs in parasitic and mutualistic symbioses, *Biol. Rev.* **44**, 17-90.

Smith, T.A. (1971) The occurrence, metabolism and functions of amines in plants, *Biol. Rev.* **46**, 201-241.

Smith, T.A. (1975) Recent advances in biochemistry of plant amines, *Phytochemistry* **14**, 865-890.

Stewart, W.D.P. (1980) Some aspects of structure and function in N_2-fixing cyanobacteria, *Annu. Rev. Microbiol.* **34**, 497-536.

Stewart, W.D.P. and Rowell, P. (1975) Effect of L-methionine-DL-sulphoximine on the assimilation of newly fixed NH_3, acetylene reduction and heterocyst production in *Anabaena cylindrica*, *Biochem. Biophys. Res. Commun.* **65**, 846-857.

Stewart, W.D.P. and Rowell, P. (1977) Modifications of nitrogen-fixing algae in lichen symbioses, *Nature* **265**, 371-372.

Stewart, W.D.P., Rowell, P. and Rai, A.N. (1983) Cyanobacteria-eukaryotic plant symbioses, *Ann. Microbiol. (Inst. Pasteur)* **134B**, 205-228.

Thomas, J., Meeks, J.C., Wolk, C.P., Shaffer, P.W., Austin, S.M. and Chien, W.S. (1977) Formation of glutamine from ^{13}N ammonia, ^{13}N dinitrogen and ^{14}C glutamate by heterocysts isolated from *Anabaena cylindrica*, *J. Bacteriol.* **129**, 1545-1555.

Wolk, C.P., Thomas, J., Shaffer, P.W., Austin, S.M. and Galonsky, A. (1976) Pathways of nitrogen metabolism after fixation of ^{13}N-labelled nitrogen gas by the cyanobacterium *Anabaena cylindrica*, *J. Biol. Chem.* **251**, 5027-5034.

Wolk, C.P., Ernst, A. and Elhai, J. (1994) Heterocyst metabolism and development, in D. Bryant (ed.), *The Molecular Biology of Cyanobacteria*, Kluwer Academic Publishers, Dordrecht, pp. 769-823.

Yoon, H-S and Golden, J.W. (1998) Heterocyst pattern formation controlled by a diffusible peptide, *Science* **282**, 891-892.

Chapter 7

CYANOBACTERIA IN SYMBIOSIS WITH HORNWORTS AND LIVERWORTS

DAVID G. ADAMS
Division of Microbiology, School of Biochemistry and Molecular Biology
University of Leeds, Leeds LS2 9JT, U. K.

1. BRYOPHYTE SYMBIOSES

The division Bryophyta consists of three classes of small, non-vascular terrestrial plant groups: the Hepaticae (liverworts), the Anthocerotae (hornworts), and the Musci (mosses). All of these form epiphytic or endophytic associations with cyanobacteria, primarily of the genus *Nostoc*. This chapter considers the liverworts and hornworts, but the moss associations are discussed in the next chapter. In their natural habitat the liverworts and hornworts grow as a prostrate gametophyte thallus a few centimeters in length, attached to the substratum by primitive roots known as rhizoids. Mature symbiotic colonies are visible as dark spots 0.5-1.0 mm in diameter within the plant tissue (Figure 1A). Only four of over 340 liverwort genera are known to develop associations with cyanobacteria, two (*Marchantia* and *Porella*) forming epiphytic associations and two (*Blasia* and *Cavicularia*) forming endophytic associations (see Meeks, 1990). Four of the six hornwort genera (*Anthoceros*, *Phaeoceros*, *Notothylas* and *Dendroceros*) form endophytic associations (see Meeks, 1990). The epiphytic associations may be more common than once thought, but they remain poorly understood (Dalton and Chatfield, 1985; Brasell *et al*., 1986), whereas the endophytic associations have been closely studied because of their adaptability to laboratory experimentation. Ridgway (1967) was the first to demonstrate the reconstitution of several hornwort associations from the separately cultured partners grown on medium solidified with agar. Rodgers and Stewart (1977) were able to reconstitute both *Anthoceros* and *Blasia* associations using moistened Perlite or sand as the growth substrate. However, hornworts such as *Anthoceros* and *Phaeoceros*, and liverworts such as *Blasia*, can be grown more conveniently in axenic shaken liquid culture (Figure 1B), with or without their symbiotic partners and can be readily re-infected with pure cultures of cyanobacteria (Figure 3; Enderlin and Meeks, 1983; Meeks, 1988; Kimura and Nakano, 1990; Meeks, 1990; Babic, 1996; West and Adams, 1997).

Successful reconstitution can be achieved not only with the original partner, but also with cyanobionts from *Gunnera*, cycads, lichens and even some free-living strains. The symbiotically-competent cyanobacteria are hormogonia-forming strains of mostly the

A.N. Rai, B. Bergman and U. Rasmussen (eds.), Cyanobacteria in Symbiosis, 117-135.

genus *Nostoc*, although *Calothrix* and *Chlorogloeopsis* strains have been shown to reconstitute the symbiosis with *Blasia* and *Phaeoceros* (West and Adams, 1997). However, reconstitution experiments such as these, which use axenic cultures of a single cyanobacterium and a single host plant, may give a false impression of the breadth of strains capable of infecting plants in the field. Indeed, in the West and Adams (1997) study of a field population of *Phaeoceros*, the cyanobacterium *Calothrix* was very rarely found as the symbiont and *Chlorogloeopsis* was never found, possibly reflecting both the relative abundance of these strains in the soil and their inability to compete with the natural populations of *Nostoc* for infection of the plant.

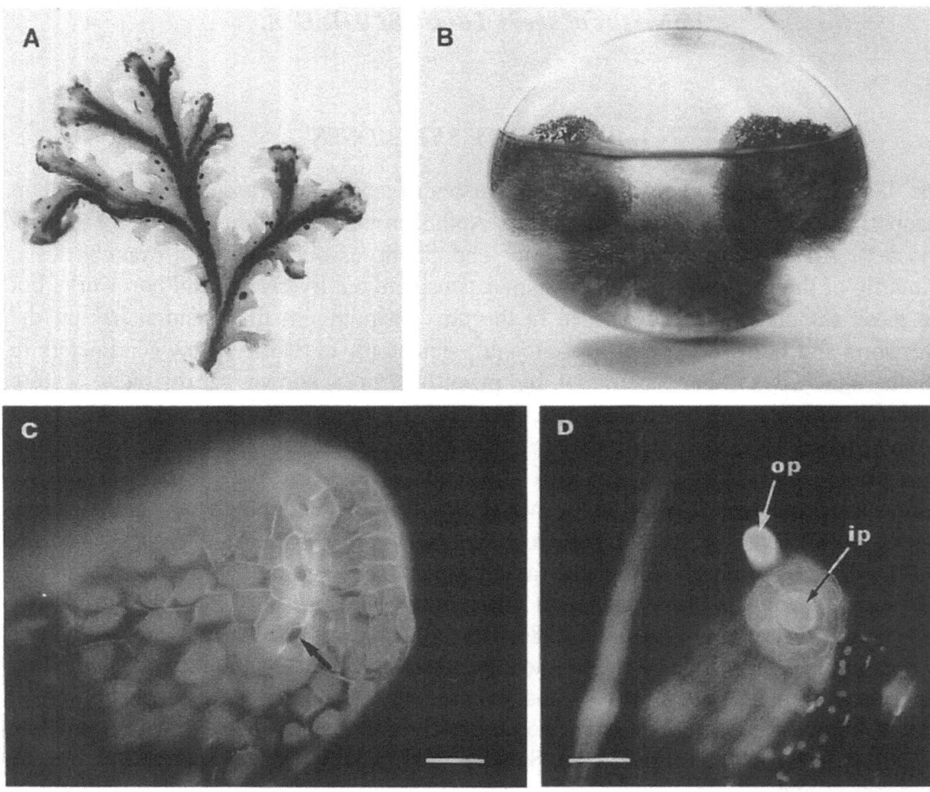

Figure 1. (A) The liverwort Blasia pusilla *showing the prominent midribs of the thallus surrounded by the dark spots of* Nostoc *colonies, which are 0.5-1 mm in diameter. (B)* B. pusilla *grown in liquid medium viewed from the underneath of a 250 ml Erlenmeyer flask. Three large masses of thallus are shown here, but for experimental purposes the thallus would be periodically fragmented to maintain a large number of smaller pieces. (C) Fluorescence photomicrograph of the hornwort* Phaeoceros *sp. showing the slit-like entrance (arrow) to the slime cavity. (D) Fluorescence photomicrogrpah of a* B. pusilla *auricle showing the inner (ip) and outer (op) papillae. (C) and (D) stained with calcofluor white. Bars 30 µm. Photographs (C) and (D) courtesy S. Babic. (A), (C) and (D) reproduced with permission from Adams (2000).*

Even competition experiments using several cyanobacteria at once (Enderlin and Meeks, 1983), do not mimic the natural environment in which many cyanobacterial strains will be present, not to mention the unknown influences of the many other bacteria and eukaryotes. However, the experiments of Enderlin and Meeks (1983) did reveal that symbiotically-competent cyanobacterial strains varied in their ability to infect *Anthoceros* and that only on very rare occasions did more than one strain infect an individual slime cavity. The results of these competition experiments implied that the infectiveness of a *Nostoc* strain was determined by more than a single property.

2. THE SYMBIONTS

For the successful establishment of plant symbioses cyanobacteria require at least two essential characteristics - the ability to develop both heterocysts, which are specialised dinitrogen-fixing cells (for review see Adams and Duggan, 1999), and hormogonia, which are motile filaments that provide a means of dispersal (Campbell and Meeks, 1989; Meeks, 1990; Johansson and Bergman, 1994; Bergman *et al.*, 1996; Meeks, 1998). Heterocysts provide fixed nitrogen for both partners, and the motile hormogonia enable the otherwise immotile cyanobacterial filaments to gain entry to the plant structures that will house the symbiotic colonies (see Section 3). Hormogonia are motile by gliding, the mechanism of which is a mystery (Adams, 2001). In members of the families Nostocaceae and Stigonemataceae, all of which are heterocystous, hormogonia formation is preceded by rapid, highly synchronous cell divisions, which occur in the absence of DNA replication and cell growth, resulting in a decrease in cell size (for reviews see Herdman and Rippka, 1988; Tandeau de Marsac, 1994; Adams, 1997). Hormogonia formation is associated with filament fragmentation and in the case of heterocystous filaments this occurs at the vegetative cell-heterocyst junction, releasing the heterocysts (which consequently lose viability). The liberated sections of filament become the motile hormogonia, which, after a period of 12-48 hours, lose motility and differentiate new heterocysts.

The formation of these hormogonia is triggered by various environmental stimuli, including dilution of liquid cultures with fresh medium or transfer to solid medium, or exposure to red light (see Herdman & Rippka, 1988; Tandeau de Marsac, 1994). Their formation can also be triggered by unidentified components of exudates from plants such as *Anthoceros* (Campbell & Meeks, 1989), *Blasia* (Knight & Adams, 1996), *Gunnera* (Rasmussen *et al.*, 1994) and wheat roots (Gantar *et al.*, 1993; Knight & Adams, 1996). In addition to being smaller than the vegetative filaments, the hormogonia of the families Nostocaceae and Stigonemataceae lack heterocysts and have a different cell shape. By contrast, in other non-heterocystous, hormogonia-forming cyanobacteria, such as *Oscillatoria*, hormogonia result from simple fragmentation of the parent trichome and, although shorter in length, resemble the original trichome in all respects, including being motile or immotile. It may be that the characteristic small-celled hormogonia formed by reductive cell division, exemplified by the genus *Nostoc*, are the only sort that can infect plants, possibly because efficient infection requires the possession of motility together with chemotaxis, and probably the ability to initiate hormogonia formation in response to plant extracellular compounds (see Section 4).

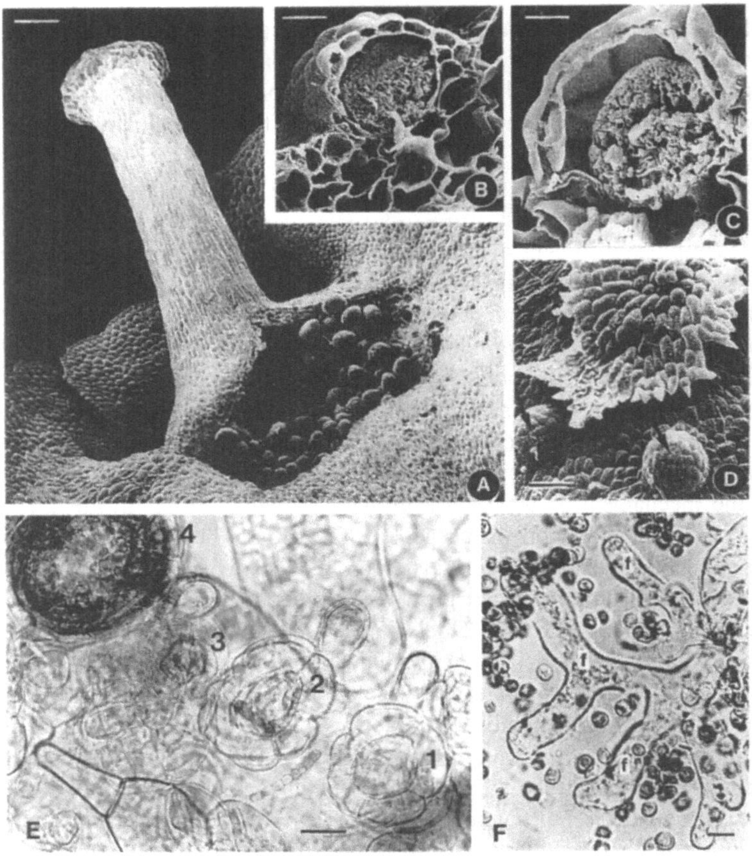

Figure 2. Cyanobacteria-liverwort symbioses. (A) - (D) Scanning electron micrographs of Blasia pusilla*. (A) Flask-shaped gemma receptacle on the ventral surface of the thallus. Individual ellipsoidal gemmae (a form of asexual propagule) can be seen where the receptacle wall has been damaged at the base. (B) and (C) Cross sections of auricles containing cyanobacterial colonies. (D) Stellate gemma (a second form of asexual propagule) on the surface of the thallus, in proximity to two auricles (arrows). (E) Light micrograph of four* Blasia pusilla *auricles (1-4). Auricle 1 is uninfected, 2 and 3 have recently become infected and contain single cyanobacterial filaments, and 4 contains a developing symbiotic colony. (F) Filamentous protrusions (f) extruded from an infected auricle, surrounded by cyanobacterial cells showing the lack of the filamentous habit that characterises the free-living state. Bars = 100 μm (A), 20 μm (B and C), 50 μm (D), 20 μm (E) and 10 μm (F). Reproduced with permission from: (A)-(D) Honnegger (1980); (E) Adams (2000); (F) Rodgers and Stewart (1977); (E) photograph courtesy S. Babic.*

The cyanobacterial symbiont in plant symbioses, including the bryophytes, is almost always a member of the genus *Nostoc*. Of 31 symbionts isolated by West and Adams (1997) from the hornwort *Phaeoceros* at several closely spaced field sites, a single strain of *Calothrix*, another hormogonia-forming member of the Nostocaceae, was identified, but the rest were *Nostoc*. The variety of strains was also compared using polymerase chain reaction (PCR) amplification of cyanobiont DNA with either arbitrary primers or primers specific for the conserved regions flanking the 16S-23S rRNA internal transcribed spacer regions (West and Adams, 1997). This showed that a wide variety of *Nostoc* strains could infect the same thallus, with symbionts in adjacent colonies on the same thallus being as likely to be different as to be the same. Some strains were found in different thalli at the same locale, but none of the symbionts was ever found among 40 free-living strains isolated from the immediate vicinity of the thalli. Nevertheless, all but one of these free-living strains was able to infect both *Phaeoceros* and *Blasia* in the laboratory. These results were later confirmed using pyrolysis mass spectrometry of the strains (West *et al.*, 1999). Broadly similar results were obtained by Costa *et al.* (2001) using the $tRNA^{Leu}$ (UAA) intron sequence to compare symbionts in *Anthoceros fusiformis* and *Blasia pusilla*. Their data confirmed that the two bryophytes were infected by many different *Nostoc* strains in the field and that an individual thallus could be infected by different strains, or by a single strain. However, they also found the same *Nostoc* strain shared by thalli growing 2,000 m apart, whereas West and Adams (1997) never found the same strain in thalli at different sites.

Although hormogonia are the infective agents in most plant-cyanobacteria symbioses, the ability to form hormogonia does not guarantee symbiotic competence. Indeed, many hormogonia-forming strains fail to establish symbioses with suitable host plants (Enderlin and Meeks, 1983; Meeks, 1990; Johansson and Bergman, 1994; Bergman *et al.*, 1996). An example is *Nostoc* ATCC 27896, which produces motile hormogonia in response to *Anthoceros* exudates, but only very rarely infects the hornwort (Campbell and Meeks, 1989). The colonies it does produce fix dinitrogen and release this to the plant, but they fail to reinfect new areas of tissue and so the infection does not persist. This may be due to an inability to sense or respond to chemotactic signals from *Anthoceros* (Meeks *et al.*, 1999). For symbionts that occupy regions of plants that receive little or no light, the ability to metabolise organic carbon sources is likely to be essential, and many cyanobionts are indeed facultative heterotrophs. Such strains are also able to commit more vegetative cells to become heterocysts, thereby reducing their ability to fix CO_2, but enhancing dinitrogen fixation capacity, and this elevated heterocyst frequency is commonly seen in the cyanobionts of bryophytes (see Section 5.2).

The formation of both endophytic and epiphytic associations requires the attachment of cyanobacteria to the host surface, and in some cyanobacteria-plant symbioses both fimbriae (pili) and extracellular polysaccharides have been implicated in the process (see Adams, 2000). However, it is not known if these are involved in the bryophyte symbioses.

3. BRYOPHYTE STRUCTURES AND THEIR INFECTION

In the bryophyte-cyanobacteria symbioses the symbionts infect existing plant structures. In the liverwort *Blasia* the cyanobacteria occupy almost spherical structures, known as auricles, on the ventral surface of the thallus (Figures 1D, 2B-E, 3A-D). By contrast, in the much thicker thallus of the hornworts *Anthoceros* and *Phaeoceros*, the cyanobacteria are found within the thallus, in slime cavities that open to the ventral surface via narrow slit-like pores (Figure 1C). The auricles of *Blasia* have two slime papillae, one of which (the inner slime papilla) partly fills the auricle itself, and the other (the outer slime papilla) arises from the thallus just outside the auricle (Figures 1D, 3A, B). The cyanobacteria enter *Blasia* auricles, and presumably hornwort slime cavities, as hormogonia (Section 2), whereupon they lose motility and differentiate heterocysts (Kimura and Nakano, 1990; Babic, 1996). In both cases the mature symbiotic colonies are 0.5-1.0 mm in diameter. Oxygen microelectrode measurements have shown that, in *Anthoceros* at least, the cavity is microaerobic and as a result, can support nitrogenase activity in a mutant of the symbiotic *Nostoc* ATCC 29133 that has a defective heterocyst envelope and is consequently unable to fix dinitrogen under aerobic conditions (Campbell and Meeks, 1992).

4. BRYOPHYTE-CYANOBACTERIUM SIGNAL EXCHANGE

As the infective agents of plant symbioses are hormogonia, a plant's chances of becoming infected will be enhanced if it can stimulate hormogonia formation in the local cyanobacterial population. Indeed, plants have been shown to produce compounds that do this. *Anthoceros punctatus* releases a low molecular mass, heat-labile product that stimulates hormogonia formation in *Nostoc* strains (Campbell and Meeks, 1989). This hormogonia-inducing factor (HIF) is not produced when the hornwort is cultured in medium containing excess NH_4^+, implying that HIF production is a result of nitrogen starvation. The importance of HIF for efficient plant infection is illustrated by a transposon mutant of the symbiotic *Nostoc* strain ATCC 29133, which shows increased sensitivity to the *Anthoceros* HIF and, as a consequence, has a much higher initial frequency of infection of the hornwort than the wild-type (Cohen *et al.*, 1994). The identity of HIF is not known although compounds with similar activity are present in *Gunnera* stem gland mucilage (Rasmussen *et al.*, 1994), wheat root exudates (Gantar *et al.*, 1993) and *Blasia* exudates (Babic, 1996; Watts *et al.*, 1999; Watts, 2001).

Because of the narrow entrances to slime cavities and auricles and their relative paucity on the thallus, efficient infection of bryophytes by hormogonia is unlikely to occur by chance; the hormogonia must be guided into the plant by chemoattraction. Indeed, there is evidence for this; when starved of combined nitrogen the liverwort *Blasia* releases extracellular signals that not only trigger hormogonia formation (Babic, 1996), but also serve as very effective chemoattractants (Knight and Adams, 1996; Watts *et al.*, 1999; Watts, 2001). In addition, the *Blasia* exudate significantly increases the speed of motility of hormogonia (Watts, 2001).

The growth of the finger-like transfer cells (Figure 2F) into symbiotic colonies in bryophytes such as *Anthoceros* and *Blasia* (see Section 5.2) may also result from the presence of signals derived from the cyanobacteria. Indeed cell division during embryogenesis in carrot is stimulated by extracts of certain cyanobacteria (Wake *et al.*, 1991, 1992). A possible source of the signals that induce such changes in the plant host may be the arabinogalactan proteins (AGPs) released by many cyanobacteria (Bergman *et al.*, 1996); such AGPs are thought to have important roles in plant growth and development (Pennell, 1992). Liverworts have also been shown to produce AGPs (Basile, 1990) and the inner and outer slime papillae of *Blasia* and the slime cavity of *Phaeoceros* stain with both Yariv reagent, which is specific for AGPs, and with anti-AGP monoclonal antibodies (Watts, 2001).

Flavonoids are another group of potential signalling molecules in cyanobacteria-plant symbioses. Flavonoids secreted by legumes are involved in the initial signalling in the symbiosis with *Rhizobium*, by binding to the transcriptional activator NodD (Fisher and Long, 1992). Seed rinse from *Gunnera*, an angiosperm that forms symbiosis with *Nostoc*, can induce expression of *nod* genes in *Rhizobium* (Rasmussen *et al.*, 1996; Bergman *et al.*, 1996; Rai *et al.*, 2000) and the flavonoid naringin induces expression of *hrmA* (see Sections 5.1 and 6.1) in *Nostoc* ATCC 29133 (Cohen and Yamasaki, 2000).

5. HOST-CYANOBIONT INTERACTIONS POST INFECTION

5.1. Cell Division Control and Hormogonia Formation

For the establishment of a stable symbiosis between a cyanobacterium and its host the growth rate of the former must match that of the host; indeed, symbiotically associated strains have doubling times considerably slower than the same free-living strains (Peters and Meeks, 1989; Hill, 1989; Braun-Howland and Nierzwicki-Bauer, 1990). For example, in *Nostoc* symbiotically associated with *Anthoceros* the doubling time can be 240 h, compared with 45 h in the free-living state (Meeks, 1990). The mechanism of this growth control is unknown, but it seems not to be nitrogen limitation, even though the host takes most of the dinitrogen fixed by its partner (see Section 5.3). It may be that the primary control is exerted on cell division of the symbiont, resulting in reduced demand for nutrients, the excess becoming available for use by the host (Hill, 1989).

The plant also appears to exert control over the biomass of its symbiotic colonies. When the *Nostoc-Anthoceros* association is grown in liquid culture, *Nostoc* can escape from existing colonies and infect new slime cavities. However, this can be prevented by the addition of penicillin which inhibits surface growth of *Nostoc*. When this is done, new colonies are prevented from developing, but existing colonies show increased size and dinitrogen fixation rate, although heterocyst frequencies do not change (Enderlin and Meeks, 1983). Addition of ammonium to the *Nostoc-Anthoceros* association leads to a decrease in dinitrogen fixation and in heterocyst frequency (Enderlin and Meeks, 1983). The rate of infection of *Anthoceros* cavities is unaffected in the presence of ammonium, but the colonies that form remain small, with low frequencies of heterocysts and a low rate of dinitrogen fixation. These observations support the idea that the metabolic nitrogen status of the *Anthoceros* regulates the size and dinitrogen-fixing capacity of the symbiotic colonies.

As well as controlling the growth rate of the cyanobiont, the host must control hormogonia formation. Having produced signals (HIF) to stimulate the development of hormogonia in potential partners (see Section 4), the plant must prevent their continued formation once infection has occurred. This is because hormogonia do not form heterocysts, and a viable symbiotic colony would not be produced if, immediately upon infection, the cyanobacteria were induced to form hormogonia once more. Evidence for a system that represses hormogonia formation has been obtained (Cohen and Meeks, 1997; Meeks, 1998) by analysis of a transposon-induced mutant of *Nostoc punctiforme* (*Nostoc* ATCC 29133) that is 50-fold more infective than the wild-type in *Anthoceros* (Cohen *et al.*, 1994). This led to the identification of two genes, *hrmA* and *hrmU*, expression of which is induced by an aqueous extract of *Anthoceros* tissue, but not by the hormogonia inducing factor (HIF) of the hornwort. The aqueous extract contains a hormogonia repressing factor (HRF) that inhibits HIF-induced hormogonia formation in wild-type *N. punctiforme*, and it seems that the mutant is more infective because it continues to respond to HIF, producing hormogonia and extending its infective state. These observations imply that the gene products of the *hrmUA* operon block hormogonium formation, perhaps by the production of an inhibitor or by catabolism of an activator (see Section 6; Cohen and Meeks, 1997).

5.2. Morphological Modifications of Bryophyte and Symbiont

In symbiosis many cyanobacteria exhibit heterocyst frequencies considerably higher than in the free-living state and this is true of the hornwort and liverwort symbioses also (Table 1; see Adams, 2000). This occurs in all cases where the host is photosynthetic, permitting the cyanobacteria to enhance their dinitrogen-fixing capabilities by increasing heterocyst numbers at the expense of vegetative cells. The resulting loss of CO_2-fixing capacity can be compensated by the supply of carbon skeletons by the host. In *Anthoceros*, and presumably all endophytic bryophyte associations, nitrogenase gene expression and heterocyst development in the symbiotically associated *Nostoc* appear to be controlled by plant signals. This conclusion is derived from studies of a *Nostoc* mutant defective in nitrate assimilation (Campbell and Meeks, 1992). In the free-living mutant strain nitrate fails to repress dinitrogen fixation and heterocyst development, but represses both when the strain is symbiotically associated with the hornwort *Anthoceros*. Ammonium is an effective repressor in both wild-type and mutant whether free-living or symbiotically associated. However, the repressive effect of nitrate in the symbiotically associated mutant does not result from the accumulation of NH_4^+, generated by the reduction of NO_3^- in *Anthoceros*, because NH_4^+ pools are the same in tissue from hornwort grown in the presence and absence of NO_3^-. In any case, ammonium *per se* does not inhibit heterocyst development. This has been clearly demonstrated by adding methione sulphoximine, an inhibitor of glutamine synthetase, to cultures of *Anabaena cylindrica* growing on medium supplemented with ammonium chloride and therefore lacking heterocysts. The inhibition of GS activity means that the ammonium can not be assimilated, resulting in nitrogen starvation and the development of heterocysts in the presence of high levels of ammonium (Stewart and Rowell, 1975).

Table 1. Summary of morphological and physiological changes in cyanobacteria symbiotically associated with hornworts and liverworts

	Hornworts	**Liverworts**
Plant structure infected	Slime cavities	Auricles
Cyanobiont	*Nostoc*	*Nostoc*[a]
Location of cyanobiont	Intercellular	Intercellular
Heterocyst frequency[b]	30-50%	30-50%
Nitrogenase specific activity	23.5	n.d.
Glutamine synthetase:		
Amount of protein[c]	~86%	n.d.
Specific activity[c]	~38%	n.d.
Form of combined N released	NH_4^+	NH_4^+
Light-dependent CO_2 fixation	12%	n.d.
Rubisco:		
Amount of protein[d]	100%	n.d.
Specific activity[d]	12%	n.d.

[a] The symbionts are *Nostoc* spp. in almost all cases; there have been rare reports of *Calothrix* spp. as symbionts (see Section 2).
[b] Heterocyst frequencies are expressed as a percentage of total cells. Typical values for free-living cyanobacteria are 4-10%.
[c] Values are for the symbiont as a percentage of the same cyanobacterium in the free-living state.
[d] Values are expressed as a percentage of those for the free-living cyanobacteria.
Table compiled from the following references: Steinberg and Meeks (1989), Meeks (1990), Rai (1990), Bergman *et al.* (1992).
n.d. = not determined, though likely to be similar to hornwort data.

The true significance of the elevated heterocyst frequencies found in bryophyte symbiotic colonies is difficult to assess for several reasons. Within the colonies heterocysts often loose their characteristic regular shape and thickened walls, making them difficult to distinguish from the vegetative cells, which are themselves enlarged and irregular in shape (Meeks, 1990). The cause of this alteration in cell shape of symbiotically associated *Nostoc* is not clear, although it has been suggested that the cells lack or partially lack the murein layer (Gorelova *et al.*, 1996). In addition to the altered cell morphology, the cell-cell junctions seem to be weakened in symbiosis so that even very gentle pressure will separate the cells from each other (Figure 2F; Meeks, 1990). The loss of the thickened heterocyst walls presumably results from the anaerobic nature of some of the symbiotic cavities (Lindblad *et al.*, 1991; Campbell and Meeks, 1992), because anaerobiosis inhibits the formation of the additional heterocyst wall layers (Rippka and Stanier, 1978). Even when heterocysts can be recognised with certainty by light or electron microscopy, their full metabolic capabilities can not be known, and at least in *Anthoceros* it seems likely that some of the heterocysts are senescent or dead (Meeks, 1990). Nevertheless, there is general correlation between the increase in heterocyst frequency of many symbiotically associated cyanobacteria and elevated rates of dinitrogen fixation.

Morphological changes are also observed in the bryophyte following infection. In both *Blasia* and *Anthoceros* branched, multicellular filaments grow from the wall of the plant and ramify within the symbiotic colony (Figure 2F), increasing the surface area of contact between the cyanobacterial colony and the bryophyte (Rodgers and Stewart, 1974, 1977; Duckett *et al.*, 1977; Renzaglia, 1982; Kimura and Nakano, 1990; Gorelova *et al.*, 1996). In *Blasia* these bryophytic filaments are derived from the inner slime papilla (Figures 1D, 3A, B) and possess transfer cell morphology, implying a modification to facilitate nutrient exchange, whereas those from *Anthoceros* lack such morphology; this difference may be related to the higher nutrient status of the field sites where the hornwort was collected (Duckett *et al.*, 1977). In *Blasia* the initial branching of the inner hair (the inner slime papilla) may close the auricle entrance (Renzaglia, 1982; Kimura and Nakano, 1990), preventing further infection.

5.3. N_2 Fixation and Transfer of Fixed Nitrogen

The rate of dinitrogen fixation in bryophyte-associated cyanobacteria is much greater than that of the same free-living strains and this broadly correlates with the increased heterocyst frequency in symbiosis (Table 1). In the *Anthoceros-Nostoc* association dinitrogen fixation is 4 to 35-fold higher than that of free-living *Nostoc* (Steinberg and Meeks, 1991). This elevated rate of dinitrogen fixation can not be supported by the lowered photosynthetic capacity of symbiotically associated cyanobacteria, and must rely on reduced carbon derived from the plant. Indeed, the rate of dark heterotrophic dinitrogen fixation in the *Anthoceros punctatus-Nostoc* association is equal to the light-dependent rate, whereas the rate of the free-living *Nostoc* is greatly reduced in the dark (Steinberg & Meeks, 1991).

In *Anthoceros* (Rodgers and Stewart, 1974; Stewart and Rodgers, 1977; Meeks *et al.*, 1985a, b) and in *Blasia* (Rodgers and Stewart, 1974; Stewart and Rodgers, 1977) dinitrogen fixed by the cyanobiont is released to the plant as ammonia (Table 1). In *Anthoceros* as little as 20% of the dinitrogen fixed is retained by the cyanobiont (Meeks *et al.*, 1985a). Initial uptake of the ammonia released by the cyanobionts in bryophytes occurs via the GS-GOGAT pathway of the host (Meeks *et al.*, 1983, 1985b; Meeks, 1990; Rai, 1990). The form in which nitrogen is transported from the symbiotic tissue to the rest of the host is not known. Ammonia release by the cyanobiont is a consequence of decreased activity of glutamine synthetase, the first enzyme in the glutamine synthetase-glutamate synthase (GS-GOGAT) pathway, which is the primary route of ammonia assimilation in cyanobacteria and in bryophytes. In *Anthoceros* the amount of GS protein in filaments of free-living and symbiotically-associated *Nostoc* is little different (Table 1), implying that the decreased GS activity is the result of an undetermined post-translational modification of the enzyme (Joseph and Meeks, 1987; Lee *et al.*, 1988; Meeks, 1990). However, Rai *et al.* (1989) found that in heterocysts of the symbiotically-associated *Nostoc*, the level of the GS protein is reduced by 50%. Indeed, it is the GS in heterocysts that is critical for assimilation of N_2-derived ammonia, and the drastic reduction in GS levels in heterocysts may explain the ammonia release by cyanobionts.

Despite supplying the host with most of the dinitrogen it fixes, the cyanobiont appears not to suffer from nitrogen deprivation, and indeed retains two nitrogen reserve materials that would be expected to be broken down in nitrogen starvation. These are the phycobiliproteins, which serve as both photosynthetic accessory pigments and nitrogen reserves in cyanobacteria and are still present in the cyanobionts of bryophytes (Rai *et al.*, 1989; Meeks, 1990), and the unique cyanobacterial nitrogen reserve, cyanophycin, a co-polymer of arginine and aspartic acid (Simon, 1987), which is also present in the cyanobionts of bryophytes (Honegger, 1980; Rai *et al.*, 1989; Meeks, 1990).

5.4. CO_2 assimilation and Transfer of Carbon

The primary route of CO_2 fixation in free-living and symbiotically-associated cyanobacteria is the Calvin cycle, with ribulose-1, 5-bisphosphate carboxylase/oxygenase (Rubisco) as the primary carboxylating enzyme (Tabita, 1994). Immediately after its separation from symbiosis, the *Nostoc* symbiont of the hornwort *Anthoceros* has a rate of light-dependent CO_2 fixation eight-fold lower than the same cyanobacterium grown in the free-living state (Table 1; Steinberg and Meeks, 1989; Meeks, 1990). The level of Rubisco protein is little different in the two cases (Steinberg and Meeks, 1989; Rai *et al.*, 1989; Meeks, 1990), implying that activity is regulated by post translational modification of the enzyme by an unknown mechanism (Steinberg and Meeks, 1989; Meeks, 1990). The cyanobiont therefore grows photoheterotrophically, receiving fixed carbon from its photosynthetic host, probably in the form of sucrose (Stewart and Rodgers, 1977; Steinberg and Meeks, 1991). In at least *Anthoceros* the symbiotically associated *Nostoc* appears not to be starved of carbon as the cells contain glycogen granules that would be degraded during a period of carbon starvation (Meeks, 1990).

6. GENETIC ANALYSIS OF THE *NOSTOC-ANTHOCEROS* ASSOCIATION

In recent years genetic techniques, including transposon mutagenesis, have been developed for the analysis of the symbiotically competent cyanobacterium *Nostoc punctiforme* strain ATCC 29133 (Cohen *et al.*, 1994, 1998). These techniques are an important advance in our ability to dissect the symbiotic interactions in the cyanobacteria-bryophyte associations and they have been used by Meeks and co-workers to identify a number of genes involved in the initial infection of *Anthoceros*.

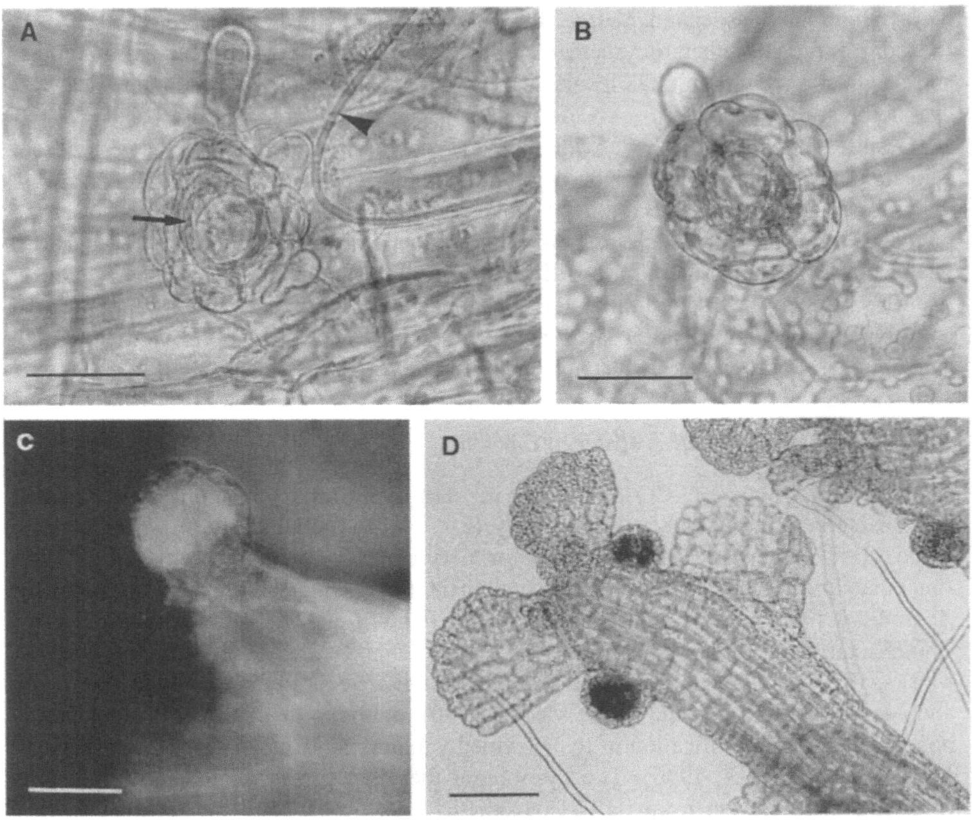

Figure 3. Photomicrographs of Blasia pusilla *auricles infected with cyanobacteria in the laboratory. (A) This auricle has just been infected by a hormogonium which is still motile (and hence out of focus) within the space (arrow) between the inner papilla and the auricle wall. Another motile hormogonium is present just outside the auricle (arrowhead). (B) An auricle several days after infection, showing a symbiotic colony beginning to develop. (C) Fluorescence image of a mature symbiotic colony showing its autofluorescence. (D) Part of a* Blasia pusilla *thallus infected in the laboratory with two different cyanobacteria. The two darker colonies in the middle contain a different* Nostoc *sp. to the paler one to the right. Bars = 30 μm (A and B), 60 μm (C), and 120 μm (D). All photographs courtesy of S. Babic. (C) reproduced with permission from Adams (2000. (B) and (D) reproduced with permission from Adams and Duggan (1999).*

Table 2. Effect of insertion mutations on the symbiotic infectiveness and effectiveness of Nostoc *ATCC 29133 strains in association with* Anthoceros punctatus

Strain	Colonies / mg dry wt / µg Chl a	nmol C_2H_2 reduced / min /: g fresh weight	nmol C_2H_2 reduced / min /: Colony x 10^{-3}	Gene induced by:
ATCC 29133 (WT)	0.21	6.3	12.4	—
UCD 328 (*hrmA*)	1.6	6.1	8.6	HRF
UCD 398 (*sigH*)	1.2	8.0	10.1	HIF
UCD 400 (*tprN*)	0.49	10.4	6.7	HIF & HRF

Table taken from Meeks *et al.* (1999). The original figures for standard deviation and number of replicates have been omitted for simplicity.
HRF = aqueous extract of *A. punctatus* containing hormogonium repressing factor identified as inducing the *hrm* operon; HIF = exudate of *A. punctatus* containing hormogonium inducing factor.

6.1. The hrm *Operon*

Transposon mutagenesis has been used to generate *Nostoc* ATCC 29133 mutants with an increased rate of initial infection of *Anthoceros* (Cohen *et al.*, 1994; Meeks *et al.*, 1999). *Nostoc* ATCC 29133 transposon mutant strain UCD 328 forms a high frequency of hormogonia in the presence of *A. punctatus* and also shows an almost eight-fold increase in the initial infection frequency of the hornwort (Table 2; Cohen and Meeks, 1997). Two open reading frames, *hrmU* and *hrmA*, were identified flanking the site of transposition (Figure 4; Meeks *et al.*, 1999). *hrmA* has no significant similarity to sequences in major databases, whereas *hrmU* has similarity to a family of NAD(P)H-oxidoreductases. Expression of *hrmUA* is induced by an aqueous extract of *A. punctatus*, but not by the hormogonium inducing factor, HIF. The aqueous extract also suppresses HIF-induced hormogonia formation in the wild-type, but not in the mutant, and therefore appears to contain a hormogonium repressing factor (HRF). Unlike HIF, HRF is not released into the growth medium, but may be released into the symbiotic cavity, ensuring that, once the cavity is infected, further hormogonium formation is suppressed, permitting the differentiation of heterocysts.

Three further open reading frames (ORFs), *hrmI*, *hrmR* and *orfK*, were subsequently identified 5' of *hrmUA* (Figure 4). These all have similarity to genes encoding proteins involved in the metabolism of gluconate and glucuronate; the significance of this in terms of hormogonium formation and symbiotic infection is not known.

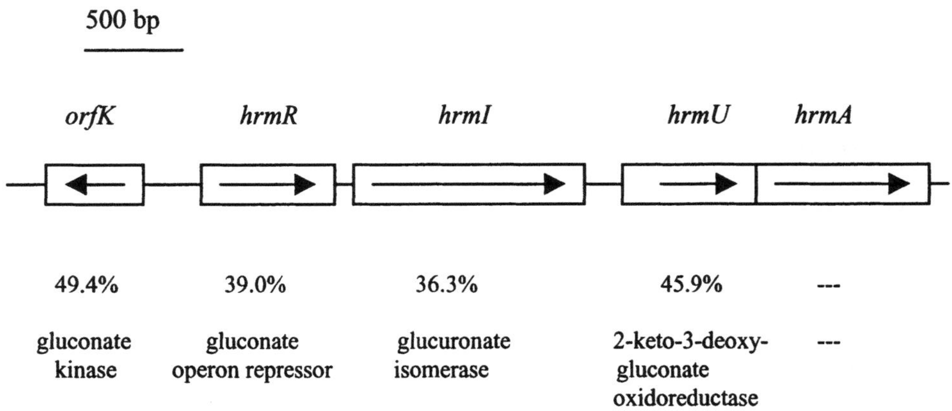

Figure 4. Map of the hrm operon, showing the percentage similarity values to amino acid sequences from BLAST analysis of double stranded nucleotide sequences. Beneath the percentage similarity values are the names of the proteins showing the greatest similarity. There are no significant sequence similarities to hrmA. *The direction of transcription is indicated by the arrows. Figure redrawn from Meeks* et al. *(1999).*

6.2. sigH *and* ctpH

The gene *sigH* encodes an alternative sigma subunit of RNA polymerase and was identified in *Nostoc* ATCC 29133 using the *Anabaena* PCC 7120 *sigB* gene as a heterologous probe (Campbell *et al.*, 1998; Meeks *et al.*, 1999). Mutant strain UCD 398, generated by the interruption of *sigH* with an antibiotic resistance cassette, has no obvious phenotype when grown in medium with or without combined nitrogen, but has a high infection phenotype (Table 2) when co-cultured with *A. punctatus*. Transcription of *sigH* is induced by *Anthoceros* HIF, but not by HRF and *hrmA* transcription is not altered in a *sigH*$^-$ mutant. Thus, although the *hrmA*$^-$ and *sigH*$^-$ mutants both have an increased infection phenotype, it seems likely that these have a different basis in the two strains (Meeks *et al.*, 1999).

Immediately 5' of *sigH* is the gene *ctpH* which encodes a protein with significant similarity to carboxy-terminal proteases of the cyanobacterium *Synechocystis* PCC 6803 (Meeks *et al.*, 1999). *Synechocystis* 6803 *ctpH* is required for processing the carboxy-terminal portion of the photosystem II D1 protein in the thylakoid lumen (Anbudurai *et al.*, 1994). However, in *Nostoc* ATCC 29133 *ctpH* seems to have a different physiological role because it is not transcribed under vegetative growth conditions, but transcription is induced by *Anthoceros* HIF. The significance of this is not understood.

6.3. tprN

The gene *tprN* encodes a protein with similarity to tetratricopeptide repeat proteins (TPR) and lies 3' of the gene *devR*, expression of which is required for heterocyst maturation (Campbell *et al.*, 1996). TPR proteins have been studied primarily in eukaryotes in which they are required for a variety of functions from cell cycle control to transcription repression and protein transport (Lamb *et al.*, 1995). Inactivation of *tprN* in *Nostoc* ATCC 29133 (producing mutant strain UCD 400) has no apparent phenotypic effect in the free-living growth state, but the mutant infects *Anthoceros* at about twice the level of the wild-type (Table 2). The gene is transcribed during vegetative growth, but the level of transcription increases following exposure to both HIF and HRF (Meeks *et al.*, 1999). The significance of this is the infection process is not known.

6.4. ntcA, hetR *and* hetF

Nostoc punctiforme (*Nostoc* ATCC 29133) strains with mutations in any of the three genes *ntcA*, *hetR* and *hetF* are unable to differentiate heterocysts. In cyanobacteria, NtcA functions as a nitrogen-dependent global regulator and controls the transcription of, among other genes, several involved in heterocyst development, including *hetR* (Herrero *et al.*, 2001; Fiedler *et al.*, 2001). *hetR* is the first heterocyst-specific gene to be expressed following combined nitrogen deprivation, and is thought to be the primary activator of heterocyst development (Wolk, 2000). HetR protein specifically accumulates in developing heterocysts in wild-type filaments of *Nostoc punctiforme*, but accumulates in all cells in a *hetF* mutant, which is unable to develop heterocysts (Wong and Meeks, 2001). The HetF protein therefore seems to be a positive activator of heterocyst differentiation, enhancing transcription of *hetR* and ensuring the localisation of HetR to developing heterocysts.

The *hetR* and *hetF* mutants (strains UCD 415 and UCD 416, respectively) both infect *Anthoceros punctatus* with a similar frequency to the wild-type, but are unable to support growth of the plant because of their inability to develop heterocysts and fix dinitrogen (Wong and Meeks, 2002). This clearly demonstrates that the ability to form a functional dinitrogen-fixing association is not essential for infection of the plant to occur. The *ntcA* mutant (strain UCD 444) forms hormogonia at 5-15% of the wild-type frequency and the hormogonia are motile and show similar morphology to those produced by the wild-type. However, rather than infecting *Anthoceros* at a reduced frequency, as might be expected, the *ntcA* mutant fails to infect at all. This non-infective phenotype can be complemented with copies of *ntcA*.

7. CONCLUDING REMARKS

The cyanobacteria-bryophyte associations provide an excellent experimental system. The hornworts *Anthoceros* and *Phaeoceros* and the liverwort *Blasia*, grow well in cyanobacterial liquid growth medium BG-11 (Rippka *et al.*, 1979) in shaken culture (Figure 1B). The plants can be freed of their symbionts and grown in medium supplemented with nitrate, and can be re-infected with the original or with novel cyanobacteria. The progress of the infection can be followed microscopically from the earliest stage, particularly in the liverwort *Blasia* in which the auricles are especially easy to visualise (Figures 1D, 2E, 3A-D). Recent developments in genetic analysis of the symbiosis (see Section 6) offer the hope of significant advances in our understanding of the cyanobacteria-bryophyte associations, which will in turn enhance our understanding of other cyanobacteria-plant symbioses. These techniques should also help to elucidate one of the most interesting aspects of these associations - the signalling between symbiont and host, including the identity of the chemical signals and of the genes they regulate, which are presently unknown.

REFERENCES

Adams, D.G. (1997) Cyanobacteria, in J. Shapiro and M. Dworkin (eds.), *Bacteria as Multicellular Organisms*, Oxford University Press, New York, pp 109-148.

Adams, D.G. (2000) Symbiotic interactions, in B. Whitton and M. Potts (eds.), *Ecology of Cyanobacteria: Their Diversity in Time and Space*, Kluwer Academic Publishers, Dordrecht, pp 523-561.

Adams, D.G. (2001) How do cyanobacteria glide? *Microbiology Today* **28**, 131-133.

Adams, D. G. and Duggan, P. (1999) Heterocyst and akinete differentiation in cyanobacteria. Tansley Review No. 107, *New Phytol* **144**, 3-33.

Anbudurai, P.R., Mor, T.S., Ohad, I., Shestakov, S.V. and Pakrasi, H.B. (1994) The *ctpA* Gene Encodes the C-Terminal Processing Protease for the D1 Protein of the Photosystem II Reaction Center Complex, *Proc Nat Acad Sci USA* **91**, 8082-8086.

Babic, S. (1996) Hormogonia formation and the establishment of symbiotic associations between cyanobacteria and the bryophytes *Blasia* and *Phaeoceros*, Ph.D. thesis, University of Leeds, England.

Basile, D.V. (1990) Morphological role of hydroxyproline containing proteins in liverworts, in R.N. Chopra and S.C. Bhatia (eds.), *Bryophyte Development: Physiology and Biochemitry*, CRC Press, Boca Raton, Florida, pp 225-243.

Bergman, B., Rai, A.N., Johansson, C. and Söderbäck, E. (1992) Cyanobacterial-plant symbioses, *Symbiosis* **14**, 61-81.

Bergman B., Matveyev, A. and Rasmussen, U. (1996) Chemical signalling in cyanobacterial-plant symbioses, *Trends Plant Sci* **1**, 191-197.

Brasell, H.M., Davies, S.K. and Mattay, J.P. (1986) Nitrogen fixation associated with bryophytes colonizing burnt sites in Southern Tasmania, Australia, *J Bryol* **14**, 139-149.

Braun-Howland, E.B. and Nierzwicki-Bauer, S.A. (1990) *Azolla-Anabaena* symbiosis: biochemistry, physiology, ultrastructure, and molecular biology, in A.N. Rai (ed.), *CRC Handbook of Symbiotic Cyanobacteria*, CRC Press, Boca Raton, Florida, pp 65-117.

Campbell, E.L. and Meeks, J.C. (1989) Characteristics of hormogonia formation by symbiotic *Nostoc* spp. in response to the presence of *Anthoceros punctatus* or its extracellular products, *Appl Environ Microbiol* **55**, 125-131.

Campbell, E.L. and Meeks, J.C. (1992) Evidence for plant-mediated regulation of nitrogenase expression in the *Anthoceros-Nostoc* symbiotic association, *J Gen Microbiol* **138**, 473-480.

Campbell, E.L., Hagen, K.D., Cohen, M.F., Summers, M.L. and Meeks, J.C. (1996) The *devR* gene product is characteristic of receivers of two-component regulatory systems and is essential for heterocyst development in the filamentous cyanobacterium *Nostoc* sp. strain ATCC 29133, *J Bacteriol* **178**, 2037-2043.

Campbell, E.L., Brahamsha, B. and Meeks, J.C. (1998) Mutation of an alternative sigma factor in the cyanobacterium *Nostoc punctiforme* results in increased infection of its symbiotic plant partner, *Anthoceros punctatus, J Bacteriol* **180**, 4938-4941.

Cohen, M.F. and Meeks, J.C. (1997) A hormogonium regulating locus, *hrmUA*, of the cyanobacterium *Nostoc punctiforme* strain ATCC 29133 and its response to an extract of a symbiotic plant partner *Anthoceros punctatus, Mol. Plant- Microbe Interactions* **10**, 280-289.

Cohen, M.F. and Yamasaki, H. (2000) Flavonoid-induced expression of a symbiosis-related gene in the cyanobacterium *Nostoc punctiforme, J Bacteriol* **182**, 4644-4646.

Cohen, M.F., Wallis, J.G., Campbell, E.L. and Meeks, J.C. (1994) Transposon mutagenesis of *Nostoc* sp. strain ATCC 29133, a filamentous cyanobacterium with multiple cellular differentiation alternatives, *Microbiol* **140**, 3233-3240.

Cohen, M.F., Meeks, J.C., Cai, Y. and Wolk, C.P. (1998) Transposon mutagenesis of heterocyst-forming filamentous cyanobacteria, *Methods Enzymol* **297**, 3-17.

Costa, J.-L., Paulsrud, P., Rikkinen, J. and Lindblad, P. (2001) Genetic diversity of *Nostoc* symbionts endophytically associated with two bryophyte species, *Appl Environ Microbiol* **67**, 4393-4396.

Dalton, D.A. and Chatfield, J.M. (1985) A new nitrogen-fixing cyanophyte-hepatic association: *Nostoc* and *Porella, Am J Bot* **72**, 781-784.

Duckett, J.G., Prasad, A.K.S.K., Davies, D.A. and Walker, S. (1977) A cytological analysis of the *Nostoc*-bryophyte relationship, *New Phytol* **79**, 349-362.

Enderlin, C.S. and Meeks, J.C. (1983) Pure culture and reconstitution of the *Anthoceros-Nostoc* symbiotic association, *Planta* **158**, 157-165.

Fiedler, G., Muro-Pastor, A., Flores, E. and Maldener, I. (2001) NtcA-dependent expression of the *devBCA* operon, encoding a heterocyst-specific ATP-binding cassette transporter in *Anabaena* spp. *J Bacteriol* **183**, 3795-3799.

Fisher, R.F. and Long, S.R. (1992) *Rhizobium*-plant signal exchange, *Nature* **357**, 655-660.

Gantar, M., Kerby, N.W. and Rowell, P. (1993) Colonization of wheat (*Triticum vulgare* L.) by N_2-fixing cyanobacteria: III. The role of a hormogonia-promoting factor, *New Phytol* **124**, 505-513.

Gorelova, O.A., Baulina, O.I., Shchelmanova, A.G., Korzhenevskaya, T.G. and Gusev, M.V. (1996) Heteromorphism of the cyanobacterium *Nostoc* sp., a microsymbiont of the *Blasia pusilla* moss, *Microbiology* (translation of *Mikrobiologiya*) **65**, 719-726.

Herdman, M. and Rippka, R. (1988) Cellular differentiation: hormogonia and baeocytes, *Methods Enzymol* **167**, 232-242.

Herrero, A., Muro-Pastor, A.M. and Flores, E. (2001) Nitrogen control in cyanobacteria, *J Bacteriol* **183**, 411-425.

Hill, D.J. (1989) The control of the cell cycle in microbial symbionts, *New Phytol* **112**, 175-184.

Honegger, R. (1980) Cytology of the cyanophyte-hornwort-symbiosis in Icelandic *Anthoceros laevis, Flora* **170**, 290-302.

Johansson, C. and Bergman, B. (1994) Reconstruction of the symbiosis of *Gunnera manicata* Linden: cyanobacterial specificity, *New Phytol* **126**, 643-652.

Joseph, C.M. and Meeks, J.C. (1987) Regulation of expression of glutamine synthetase in a symbiotic *Nostoc* strain associated with *Anthoceros punctatus, J Bacteriol* **169**, 2471-2475.

Kimura, J. and Nakano, T. (1990) Reconstitution of a *Blasia-Nostoc* symbiotic association under axenic conditions, *Nova Hedwigia* **50**, 191-200.

Knight, C.D. and Adams, D.G. (1996) A method for studying chemotaxis in nitrogen-fixing cyanobacterium-plant symbioses, *Physiol Molec Plant Pathol* **49**, 73-77.

Lamb, J.R., Tugendreich, S. and Hieter, P. (1995) Tetratrico peptide repeat interactions: to TPR or not to TPR? *Trends in Biochemical Sciences* **20**, 257-259.

Lee, K.Y., Joseph, C.M. and Meeks, J.C. (1988) Glutamine synthetase specific activity and protein concentration in symbiotic *Anabaena* associated with *Azolla caroliniana, Antonie van Leeuwenhoek* **54**, 345-355.

Lindblad, P., Atkins, C.A. and Pate, J.S. (1991) N_2-fixation by freshly isolated *Nostoc* from coralloid roots of the cycad *Macrozamia riedlei* (Fisch. ex Gaud.) Gardn., *Plant Physiol* **95**, 753-759.

Meeks, J.C. (1988) Symbiotic associations, *Meth Enzymol* **167**, 113-121.

Meeks, J.C. (1990) Cyanobacterial-bryophyte associations, in A.N. Rai (ed.), *CRC Handbook of Symbiotic Cyanobacteria*, CRC Press, Boca Raton, Florida, pp 43-63.

Meeks, J.C. (1998) Symbiosis between nitrogen-fixing cyanobacteria and plants, *Bioscience* **48**, 266-276.

Meeks, J.C., Enderlin, C.S., Wycoff, K.L., Chapman, J.S. and Joseph, C.M. (1983) Assimilation of $^{13}NH_4^+$ by *Anthoceros* grown with and without symbiotic *Nostoc, Planta* **158**, 384-391.

Meeks, J.C., Enderlin, C.S., Joesph, C.M., Chapman, J.S. and Lollar, M.W.L. (1985a) Fixation of $[^{13}N]N_2$ and transfer of fixed nitrogen in the *Anthoceros-Nostoc* symbiotic association, *Planta* **164**, 406-414.

Meeks, J.C., Steinberg, N., Joseph, C.M., Enderlin, C.S., Jorgensen, P.A. and Peters, G.A. (1985b) Assimilation of exogenous and dinitrogen-derived $^{13}NH_4^+$ by *Anabaena azollae* separated from *Azolla caroliniana* Willd., *Arch Microbiol* **142**, 229-233.

Meeks, J.C., Campbell, E., Hagen, K., Hanson, T., Hitzeman, N. and Wong, F. (1999). Developmental alternatives of symbiotic *Nostoc punctiforme* in response to its symbiotic partner *Anthoceros punctatus,* in G.A. Peschek, W. Löffelhardt and G. Schmetterer (eds.), *The Phototrophic Prokaryotes*, Kluwer Academic/Plenum Publishers, New York, pp 665-678.

Pennell, R.I. (1992) Cell surface arabinogalactan proteins, arabinogalactans and plant development, in J.A. Collow and J.R. Green (eds.), *Perspectives in Plant Cell Recognition*, Cambridge University Press, Cambridge, pp 105-121.

Peters, G.A. and Meeks, J.C. (1989) The *Azolla-Anabaena* symbiosis: basic biology, Annu Rev Plant Physiol **40**, 193-210

Rai, AN (1990) Cyanobacteria in Symbiosis, in A.N. Rai (ed.), *CRC Handbook of Symbiotic Cyanobacteria*, CRC Press, Boca Raton, Florida, pp 1-7.

Rai, A.N., Borthakur, M., Singh, S. and Bergman, B. (1989) *Anthoceros-Nostoc* symbiosis: immunoelectronmicroscopic localization of nitrogenase, glutamine synthetase, phycoerythrin and ribulose-1,5-bisphosphate carboxylase/oxygenase in the cyanobiont and the cultured (free-living) isolate *Nostoc* 7801, *J Gen Microbiol* **135**, 385-395.

Rai, A.N., Söderbäck, E. and Bergman, B. (2000) Tansley Review No. 116, Cyanobacterium-plant symbioses, *New Phytol* **147**, 449-481.

Rasmussen, U., Johansson, C., and Bergman, B. (1994) Early communication in the *Gunnera-Nostoc* symbiosis: Plant-induced cell differentiation and protein synthesis in the cyanobacterium, *Molecular Plant-Microbe Interactions* **7**, 696-702.

Rasmussen, U., Johansson, C., Renglin, A., Petersson, C. and Bergman, B. (1996) A molecular characterization of the *Gunnera-Nostoc* symbiosis: comparison with *Rhizobium-* and *Agrobacterium-*plant interactions, *New Phytol* **133**, 391-398.

Renzaglia, K.S. (1982) A comparative developmental investigation of the gametophyte generation in the Metzgeriales (Hepatophyta), *Bryophytorum Bibliotheca* **24**, 1-238.

Ridgway, J.E. (1967) The biotic relationship of *Anthoceros* and *Phaeoceros* to certain cyanophyta, *Ann Missouri Bot Gdn* **54**, 95-102.

Rippka, R. and Stanier, R.Y. (1978) The effects of anaerobiosis on nitrogenase synthesis and heterocyst development by Nostocacean cyanobacteria, *J Gen Microbiol* **105**, 83-94.

Rippka, R., Deruelles, J., Waterbury, J.B., Herdman, M. and Stanier, R.Y. (1979) Generic assignments, strain histories and properties of pure cultures of cyanobacteria, *J Gen Microbiol* **111**, 1-61.

Rodgers, G.A. and Stewart, W.D.P. (1974) Physiological interrelations of the blue-green alga *Nostoc* with the liverworts *Anthoceros* and *Blasia*, *Brit Phycol J* **9**, 223.

Rodgers, G.A. and Stewart, W.D.P. (1977) The cyanophyte-hepatic symbiosis. I. Morphology and physiology, New Phytol **78**, 441-458.

Simon, R.D. (1987) Inclusion bodies in the cyanobacteria, in P. Fay and C. Van Baalen (eds.), *The Cyanobacteria*, Elsevier Science Publishers, Amsterdam, pp 199-225.

Steinberg, N.A. and Meeks, J.C. (1989) Photosynthetic CO_2 fixation and ribulose bisphosphate carboxylase/oxygenase activity of *Nostoc* sp. strain UCD 7801 in symbiotic association with *Anthoceros punctatus, J Bacteriol* **171**, 6227-6233.

Steinberg, N.A. and Meeks, J.C. (1991) Physiological sources of reductant for nitrogen fixation activity in *Nostoc* sp. strain UCD 7801 in symbiotic association with *Anthoceros punctatus, J Bacteriol* **173**, 7324-7329.

Stewart, W.D.P. and Rogers, G.A. (1977) The cyanophyte-hepatic symbiosis. II. Nitrogen fixation and the interchange of nitrogen and carbon, *New Phytol* **78**, 459-471.

Stewart, W.D.P. and Rowell, P. (1975) Effects of L-methionine-DL-sulphoximine on the assimilation of newly fixed NH_3, acetylene reduction and heterocyst production in *Anabaena cylindrica, Biochem Biophys Res Commun* **65**, 846-856.

Tabita, F.R. (1994) The biochemistry and molecular regulation of carbon dioxide metabolism in cyanobacteria, in D.A. Bryant (ed.), *The Molecular Biology of Cyanobacteria*, Kluwer Academic Publishers, Dordrecht, pp 437-467.

Tandeau de Marsac, N. (1994) Differentiation of hormogonia and relationships with other biological processes, in D.A. Bryant (ed.), *The Molecular Biology of Cyanobacteria*, Kluwer Academic Publishers, Amsterdam, pp 825-842.

Wake, H., Umetsu, H., Ozeki, Y., Shimomura, K. and Matsunaga, T. (1991) Extracts of marine cyanobacteria stimulated somatic embryogenesis of *Daucus carota* L., *Plant Cell Rep* **9**, 655-658.

Wake, H., Akasaka, A., Umetsu, H., Ozeki, Y., Shimomura, K. and Matsunaga, T. (1992) Promotion of plantlet formation from somatic embryos of carrot treated with a high molecular weight extract from a marine cyanobacterium, *Plant Cell Rep* **11**, 62-65.

Watts, S. (2001) Signalling in the *Nostoc*-plant symbiosis, PhD thesis, University of Leeds, UK.

Watts, S.D., Knight, C.D. and Adams, D.G. (1999) Characterisation of plant exudates inducing chemotaxis in nitrogen-fixing cyanobacteria, in G.A. Peschek, W. Löffelhardt and G. Schmetterer (eds.), *The Phototrophic Prokaryotes*, Kluwer Academic/Plenum Publishers, New York, pp 679-684.

West, N. and Adams, D.G. (1997) Phenotypic and genotypic comparison of symbiotic and free-living cyanobacteria from a single field site, *Appl Environ Microbiol* **63**, 4479-4484.

West, N.J., Adams, D.G., Sisson, P.R., Freeman, R. and Hawkey, P.M. (1999). Pyrolysis mass spectrometry analysis of free-living and symbiotic cyanobacteria, *Antonie van Leeuwenhoek* **75**, 201-206.

Wolk, C.P. (2000) Heterocyst formation in *Anabaena*, in Y.V. Brun and L.J. Shimkets (eds.), *Prokaryotic Development*, American Society for Microbiology, Washington, DC, pp 83-104.

Wong, F.C.Y. and Meeks, J.C. (2001) The *hetF* gene product is essential to heterocyst differentiation and affects HetR function in the cyanobacterium *Nostoc punctiforme*, *J Bacteriol* **183**, 2654-2661.

Wong, F.C.Y. and Meeks, J.C. (2002) Establishment of a functional symbiosis between the cyanobacterium *Nostoc punctiforme* and the bryophyte *Anthoceros punctatus* requires genes involved in nitrogen control and initiation of heterocyst differentiation, *Microbiology* **148**, 315-323.

Chapter 8

ASSOCIATIONS BETWEEN CYANOBACTERIA AND MOSSES

BJØRN SOLHEIM[1] AND MATTHIAS ZIELKE[1, 2]
[1]*Department of Biology, University of Tromsø, N-9037 Tromsø, Norway*
[2]*The University Courses on Svalbard–UNIS, N-9171 Longyearbyen, Norway*

1. INTRODUCTION

Cyanobacteria can form associations with species from most groups of Bryophytes. The Bryophytes contains three classes:

Hepatophyta (liverworts)
Anthocerophyta (hornworts)
Bryophyta (mosses or musci)

In the first two classes, Hepatophyta (liverworts) and Anthocerophyta (hornworts), true symbiotic associations between the bryophytes and cyanobacteria are found. These symbioses are described in the previous chapter. The class *Bryophyta* (mosses or musci), is divided in three sub-classes:

Sphagnidae (*Sphagnum* or peat mosses)
Andreaeidae (granite or rock mosses)
Bryidae (true mosses)

Associations between members of Bryophyta (mosses or musci) and cyanobacteria are found, but special symbiotic structures comparable to those in liverworts and hornworts have never been described. The associations probably range from relationships that are simply fortuitous, through more or less specific epiphytic associations, to intracellular colonisation of dead moss cells by the cyanobacteria. We know very little about how the associations are formed, how the cyanobacteria infect or colonise the mosses, if special genes are involved, or if nutrients are exchanged between the two partners. In this chapter we will use the terms epiphytic associations and intracellular colonisation to describe associations between cyanobacteria and mosses.

Except for a few studies, our knowledge about cyanobacteria-moss associations is based on work from polar or sub-polar environments. In the more extreme terrestrial polar habitats, cyanobacteria constitute not only the dominant phototrophs and nitrogen fixers, but also most of the microbial ecosystem biomass (Vincent, 2000). In the areas covered by vegetation, cyanobacteria in association with mosses are the main source of fixed nitrogen and are important for the productivity of the ecosystem. Cyanobacteria-

A.N. Rai, B. Bergman and U. Rasmussen (eds.), Cyanobacteria in Symbiosis, 137-152.

moss associations have the same ecological role in polar habitats as the nitrogen fixing symbioses have in more temperate and warm habitats.

2. FORMATION AND CYTOLOGY OF THE ASSOCIATION

2.1 Dispersal of Cyanobacteria to Mosses

Wind dispersal of terrestrial algae at Signy Island was studied by Broady (1979b) from March to December in 1973. Dispersal of algae and plant debris were found at all sampling sites lacking snow cover in the lowland during the summer months. Broady concluded that the quantities of terrestrial algae in the air at Signy Island were large enough to distribute most taxa to all available habitats on the island and thus provide the opportunity for colonisation. By using spore traps and culturing, Marshall and Chamers (1997) found cyanobacterial propagules produced from diverse vegetation with mosses and lichens at the end of the growing season when the vegetation was drying up. They also found propagules in air mass from South America and suggested that long-distance transport potentially provided inocula for colonisation of Antarctica as regional warming continued to expose fresh habitats. Gordon *et al.* (2000) studied the phylogenetic diversity of bacteria and cyanobacteria colonising sediment particles in permanent ice cover of an Antarctic lake by analyses of 16S rRNA genes amplified from environmental DNA. By comparing with 16S rRNA genes from terrestrial sites in the area they found that the lake ice microbial community appeared to be dominated by organisms originating from the surrounding region. In addition to dispersal by wind, dispersal also takes place by means of melt-water streams (Broady, 1977), in collembolan guts and faeces (Broady, 1979c; Birkemoe and Liengen, 2000), and by larger animals and humans. On the newly formed volcanic island Surtsey in Island most cayanobacterial propagules were introduced by wind dispersal from the mainland, with the exception of, *Nodularia* sp. (Schwabe, 1974). While the other Nostocacean cyanobacteria were found associated with mosses all over the island, *Nodularia* was found only at the resting sites for birds. Schwabe (1974) concluded that the dispersal of *Nodularia* was only by birds.

2.2. Establishing the Association

The taxonomy of algae growing as epiphytes on mosses in Antarctica has been studied in great detail (Broady, 1987; Ohtani and Kanda, 1987; Fumanti *et al.*, 1997; Alfinito *et al.*, 1998). Only a limited number of cyanobacterial taxa were found and of these only two, *Phormidium frigidum* Fritsch and *Nostoc commune* Vaucher were abundant; frequently, *Microcystis parasitica* Kützing was also found (Alfinito *et al.*, 1998). Jordan *et al.* (1978) also found *Nostoc* sp. to be dominant on leaves of mosses in the Arctic, although *Anabena* sp. and *Oscillatoria* sp. were also observed. For a cyanobacterium to be a true epiphyte it must be able to colonise new leaf areas formed by seasonal growth of its host plant, either by passive transport or by active movement. All types of cyanobacteria can be transported passively, mainly by water, while only a few have developed motile stages in their life cycle. Chemotaxis and active infection of symbiotic structures in host plants by cyanobacteria are discussed elsewhere (Rai *et al.*, 2000; see also the previous chapter and the chapter on *Gunnera*). An uneven distribution of epiphytic cyanobacteria (Broady, 1979a; Smith, 1984) and nitrogen fixation activity (Smith, 1984) were found along moss plants. The same pattern was found for the intracellular colonisation of mosses by cyanobacteria (Granhall and v. Hofsten, 1976; Basilier *et al.*, 1978). In all cases, highest number of cyanobacteria were found in the living green part of the moss and a lower number at the apex of the plant. In a thorough study on Signy Island, Broady (1979a) followed the vertical micro-distribution of phototrophs, including three species of cyanobacteria, along moss plants over three years. He made a model for growth and life cycle of *Nostoc muscorum* on moss shoots. At the beginning of the summer, thalli with heterocysts were located near the previous year's shoot apices. When the moss started to grow the trichomes fragmented and motile hormogones were released and moved up the growing moss shot. The hormogones reached the site of the old terminal bud and then produced colonies with heterocysts. The new bud had few or no cyanobacteria. In this way the cyanobacteria colonise new leaf areas of the host plant the year after it has been formed. In the older part of the moss, both moss tissue and cyanobacterial trichomes died and released nutrients to the environment. Broady (1979a) concluded that *N. muscorum* possessed a life cycle with a motile stage that enabled it to maintain a population in the regions with higher light intensity. Fragmentation and hormogonia formation was possibly the result of a developmental stimulus of the higher summer light levels after snowmelt.

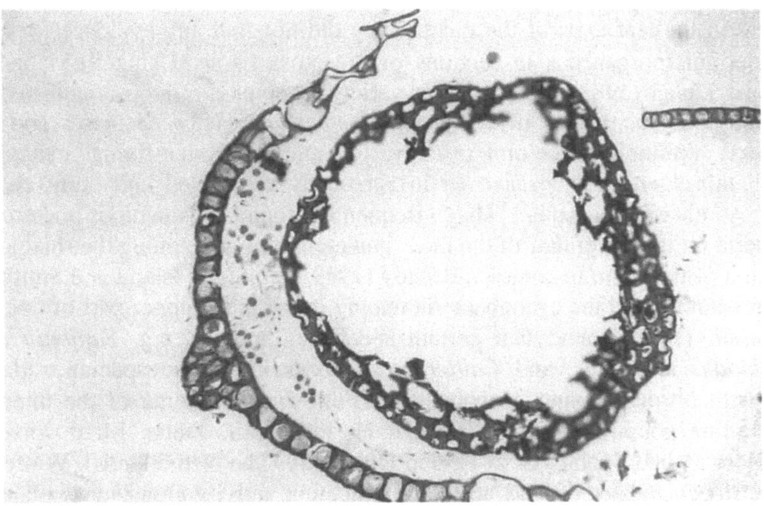

Figure 1. Section perpendicular to the stem of Calliergon richardsonii. *Cyanobacteria are concentrated between the stem and the leaf. The central part of the stem is missing due to inadequate infiltration of the embedding material. Sample fixed and embedded in Epon-Araldite, sectioned and stained with toluidine blue. Light microscopy, 150x (H. Vigstad, unpublished).*

Granhall and v. Hofsten (1976) and Granhall and Selander (1973) found *Nostoc* sp. living epiphytically on, as well as intracellularly in, *Sphagnum* mosses. The water-filled hyaline cells of *Sphagnum* mosses seem to constitute a complex microhabitat for various microorganisms (Granhall and v. Hofsten, 1976). The intracellular colonisation takes place through pores in the hyaline cells. Granhall and v. Hofsten (1976) found *Nostoc* sp. intracellularly in *Sphagnum lindebergii* and *S. riparium*, but not in *S. balticum*, *S. fuscum* or *S. annulatum* (*jensenii*). The determining factor for intracellular or epiphytic colonisation of the *Sphagnum* moss seemed to be the pH of the mire. Cyanobacteria could escape the unfavorable pH by entering the less acidic environment of hyaline cells. At pH 4.9 and 4.2, *S. riparium* and *S lindeberii* were colonised intracellularly. At pH 3.8, the value in the mire where *S. balticum*, *S. fuscum* and *S. annulatum* (*jensenii*) were growing, cyanobacteria were totally absent. At higher pH values (minerotrophic condition), only epiphytic and free-living cyanobacteria were found in association with *Sphagnum* mosses (Granhall and Selander, 1973).

2.3 Cytology of the Association

2.3.1. Epiphytic Associations

Few papers have been published on the cytology of the association between cyanobacteria and mosses. Jordan *et al.* (1978), studying a terrestrial environment in the high Canadian arctic, found mainly *Nostoc* sp. on the mosses studied, although *Anabena* sp. and *Oscillatoria* sp. were also present. Large globular colonies of *Nostoc* sp. were often found in the leaf axils of the moss. They did not find any evidence for inter- or intra-cellular microorganisms in sections of the moss tissue. Using SEM and TEM, Scheirer and Dolan (1983) found a great variety of epiphytic microorganisms, but no cyanobacteria, on the leaves of *Polytrichum commune* Hedw. Scheirer and Brasell (1984) used epifluorescence microscopy to study nitrogen-fixing cyanobacteria associated with *Funaria hygrometrica* in a recently clearfelled and burnt eucalyptus forest of southern Tasmania. They frequently found filamentous heterocystous cyanobacteria on the lower half of the moss gametophores and among the rhizoids at the basal portion of the stem. In contrast, Broady (1979a) on Signy Island and Smith (1984) on Marion Island found the cyanobacteria mainly in the green upper part of the mosses. Solheim *et al.* (1996) found that certain species of mosses, e.g. *Sanionia uncinata* (*Drepanocladus uncinatus*) and *Calliergon richardsonii*, were especially adapted to associate with nitrogen fixing cyanobacteria. Following are some of the unpublished results from our group (Solheim, B., Vigstad, H., Røberg, S., Zielke, M. at University of Tromsø, Norway and Spaink, H. at Leiden University, The Netherlands). We found the distribution of cyanobacteria and acetylene reduction activity along moss plants to be similar to that reported by Broady (1979a) and Smith (1984). The epiphytic association between different species of mosses from the Arctic (Spitsbergen, Norway) and the Sub-arctic (Abisko, Sweden) were studied by fluorescence and laser confocal microscopy. Like Scheirer and Brasell (1984), we found epifluorescence microscopy to be a rapid and easy technique to localise cyanobacteria along the moss plants. We found cyanobacteria mostly along the leaves of, but occasionally intracellularly in, *S.*

uncinata. Differrent morphological types of cyanobacteria were also found on the leaves of *C. richardsonii.* However, most of the time the cyanobacteria were located behind moss tissue and could only be seen as bright red spots or areas of fluorescent light. The only method that could be used on living material to distinguish and accurately localise the cyanobacteria on the moss was laser confocal microscopy. Figure 1 shows the cyanobacteria between the stem and a leaf of *C. richardsonii.* The hollow stem is an artefact, it was difficult to find fixation and embedding techniques that both gave good staining of cyanobacteria and preserved the soft tissue in the middle of the stem. At higher magnification (Fig. 2), it is evident that the cyanobacteria are inside a matrix between the stem and the leaf. This matrix probably stabilises the association and makes transfer of nutrient between the organisms more efficient. By using $^{15}N_2$-gas, Alexander *et al.* (1974) showed a rapid transfer of fixed nitrogen from cyanobacteria to mosses and other plants. By scanning electron microscopy (SEM) it was possible to see a high number of heterotrophic bacteria (Fig. 3) constituting an important part of the epiphytic microbial community. Heterotrophic bacteria could also be seen in direct contact with cyanobacteria (Fig. 4). Most of the observed cyanobacteria were filamentous with heterocysts (Fig. 2), but other types were also observed.

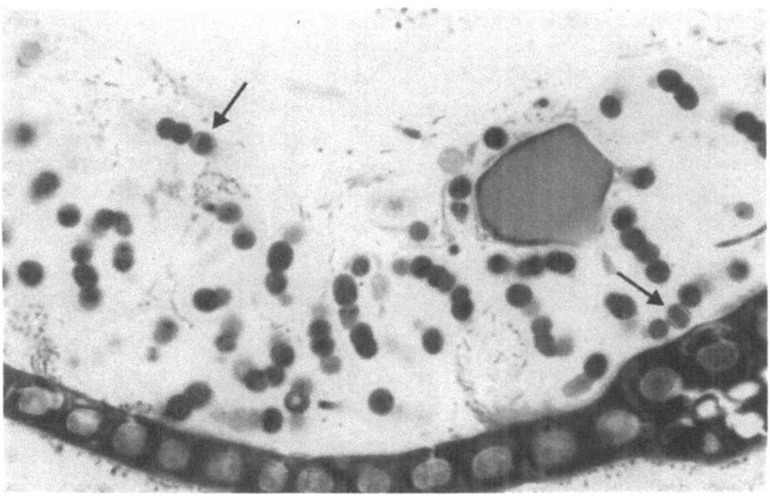

Figure 2. Same sample as Fig. 1, but at a higher magnification (300x). Arrows indicate heterocysts. Note the matrix immobilising the cyanobacteria close to the moss leaf.

Figure 3. Scanning electron photomicrograph of heterotrophic bacteria growing as epiphytes on a leaf of Sanionia uncinata *(H. Vigstad, unpublished results).*

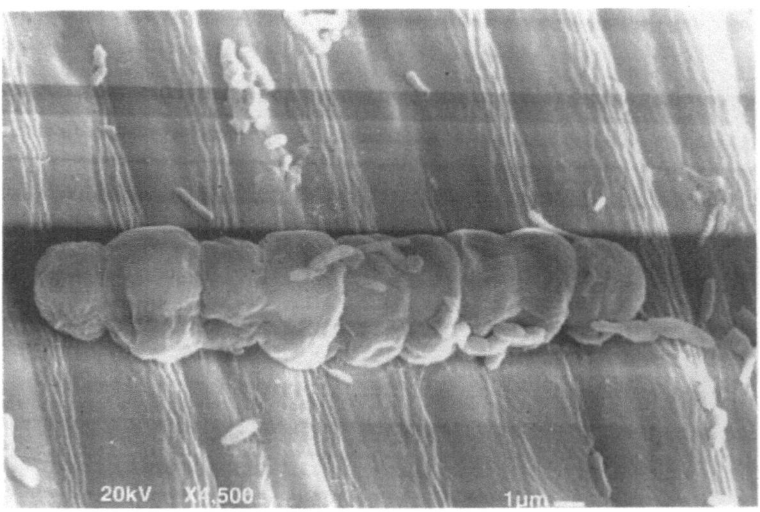

Figure 4. Scanning electron photomicrograph of cyanobacteria growing as epiphytes on a leaf of Sanionia uncinata. *Heterotrophic bacteria are associated with cyanobacteria as well as with the host plant (S. Roberg, unpublished results).*

2.3.2. Intracellular Colonisation

Intracellular colonisation of hyaline cells in *Sphagnum* sp. by cyanobacteria was reported by Stewart (1966), but Granhall and v. Hofsten (1976) are believed to be the first authors to describe this association between cyanobacteria and mosses in great detail using electron microscopy. They found cyanobacterial vegetative cells and heterocysts in the hyaline cells of *S. riparium* (Fig. 5). Cell aggregates were generally embedded in mucilaginous material, presumably produced by the cyanobacteria. In addition to cyanobacteria, typical heterotrophic bacteria, *Methanosarcina*-like bacteria, and fungal mycelia were found inside the hyaline cells. Based on observations that *Sphagnum* mosses from areas with cyanobacteria grew much faster than mosses from areas devoid of cyanobacteria (Sonesson, 1973) and their own work, Granhall and v. Hofsten (1976) suggested a model of chemical interactions among some intracellular organisms in *Sphagnum* mosses (Fig. 6). In addition to exchange of nutrients between moss cells and cyanobacteria synergistic interrelations are also likely to take place between the cyanobacteria and other bacteria in the enclosed space in the hyaline cells. The bacteria can use nitrogenous compounds and carbohydrates excreted by cyanobacteria, while at the same time, bacterial respiration and removal of nitrogen would stimulate nitrogen fixation of the cyanobacteria by increasing CO_2 levels and reducing O_2 levels. The products of nitrogen fixation, hydrogen and ammonium, can be used as substrate by methane-producing bacteria.

3. GEOGRAPHICAL DISTRIBUTION

3.1. Polar Regions

There are several approaches to dividing the polar areas into different climatic zones. In this review we use the term "polar areas" without distinction between these zones. Polar areas in the Arctic include North Alaska and Canada with the islands of the Northwest Territories, the northern half of Greenland, Svalbard and Novaya Zemlya. Polar areas in the Southern Hemisphere are limited to the Continental Antarctic. These regions are not only characterised by their extremely harsh climatic conditions (cold, drought, and low soil nutrient content) but also by the severe seasonal changes. Furthermore, large areas are covered by ice, and areas covered by vegetation can often be found only at the margins of these hostile landmasses. Apart from a few exceptions, the soils in ice-free areas are characterised by a low nutrient content, particularly nitrogen and phosphorus (Vincent, 1988; Lennihan *et al.*, 1994), which limit the terrestrial plant life. Lichens, liverworts and mosses cope best with these environmental conditions and play a dominant role in most polar ecosystems (Aleksandrova, 1988; Longton, 1988; Elvebakk, 1994). As described above, many of the mosses are able to harbour cyanobacteria confering high nitrogen fixation activity to the vegetation. Studies from the Arctic (Karagatzides *et al.*, 1985; Solheim *et al.*, 1996) describe *Bryum* sp., *Drepanocladus* (*Sanionia*) sp., *Calligeron* sp., *Dupontia* sp., and *Grimmia* sp., as the most abundant host plants for cyanobacteria. In the Antarctic, Davey (1983) found epiphytic *Nostoc* on all mosses at the Vestfold Hills. In a contiuous cover of moss-*Nostoc* the bryophytes included a couple of *Bryum* sp., *Grimmia* sp. and *Sarconeurum* sp. (Davey and Marchant, 1983). Nakatsubo and Ino (1986, 1987) collected cyanobacteria-moss samples along the coast of East Antarctica and found *Bryum* sp., *Grimmia* sp., and *Ceratodon* sp. The most dominant cyanobacteria were members of the *Nostoc-Anabaena* group and *Calothrix* sp.

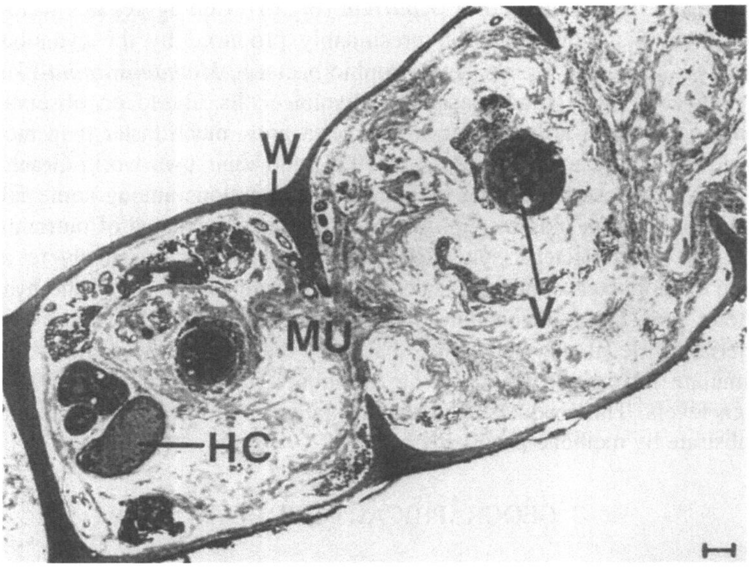

Figure 5. Nostoc *in hyaline cells of* Sphagnum riparium. *V, vegetative cell; HC, heterocyst; MU, mucilaginous material; W, moss cell wall. Bar, 1 μm. From Granhall and v. Hofsten (1976) with permission.*

3.2. Sub-Polar Regions

The Sub-polar regions, commonly called tundra, surround the Polar regions like a belt. In the Southern Hemisphere they include the Antarctic islands in the southern parts of the Indian, Pacific and Atlantic Ocean as well as western coast of Patagonia. In the Northern Hemisphere, the Sub-arctic regions roughly have a southern extension along the North Coast of North America, South Cape of Greenland, the northern coast of Norway, northern parts of Finland, and the northern Eurasian and Asian coastline. However, due to several similarities in environmental characteristics many Alpine regions such as parts of the Rocky Mountains in USA and Canada, northern Sweden, and the plateau of the Hardangervidda in Norway, are also classified as tundra.

The climate conditions in Sub-polar and Alpine regions can be harsh and extreme, but they are generally more moderate than in the Polar regions and provide more favourable environmental conditions for the biota. Although extensive grass and dwarf shrub heaths are common in the Sub-polar regions, mosses and other bryophytes are important and often even dominating components of the sub-polar and alpine vegetation. In contrast to the Polar regions, where *Sphagnum* spp. are rare, this moss is abundant in many mires in the Sub-polar regions (Aleksandrova, 1988; Longton, 1997) and often found associated with nitrogen fixing cyanobacteria (Granhall and Selander, 1973; Granhall and v. Hofsten, 1976; Basilier *et al.*, 1978). Nitrogen-fixing associations between cyanobacteria and *S. uncianata* growing in Alpine grassland in northern Finland (Solheim, unpublished results), and *Hylocomium splendens* growing in a sub-arctic birch forest (Solheim *et al.*, 2002), have been described. In the most hospitable areas of the Sub-polar and Alpine regions, symbiotic nitrogen fixation has been reported for several legumes (*Oxytropis* spp. and *Astragalus alpinus*) and at least two *Dryas*-species (Alexander *et al.*, 1978; Waughman *et al.*, 1981; Karagatzides *et al.*, 1985). Nevertheless, nitrogen fixation by cyanobacteria-moss associations has been found to contribute significantly to the nitrogen economy of Sub-polar and Fennoscandia tundra ecosystems (Granhall and Basilier, 1973; Granhall and Selander, 1973; Granhall and Lid-Torsvik, 1975) as well as to Alpine ecosystems (Englund, 1976; Alexander, *et al.*, 1978; Wojchiechowski and Heimbrook, 1984). Hardly any nitrogen fixation by heterotrophic soil bacteria has been found (Granhall and Selander, 1973; Smith, 1985). In black spruce forest in Alaska which is the most widespread vegetation type in the Alaskan taiga, moss-cyanobacteria associations were found to be the largest nitrogen-fixing entity based on plant cover (Billington and Alexander, 1978). In the Subantarctic Marion (Smith and Ashton, 1981; Smith and Rusell, 1982; Smith, 1984), Macquarie (Line, 1992) and Signy islands (Broady, 1979a) a large number of different species of mosses were found to associate mainly with filamentous cyanobacteria.

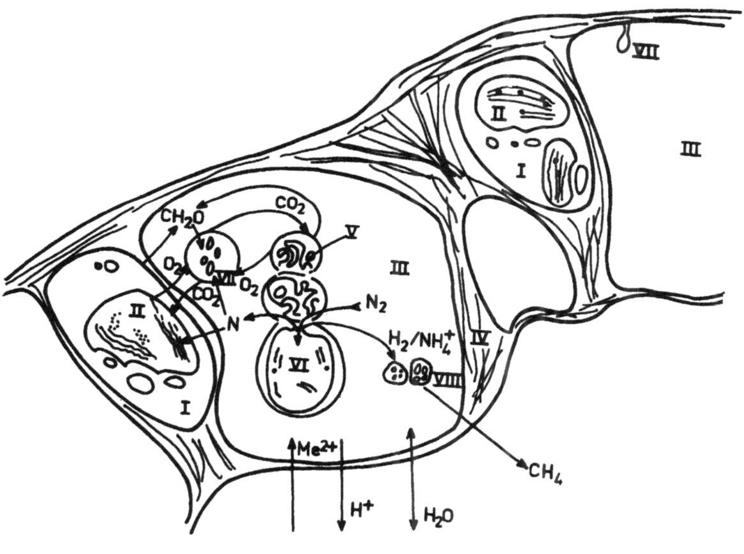

Figure 6. Suggested chemical interactions among intracellular organisms in Sphagnum *mosses. Chemical compounds as per convention. N, nitrogen compounds; I, chlorophyllous moss cell; II, chloroplast; III, hyaline cell; IV, moss cell wall; V, cyanobacterial vegetative cell; VI, heterocyst; VII, bacteria; VIII, methane-producing bacteria. From Grabhall and v. Hofsten (1976) with permission.*

3.3 Temperate and Tropical Regions

The major part of the earth's terrestrial habitats are situated in Temperate and Tropical regions, but only a few studies have been carried out to investigate moss-associated cyanobacteria and their nitrogen fixation activity in these regions. This may be due to the fact that Temperate and Tropical regions are often dominated by cultural landscape and large areas with moss vegetation are not common. Furthermore, the arid and dry regions do not favour growth of mosses. However, some reports on nitrogen fixation by moss-associated cyanobacteria in temperate and tropical ecosystems do exist. Studied ecosystems include coniferous forests and mires in Scandinavia, North American grasslands, and tropical and temperate forests in India and Tasmania. Granhall and Lindberg (1978) described intracelluar nitrogen-fixing cyanobacteria within hyaline cells of *Sphagnum* sp., whereas Basilier (1979) reported epiphytic cyanobacteria on *Drepanocladus* sp., *Calliergon* sp., and *Sphagnum* sp. In both these studies *Nostoc* was found to be most common cyanobacterium. Nitrogen fixation by moss-associated cyanobacteria was also described for temperate grassland ecosystems (Vlassak *et al.*, 1973; Reddy and Giddens, 1981). The identified cyanobacteria belong predominantly to the genera *Nostoc* and *Anabena, Oscillatoria* and *Lyngbya* were also found. The cyanobacteria were mostly observed on *Bryum* mosses, *Weisia controversa* and *Ceratodon purpureus*. Nitrogen-fixing cyanobacteria-moss associations and their role in the succession of a clearfelled and slash-burnt mixed forest in southern Tasmania has been reported (Scheirer and Brasell, 1984; Brasell *et al.,* 1986). They found *Nostoc* and *Anabena* spp. as nitrogen-fixing epiphytes on the mosses *Funaria hygrometrica* and *C. prupureus,* and to a lesser extent on *Polytrichum juniperinum* and *Campylopus introflexus*. These associations are also suggested to be the earliest colonisers of the burnt areas. *Nostoc* sp. and species of *Chroococcus*, *Lyngbya*, *Scytonema*, and *Phormidium* have been found on mosses in the tropical montane rainforests of the Western Ghats of Kerala, India (Madhusoodanan and Dominic, 1996).

4. ECOLOGICAL SIGNIFICANCE

Although the values for nitrogen fixation in polar regions are generally lower than the ones measured in temperate and tropical environments, they represent a major part of the nitrogen input in terrestrial polar ecosystems (Croome, 1973; Alexander, 1974; Alexander *et al.* 1974; Henry and Svoboda, 1986; Lennihan and Dickson, 1989; Chapin and Bledsoe, 1992). This may be ascribed to the fact that the input of nitrogen by precipitation is low and nitrogen fixation by heterotrophic soil bacteria is either limited or absent due to low soil temperature or nutrient availability (Jordan *et al.*, 1978; Smith, 1985; Christie, 1987). Although free-living edaphic cyanobacteria have a high nitrogen-fixing potential, their contribution to the total nitrogen input might be less significant. This is because of their limited (localised) distribution and the fact that they easily dry up due to their exposed position on the soil surface. Nitrogen-fixing symbioses between soil bacteria and legumes or *Dryas octopetala* as reported for Sub-arctic and Alpine tundra, are not found in the Arctic and Antarctic (Henry and Svoboda, 1986). However, cyanobacterial nitrogen fixation is subject to environmental constraints and shows wide spatial and temporal variation (Chapin and Bledsoe, 1992). Chapin and Bledsoe (1992) estimated that nitrogen fixation in the Arctic ranges from 19 to 255 mg N m^{-2} yr^{-1}, and Vincent (1988) estimated it to be 10 to 192 mg N m^{-2} yr^{-1} in the Antarctic. Further, moss-associated cyanobacteria provide 2-58% of the nitrogen input to arctic ecosystems and 42-84% to Antarctic moss vegetation (Dodds *et al.*, 1995). In the Arctic, large populations of herbivorous animals harvest a considerable part of the annual plant production. Geese especially export nitrogen and other nutrients out of the ecosystem by the autumn migration of their newly produced goslings. Bazely and Jefferies (1989) found that the nitrogen exported out of the system was compensated by the increased nitrogen fixation by free-living cyanobacteria in salt marshes in Northern Canada. Zacheis *et al.* (2001) concluded that increased nitrogen availability in areas grazed by geese in an Alaskan salt marsh was primarily due to the increased mineralisation of organic matter although increased nitrogen fixation also made a contribution. At Spitsbergen (Norway), Loonen and Solheim (1998) found significantly higher nitrogen fixation by cyanobacteria-moss associations in vegetation grazed by geese than in vegetation that had not been grazed for 5-6 years. Intensively grazed areas in the Arctic would probably not be able to sustain the current populations of migratory geese without the extra input of nitrogen from biological nitrogen fixation.

The importance of nitrogen fixation by moss-associated cyanobacteria in Temperate and Tropical regions has been less investigated than in Polar and Sub-polar regions. In contrast to most Polar and Sub-polar regions, where moss-associated nitrogen fixation is the major contributor to the nitrogen cycle, in temperate and tropical regions this may be the case only in specific habitats. Basilier (1979) found that only cyanobacterial nitrogen fixation was associated with the *Sphagnum* mosses in open mires. In the temperate coniferous forests, *Sphagnum* mosses did not associate with cyanobacteria and only a very low nitrogenase activity could be detected due to heterotrophic bacteria. In a similar ecosystem, Granhall and Lindberg (1978) reported high nitrogen fixation by *Sphagnum*-cyanobacteria associations in mires with values up to 37.9 kg N ha^{-1} yr^{-1}, exceeding the values reported by Todd *et al.* (1978) from a deciduous forest in south-eastern United States. In a North American grassland ecosystem, nitrogen fixation by different moss-cyanobacteria associations were highly dependent on moisture and light (Vlassak *et al.*, 1973). Samples of the moss *C. purpureus* with epiphytic cyanobacteria had an estimated nitrogen fixation activity of 25-110 ng N g^{-1} dry matter h^{-1} depending on the moisture content. The importance of nitrogen fixation by moss-associated cyanobacteria in the succession of burnt forests in southern Tasmania was studied by Brasell *et al.* (1986). Cyanobacteria were the dominant diazotrophs since the low soil organic matter content, after the burning of the forests, limited the nitrogen fixation by heterotrophs. Nitrogen fixation by cyanobacteria associated with mosses played a major role in succession processes after burning. Burning itself did not stimulate nitrogen fixation; in mature forests nitrogen fixation was just as high as at the burnt sites. The estimated values of nitrogen fixation increased from 43 mg N m^{-2} in the second year after burning, to 99 and 152 mg N m^{-2} in the third and fourth year after burning, respectively, probably reflecting growth of the cyanobacterial population. No seasonal variation in activity could be determined. This was probably due to the mild and relatively stable climate in southern Tasmania. However, as reported for other regions, activity showed a clear moisture and temperature dependence under laboratory conditions.

The influence of abiotic factors on cyanobacterial nitrogen fixation in Polar regions has been subject of many studies (Jordan *et al.* 1978; Davey, 1983; Davey and Marchant, 1983; Christie, 1987; Nakatsubo and Ino, 1987; Chapin *et al.* 1991; Lennihan *et al.* 1994; Liengen and Olsen, 1997; Liengen, 1999; Dickson, 2000; Solheim *et al.*, 2002; Zielke *et al.*, 2002). Except for some very site-specific variations, the studies above concluded that moisture, temperature, light and nutrient availability were the most important environmental factors for cyanobacterial nitrogen fixation in the Arctic and Antarctic. Soil moisture is one of the most important environmental factors controlling the nitrogen fixation activity of cyanobacteria-moss associations. Results from *in situ* studies conducted in different types of polar and sub-polar habitats and from laboratory experiments, have shown a strong correlation between nitrogen fixation rates and soil moisture (Alexander *et al.*, 1974; Davey, 1983; Davey and Marchant 1983; Wojchiechowski and Heimbrook, 1984; Gold and Bliss, 1995; Liengen and Olsen, 1997; Dickson, 2000; Zielke *et al.*, 2002). Billington and Alexander (1978) showed that in the Alaskan taiga nitrogenase activity was related to moisture, temperature and day length. They found that moisture appeared to be the major abiotic controller, but was at times overshadowed by temperature and/or day length. The nitrogen fixation activity of epiphytic cyanobacteria in the moss vegetation responded strongly to different temperature regimes. Although the moss-associated cyanobacteria fix nitrogen even at 0°C, they do optimally at temperatures between 20 and 30°C (Alexander *et al.*, 1974; Davey and Marchant, 1983; Vincent, 1988; Chapin *et al.*, 1991; Zielke *et al.*, 2002). Consequently, they have to be classified as psychrotolerants rather than as psychrophiles.

Photosynthesis is the main source of energy (ATP) and reducing power for nitrogen fixation in cyanobacteria, and the influence of light quality and intensity on photosynthesis might have an indirect effect on nitrogen fixation activity. Studies on samples of moss-associated cyanobacteria from polar environments (Alexander *et al.*, 1974; Chapin and Bledsoe, 1992; Vincent *et al.*, 1993; Lennihan *et al.*, 1994; Zielke *et al.*, 2002) show a clear light-dependent response of nitrogen fixation activity. The reported point of light saturation was 100-150 μmol m^{-2} s^{-1} Photosynthetic Photon Flux Density (PPFD) at 12-14°C. This temperature is often reached in the vegetation layer during the growing season (Davey and Marchant, 1983). Due to the shadowing effect by plant tissues and dead plant material, the actual PPFD available to cyanobacteria is lower. However, in the Arctic, with continuous light during the whole growing season, low light intensity is probably not an important limiting factor for nitrogen fixation. Furthermore, some cyanobacteria continue to fix nitrogen for several hours when transferred from light into the darkness. This indicates that the nitrogen fixation process can be driven by stored energy or energy derived solely from respiration (Coxson and Kershaw, 1983; Maryan *et al.*, 1986; Fritz-Sheridan, 1988). Light quality, especially the amount of UV-B radiation can severely inhibit nitrogen fixation in the Arctic. Solheim *et al.* (2002) found a 50% reduction in nitrogen fixation activity in a cyanobacteria-moss community at Spitsbergen (Norway) that had been under enhanced UV-B radiation, equivalent to a 15% reduction in the ozone layer, during the growing season for 3-4 years. However, cyanobacteria-moss communities from Abisko (Sweden) did not show such reduction of nitrogen fixation activity.

The availability of mineral forms of N, P, Co and Mo might have a significant effect on the nitrogen fixation in moss vegetation. While higher levels of P (component of ATP), Co, and Mo (components of the nitrogenase enzyme complex) positively correlate with cyanobacterial nitrogen fixation, increased availability of nitrogen (as either NO_3^- or NH_4^+) generally leads to a lower nitrogen fixation activity (Chapin *et al.*, 1991; Solheim *et al.*, 1996).

Soil moisture and temperature appear to be the most important factors affecting the terrestrial nitrogen fixation in the Polar regions. A major part of the seasonal and spatial variations in nitrogen fixation results from changes in temperature and moisture regime (Alexander *et al.*, 1974; Davey and Marchant, 1983). Despite ambient summer temperatures scarcely exceeding 10°C in the Arctic and 0°C in the Antarctic, the light absorbing and isolating property of the vegetation often results in temperatures 5 to 10°C higher in the upper part of the vegetation compared to the air temperatures. However, since the temperature in the vegetation layer does not vary significantly on a spatial basis this factor might be involved mostly in the seasonal variations in nitrogen fixation. In contrast, much of the spatial and the seasonal variations in cyanobacterial nitrogen fixation can be attributed to local and temporal differences in soil moisture (Alexander and Schell, 1973; Henry and Svoboda, 1986; Chapin *et al.*, 1991). Davey and Marchant (1983) report almost simultaneous seasonal progressions for soil surface temperature, soil moisture and nitrogen fixation activity. The Arctic and the Antarctic are characterised by continuos daylight during the 24-h photoperiod of their respective summers. Considering this fact, combined with the relatively low light requirements of moss-associated cyanobacteria and their ability to overcome short periods of insufficient light, light is probably not a limiting factor for cyanobacterial nitrogen fixation in the vegetation layer. Global climate change will obviously influence the cyanobacteria-moss associations and the nitrogen fixation activity in the ecosystem. This aspect has been reviewed in detail by Chapin and Bledsoe (1992).

ACKNOWLEDGEMENT

We thank Stian Røberg, Herman Spainak and Hege Vigstad for use of unpublished pictures and Andy Dean for correcting the English.

REFERENCES

Alexander, V. (1974) A synthesis of the IBP Tundra biome circumpolar study of nitrogen fixation, in A. Holding, J.O.W. Heal, S.F. Maclean and P.W. Flanagan (eds.), *Soil organisms and Decomposition in Tundra*, Tundra Biome Steering Committee, Stockholm, pp. 109-121.

Alexander, V., Billington, M., and Schell, D. (1974) The influence of abiotic factors on nitrogen fixation rates in the Barrow, Alaska, arctic tundra, *Rep. Kevo Subarct. Res. Sta.* **11**, 3-11.

Alexander, V., Billington, M., and Schell, D.M. (1978) Nitrogen fixation in Arctic and Alpine tundra, in L.L. Tieszen (ed.),*Vegetation and Production Ecology of an Alaskan Arctic Tundra, vol. 29. Ecological Studies*, Springer Verlag, New York, pp. 539-558.

Alexander, V. and Schell, D.M. (1973) Seasonal and spatial variation of nitrogen fixation in the Barrow, Alaska, tundra, *Arct. Alp. Res.* **5**, 77-88.

Aleksandrova, V.D. (1988) *Vegetation of the Soviet Polar Deserts*. Cambridge University Press, Cambridge.
Alfnito, S., Fumanti, B., and Cavacini, P. (1998) Epiphytic algae on mosses from North Victoria Land (Antarctica), *Nova Hedwigia* **66**, 473-480.
Basilier, K. (1979) Moss-associated nitrogen fixation in some mire and coniferous forest environments around Uppsala, Sweden, *Lindbergia* **5**, 84-88
Basilier, K., Granhall, U., and Stenström, T.-A. (1978) Nitrogen fixation in wet minerotrophic moss communities of a subarctic mire, *Oikos* **31**, 236-246.
Bazely, D.R. and Jefferies, R.L. (1989) Lesser snow geese and the nitrogen economy of a grazed salt march, *J. Ecol.* **77**, 24-34.
Billington, M. and Alexander, V. (1978) Nitrogen fixation in a black spruce (*Picea mariana* [Mill] B.S.P.) forest in Alaska, *Ecol. Bull.* **26**, 209-215.
Birkemoe, T. and Liengen, T. (2000) Does collembolan grazing influence nitrogen fixation by cyanobacteria in the high Arctic, *Polar Biol.* **23**, 589-592.
Brasell, H.M., Davies, S.K., and Mattay, J.P. (1986) Nitrogen fixation associated with bryophytes colonizing burnt sites in Southern Tasmania, Australia, *J. Bryol.* **14**, 139-149.
Broady, P.A. (1977) The Signy Island terrestrial reference site: VII. The ecology of the algae of site 1, a moss turf, *Br. Antarct. Surv. Bull.* **45**, 47-62.
Broady, P.A. (1979a) The Signy Islands terrestrial reference sites: IX. The ecology of the algae of site 2. a moss carpet, *Br. Antarct. Surv. Bull.* **47**, 13-29.
Broady, P.A. (1979b) Wind dispersal of terrestrial algae at Signy Island, South Orkney Islands, *Br. Antarct. Surv. Bull.* **48**, 99-102.
Broady, P.A. (1979c) Feeding studies on the collembolan *Cryptopygus antarcticus* Willem at Signy Island, South Orkney Islands, *Br. Antarct. Surv. Bull.* **48**, 37-46.
Broady, P.A. (1987) A floristic survey of algae at four locations in Northern Victoria Land, *N. Z. Antarct. Rec.* **7**, 8-19.
Chapin, D.M. and Bledsoe, C.S. (1992) Nitrogen fixation in arctic plant communities, in F.S. Chapin III, R.L. Jefferies, J.F. Reynolds, G.R. Shaver and J. Svoboda (eds.), *Arctic Ecosystems in a Changing Climate. An Ecophysiological Perspective*, Academic Press, Inc., San Diego, pp.301319.
Chapin, D.M., Bliss, L.C., and Bledsoe, L.J. (1991) Environmental regulation of nitrogen-fixation in a high arctic lowland ecosystem, *Can. J. Bot.* **69**, 2744-2755.
Christie, P. (1987) Nitrogen in two constrasting Antarctic bryophyte communities, *J. Ecol.* **75**, 73-93.
Coxson, D.S. and Kershaw, K.A. (1983) The pattern of *in situ* summer nitrogenase activity in terrestrial *Nostoc commune* from *Stipa-Bouteloa* grass, *Can. J. Bot.* **61**, 2686-2693.
Croome, R.L. (1973) Nitrogen fixation in the algal mats on Marion Island, *S-Afr. Tydskr. Navors. Antarkt.* **3**, 64-67.
Davey, A. (1983) Effects of abiotic factors on nitrogen fixation by blue-green algae in Antartica, *Polar Biol.* **2**, 95-100.
Davey, A. and Marchant, H.J. (1983) Seasonal variation in nitrogen fixation by *Nostoc commune* Vaucher at the Vestfold Hills, Antarctica, *Phycologia* **22**, 377-385.
Dickson, L.G. (2000) Constraints to nitrogen fixation by cryptogamic crusts in a polar desert ecosystem, Devon Island, N.W.T., Canada, *Arc. Antarc. Alp. Res.* **32**, 40-45.
Elvebakk, A. (1994) A survey of plant associations and alliances from Svalbard, *J. Veg. Sci.* **5**, 791-802.
Englund, B. (1976) Nitrogen fixation by free-living microorganisms on the lava field of Heimaey, Iceland, *Oikos* **27**, 428-432.
Fritz-Sheridan, R.P. (1988) Physiological ecology of nitrogen fixing blue-green algal crusts in the upper-subalpine life zone, *J. Phycol.* **24**, 302-309.
Fumanti, B., Cavacini, P., and Alfinito, S. (1997) Benthic algae mats of some lakes of Inexpressible Island (Northern Victoria Land, Antarctica), *Polar Biol.* **7**, 97-113.
Dodds, W.K., Gudder, D.A., and Mollenhauer, D. (1995) The ecology of *Nostoc*, *J. Phycol.* **31**, 2-18.
Gold, W.G. and Bliss, L.C. (1995). Water limitations and plant community-development in a polar desert. *Ecology* **76**, 1558-1568.
Gordon, D.A., Priscu, J., and Giovannoni, S. (2000) Origin and phylogeny of microbes living in permanent Antarctic lake ice, *Microb. Ecol.* **39**, 197-202.
Granhall, U. and Basilier, K. (1973) Nitrogen fixation in tundra moss communities, Progress Report 1972, *Swedish IBP Tundra Biome Project Technical Report* **14**, 174-190.
Granhall,U. and v. Hofsten, A. (1976) Nitrogenase activity in relation to intracellular organisms in *Sphagnum* mosses, *Physiol. Plant.* **36**, 88-94.

Granhall, U. and Lid-Torsvik, V. (1975) Nitrogen fixation by bacteria and free-living blue-green algae in tundra areas, *Ecol. Stud.* **16**, 305-315.

Granhall, U. and Lindberg, T. (1978) Nitrogen fixation in some coniferous forest ecosystems, *Ecol. Bull.* **26**, 178-192.

Granhall, U. and Selander, H. (1973) Nitrogen fixation in a subarctic mire, *Oikos* **24**, 8-15

Henry, G.H.R. and Svoboda, J. (1986) Dinitrogen fixation (acetylene reduction) in High Arctic segde meadow communities, *Arct. Alp. Res.* **18**, 181-187.

Jordan, D.C., McNicol, P.J., and Marshall, M.R. (1978). Biological nitrogen fixation in the terrestrial environment of a high Arctic ecosystem (Truelove Lowland, Devon Island, N.W.T.), *Can. J. Microbiol.* **24**, 643-649.

Karagatzides, J.D., Lewis, M.C. and Schulmann, H.M. (1985) Nitrogen fixation in the high artic tundra at Scarpa Lake, Northwest Territories, *Can. J. Bot.* **63**, 974-979.

Lennihan, R. and Dickson, L.G. (1989) Distribution, abundance and physiological aspects of *N. commune* in a high arctic ecosystem, *J. Phycol. (Suppl.)* **25**, 16.

Lennihan, R., Chapin, D.M., and Dickson, L.G. (1994) Nitrogen fixation and photosynthesis in high arctic forms of *Nostoc commune*, *Can. J. Bot.* **72**, 940-945.

Liengen, T. (1999) Environmental factors influencing the nitrogen fixation activity of free-living terrestrial cyanobacteria from a high arctic area, Spitsbergen, *Can. J. Microbiol.* **45**, 573-581.

Liengen, T. and Olsen, R.A. (1997) Seasonal and site-specific variations in nitrogen fixation in a high arctic area, Ny-Ålesund, Spitsbergen, *Can. J. Microbiol.* **43**, 759-769.

Line, M.A. (1992) Nitrogen fixation in the sub-Arctic Macquarie Island, *Polar Biol.* **11**, 601-606.

Longton, R.E. (1988) *The Biology of Polar Bryophytes and Lichens*, Cambridge University Press, Cambridge.

Longton, R.E. (1997) The role of bryophytes and lichens in polar ecosystems, in S.J. Woodin and M. Marquis eds. *Ecology of Arctic Enivironments*, vol. 13. *Special Publication of the British Ecological Society*, Blackwell Science Ltd., Oxford.

Loonen, M.J.J.E. and Solheim, B. (1998) Does arctic vegetation change when grazed by barnacle geese? A pilot study, in F. Mehlum, J.M. Black and J. Madsen (eds.), *Research on Arctic Geese. Nor. Polarinst. Skr.* **200**, 99-103

Madhusoodanan, P.V. and Dominic, T.K. (1996) Epiphytic cyanobacteria on mosses from the Western Ghats of Kerala, *J. Econ. Tax. Bot.* **20**, 355-359.

Marshall, W.A. and Chalmers, M.O. (1997) Airborne dispersal of antarctic terrestrial algae and cyanobacteria, *Ecography* **20**, 585-594.

Maryan, P.S., Eady, R.R., Chaplin, A.E., and Gallon, J.R. (1986) Nitrogen fixation by *Gloeothece* sp. PCC 6909: Respiration and not photosynthesis supports nitrogenase activity in the light, *J. Gen. Microbiol.* **132**, 789-796.

Nakatsubo, T. and Ino, Y. (1986) Nitrogen cycling in an Antarctic ecosystem. 1. Biological nitrogen fixation in the vicinity of Syowa Station, *Mem. Natl. Inst. Polar Res. Ser.* **37**,1-10.

Nakatsubo, T. and Ino, Y. (1987) Nitrogen cycling in an Antarctic ecosystem. 2. Estimation of the amount of nitrogen fixation in a moss community on East Ongul Island, *Ecol. Res.* **2**, 31-40.

Ohtani, S. and Kanda, H. (1987) Epiphytic algae on moss community of *Grimmia lawiana* around Syowa Station, Antarctica, *Proceeding NIPR Symposium. Polar Biol.* **1**, 255-264.

Rai, A.N., Söderback, E. And Bergman, B. (2000) Tansley Review No. 116. Cyanobacterium-plant symbioses, *New Phytol.* **147**, 449-481.

Reddy, G.B. and Giddens, J. (1981) Nitrogen fixation by moss-algal associations in grassland, *Soil Biol. Biochem.* **13**, 537-538.

Scheirer, D.C. and Dolan, H.A. (1983) Bryophyte leafe epiflora: An SEM and TEM study of *Polytrichum commune* Hedw, *Am. J. Bot* **70**, 712-718.

Scheirer, D.C. and Brasell, H.M. (1984) Epifluorescence microscopy for the study of nitrogen fixing blue-green algae associated with *Funaria hygrometrica* (Bryophyta), *Am. J. Bot.* **71**, 461-465.

Schwabe, G.H. (1974) Nitrogen fixing blue-green algae ae pioneer plants on Surtsey 1968-1973, *Surtsey Res. Progress Report* **7**, 22-25.

Smith, V.R. (1984) Effects of abiotic factors on acetylene reduction by cyanobacteria epiphytic on moss at a subarctic island, *Appl. Environ. Microbiol.* **48**, 594-600.

Smith, V.R. (1985) Heterotrophic acetylene reduction in soil at Marion island, in Siegfried, W.R., Condy, P.R., Laws, R.M. (eds), *Antarctic Nutrient Cycles and Food Webs*, Springer, Berlin, Heidelberg, New York, pp. 185-191.

Smith, V.R. and Ashton, P.J. (1981) Bryophyte-cyanobacteria associations on sub-Arctic Marion Island: are they important in nitrogen fixation, *S-Afr. Tydskr. Navors. Antarkt.* **10/11**, 24-26.

Smith, V.R. and Russell, S. (1982) Acetylene reduction by bryophyte-cyanobacteria associations on a Subantarctic island, *Polar Biol.* **1,** 153-157.

Solheim, B., Endal, A., and Vigstad, H. (1996) Nitrogen fixation in Arctic vegetation and soils from Svalbard, Norway, *Polar Biol.* **16,** 35-40.

Solheim, B., Johanson, U., Callaghan, T.V., Lee, J.A., and Gwynn-Jones, D. (2002) The nitrogen fixation potential of arctic cryptogam species are influenced by enhanced UV-B radiation, *Oecologia* (in press).

Sonesson, M. (1973) Studies in production and turnover of bryophytes at Stordalen 1972, in Sonesson, M. (ed.), *Progress Report 1972. Swedish IBP Tundra Project Technical Report* **14**, 66-75.

Stewart. W.D.P. (1966) *Nitrogen Fixation in Plants*. Athlone Press, London.

Todd, R.L., Meyer, R.D., and Waide, J.B. (1978) Nitrogen fixation in a deciduous forest in the south-eastern United States, *Ecol. Bull.* **26**, 172-177.

Vincent, W.F. (1988) *Microbial Ecosystems of Antarctica*, Press Syndicate of the University of Cambridge, Cambridge.

Vincent, W.F. (2000) Cyanobacterial dominance in the polar regions, in Whitton, B.A. and Potts, M. (eds.), *The Ecology of Cyanobacteria. Their Diversity in Time and Space*, Kluwer Academic Publishers, Dordrecht, pp. 321-340.

Vincent, W.F., Castenholz, R.W., Downes, M.T. and Howard-Williams, C. (1993) Antarctic cyanobacteria: Light, nutrients, and photosynthesis in the microbial mat environment, *J. Phycol.* **29,** 45-755.

Vlassak, K., Paul, E.A., and Harris, R.E. (1973) Assessment of biological nitrogen fixation in grassland and associated sites, *Plant Soil* **38**, 637-649.

Waughman, G.J., French, J.R.J., and Jones, K. (1981) Nitrogen fixation in some terrestrial environments, in Broughton, W.J. (ed.), *Nitrogen fixation. Volume 1. Ecology*, Clarendon Press, Oxford, pp. 135-192.

Wojchiechowski, M.F. and Heimbrook, M.E. (1984) Dinitrogen fixation in alpine tundra, Niwot Ridge, Front Range, Colorado, USA, *Arct. Alp. Res.* **16**, 1-10.

Zacheis, A., Ruess, R.W. and Hupp, J.W. (200) Nitrogen dynamics in an Alaskan salt marsh following spring use by geese, *Oecologia* **130**, 600-608.

Zielke, M., Ekker, A.S., Olsen, R.A., Spjelkavik, S. and Solheim, B. (2002) The influence of abiotic factors on biological nitrogen fixation in different types of vegetation in the High Arctic, Svalbard, *Arct. Antarct. Alp. Res.* (in press).

Chapter 9

AZOLLA – ANABAENA SYMBIOSIS

SIGAL LECHNO-YOSSEF AND SANDRA A. NIERZWICKI-BAUER
Department of Biology, Rensselaer Polytechnic Institute
Troy, NY 12180 USA

1. INTRODUCTION

The *Azolla-Anabaena* symbiosis is a mutualistic association between the water fern *Azolla*, the nitrogen-fixing cyanobacterium *Anabaena*, and endosymbiotic bacteria. This association has gained attention, because of its potential as a biofertilizer and supplemental animal feed. In recent review articles (Van Hove and Lejeune, 1996; Lejeune *et al.*, 1999) ten useful characteristics attributed to this association were described. Five of these, which were considered to be unquestionable, were the capacity to fix atmospheric nitrogen, high productivity, high protein content and a depressive influence on both aquatic weeds and NH_3 volatilization (Lejeune *et al.*, 1999). These applied aspects related to the utility of this symbiotic association are being reviewed in the next chapter. The focus of this chapter is advancements in our knowledge of basic characteristics of the *Azolla-Anabaena* association, through biochemical, physiological, ultrastructural and molecular biological studies, primarily those carried out over the last ten years which have not already been extensively reviewed. The reader is also directed to several earlier reviews for information on basic *Azolla* research into the 1990's (Braun-Howland and Nierzwicki-Bauer, 1990a; Plazinski, 1990; Van Hove and Lejeune, 1996; Wagner, 1997; Bushnell and Nierzwicki-Bauer, 1999; Lejeune *et al.*, 1999). Highlights included in this review include but are not limited to abscission and vegetative propagation and growth of *Azolla*, sporulation, advancements in understanding the leaf cavity, taxonomic research on the symbiosis, characterization of the endosymbiotic bacteria, and other advancements of knowledge using molecular techniques. The chapter concludes with suggested future research to further improve our understanding of this important symbiosis.

2. *AZOLLA* PHYSIOLOGY

2.1. Abscission and Vegetative Propagation

Abscission of *Azolla* sp. has been studied by Uheda and colleagues (Uheda and Kitoh, 1994; Uheda *et al.*, 1994; Uheda *et al.*, 1995b; Uheda *et al.*, 1999; Uheda and

A.N. Rai, B. Bergman and U. Rasmussen (eds.), Cyanobacteria in Symbiosis, 153-178.

Nakamura, 2000). Shedding of roots and branches was related to environmental and physiological factors. This enables the plant to reproduce (via vegetative fragmentation) and be carried with air currents and/or movement of the water surface to new growth locations. This vegetative fragmentation is the major form of reproduction in *Azolla*.

In *Azolla filiculoides* plants, inhibitors of respiration were shown to cause rapid shedding of roots (Uheda and Kitoh, 1994) and abscission of branches (Uheda *et al.*, 1995b). The events in the process of rapid abscission are similar in the abscission layers of roots and branches. Abscission layers in roots and branches of the intact plant are composed of large cells in the base of the root (Uheda and Kitoh, 1994) and of large flattened cells in the surface of the stem (Uheda *et al.*, 1995b). These large cells expand following treatment with inhibitors of respiration. The expansion is due to absorption of water and may be inhibited by mannitol, which decreases the turgor pressure in the cells. This expansion enables the separation of cells and eventually shedding of the roots or fragmentation of the stem. The expansion and separation of cells in roots is inhibited by acidic pH, and enhanced by neutral pH, suggesting that a change of pH in the vicinity of the roots is involved in the process (Uheda and Kitoh, 1994). In a buffer of neutral or alkaline pH, there was separation of large cells from detached roots even in the absence of chemicals that inhibit respiration (Uheda and Kitoh, 1994; Uheda *et al.*, 1994). Changes in ion flux as a result of alkalization may be related to membrane permeability and the absorption of water into the large cells. Apparently, protein and RNA synthesis are not involved in the rapid abscission, since inhibitors of these processes do not affect shedding of roots (Uheda and Kitoh, 1994), expansion and separation of the large cells in the roots (Uheda *et al.*, 1994), and rapid abscission of branches (Uheda *et al.*, 1995b). However, existing enzymes are probably involved in decomposition of the middle lamella and primary walls of the cells in the abscission layer, since pretreatment of detached roots with papain inhibited separation of cells from the base of the roots (Uheda *et al.*, 1994). The effect of sodium azide treatment on the middle lamella and the cell walls of the large flattened cells in abscising branches are demonstrated by Uheda *et al.* (1995b). In untreated plants the process of abscission of branches starts early in the development of the plant and is dependent on age, which enables the rapid abscission in the presence of chemical or physiological stimulation such as sodium azide. The known events that are involved in rapid abscission as a result of treatment with inhibitors of respiration are summarized below in Figure 1.

Ethylene treatment induces separation of an *Azolla* frond into four to five fragments. However, this abscission process is different than the rapid abscission that results from sodium azide treatment. The differences are described by (Uheda and Nakamura, 2000), and summarized in Table 1.

While ethylene-induced abscission is associated with increased cellulase and polygalacturonase activities that degrade cell wall and unesterified pectin in the middle lamella, in sodium azide-induced abscission there is most likely activation of different enzymes that are already synthesized in the abscission layer.

Rapid abscission occurred in eight *Azolla* strains from six species in response to transient high temperature. The magnitude of rapid abscission and the temperature and time that caused abscission were different for the different strains. Effective treatment time ranged from 30 to 120 minutes, and effective temperature that caused abscission ranged from 40 to 44°C. In *A. filiculoides* and *A. microphylla* the abscised branches were alive and proliferative when the temperature treatment time was short. This suggests that this type of abscission may be involved in survival of the plant under stressful conditions (Uheda *et al.*, 1999).

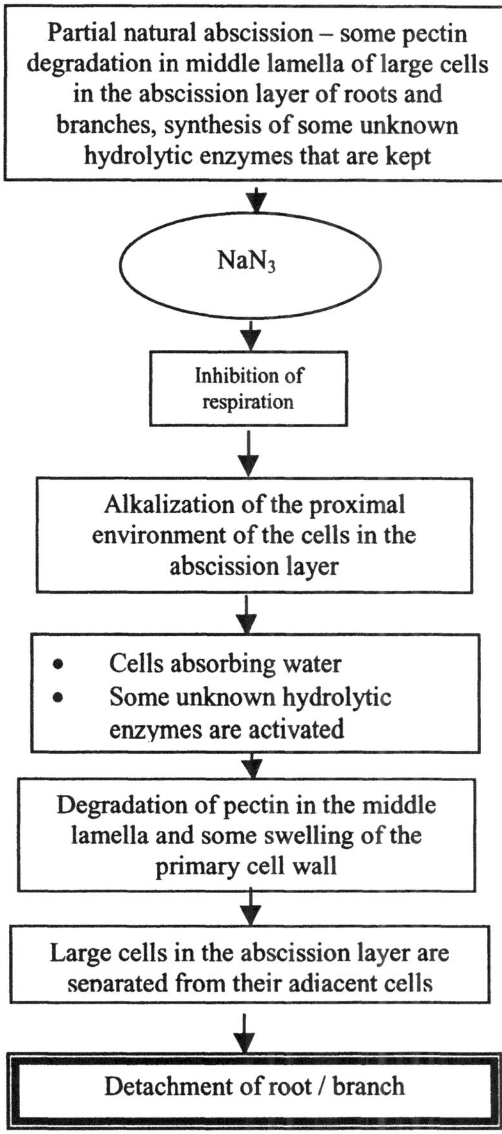

Figure 1. Events in the rapid abscission of roots and branches in A. filiculoides *in response to* NaN_3 *treatment. Summarized from Uheda and Kitoh, 1994; Uheda* et al., *1994; Uheda* et al., *1995b; and Uheda and Nakamura, 2000.*

Table 1. Differences in the effects of ethylene and sodium azide treatments on abscission of Azolla *branches*

Property tested	**Ethylene treatment**	**Sodium azide treatment**
Time to full response	20 – 24 hr	30 min
Location of abscised branches relative to the tip	5-10	4-9
Cellulase activity	Increase relative to control	No increase
Polygalacturonase activity	Increase relative to control	No increase
Cycloheximide effect	Inhibition	No effect
Actinomycin D effect	Low or no effect	No effect
Primary wall	Breakdown and modification	Little or no change
Middle lamella	Remained	Breakdown
Unesterified pectin in middle lamella	Breakdown	Breakdown
Methyl esterified pectin in middle lamella	Remained	Breakdown

2.2. Vegetative Growth

The factors affecting growth of *Azolla* have been recently reviewed (Singh and Singh, 1997). These factors include genotype; high and low temperatures; light intensity, quality and photoperiod; water management, when grown in the field; pH; salinity; availability of nutrients, mostly phosphorus; and influence of pests and diseases. Recent research on temperature and light intensity, phosphorus requirement and uptake, and the influence of salt are described below.

2.2.1. Temperature and Light Intensity

Temperature is clearly one of the most important environmental variables governing the growth and distribution of *Azolla* in aquatic environments. It continues to be shown that different *Azolla* strains and species have different temperature sensitivities (Uheda *et al.,* 1999). Additionally it has been shown that the optimal growth temperature is affected by the light intensity. Results of both field and laboratory studies (carried out before 1990) on the effects of temperature and light intensity on *Azolla* were extensively reviewed and summarized in tabular form by Braun-Howland and Nierzwicki-Bauer (1990a). Since that time, Janes (1998a) has carried out the most comprehensive study that adds to our understanding of the effects of light intensity, temperature and seasonal effects on *Azolla* growth. Studying *A. filiculoides* it was determined that in the laboratory normal vegetative growth could continue at 5°C. Also it was shown that adult plants could survive after short-term (18hr) exposure to temperatures as low as –2°C, but not –4°C or lower. In the field, three phenotypic forms of *A. filiculoides* were described (Janes, 1998a):

* *The survival form* – Appears in the winter. Plants are red, small, with few branches, slow-growing.

* *The colonizing form* – Appears in spring and summer, in free water, where there is no space limitation. The plants are compact and green with a very fast growth rate. Sporulation is sometimes observed at this stage.

* *The mat form* – Appears in mid- to late-summer. Large green plants grow in mats that are composed of many layers. The growth is slower than in the colonizing form, but extensive sporulation is observed.

Seasonal effects on the growth of several *Azolla* hybrids and wild types have also been examined (Gopalaswamy and Kannaiyan, 2000). In that study the environmental conditions that were suggested to be responsible for differences in growth were temperature, relative humidity, wind velocity and evaporation.

2.2.2. Phosphorus Requirement

All *Azolla* species require minimal concentrations of phosphorus (P) for growth. However, there have been differing reports regarding the levels of P necessary for sustained growth. In the laboratory a concentration of approximately 0.06 ppm (2 μM) is reported as being sufficient. Based on field studies it has been suggested that a range from 0.3 to 1 ppm (10 to 33 μM) is necessary. A number of studies have focused on assessing the growth response of different *Azolla* species to different concentrations of P, including the effect of low P on growth (Kushari and Watanabe, 1991; Kushari and Watanabe, 1992). Based on these studies, it appears that *A. pinnata* is best suited for growth under conditions with low P. Additionally, it can be concluded that there is variation among species in the level of phosphorus they require to achieve growth, thus, a selection of species with lower requirements for application in the field is possible. A more detailed review of these results is provided in Bushnell and Nierzwicki-Bauer (1999).

Rates of P uptake and efflux in *Azolla*, grown either in medium containing high or low P, have also been measured using radioactive ^{32}P as a tracer (Bieleski and Lauchli, 1992). Results of this study revealed that *Azolla* has two different uptake systems for low and high concentrations of P, similar to those proposed for other water plants.

2.2.3. Influence of Salt

As salinization of agricultural soils and irrigation water increases, an effort to assess the tolerance of *Azolla* to salt has been examined (Rai and Rai, 1999; Rai and Rai, 2000). In these studies, it has been shown that *A. pinnata* is extremely sensitive to NaCl. The dry weight accumulation of plants was decreased in salt concentrations as low as 10 mmol L^{-1}. While the growth of the plants was inhibited under all of the salt concentrations studied, their appearance was not affected under NaCl concentrations of 10 and 20 mmol L^{-1}. By growing plants in 20 mmol L^{-1} NaCl for 18 days, it was possible to adapt the plants to salinity. Adapted plants were capable of growth in 60 mmol L^{-1} NaCl at a similar rate to control plants. The increase in salt-tolerance of the plants was not related to the age of the culture. Adapted plants grown in 60 mmol L^{-1} NaCl contained higher Na^+ ion concentrations than control plants, but lower concentrations than in the 60 mmol L^{-1} NaCl-exposed plants. Adapted and stressed plants had lower K^+ and higher Ca^{+2} ion concentrations than the control. These results suggest a physiological adaptation to the salt stress (Rai and Rai, 1999).

The effect of salt stress on photosynthetic processes and respiration in NaCl-adapted and unadapted *A. pinnata* has also been studied (Rai and Rai 2000). Chlorophyll *a* levels did not change during 48 hours in control plants and adapted plants but it declined in the 60 mmol L^{-1} NaCl-stressed plants. Similarly, in control and adapted plants there was no change in the chlorophyll *b* content, while the stressed plants showed oscillation in chlorophyll *b* levels during 48 hours of measurements. Adapted plants displayed a higher O_2 evolution rate than control and stressed plants. Adapted and control plants had similar activities of PSII and the entire electron transfer chain, while the PSI activity was lower in adapted plants than in the control. In stressed plants there was enhanced PSI and PSII activity to that of the control plants, but decreased whole chain electron transport activity. The authors suggest that these results indicate that the salt induced damage was located near the linkage of the PSII and PSI reaction centers, probably at Q_b and/or the Cyt $_{b6\text{-}f}$ complex. Additionally, stressed plants showed an increased respiration rate in comparison to control and adapted plants (Rai and Rai, 2000).

2.3. *Sporulation*

Sporulation is the sexual reproduction process in *Azolla*. Although normally *Azolla* displays vegetative reproduction via fragmentation, knowledge of the sporulation process is important for several applications such as:

- *Azolla* as a weed in Britain - Possible mechanisms of reproduction and population control (Janes, 1998a).
- *Azolla* as a biofertilizer - Ways to sustain the population during off-seasons in rice fields.
- Breeding between different *Azolla* strains and species- To form new, better-fit strains.

Different *Azolla* species and strains display different sensitivities to environmental factors such as photoperiod, light intensity, temperature, pH, and phosphate concentration that can influence sporulation. Kar *et al.* (1999) observed differences between strains related to the season of sporulation, while Janes (1998b) has examined *Azolla filiculoides* which sporulates in Britain under natural conditions in the summer and fall, with sporulation often being associated within or with the formation of thick mats. Some strains of *Azolla* (such as CA3001) have never been observed to sporulate regardless of the specific growth conditions (Marsh *et al.*, 1998). For other *Azolla* species (that have been observed to sporulate) the exact factors involved with sporulation and spore germination are still not fully understood but progress is being made. The megasporocarp production is affected by phosphate concentration and plant abundance.

In particular, the influence of phosphorus treatment on sporulation and growth of *Azolla* has been further explored (Kar *et al.*, 2000; Kar *et al.*, 2001; Janes, 1998b). While application of phosphorus is often necessary for maximum *Azolla* biomass production, the phosphorus can inhibit/decrease the sporulation rate and the number of sporocarps per plant (Kar *et al.*, 2000; Kar *et al.*, 2001). It should be noted however that there appears to be a differential response of *Azolla* species to phosphorus for sporulation. For example, the application of phosphorus at 1 mg L^{-1} and 10 mg L^{-1} actually stimulated sporulation in *A. filiculoides* (Janes, 1998b). This result may have been influenced by the association of sporulation with thick mats that have reduced growth in *A. filiculoides*, or possibly due to genetic variability among different *Azolla* species/strains. Additionally, Kar *et al.* (2000) observed that in a strain of *A. caroliniana*, sporulation frequency and sporocarp numbers remained unaffected with additions of phosphate at 2500 µg L^{-1} to 20,000 µg L^{-1}.

The effect of application of gibberellic acid (GA) on sporulation has also been studied. An increased sporulation rate was observed for *A. microphylla, A. pinnata* and *A. caroliniana* when sprayed once with GA in concentrations of 2.5 – 100 µg L^{-1}, and grown without phosphate, in the field or in a nethouse. The level of this effect varied for the different species and strains of *Azolla* tested (Kar *et al.*, 1999). Additionally, a GA spray (of 2.5 µg L^{-1}) combined with phosphate treatment or enrichment of the inoculum with phosphate (by growing the plants under high phosphate concentration before the inoculation) increased both sporulation frequency and biomass accumulation compared to the control (Kar *et al.*, 2001).

In an attempt to define the physiological events associated with sporulation Marsh *et al.* (1998) compared the amino acid and polyamine content in species that are sporulating under different light conditions, or those that are not sporulating, under the experimental conditions. They found that an increase in culture age and plant density (achievement of surface cover after inoculation) in *A. caroliniana*, *A. mexicana* and *Azolla* sp. was accompanied by an increase in glutamine concentrations. Apparently, glutamine is the major form of nitrogenous storage in the plant in conditions of increased plant density and reduced growth. This increase in the amino acid pool, however, occurred in all species tested and is not specifically associated with sporulation. On the contrary, differences in the ratio between the diamine putrescine and the polyamine spermidine were detected between the different species. In one of the species, *Azolla* sp., that sporulates only under constant light conditions, this ratio change appears to be associated with sporulation.

2.4. Sexual Hybridization and Cyanobiont Exchange Between Different Azolla *Species*

In sporulating species, there have been some successful attempts of sexual hybridization between different *Azolla* species, as well as the formation of new combinations of *Azolla – Anabaena* (reviewed by Watanabe, 1994 and Watanabe and Van Hove, 1996). For example, *Anabaena* from *A. microphylla* (MI4031) was successfully introduced into *A. filiculoides* (FI1034) by exchange of the indusium cap of the megaspore (Lin *et al.*, 1989). Successful sexual hybridizations between *A. microphylla* (megasporocarp) and *A. filiculoides* (microsporocarp) (Do *et al.*, 1989; Wei *et al.*, 1988), between *A.*

filiculoides (megasporocarp) and *A. microphylla* (microsporocarp) (Watanabe *et al.*, 1993) and between *A. mexicana* and *A. microphylla* (Zimmerman *et al.*, 1991) have also been reported.

2.5. Spore Germination

Although much information still needs to be elucidated, the influence of temperature, light exposure, and storage on spore germination has recently been reported. In one study (Janes, 1998b), spores of *A. filiculoides*, when freshly collected, germinated at 15°C and 20°C, but not at 5°C. Pre-incubation of the spores at 5°C for longer than one month reduced germination (at 20°C). However, freezing (at –10 ° C) for up to 19 days did not affect germination. Light exposure, even for as short as 4 hours was shown to be required for germination. Only 1% of spores germinated after 3 years of storage in water in the darkness (Janes, 1998b). A great reduction in viability of spores was also observed after storage for longer than 1 year in a different study (Shanmugasundaram and Kannaiyan, 1992). These authors have also reported that pre-soaking dried spores of *A. microphylla* in GA (100 ppm) provided a means to increase the germination rate by 50% (Shanmugasundaram and Kannaiyan, 1992).

3. THE LEAF CAVITY

The leaf cavity is an extracellular compartment in the dorsal lobe of the leaf. In mature leaves, the microsymbionts – cyanobacteria and bacteria are located in the periphery of the leaf cavity. The microbionts are enclosed in a mucilaginous material between internal (Nierzwicki-Bauer *et al.*, 1989) and external envelopes (Uheda and Kitoh, 1991). The inner envelope is best preserved by potassium permanganate fixation, suggesting that it is rich in lipids. However, it does not have a tripartite structure typical for a membrane (Nierzwicki-Bauer *et al.*, 1989). The external three-layered envelope is resistant to treatment with cell-wall degrading enzymes, protease, lipase, sodium hydroxide and nitric acid (Uheda and Kitoh, 1991). An array of chemical treatments that affect different substances in plant cells – proteins, cell membranes, cellulose, wall polysaccharides, pectins, lignins, wax, soluble lipids, sporopollenin and cutinic and suberic substances, was carried out in order to verify the chemical composition of the external envelope. The only treatment that degraded isolated external envelopes was hot alkali methanol. This treatment specifically degrades cutinic and suberic substances (de Roissart *et al.*, 1994). Both, the inner and the outer envelopes were observed in *Anabaena*-containing, and *Anabaena*-free *Azolla*. This finding excludes the involvement of the cyanobionts in the formation of the envelopes. However, in both of these studies, the cyanobiont-free plants contained symbiotic bacteria, that cannot be excluded as playing a role in the synthesis of the envelopes.

A pore is located in the adaxial epidermis of the leaf cavity. The anatomy and the function of the pore were studied by Veys *et al.* (1999, 2000). Two cell layers surround the pore. The layer inside the pore is made of teat-shaped cells extended from the adaxial epidermis. The other layer corresponds to the inner epidermis, which lines the inside of the cavity. Three to four tiers of teat cells form a cone-like pore with an average diameter of 80 μm at the base. The teat cells sometimes overlap but do not touch each other, and a continuous lumen of up to 15 μm was observed. No mucilage was observed in the pore, and a gas exchange between the cavity and the external environment seems possible. The pore opening is larger in younger leaves, before the development of the teat cells. Mature teat-cells surrounding the pore are highly vacuolated and their cytoplasm has high organelle density (Veys *et al.*, 1999). In addition, teat-cells have denser wax deposits than the surrounding epidermal cells (Veys *et al.*, 2000). The authors suggest, based on the morphology of the teat-cells, that these cells function as a physical barrier to prevent particles and organisms from entering the cavity, and microbionts from exiting. The thick wax which coats the teat-cells may function as a water repellant, as well as a repellant for the mucilage that contains the microbionts, supporting the hypothesis that the pore is involved in the maintenance of the symbiotic association. In addition, the organelle rich cytoplasm suggests high metabolic activity and a secretory role of the teat-cells. Cell-wall projections in the base of the teat-cells are rich in pectin, callose and proteins. Recently, it has been demonstrated that *Azolla* teat-cell projections are actively produced and compounds from them are excreted by an exocytotic mechanism (Veys *et al.*, 2002). This differs fundamentally from other types of previously described cell wall projections. It was originally suggested that these projections are involved in a chemical defense against invasion of organisms into the leaf cavity, as well as water repulsion (Veys *et al.*, 1999). Experiments demonstrating that some stress factors promote the formation of projections in *Azolla* add additional support to the hypothesis of a defense role for the teat-cell projections (Veys *et al.*, 2002).

Pores in the leaf cavities of species from *Azolla* section are morphologically different than those of the *Rhizosperma* section. Species in the *Azolla* section have a greater number of teat-cells than *Rhizosperma*, and have the cell-wall projections that are not apparent in *Rhizosperma*, suggesting a better physical protection of the cavity from invasion in these species (Veys *et al.*, 2000). It should be noted however that some stress conditions were able to result in the emergence of such projections even in a member of the *Rhizosperma* section (Veys *et al.*, 2002). The pore region is free from *Anabaena* and hair-cells (Veys *et al.*, 2000). In the region of the pore there is continuity of the inner one-layered envelope, and the outer three partite envelope. In this region the mucilage harboring the symbionts is very thin (Veys *et al.*, 2000). The outermost layer in the outer envelope represents the thin cuticle of the epidermis cells surrounding the pore.

3.1. Conditions in the leaf Cavity

In *A. filiculoides* there is a gradient of ion concentrations and enzymatic activity along the main axis of the plant (Grilli Caiola *et al.*, 1989; Canini *et al.*, 1990; Canini *et al.*, 1991; Canini *et al.*, 1992a,b; Albertano *et al.*, 1993; Canini *et al.*, 1993). The findings of these studies are summarized in Table 2. In these studies the main axis of the plant was divided into 5 regions – The apex, Region I (1st – 4th leaves), II (5th – 8th leaves), III (9th – 12th leaves) and IV (13th to 16th leaves). Ions and oxygen were measured *in situ* using microelectrodes and microprobe, respectively. Canini *et. al.* (1990) found that nitrogenase activity was not detected in the apex, increased between regions I and III and decreased in region IV. The nitrogenase activity corresponded to the heterocyst frequencies in these regions. Ammonium concentration was low in the young regions (0.8 to 2 mM) and high in region IV (6 mM), suggesting that heterocyst development is not inhibited under a certain concentration of ammonium, and that there is a mechanism of ammonium removal from leaf cavities that have high ammonium production. A suggested mechanism for the removal of ammonium is its storage in the vegetative cells of *Anabaena* as cyanophycin granules. The percentage of vegetative cells that contained cyanophycin granules was low in the apex and in the base of the axis and high in the mid section (Canini *et al.*, 1990). Other possible explanations include uptake by branched hair cells of the plant in the leaf cavity, and/or utilization by the bacteria.

Oxygen concentration was measured inside the leaf cavities using a microprobe. These measurements showed that in all leaf cavities, including in the apex the oxygen concentration was lower than in aerated water. The oxygen concentration was lower in the midsection, where nitrogenase activity is high than in the apex and the base of the plant. Additional studies indicated that respiration contributed more than photosynthesis in arriving at the lower O_2 concentration in leaves in region III (Grilli Caiola *et al.*, 1989). The measured oxygen concentrations in this study are in agreement with the measured O_2 concentration in gas bubbles isolated from the leaf cavities of *A. rubra* (Uheda *et al.*, 1995a). It is important to note that the concentration of O_2 in the leaf cavities of *Azolla* is much higher than that of the root nodules of legumes (1.2 – 12% of water-saturated soil; reviewed by Day and Copeland, 1991; Delgado *et al.*, 1998). The higher oxygen concentration in the leaf cavity seems permissible due to the presence of heterocysts possessing envelope layers that serve as an additional protection for nitrogenase from oxygen. In addition, there is a diverse population of symbionts in the leaf cavity, which may enable higher respiration contributing high ATP for potential use in association with the nitrogenase activity.

In the basal leaves, higher concentrations of cations (calcium, potassium and sodium) were found, while the concentration of chloride ions, and the pH were lower in older leaves than in other groups of leaves (Canini *et al.*, 1992b). The authors suggest that a more neutral pH could be important for nitrogenase activity since the pH is higher in leaves where a higher rate of nitrogen fixation is taking place. The concentration of calcium changed most dramatically in relationship to leaf age. The authors suggest that the high concentrations of calcium and sodium in the basal leaves are due to inefficient utilization of these ions by old and disintegrating hair cells, and that these high concentrations are a possible reason for the observed decreased activity of nitrogenase in the basal leaves (Canini *et al.*, 1992b).

Three pools of calcium were found in different symbionts of the leaf cavity using fluorescence signal (due to formation of a complex of chlorotetracycline (CTC) with calcium), electron spectroscopic imaging (ESI) and electron energy loss spectroscopy (EELS) techniques (Canini *et al.*, 1993). Loosely membrane-bound and ionic calcium could be chelated by EGTA and was found in *Azolla* hair cells and in the cytoplasm of vegetative cells of *Anabaena azollae* in basal regions of the fronds. In terminal cells of hairs and in cyanophycin granules and carboxysomes in *Anabaena azollae,* tightly bound calcium were found. A third pool of calcium was found associated with cells walls of simple and branched hair cells of *Azolla* and within the heterocyst envelope of the cyanobionts. The third pool of calcium is suggested to be involved in maintenance of cell wall rigidity in the plant cells, and reduced permeability of the heterocyst to oxygen in the cyanobiont. Calcium was also found in cell walls and within the cytoplasm of bacterial cells (Canini *et al.*, 1993).

The distribution of nitrogen compounds in leaf cavities of different ages was also studied using ESI and EELS (Albertano *et al.*, 1993). Using these techniques it is possible to detect nitrogen in proteins and insoluble components, but not soluble forms such as ammonia that are washed out of the sample during the preparation. In the apical leaves, nitrogen is found mostly in the terminal cells of primary branched hairs (PBH) and in vegetative cells of *Anabaena azollae*. It is suggested that the PBH are involved in transfer of nitrogen compounds to the apical colony. Nitrogen compounds in leaves of group I and II were found mostly in secondary branched hairs. While in leaves of group III, where the nitrogen fixation rate is the highest, nitrogen compounds were found in the cyanobionts and the bactobionts, and less in the plant hair cells. The opposite is true for the basal leaves, in which less nitrogen is found in the microbionts with more in the plant hair cell. These results support the ability of all of the partners in the symbiosis to efficiently utilize nitrogen fixed by the cyanobionts.

4. TAXONOMIC RESEARCH AND ENHANCEMENT OF *AZOLLA*

An effort to classify all known *Azolla* species has been in progress over the past two decades. Traditionally, *Azolla* has been classified based on the morphology of the spores. This method of classification is sometimes problematic, because some *Azolla* species do not form spores, or only form spores under certain conditions. Classification using spores can also lead to some misconceptions about the taxonomic relationships between species in the genus. The presence of the IRRI collection with 564 accessions (in 1999) and the more recent availability of molecular approaches have increasingly facilitated the classification effort (reviewed by Watanabe and Van Hove, 1996).

The collection of species from different regions in the world and the formation of new variants by mutations, sexual hybridization and spore manipulation (to transfer cyanobionts between species) will enable the production and selection of *Azolla* species that are better fit for use as a biofertilizer. The desirable characteristics are increased sporulation and spore germination, increased growth rate, higher nitrogen fixation, better adaptation to environmental condition such as temperature, light, pH, availability of nutrients (especially phosphorus) and increased resistance against pests and diseases. Experiments in hybridization, transfer of the symbionts between species, and re-inoculation of *Anabaena* – free *Azolla* have been reviewed elsewhere (Watanabe, 1994; Watanabe and Van Hove, 1996; Bushnell and Nierzwicki-Bauer, 1999). Currently, these experiments have showed only limited success.

4.1. Taxonomy of the Cyanobacteria

Attempts to classify the cyanobionts from different *Azolla* species have been carried out in recent years using a variety of molecular markers.

Table 2. Gradient in conditions along the stem axis of A. filiculoides. ND, not detected

Measured Property	Apex	Leaf 1-4	Leaf 5-8	Leaf 9-12	Leaf 13-16	Growth medium	Ref.
Ammonium (mM)	0.8 ± 0.05	~2	0.6 ± 0.05	0.8 ± 0.06	6 ± 0.2	0.08	Canini *et al.*, 1990
Nitrogenase (pmol C_2H_4. h^{-1} per leaf cavity)	ND	275 ± 17	390 ± 37	453 ± 27	90 ± 8		"
Heterocyst frequency (%)	5 ± 0.4	8 ± 0.6	15 ± 0.3	19 ± 0.4	12 ± 0.5		"
% vegetative cells containing cyanophycin)	~45	~80	~85	~85	~60		"
O_2 (atm)	0.18 ± 0.014	0.16 ± 0.003	0.157 ± 0.003	0.174 ± 0.005	0.18 ± 0.01	0.21	Grilli Caiola *et al.*, 1989
Calcium (M)	2-4 x 10^{-5}	2-4 x 10^{-5}	1-2 x 10^{-4}	1-2 x 10^{-4}	3 x 10^{-3}	10^{-2}	Canini *et al.*, 1992b
Sodium (M)	1.5 x 10^{-4}	1.5 x 10^{-4}	2.5 x 10^{-4}	3.5 x 10^{-4}	1 x 10^{-3}	6 x 10^{-2}	"
Potassium (M)	6 x 10^{-3}	3-5 x 10^{-4}	3-5 x 10^{-4}	2 x 10^{-3}	8 x 10^{-3}		"
Chloride (M)	0.45	0.45	0.45	0.35	0.25	1	"
pH	6-6.5	6–6.5	6–6.5	6–6.5	5.7	5.5	"

Hybridization studies using probes against different *nif* genes were not capable of differentiating between *Anabaena* from *A. caroliniana, A. microphylla* and *A. mexicana*. However, using these probes species from the *Euazolla* section clustered separately from those in the *Rhizosperma* section. Within *Rhizosperma*, a differentiation between *A. pinnata* and *A. nilotica* was observed (Van Coppenolle *et al.*, 1995). In contrast, DNA probes that were randomly cloned from the genomes of extracted cyanobionts of *A. caroliniana, A. rubra* and *A. microphylla*, showed extensive variation in RFLP. A combination of these probes could distinguish between the cyanobionts of all *Azolla* species, and showed intraspecific variation among the cyanobionts from *A. filiculoides, A. caroliniana* and *A. pinnata* (Plazinski *et al.*, 1990b).

Zheng *et al.* (1999) classified the *Azolla* cyanobionts of 18 accessions using PCR-fingerprinting based on short tandemly repeated repetitive (STRR) sequences in the genome of heterocystous filamentous cyanobacteria. In this study, a clear distinction between cyanobionts from the section *Euazolla* and the section *Rhizosperma* was observed. In a phylogenetic analysis based on the PCR fingerprinting *A. filiculoides* formed a separate cluster between *Euazolla* and *Rhizosperma*. The STRR approach differentiated between cyanobacteria associated with all *Azolla* species with one exception. The cyanobacteria associated with *A. microphylla* and *A. caroliniana* gave identical patterns (Zheng *et al.*, 1999). Both, Zheng *et al.* (1999) and Van Coppenolle *et al.* (1995) could not distinguish amongst the cyanobacteria extracted from different strains of the same species. In contrast, a distinction between *Azolla - Anabaena* complexes from the same species was achieved using random amplified polymophic DNAs (RAPDs) (Van Coppenolle *et al.*, 1993). It is not clear whether the observed difference in these approaches to identify intraspecific differences is due to the specificity of the markers, or a lower variability among the cyanobionts rather than the host plants. A RAPD approach was also used to differentiate between different *Anabaena azollae* isolates, as well as to compare them with extracted cyanobionts (Kim *et al.*, 1997). This study showed variation in the RAPD pattern among the examined cyanobacteria. However, no comparison was made among cyanobacteria extracted from different strains of the same species.

Using a different approach, the taxonomy of the cyanobacteria based on fatty acid composition was carried out. Results clustered the cyanobionts of *A. rubra* within the *Rhizosperma* section, which is in contrast to the classical systematics of the genus based on spore structure. In this study, again, no intraspecific differences were observed (Caudales *et al.*, 1995).

In summary, a variety of different molecular techniques have been used in attempts to better classify the cyanobionts associated with different *Azolla* species and different strains within the same *Azolla* species. In general it has been possible, using different specific molecular techniques, to distinguish the cyanobacteria in association with the *Euazolla* from those in association with the *Rhizosperma* section. Distinction between the cyanobacteria within different *Azolla* species within each of the major sections has also been for the most part successful. Nevertheless, the ability to accurately detect differences amongst cyanobacteria within different strains of the same *Azolla* species has proven difficult. This is likely to be either due to limitations of the current techniques being used, or a very high degree of similarity of the cyanobacteria from different *Azolla* strains of the same species.

5. ENDOSYMBIOTIC CYANOBACTERIA

5.1. Isolation of Cyanobacteria

Many research groups have carried out a variety of procedures in attempts to isolate cyanobionts from the leaf cavities of *Azolla* and culture them under free-living conditions (Braun-Howland and Nierzwicki-Bauer, 1990a). In all instances, morphological and molecular characteristics of these isolates have been different in some way from freshly extracted cyanobacteria (*Anabaena azollae*) of different *Azolla* species. Isolates from *A. caroliniana* (Newton and Herman, 1979), *A. filiculoides* (Tel-Or *et al.*, 1983), *A. mexicana* and *A. pinnata* (Gebhardt and Nierzwicki-Bauer, 1991) had different morphology and different RFLP patterns (using the genes *glnA, rbcS* and *psbA* as probes) from the freshly extracted cyanobionts (Gebhardt and Nierzwicki-Bauer, 1991). On the other hand, isolates from *A. filiculoides* and *A. microphylla* were obtained, which reacted with antibodies raised against symbiotic cyanobionts from the same *Azolla* species (Tang *et al.*, 1990). These isolates required supplementation of fructose, casamino acids, combined nitrogen as well as low oxygen atmosphere (1%), and showed only a very limited growth.

Immobilization of free-living isolates in different media such as polyurethane (PU), hydrophilic polyvinyle (PV), and alginate caused changes in morphological and physiological properties of the cells. Immobilized cells were more similar to the symbiotic cells than non-immobilized cells (Shi *et al.*, 1987; Kannaiyan *et al.*, 1994; Rajini and Subramanian, 1996; Rajini and Subramanian, 1997). Immobilized cells possessed a different surface structure, had a 2-fold increase in heterocyst frequency and a 5-fold increase in nitrogenase activity in comparison to when in a non-immobilized state (Shi *et al.*, 1987). In addition, immobilized cells secreted more ammonia than free-living cultures (Kannaiyan *et al.*, 1994). The ultrastructure of immobilized *Anabaena azollae* as well as *A. variabilis* cells resembled that of symbiotic *Anabaena azollae*. Immobilized and symbiotic vegetative cells displayed a denser cytoplasm with more uniformly dispersed thylakoid membranes than the free-living ones. In heterocysts, the intracellular membranes were more diffused in the immobilized cells, than in the free-living ones (Rajini and Subramanian, 1996). However, immobilization differentially affected carbon fixation of these two species. Immobilized *A. variabilis* cells fixed 160% more carbon, while immobilized *Anabaena azollae* cells showed a slight reduction (2%) in carbon fixation compared to the non-immobilized cells (Rajini and Subramanian, 1997). Immobilization in synthetic polymers is believed to mimic conditions within the leaf cavity of *Azolla*, in which the microbionts are embedded in polysaccharide mucilage. A molecular characterization of the immobilized cells was not carried out. However, at least some of the effects of immobilization are common to free-living *A. variabilis* and to the *Anabaena azollae* isolates, suggesting physiological adaptation of the tested species. It remains unclear from these experiments if an immobilization technique may be used for the isolation of the major symbionts from *Azolla* spp.

A 6 Kb region containing the *nifB* operon and the *nifH* gene of *Anabaena azollae* isolated from *A. caroliniana* (Newton and Herman, 1979) was sequenced and compared to previously published sequences of free-living cyanobacteria (Jackman and Mulligan, 1995). The coding regions were 93% to 97% homologous between *Anabaena azollae* and *Anabaena* sp. PCC 7120 at the nucleotide level. In contrast, the degree of homology was much lower when the intergenic regions were compared (52% to 88% homology). Short tandemly repeated repetitive (STRR) sequences are found in the intergenic region, and six new STRR sequences were found in the *Anabaena azollae* operon. The arrangement, types, and number of STRR sequences differ between *Anabaena azollae* and *Anabaena* 7120. A comparison between *Anabaena azollae* from *A. caroliniana* to other free-living and freshly isolated *Anabaena azollae* was conducted (Jackman and Mulligan, 1995).

5.2. Plasmids

Large indigenous plasmids in a size range of 51 to 144 kbp were found in *Anabaena azollae* extracted from *A. caroliniana, A. filiculoides, A. mexicana, A. rubra, A. microphylla, A. nilotica* and *A. pinnata* (Plazinski *et al.*, 1991). Plasmids from different species of *Azolla* showed sequence similarity in hybridization experiments, but also some sequence uniqueness. Hybridization with probes of some symbiotic genes from *Rhizobium* (*exoX, exoY, nod* box, and *nodMN)* produced positive signals as well. A sequence from the plasmid of *Anabaena* extracted from *A. filiculoides,* complimented Exo^- mutation in *Rhizobium* spp. NGR234, leaving open the possibility of some involvement of the cyanobacteria in the production of polysaccharide in the leaf cavity and maintaining the symbiosis.

A plasmid in *Anabaena* from *A. pinnata* was found to contain the genes for glutathione dependent formaldehyde dehydrogenase (*gdfaldh*) and S-formylglutathione hydrolase (*fgh*). These two enzymes are involved in detoxification of exogenous or endogenous formaldehyde in non-methylotrophic organisms (Shaw *et al.*, 1998; Shaw *et al.*, 2001). Since free-living *Anabaena* strains have not been reported to have these genes, the authors suggest that *Anabaena azollae* acquired the genes in order to utilize methanol produced by the plant. The source of the gene is not clear, but the possibility that it was acquired from bacteria in the leaf cavity cannot be excluded.

5.3. Physiology and Molecular Studies

5.3.1. Nitrogenase

Nitrogenase is the enzyme that is responsible for N_2-fixation in microorganisms. The enzyme is composed of two proteins. The iron protein (Fe-protein, also called dinitrogenase reductase) is a α_2 homodimer of about 64 kDa, and is encoded by the *nif*H gene. The molybdenum-iron protein (MoFe-protein, also called dinitrogenase) is a $\alpha_2\beta_2$ heterotetramer of about 240 kDa, in which the α subunits are encoded by *nif*D and the β subunits by *nif*K (reviewed by Burgess and Lowe, 1996). The Fe-protein of the nitrogenase enzyme from *Anabaena azollae* extracted from *A. filiculoides* has two isoforms. A 30-kDa isoform is expressed during 18-hour dark incubation of the plants, stays stable during 20 minutes of re-illumination, and increases in quantity after longer illumination periods. Under a dark incubation, however, nitrogenase activity (measured as acetylene reduction rate) is not detected. Recovery of nitrogenase activity after re-illumination corresponds to the appearance of a 36-kDa isoform of the nitrogenase within 10 minutes of illumination (Bar and Tel-Or, 1994).

5.3.2. Protection of Nitrogenase from Reactive Oxygen

The enzyme superoxide dismutase (SOD) converts superoxide radicals (O_2^-), formed in electron transfer processes such as respiration and photosynthesis, into molecular oxygen and hydrogen peroxide. The enzyme has several isozymes that are different in their metal ligand. The different isozymes are involved in the protection of different systems from damage caused by superoxide radicals. In vegetative cells, SOD protects mostly the photosynthetic systems, membranes and proteins, while in heterocysts, it is believed to protect the enzyme nitrogenase. In *Anabaena azollae* from *A. filiculoides* three forms of SOD were found, the Fe-SOD was the major form in both vegetative cells and heterocysts. In vegetative cells, manganese-SOD (Mn-SOD), iron-SOD (Fe-SOD) and hybrid iron-manganese (Hy-SOD) isozymes were found. The Mn-SOD was not found in heterocysts. The total activity of SOD in the heterocysts, measured by activity staining on a polyacrylamide gel, was 15% lower than that in the vegetative cells. The activity of Fe-SOD in heterocysts was 25% of its activity measured in vegetative cells, while the activity of Hy-SOD in heterocysts was comparable to that in vegetative cells. (Canini *et al.*, 1991).

The quantity of Fe-SOD protein was assayed using immunogold labeling with antibodies raised against the Fe-SOD protein of *A. cylindrica*. In vegetative cells of *Anabaena azollae* the level of the protein (expressed as number of gold particles. μm^{-2}) increased slightly from the apical leaves to mature leaves (leaf 5-8) and kept constant along the rest of the stem axis. In contrast, in heterocysts, the level of the protein in the basal leaves was about half of that found in leaves of group II (5-8), in which the nitrogenase activity is highest. The density of gold particles was about 30% higher in heterocysts of leaves from group II than in vegetative cells. This finding is in contrast to the measurement of SOD activity in the whole plant that was higher in vegetative cells than in heterocysts (Canini *et al.*, 1991). The difference between the two studies may have resulted from the difference in leaf age, or from regulation of the activity of the protein. In vegetative cells, gold particles were found in the thylakoid membrane and the outer membrane, while in heterocysts, higher densities of gold particles were found in the cytoplasm than within the honeycomb regions (Canini *et al.*, 1992a). The cellular localization of Fe-SOD in vegetative cells supports its role in protection of the photosynthesis and respiration systems, and the removal of superoxide ions that are formed in this process. In heterocysts, the presence of Fe-SOD in proximity to nitrogenase suggests it has a role in protection of that enzyme from superoxide radicals that are formed inside the heterocysts.

5.3.3. The 32-KDa Protein

The 32-KDa protein (D1) is encoded by *psbA* genes. This protein is involved in oxygenic photosynthesis in plants and cyanobacteria. Preliminary experiments showed a 10-fold higher transcription of *psbA* genes in *Anabaena azollae* from *A. caroliniana* compared to the free-living *Anabaena* 7120 (Nierzwicki-Bauer and Haselkorn, 1986). Based on these observations, the authors suggest that the gene product has an additional function apart from its involvement in the photosynthesis. Immunogold labeling with antibodies against three different epitopes of the protein detected the 32-KDa protein in vegetative cells, heterocysts and akinetes of *Anabaena azollae* in *A. caroliniana*. The presence of the protein in heterocysts is unexpected since these cells do not have active photosystem II (PSII). These results further support the earlier suggestion that in *Anabaena azollae* the 32-KDa protein plays another role in addition to its involvement in the photosynthetic process (Braun-Howland and Nierzwicki-Bauer, 1990b).

5.3.4. RuBP Carboxylase (RuBisCO)
RuBisCO is the first enzyme involved in assimilation of CO_2 in the Calvin Cycle of photosynthesis. The enzyme in plants and cyanobacteria is encoded for by two genes – *rbcS* encodes the small subunit and *rbcL* encodes the large subunit of the protein. Using *in situ* and Northern blot hybridization, the presence of transcripts for this enzyme was found in vegetative cells as well as in heterocysts of *Anabaena azollae* from *A. pinnata* and *A. caroliniana.* Since the presence of *rbcLS* transcripts in the heterocysts was unexpected, the authors suggest that it will be important to carry out immunogold labeling to determine if the protein is also present in heterocysts (Madan and Nierzwicki-Bauer, 1993).

5.3.5. Influence of Salt
It has been observed that in addition to the impact of salt exposure on the *Azolla* plant, there are also effects on the cyanobionts (Rai *et al.*, 2001). Exposure of the *Azolla-Anabaena* association to a salt concentration of 60 mmol L^{-1} for 48 hours caused increased heterocyst frequencies. Maximum heterocyst frequencies were observed in leaf numbers 13 and 14 in the control plants, versus leaves 9-11 in plants exposed to 60 mmol L^{-1} NaCl. However, in salt exposed plants there was no correlation between heterocyst frequency and nitrogenase activity in leaves of different ages. Sodium chloride is believed to affect nitrogenase assembly, its protection from oxygen and the availability of electrons for its reaction. Exposure of plants to a 40 mmol L^{-1} NaCl concentration, only caused a reduction of 39% in nitrogenase activity. Overall, salt had a higher effect on the growth of the system than on nitrogenase activity, suggesting that there is some protection of the cyanobiont from salt stress within the host leaf cavity.

6. ENDOSYMBIOTIC BACTERIA

While the presence of bacteria associated with the cyanobionts of *Azolla* sp. has been recognized for many years, it still remains unclear as to whether the bacteria have specific functions and are essential to the symbiosis. The majority of research on the endosymbiotic bacteria has employed traditional microbiological, biochemical and physiological techniques for the identification of bacteria isolated from different *Azolla* species and cultures. Unfortunately, using these techniques, it is possible to overlook bacteria that are difficult to culture, or present in only small quantities within leaf cavities. In addition, bacteria that are grown in culture may be physiologically different than when residing within the leaf cavity. In recent years, the development and utilization of modern molecular techniques for the identification of bacteria has enhanced the study of bacteria in the symbiosis. These newer techniques have made it possible to study bacteria without the need of first growing them in culture, as well as to examine them *in situ*, within leaf cavities. A review of transmission electron micrograph observations, studies that have focused on identification of bacteria, and experiments aimed at elucidation of possible functions of the bacteria are presented below.

6.1. Transmission Electron Microscopy (TEM) Observations

The first TEM observation within the leaf cavity was obtained in 1964 (Grilli, 1964). In the past decade bacteria have been observed within the leaf cavities of all *Azolla* species examined, as well as in megaspores. The most extensive TEM examination of bacteria within leaf cavities was carried out in *A. caroliniana* and *A. mexicana*.

In *A. caroliniana*, five different morphotypes of bacteria were observed. In this study it was demonstrated that the size of the bacterial population and the quantitative relationships among the different morphotypes were dependent on the leaf age. Only one of the morphotypes had the characteristic appearance of Gram-positive bacteria. The Gram-positive morphotype appeared in a very low abundance (Nierzwicki-Bauer and Aulfinger, 1991). The correlation between leaf age and bacterial population size is in agreement with earlier isolation experiments (Petro and Gates, 1987). In *A. mexicana*, three morphotypes were observed in *Anabaena*-free plants, and two morphotypes in the symbiotic plants. One of the morphotypes in *A. mexicana* appears similar to one of the morphotypes from *A. caroliniana* (Nierzwicki-Bauer and Aulfinger, 1990).

Gram-negative bacteria were observed in germinating sporelings of *Azolla microphylla* and in the megaspores (Aulfinger *et al.*, 1991). Bacteria having coryneform or rod morphology typical of the genus *Arthrobacter* have also been observed in the megaspores, as well as in mature leaves, of *A. filiculoides* (Carrapico, 1991). The existence of bacteria within the megaspore, where cyanobacterial akinetes are also present, demonstrates that the bacteria are actively maintained by the plant and suggests that the bacteria play a role in the symbiosis.

6.2. Identification of the Bactobionts

Bacteria have been isolated from leaves of different *Azolla* species (see references in Table 3), as well as from the megasporocarps and microsporocarps of *A. filiculoides* (Forni *et al.,* 1990). In all of the described studies the plants were surface sterilized with hypochlorite prior to the isolation. The techniques used to extract bacteria from the leaves were variable and included the use of a microcapillary tube to extract microbionts directly from the leaf cavity, or crushing the plants using different techniques. The growth media for culturing the isolates varied in different laboratories. In most instances identification was carried out using biochemical techniques and nutritional requirements of the isolates. The results of different isolation experiments are summarized in Table 3. In most of the studies, bacteria were identified as *Arthrobacter* spp. or a relative, Gram-positive *Corynebacterium* sp. The presence of Gram-positive, *Arthrobacter*-like bacteria within the leaf cavity of *A. filiculoides* was verified using immunogold labeling, with antibodies raised against free-living *Arthrobacter globiformis* and against isolates from the same species (Leonardi *et al.*, 1993). However, the Gram-negative bacterium *Agrobacterium* sp. was isolated from *A. filiculoides* (Plazinski *et al.*, 1990a; Serrano *et al.*, 1999), *A. microphylla* (Shannon *et al.*, 1993) and *A. caroliniana* (S. Lechno-Yossef and S.A. Nierzwicki-Bauer, unpublished results). Also, bacteria belonging to the β-proteobacteria subgroup were isolated from *A. caroliniana* (Newton and Herman, 1979; Lechno-Yossef and Nierzwicki-Bauer, unpub.

results), and bacteria from the flavobacter-bacteroid group were isolated from *A. pinnata* (Serrano *et al.*, 1999). All of the isolated bacteria belong to groups of soil bacteria that are normally associated with plants. It appears that the results of isolation experiments depend on the isolation technique, the growth conditions, as well as the relative abundance of the bacterial species within the leaf cavity. It should be noted, however, that some slow-growing bacteria might be overlooked in these types of experiments. In addition, if bacteria have adapted to growth within the leaf cavity, it may be hard or impossible to isolate them without additional information on specific nutritional and growth requirements.

Fortunately using molecular techniques, such as sequencing 16S rRNA genes from bacteria extracted directly from the plants, (without isolation) it is now possible to obtain additional information about the identity of bacteria residing within the leaf cavities of *Azolla* species, without the need to first isolate and culture them. Using this approach it was found that in addition to the four bacterial isolates that were successfully cultured from *A. caroliniana*, bacteria from the flavobacter-bacteroid group, and *Bradyrhizobium* sp. are present in the leaf cavity, albeit probably in very small quantities. In these experiments, the presence of bacteria closely related to *Arthrobacter* sp. has recently been demonstrated using PCR primers specific to that group of bacteria. The use of specific primers and PCR amplification was the only method that permitted the detection of these sequences, suggesting that these bacteria may only be present in extremely small quantity in the *Azolla* species that were examined (Lechno-Yossef and Nierzwicki-Bauer, unpubl. results). These results are in agreement with the microscopic observation of only a small percentage of the bacteria belonging to this same species (Nierzwicki-Bauer and Aulfinger, 1991).

The isolation of a *Rhizobium* sp. (Isolate CA3 from *A. caroliniana*) has not been described before. This isolate has a 16S rDNA sequence identical to a sequence obtained by PCR amplification of the gene from freshly extracted bacteria from the same species. The morphology of the isolate is pleomorphic (Figure 2), and is different from the bacteria observed in the leaf cavity of that species (Nierzwicki-Bauer and Aulfinger 1991). However, because of the pleomorphism of the isolate, it is possible that it appears as a different morphotype when residing within the leaf cavity of the plant (Lechno-Yossef and Nierzwicki-Bauer, unpubl. results).

6.3. Suggested Roles of the Endosymbiotic Bacteria

The fact that bacteria are observed in all of the *Azolla* spp., as well as in the spores (Carrapico, 1991; Forni *et al.*, 1990; Nierzwicki-Bauer and Aulfinger, 1991) and also in *Anabaena*-free *Azolla* species (Nierzwicki-Bauer and Aulfinger, 1990), suggests that the bacteria may play an important role in the symbiosis. Nevertheless, direct study of the role of the bacteria has been difficult using traditional techniques since they require *a priori* assumptions that can be tested experimentally. Also, it is not clear if free-living isolates express the same characteristics in culture as they do in the association. In light of the above, only limited progress has been made towards elucidating potential functions of bacteria in the symbiosis. The work to date, which is described below, suggests that the bacteria play an important role in the symbiosis. However additional experimentation is still needed to conclusively demonstrate the function(s) of bacteria in this symbiosis.

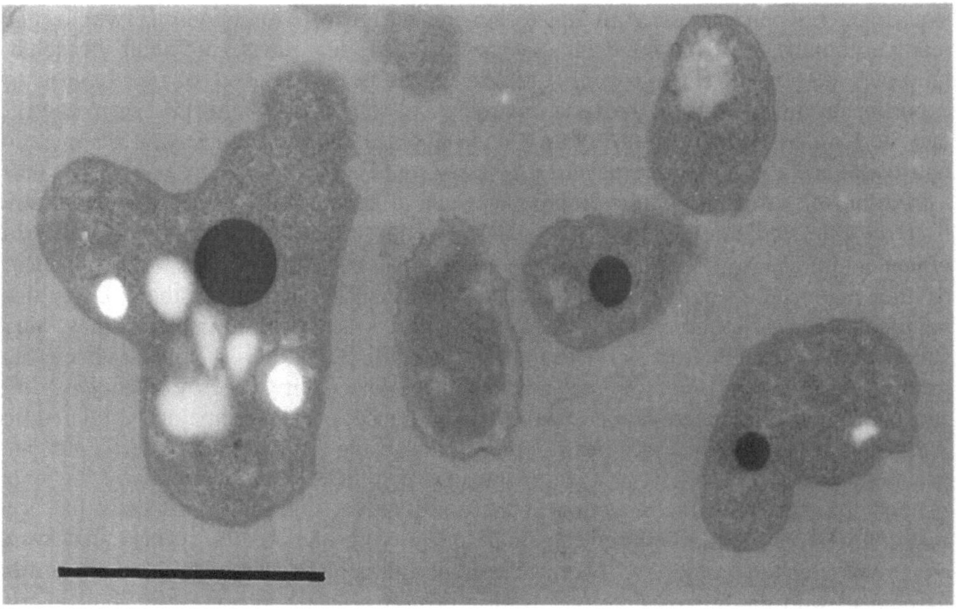

Figure 2. TEM image of isolate CA3 from A. caroliniana *(Lechno-Yossef and Nierzwicki-Bauer). Bar, 1 μm.*

Immunoelectron microscopy has been used to examine the possibility that bacteria possess the ability to fix nitrogen. Using antibodies against the Fe and the MoFe proteins of nitrogenase and immunogold labeling proteins were identified in a subset of the bacteria in the leaf cavities of *A. caroliniana* and *A. filiculoides* (Lindblad *et al.*, 1991). These observations suggest that some of the bacteria may be involved in nitrogen fixation in the leaf cavity. However, in our unpublished experiments and in a study by Forni *et al.* (1989) bacterial isolates did not show nitrogenase activity using acetylene reduction assays, suggesting that if the bacteria are carrying out nitrogen fixation *in situ*, they may require specific conditions (not provided under the experimental conditions used). Similarly, when grown in a free-living state, symbiotic *Rhizobia* have also not exhibited nitrogenase activity (reviewed by Day and Copeland, 1991).

Two other studies have examined *Arthrobacter* spp. isolated from different *Azolla* spp. In these studies it was shown that the isolates have the ability to synthesize the plant hormone auxin in culture, when supplied with the precursor, tryptophan (Forni *et al.*, 1992b). Other bacteria that are associated with plants, or grown in the rhizosphere, have also been shown to produce phytohormones (reviewed by Costacurta and Vanderleyden, 1995). It is believed that the bacterial symbionts in *Azolla* are involved in auxin production.

Table 3. Identification of bacteria isolated from different species of Azolla

Azolla **Source #**	Isolate Identity	Identification Method	Ref.
A. filiculoides accessions: ADUL FI139[a] (FI 1517)[b], FI162 (FI 1527), FI170 (FI 1531), FI174 (FI 1535), FI NA	28 different isolates identified as *Arthrobacter* sp, *A. globiformis* and *A. nicotianae*	Biochemical	Forni *et al.*, 1990
A. caroliniana (from G. Peters, no accession number).	*Alcaligenes faecalis* *Caulobacter fusiformis*	Slime forming Yellow-pigment technique of ID is not described	Newton and Herman, 1979
Wild *A. filiculoides* - Portugal	*Agrobacterium radiobacter, Staphylococcus sp., Rhodococcus sp, R. equi, Corynebacterium jeikeium*	BIOLOG and API	Serrano *et al.*, 1999
Wild *A. pinnata* -Portugal	*Staphylococcus* sp, *Weeksella zoohelcum*	BIOLOG and API	,,
Wild *A. filiculoides* - Australia	*Agrobacterium* sp. strain AFSR-1	API	Plazinski *et al.*, 1990a
A. pinnata PI0097 and A. filiculoides FI1036 both *Anabaena*-free	No isolates		,,
A. caroliniana, A. microphylla, A. mexicana A. filiculoides (no accession numbers; from G. Peters)	*Arthrobacter* sp. Maybe *Agrobacterium* (AM-5 from A. *microphylla*)	Gram stain, morphology and DNA base composition.	(Shannon *et al.*, 1993)
A. caroliniana, A. microphylla, A. mexicana, A. filiculoides, A. pinnata. (No accession numbers; from U of Naples, Italy)	*Arthrobacter globiformis, A. nicotianae, A. autescens, A. cristallopoietes* or *A. pascens*	Biochemical	Forni *et al.*, 1989
Four *Azolla* species, (not mentioned which; from C.W. Rains)	*Arthrobacter globiformis* Conn and Dimick	Biochemical	Wallace and Gates, 1986
A. caroliniana (No accession number; from G. Peters)	*Arthrobacter globiformis* Conn and Dimick or *Corynebacterium* Lehmann and Neumann	Biochemical	Gates *et al.*, 1980
A. caroliniana CA3001	*Arthrobacter* sp.	Morphology, cell wall structure	Nierzwicki-Bauer and Aulfinger, 1991
A. caroliniana CA3001	CA1 – *Variovorax* sp. CA2 – *Agrobacterium* sp. CA3 – *Rhizobium* sp. CA4 – *Frateuria* sp.	16S rRNA gene sequences and phylogenetic affiliation	Author's unpubl. results

[a] *Azolla* spp. accession numbers in the ADUL culture collection.
[b] ( ) *Azolla* spp. accession numbers in the IRRI culture collection.

Finally, Forni *et al.* (1992a) found that cultured *Arthrobacter* isolates (in medium supplemented with 50mg/ml yeast extract and 10mg/ml glucose) from *A. filiculoides* Lam. and *A. caroliniana* Willd. produce mucilaginous matrices. The mucilaginous matrices produced by bacterial isolates, as well as those produced in the leaf cavities of *A. filiculoides* Lam., were examined using stains specific for: polysaccharides (PS); carboxylic PS; sulphated PS; proteins; and lipids. In all cases, the mucilage staining patterns were similar, which led the authors to suggest that the mucilaginous matrix in the leaf cavities of *Azolla* is produced by the bactobionts, and not only the cyanobionts or the plant (Forni *et al.*, 1992a). The composition of polysaccharides in the leaf cavity of different *Azolla* species was also studied by HPLC (Forni *et al.*, 1998). In the presence of bactobionts and cyanobionts it was composed of glucose, galactose and fucose. When bactobionts were eliminated by antibiotic treatment, decreased levels of glucose and galactose were present in the PS, also supporting the involvement of bactobionts in their production (Forni *et al.*, 1998).

7. FUTURE STRATEGY

Our fundamental understanding of interactions between *Azolla*, the cyanobacteria, and the bacteria remains at an early stage. More information on the diversity and function of the bacteria should further our understanding of the intricate interactions of the association. Also, a more detailed examination of the biochemical pathways and transport of different nutrients and compounds between the partners are necessary to enhance our understanding of the association. A current trend in the study of the symbiosis is the application of *in situ* techniques (currently immunolocalization) as opposed to separation techniques. These techniques appear to give more accurate results regarding the activity site of different proteins, and the exchange of nutrients between the symbionts. The development of techniques for intensive screening of genes and gene expression in the association, techniques for *in situ* detection of transcripts and proteins in the symbionts, as well as techniques for gene knock-out and reintroduction of genetically modified bacteria and/or cyanobacteria into the association is now highly desirable.

There also remains a need for further basic research on the precise factors that induce sporulation. This work is critical to enhance the production of *Azolla*, as well as to facilitate the creation of new *Azolla* hybrids with improved characteristics such as increased nitrogen fixing capabilities, increased tolerance to high and low temperatures, and increased resistance to pests.

It is envisioned that ultimately, through a combination of basic and applied research, a more complete understanding of the intriguing and important *Azolla-Anabaena* symbiosis will be obtained.

REFERENCES

Albertano, P., Canini, A. and Caiola, M.G. (1993) Subcellular distribution of nitrogen compounds in *Azolla* and *Anabaena* by ESI and EELS analysis, *Protoplasma* **173**, 158-169.

Aulfinger, H., Braun-Howland, E.B., Kannaiyan, S. and Nierzwicki-Bauer, S.A. (1991) Ultrastructural changes of the endosymbionts of *Azolla microphylla* during megaspore germination and early plantlet development, *Can. J. Bot.* **69**, 2489-2496.

Bar, E. and Tel-Or, E. (1994) Effect of light and oxygen on nitrogenase activity and dinitrogenase reductase (Fe-Protein) content in *Azolla-Anabaena* association., *J. Plant Physiol.* **144**, 438-443.

Bieleski, R.L. and Lauchli, A. (1992) Phosphate uptake, efflux and deficiency in the water fern, *Azolla, Plant Cell Environ.* **15**, 665-673.

Braun-Howland, E.B. and Nierzwicki-Bauer, S.A. (1990a) *Azolla-Anabaena* symbiosis: biochemistry, physiology, ultrastructure and molecular biology, in A.N. Rai (ed.), *CRC Handbook of Symbiotic Cyanobacteria*, CRC Press, Boca Raton, FL, pp. 65-117.

Braun-Howland, E.B. and Nierzwicki-Bauer, S.A. (1990b) Occurrence of the 32 kDa Qb-binding protein of photosystem II in vegetative cells, heterocysts and akinetes of *Azolla caroliniana* cyanobionts, *Planta* **180**, 361-371.

Burgess, B. and Lowe, D. (1996) Mechanisms of molybdenum nitrogenase, *Chem. Rev.* **96**, 2983-3011.

Bushnell, T.P. and Nierzwicki-Bauer, S. (1999) Basic and applied studies of the *Azolla-Anabaena* symbiosis, in M. Jha (ed.), *Current Trends in Life Sciences*, Today & Tomorrow's Printers & Publishers, New Delhi, India, pp. 165-189.

Canini, A., Caiola, M.G. and Mascini, M. (1990) Ammonium content, nitrogenase activity and heterocyst frequency within the leaf cavities of *Azolla filiculoides* Lam, *FEMS Microbiol. Lett.* **71**, 205-210.

Canini, A., Galiazzo, F., Rotilio, G. and Caiola, M.G. (1991) Superoxide dismutase in the symbiont *Anabaena azollae* Strasb, *Plant Physiol.* **97**, 34-40.

Canini, A., Bergman, B., Civitareale, P., Rotilio, G. and Caiola, M.G. (1992a) Localization of iron superoxide dismutase in the cyanobiont of *Azolla filiculoides* Lam, *Protoplasma* **169**, 1-8.

Canini, A., Caiola, M.G., Bertocchi, P., Lavagnini, M.G. and Mascini, M. (1992b) Ion determinations within *Azolla* leaf cavities by microelectrodes, *Sens. Actuator B-Chem.* **7**, 431-435.

Canini, A., Albertano, P. and Caiola, M.G. (1993) Subcellular localization of calcium in *Azolla-Anabaena* symbiosis by chlorotetracycline, ESI and EELS, *Bot. Acta* **106**, 146-153.

Carrapico, F. (1991) Are bacteria the 3rd partner of the *Azolla-Anabaena* symbiosis? *Plant Soil* **137**, 157-160.

Caudales, R., Wells, J.M., Antoine, A.D. and Butterfield, J.E. (1995) Fatty acid composition of symbiotic cyanobacteria from different host plant (*Azolla*) species - evidence for coevolution of host and symbiont, *Int. J. Syst. Bacteriol*, **45**, 364-370.

Costacurta, A. and Vanderleyden, J. (1995) Synthesis of phytohormones by plant-associated bacteria, *Crit. Rev. Microbiol.* **21**, 1-18.

Day, D. and Copeland, L. (1991) Carbon metabolism and compartmentation in nitrogen-fixing legume nodules, *Plant Physiol. Biochem.* **29**, 185-201.

de Roissart, P., Jucqued, C., Watetkeyn, L., Berghmans, P. and Van Hove, C. (1994) First evidence for the cutinic nature of the envelope at the interface of *Azolla* and its endophytes, in N. A. Hegazi, M. Fayez and M. Monib (ed.), *Nitrogen Fixation with Non-legume*, The American University in Cairo Press, Cairo, pp. 133-138.

Delgado, M.J., Bedmar, E.J. and Downie, J.A. (1998) Genes involved in the formation and assembly of rhizobial cytochromes and their role in symbiotic nitrogen fixation, *Adv. Microb. Physiol.* **40**, 191-231.

Do, V.C., Watanabe, I., Zimmerman, W.J., Lumpkin, T.A. and de Waha Gaillonville, T (1989) Sexual hybridization among *Azolla* species, *Can. J. Bot.* **67**, 3482-3485.

Forni, C., Grilli Caiola, M. and Gentili, S. (1989) Bacteria in the *Azolla - Anabaena* symbiosis, in F.A.Skinner, R.M. Boddey and I. Fendrik (eds.), *Nitrogen Fixation with Non-Legumes*, Kluwer Academic Publishers, Dordrecht, The Netherlands, pp. 83-88.

Forni, C., Gentili, S., Van Hove, C. and Grilli Caiola, M. (1990) Isolation and characterization of the bacteria living in the sporocarps of *Azolla filiculoides* Lam, *Ann. Microbiol.* **40**, 235-243.

Forni, C., Haegi, A., Delgallo, M. and Caiola, M.G. (1992a) Production of polysaccharides by *Arthrobacter globiformis* associated with *Anabaena azollae* in *Azolla* leaf cavity, *Fems Microbiology Letters* **93**, 269-274.

Forni, C., Riov, J., Caiola, M.G. and Tel-Or, E. (1992b) Indole-3-acetic acid (IAA) production by *Arthrobacter* species isolated from *Azolla, J. Gen. Microbiol.* **138**, 377-381.

Forni, C., Haegi, A. and Del Gallo, M. (1998) Polysaccharide composition of the mucilage of *Azolla* algal packets, *Symbiosis* **24**, 303-313.

Gates, J.E., Fisher, R.W. and Candler, R.A. (1980) The occurrence of corynoform bacteria in the leaf cavity of *Azolla, Arch. Microbiol.* **127**, 163-165.

Gebhardt, J.S. and Nierzwicki-Bauer, S.A. (1991) Identification of a common cyanobacterial symbiont associated with *Azolla* spp. through molecular and morphological characterization of free-living and symbiotic cyanobacteria, *Appl. Environ. Microbiol.* **57**, 2141-2146.

Gopalaswamy, G. and Kannaiyan, S. (2000) Influence of seasonal factors on the growth and biomass production of *Azolla* hybrids, *Indian J. Agric. Sci.* **70**, 691-693.

Grilli Caiola, M., Canini, A. and Moscone, D. (1989) Oxygen concentration, nitrogenase activity and heterocyst frequency in the leaf cavities of *Azolla filiculoides* Lam, *FEMS Microbiol. Lett.* **59**, 283-288.

Grilli, M. (1964) Infrastructure di *Anabaena azollae* vivente nelle foglioline di *Azolla caroliniana*, *Ann. Microbiol.* **14**, 69-90.

Jackman, D.M. and Mulligan, M.E. (1995) Characterization of a nitrogen fixation (*nif*) gene cluster from *Anabaena azollae* 1a shows that closely related cyanobacteria have highly variable but structured intergenic regions, *Microbiol.* **141**, 2235-2244.

Janes, R. (1998a) Growth and survival of *Azolla filiculoides* in Britain - I. Vegetative reproduction, *New Phytol.* **138**, 367-375.

Janes, R. (1998b) Growth and survival of *Azolla filiculoides* in Britain - II. Sexual reproduction, *New Phytol.* **138**, 377-384.

Kannaiyan, S., Rao, K.K. and Hall, D.O. (1994) Immobilization of *Anabaena azollae* from *Azolla filiculoides* in polyvinyl foam for ammonia production in a photobioreactor system, *World J. Microbiol. Biotechnol.* **10**, 55-58.

Kar, P.P., Mishra, S. and Singh, D.P. (1999) Influence of gibberellic acid on the sporulation of *Azolla caroliniana*, *Azolla microphylla* and *Azolla pinnata*, *Biol. Fertil. Soils* **29**, 424-429.

Kar, P.P., Mishra, S. and Singh, D.P. (2000) Variability in *Azolla* sporulation in response to phosphorus application, *Biol. Fertil. Soils* **32**, 458-462.

Kar, P.P., Mishra, S. and Singh, D.P. (2001) Influence of different phosphorus management strategies on the sporulation and growth of *Azolla*, *Exp. Agric.* **37**, 53-64.

Kim, J.J.H., Krawczyk, K., Lorentz, W.P. and Zimmerman, W.J. (1997) Fingerprinting cyanobionts and hosts of the *Azolla* symbiosis by DNA amplification, *World J. Microbiol. Biotechnol.* **13**, 97-101.

Kushari, D.P. and Watanabe, I. (1991) Differential responses of *Azolla* to phosphorus deficiency .I. Screening methods in quantity controlled condition, *Soil Sci. Plant Nutr.* **37**, 271-282.

Kushari, D.P. and Watanabe, I. (1992) Differential responses of *Azolla* to phosphorus deficiency .II. Screening method under concentration controlled conditions, *Soil Sci. Plant Nutr.* **38**, 65-73.

Lejeune, A., Cagauan, A. and Van Hove, C. (1999) *Azolla* research and development: Recent trends and priorities, *Symbiosis* **27**, 333-351.

Leonardi, D., Canini, A. and Forni, C. (1993) Immunological comparison between *Arthrobacter* isolates and bacteria living in *Azolla filiculoides* Lam, *Symbiosis* **15**, 269-279.

Lin, C., Liu, Z.Z., Zheng, D.Y., Tang, L.F. and Watanabe, I. (1989) Re-establishment of symbiosis to *Anabaena* free *Azolla*, *Sci. China Ser. B-Chem.* **32**, 551-559.

Lindblad, P., Bergman, B. and Nierzwicki-Bauer, S.A. (1991) Immunocytochemical localization of nitrogenase in bacteria symbiotically associated with *Azolla* spp, *Appl. Environ. Microbiol.* **57**, 3637-3640.

Madan, A.P. and Nierzwicki-Bauer, S.A. (1993) *In situ* detection of transcripts for ribulose-1,5-bisphosphate carboxylase in cyanobacterial heterocysts, *J. Bacteriol.* **175**, 7301-7306.

Marsh, B.H., Corbin, J.L. and Peters, G.A. (1998) Changes in soluble amino acid and polyamine composition associated with increasing plant density and the onset of sporulation in *Azolla*, *Symbiosis* **24**, 315-326.

Newton, J.W. and Herman, A.I. (1979) Isolation of cyanobacteria from the aquatic fern *Azolla*, *Arch. Microbiol.* **120**, 161-165.

Nierzwicki-Bauer, S.A. and Aulfinger, H. (1990) Ultrastructural characterization of eubacteria residing within leaf cavities of symbiotic and cyanobiont-free *Azolla mexicana*, *Curr. Microbiol.* **21**, 123-129.

Nierzwicki-Bauer, S.A. and Aulfinger, H. (1991) Occurrence and ultrastructural characterization of bacteria in association with and isolated from *Azolla caroliniana*, *Appl. Environ. Microbiol.* **57**, 3629-3636.

Nierzwicki-Bauer, S.A. and Haselkorn, R. (1986) Differences in mRNA levels in *Anabaena* living freely of in symbiotic association with *Azolla*, *EMBO J.* **5**, 29-35.

Nierzwicki-Bauer, S.A., Aulfinger, H. and Braun-Howland, E.B. (1989) Ultrastructural characterization of an inner envelope that confines *Azolla* endosymbionts to the leaf cavity periphery, *Can. J. Bot.* **67**, 2711-2719.

Petro, M.J. and Gates, J.E. (1987) Distribution of *Arthrobacter* sp. in the leaf cavities of 4 species of the N-fixing *Azolla* fern, *Symbiosis* **3**, 41-48.

Plazinski, J. (1990) The *Azolla-Anabaena* symbiosis, in P. M. Gresshoff (ed.), *Molecular Biology of Symbiotic Nitrogen Fixation*, CRC Press, Boca Raton, Florida, pp. 51-75.

Plazinski, J., Taylor, R., Shaw, W., Croft, L., Rolfe, B.G. and Gunning, B.E.S. (1990a) Isolation of *Agrobacterium* sp. strain from the *Azolla* leaf cavity, *FEMS Microbiol. Lett.* **70**, 55-59.

Plazinski, J., Zheng, Q., Taylor, R., Croft, L., Rolfe, B.G. and Gunning, B.E.S. (1990b) DNA probes show genetic variation in cyanobacterial symbionts of the *Azolla* fern and a closer relationship to free-living *Nostoc* strains than to free-living *Anabaena* strains, *Appl. Environ. Microbiol.* **56**, 1263-1270.

Plazinski, J., Croft, L., Taylor, R., Zheng, Q., Rolfe, B.G. and Gunning, B.E.S. (1991) Indigenous plasmids in *Anabaena azollae* - their taxonomic distribution and existence of regions of homology with symbiotic genes of rhizobium, *Can. J. Microbiol.* **37**, 171-181.

Rai, A.K. and Rai, V. (2000) Response of NaCl-adapted and unadapted *Azolla pinnata - Anabaena azollae* complex to salt-stress: Partial photosynthetic processes and respiration, *Symbiosis* **29**, 249-261.

Rai, V. and Rai, A.K. (1999) Growth behaviour of *Azolla pinnata* at various salinity levels and induction of high salt tolerance, *Plant Soil* **206**, 79-84.

Rai, V., Tiwari, S.P. and Rai, A.K. (2001) Effect of NaCl on nitrogen fixation of unadapted and NaCl-adapted *Azolla pinnata - Anabaena azollae*, *Aquat. Bot.* **71**, 109-117.

Rajini, V.S. and Subramanian, G. (1996) Ultrastructural differences between free-living and foam-immobilized cyanobacterial systems, *Cytobios* **88**, 71-77.

Rajini, V.S. and Subramanian, G. (1997) The effect of immobilization on carbon flow through *Anabaena variabilis* and *A. azollae*, *Photosynthetica* **34**, 137-139.

Serrano, R., Carrapico, F. and Vidal, R. (1999) The presence of lectins in bacteria associated with the *Azolla - Anabaena* symbiosis, *Symbiosis* **27**, 169-178.

Shanmugasundaram, R. and Kannaiyan, S. (1992) Effect of growth regulator and fungicide combination on germination, fertilization and viability of the dried sporocarps of *Azolla microphylla*, *Phykos* **31**, 61-67.

Shannon, B.T., Gates, J.E. and McCowen, S.M. (1993) DNA base composition of eubacteria isolated from 4 Species of *Azolla*, *Symbiosis* **15**, 165-175.

Shaw, W.H., Arioli, T. and Plazinski, J. (1998) Cloning and sequencing of a S-Formylglutathione hydrolase (FGH) gene from the cyanobacterium *Anabaena azollae* (Accession No AF035558) (PGR98-024), *Plant Physiol.* **116**, 868.

Shaw, W.H., Muirhead, A., Arioli, T. and Plazinski, J. (2001) A glutathione-dependent formaldehyde dehydrogenase gene of symbiotic cyanobacterium *Anabaena azollae*. *Symbiosis*, **30**, 41-55.

Shi, D.-J., Brouers, M., Hall, D.O. and Robins, R.J. (1987) The effect of immobilization on the biochemical, physiological and morphological features of *Anabaena azollae*, *Planta* **172**, 298-308.

Singh, P. and Singh, D. (1997) *Azolla - Anabaena* symbiosis, in K. Ddarwal (ed.), *Biotechnological Approaches in Soil Microorganisms for Sustainable Crop Production*, Scientific Press, Jodhpur, pp. 93-107.

Tang, L.F., Watanabe, I. and Liu, C.C. (1990) Limited multiplication of symbiotic cyanobacteria of *Azolla* spp. on artificial media, *Appl. Environ. Microbiol.* **56**, 3623-3626.

Tel-Or, E., Sandovsky, T., Kobiler, D., Arad, C. and Weinberg, R. (1983) The unique symbiotic properties of *Anabaena* in the water fern *Azolla*, in G.C. Papageorgiou and L. Packer (eds.), *Prokaryotes: Cell Differentiation and Function*. Elsevier Science, New York, pp. 303-314.

Uheda, E. and Kitoh, S. (1991) Electron microscopic observations of the envelopes of isolated algal packets of *Azolla*, *Can. J. Bot.* **69**, 1418-1419.

Uheda, E. and Kitoh, S. (1994) Rapid shedding of roots from *Azolla filiculoides* plants in response to inhibitors of respiration, *Plant Cell Physiol.* **35**, 37-43.

Uheda, E., Nakamura, S. and Kitoh, S. (1994) Events associated with the rapid separation of cells from detached roots of *Azolla filiculoides* depend on pH, *J. Exp. Bot.* **45**, 1451-1457.

Uheda, E., Kitoh, S., Dohmaru, T. and Shiomi, N. (1995a) Isolation and analysis of gas bubbles in the cavities of *Azolla* leaves, *Physiol. Plant.* **93**, 1-4.

Uheda, E., Nakamura, S. and Kitoh, S. (1995b) Aspects of the very rapid abscission of *Azolla* branches - anatomy and possible mechanism, *Int. J. Plant Sci.* **156**, 756-763.

Uheda, E., Kitoh, S. and Shiomi, N. (1999) Response of six *Azolla* species to transient high-temperature stress, *Aquat. Bot.* **64**, 87-92.

Uheda, E. and Nakamura, S. (2000) Abscission of *Azolla* branches induced by ethylene and sodium azide, *Plant Cell Physiol.* **41**, 1365-1372.

Van Coppenolle, B., Watanabe, I., Van-Hove, C., Second, G., Huang, N. and McCouch, S.R. (1993) Genetic diversity and phylogeny analysis of *Azolla* based on DNA amplification by arbitrary primers, *Genome* **36**, 686-693.

Van Coppenolle, B., McCouch, S.R., Watanabe, I., Huang, N. and Van-Hove, C. (1995) Genetic diversity and phylogeny analysis of *Anabaena azollae* based on RFLP detected in *Azolla-Anabaena azollae* DNA complexes using *nif* gene probes, *Theor. Appl. Genet.* **91**, 589-597.

Van Hove, C. and Lejeune, A. (1996) Does *Azolla* have any future in agriculture? in M. Rahman (ed.), *Biological Nitrogen Fixation Associated with Rice Production*, Kluwer Academic Publishers, Dordrecht, The Netherlands, pp. 83-94.

Veys, P., Waterkeyn, L., Lejeune, A. and Van Hove, C. (1999) The pore of the leaf cavity of *Azolla*: Morphology, cytochemistry and possible functions, *Symbiosis* **27**, 33-57.

Veys, P., Lejeune, A. and Van Hove, C. (2000) The pore of the leaf cavity of *Azolla*: Interspecific morphological differences and continuity between the cavity envelopes, *Symbiosis*, **29**, 33-47.

Veys, P., Lejeune, A. and Van Hove, C. (2002) The pore of the leaf cavity of *Azolla*: Teat-cell differentiation and cell wall projections, *Protoplasma* (in press).

Wagner, G.M. (1997) *Azolla*: A review of its biology and utilization, *Bot. Rev.* **63**, 1-26.

Wallace, W.H. and Gates, J.E. (1986) Identification of eubacteria isolated from leaf cavities of 4 species of the N-fixing *Azolla* fern as *Arthrobacter* Conn and Dimmick, *Appl. Environ. Microbiol.* **52**, 425-429.

Watanabe, I. (1994) Genetic enhancement and *Azolla* collection problems in applying *Azolla-Anabaena* symbiosis, in N.A. Hegazi, M. Fayez and M. Monib (eds.), *Nitrogen Fixation with Non-Legumes*, The American University in Cairo Press, Cairo, pp. 437-450.

Watanabe, I., Lapis-Tenorio, M.T., Ventura, T.S. and Padre, B.C. (1993) Sexual hybrids of *Azolla filiculoides* with *A. microphylla*, *Soil Sci. Plant Nutr.* **39**, 669-676.

Watanabe, I. and Van Hove, C. (1996) Phylogenetic, molecular, and breeding aspects of *Azolla-Anabaena* symbiosis, in J.M. Camus, M. Gibby and R.J. Jones (eds.), *Pteridology in Perspective*, Royal Botanic Gardens, Kew, pp. 611-619.

Wei, W.X., Jin, G.Y., Zhang, N. and Chen, J. (1988) Studies of hybridization in *Azolla*, in K.H.Singh and K.U. Kramer (eds.), *Proc. International Symp. Systematic Pteriodology*, China Sci. Tech. Press, Beijing, pp. 135-139.

Zheng, W.W., Nilsson, M., Bergman, B. and Rasmussen, U. (1999) Genetic diversity and classification of cyanobacteria in different *Azolla* species by the use of PCR fingerprinting, *Theoretical and Applied Genetics* **99**, 1187-1193.

Zimmerman, W.J., Watanabe, I., Ventura, T., Payawal, P. and Lumpkin, T.A. (1991) Aspects of the genetic and botanical status of neotropical *Azolla* species, *New Phytol.* **119**, 561-566.

Chapter 10

APPLIED ASPECTS OF *AZOLLA-ANABAENA* SYMBIOSIS

CHARLES VAN HOVE and ANDRÉ LEJEUNE
Laboratory of Plant Biology, Faculty of Sciences, Catholic University of Louvain, Place Croix du Sud 5 (bte 14), B-1348 Louvain-la-Neuve, Belgium

1. INTRODUCTION

The *Azolla-Anabaena* symbiosis has given rise to abundant research, basic as well as applied, due to its remarkable sophistication (see previous chapter by Lechno-Yossef and Nierzwicki-Bauer) and the diversity of possible applications, particularly in agriculture and animal husbandry. In some areas of China and Vietnam farmers have indeed recognized the usefulness of *Azolla* as feed and as green manure since centuries, perhaps as far back as 2000 years (Liu, 1979; Lumpkin and Plucknett, 1982; Shi and Hall, 1988a). Peruvians have also highlighted the use of *Azolla* as a poultry feed since 1710 (Feuillée, 1725).

Figure 1 presents the level of *Azolla* research as illustrated by the number of scientific papers in the author's data bank on *Azolla*. Research on *Azolla* seems to have remained very modest until 1960. A peak in number of publications during the Sixties, is linked to the promotion of applied research by the Chinese and Vietnamese governments at that time, aimed at extending the use of *Azolla* which was restricted to Southern China and North Vietnam until then. The major second peak coincides with the first oil crisis, in 1973, which induced international interest for alternatives to the high energy-consuming production of chemical N-fertilizer by the Haber-Bosch process. Even if the references from the most recent years are underestimated (it takes time to collect papers published in non-international journals), the general tendency clearly indicates a decline of research devoted to *Azolla*. Figure 1 also mentions the papers which first described the differrent applications of *Azolla*.

An excellent comprehensive description of *Azolla* utilization in agriculture, essentially based on the Chinese and Vietnamese experience, has been presented by Lumpkin and Plucknett (1982). Further general information on the applied aspects of *Azolla* research are found in Nierzwicki-Bauer (1990), Roger (1993), Rai *et al.* (1996), Van Hove and Lejeune (1996), Wagner (1997), Lejeune *et al.* (1999), Mandal *et al.* (1999).

The present chapter will therefore deal only with the most recent advances related to applied aspects of *Azolla*. Section 2 deals with the agricultural utilization of *Azolla*, and includes classical topics (such as fertilization and feed) as well as the newer topics on

A.N. Rai, B. Bergman and U. Rasmussen (eds.), Cyanobacteria in Symbiosis, 179-193.

the nitrogen saving effect, the influence on methane emission, and the role of *Azolla* in integrated farming systems. A brief account of *Azolla* selection and *Azolla* sporulation are then presented. More recent topics also include *Azolla* as an invasive weed (section 3) and its potential role in phytoremediation (section 4). The later sections concern with less documented topics (section 5: *Azolla* and tropical diseases, section 6: *Azolla* in space) or topics connected to but somewhat out of the frame of this review (section 7: "*Anabaena azollae*" in biotechnology). We conclude (section 8) with a few suggestions for future research.

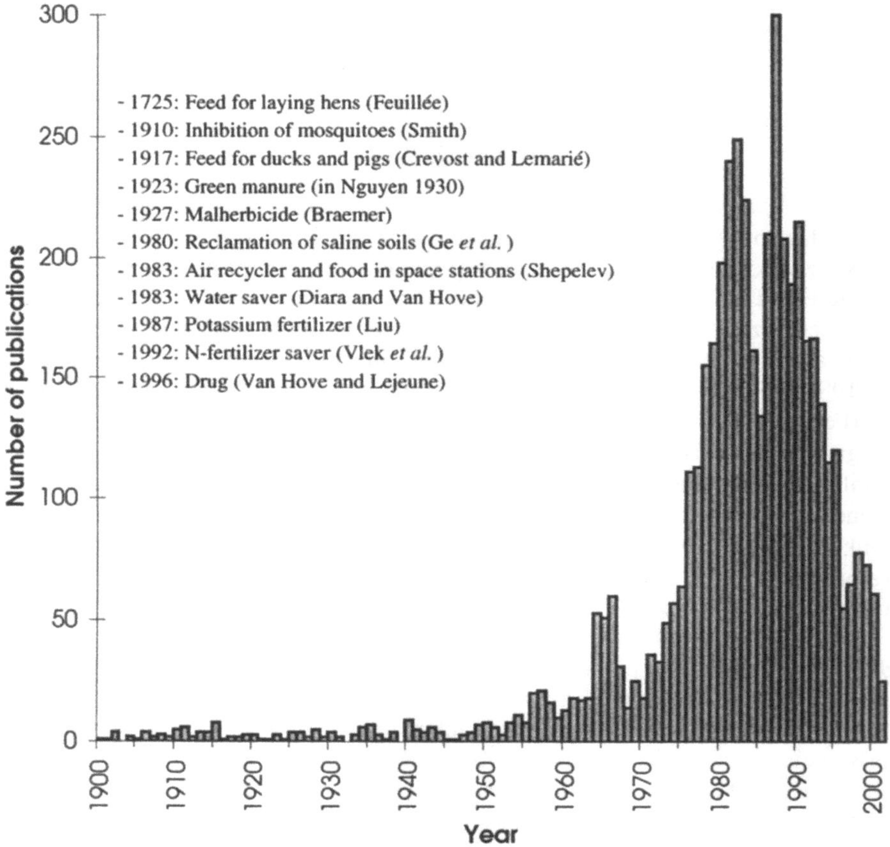

Figure 1. Number of publications related to Azolla *since 1900 as indexed in the authors' data bank (the 41 papers published between 1725 and 1899 are omitted). References of the papers that first mentioned or suggested a particular application of* Azolla *are shown.*

2. *AZOLLA* IN AGRICULTURE

2.1. Fertilizer

Azolla has been used as biofertilizer for rice cultivation since long, and its contribution to the nitrogen economy of rice fields is well established (see Nierzwicki-Bauer, 1990). Ladha *et al.* (2000) have presented data from a 14-year double-crop rice experiment at the International Rice Research Institute, Philippines. This experiment involved comparison of the long-term effects of four treatments (no addition of nitrogen, application of urea, application of *Sesbania* and application of *Azolla*) on yield, N-balance, total-N and available N pools. It confirms the positive influence of *Azolla*, with average grain yields of 5 and 6.4 t. ha^{-1} due to urea and green manure, respectively, while in the control the yields were 4 and 4.4 t. ha^{-1}. This study also indicated that the grain yield declines with time, even in the treatments involving *Azolla* or *Sesbania,* despite significant total soil N increase resulting from these treatments. This is attributed to a decline in physiological N use efficiency, the reasons for which remain unknown.

On the other hand *Azolla* seems to interfere with the fertilization process in rice fields on lateritic soils where the vailability of Cu and Zn is often limited due to their fixation on Fe and Al oxides. When *Azolla* or other green fertilizers are incorporated into the soil, the availability of Cu and Zn is further decreased due to the chelation by organic matter (Mandal *et al.*, 1997).

As for other crops, preliminary results (Plessner *et al.*, 1998) indicate that *Azolla* powder, iron-enriched (4.4% w/w.) by treatment with a solution of $FeSO_4$, could be useful as a slow-release biofertilizer for hydroponic cultures of cucumbers.

2.2. Nitrogen Saver in Rice Fields

The low efficiency of urea-N, mainly due to ammonia volatilisation which often accounts for 50% or more of the applied N within two weeks of its application, is well documented (Vlek and Craswell, 1981; Fillery and Vlek, 1986; Mandal *et al.*, 1999). Since the high rice productivities presently expected cannot be sustained by biological nitrogen fertilization alone, finding ways for minimizing these losses is essential. Only partial answers have been found to this problem (split application, deep placement of the fertilizer, use of slow-release fertilizer, inhibition of urease or of nitrification). The role of *Azolla* in curbing ammonia volatilisation in flooded rice environments, first suggested by Kröck *et al.* (1988), has been supported by experimental data (see Lejeune *et al.*, 1999) including ^{15}N tracer under field (de Macale *et al.*, 1997; de Macale *et al.*, 2002) and greenhouse (Cisse, 2001) conditions.

2.3. Methane Emission

Paddy fields significantly contribute to methane emission and thereby to the greenhouse effect (Aulakh *et al.*, 2001; Le Mer and Roger, 2001). Chen *et al.* (1997) observed a stimulation of CH_4 (and N_2O) emission in the field when *Azolla* was intercropped with rice and suggested that this could results from *Azolla* roots and decomposition of dead *Azolla*. In pot experiments, Ying *et al.* (2000) showed a stimulation of CH_4 emission by *Azolla* in the absence of rice plants. They attributed this increase in CH_4 emmission to the decreased O_2 concentration in water and increased NH_4^+ concentration in soil, caused by *Azolla*. According to King and Schnell (1994), increased NH_4^+ concentration inhibits CH_4 oxidation. However, Ying *et al.* (2000) did not find the stimilation of CH_4 emission by *Azolla* when the experiments were conducted in the presence of rice plants. Probably, *Azolla* stimulated rice root development and thereby the CH_4 oxidation in the rhizosphere. They also showed that *Azolla* facilitates CH_4 transfer from water to air. In contrast, from their field experiments on dual crops, the same authors suggested a probable increase in methane emission due to *Azolla*. Bharati *et al.* (2000) observed that *Azolla* intercropping decreased CH_4 emission due to an increased oxygen concentration in water. This contrasts with other well-documented data indicating a decrease in O_2 concentration due to *Azolla* (Kröck *et al.*, 1988; Gratwicke and Marshall, 2001; unpublished data from the authors lab). Such an effect is easily explained by the inhibition of photosynthetic activity in algae under an *Azolla* mat. In another paper which includes some of the same authors (Adhya *et al.*, 2000), it was concluded that incorporation of *Azolla* stimulates CH_4 emission, although to a lesser extent than the stimulation caused by other organic amendments (*Sesbania* or compost). According to Tang *et al.* (2000) integrated farming systems including rice, fish and *Azolla* decrease CH_4 emission. More research is clearly required for evaluating the impact of various *Azolla* cultural practices on CH_4 emission in rice fields.

2.4. Feed

There is still a need for better information on various qualities of *Azolla* required for its utilization as feed, (see Lejeune *et al.*, 1999). Only a few significant papers have been published recently.

Substitution of wheat by *Azolla* powder for feeding rabbits gave encouraging preliminary results (Feng, 1994) but this needs further study.

Azolla as a good digestible nitrogen source for pigs (Ly and Preston, 2001) as well as the nutritive value and the positive effect of *Azolla* on yolk colour for laying hens (Khatun *et al.*, 1999), has been confirmed. The carotene content in fresh samples of six *Azolla* strains belonging to different species has been analysed at four stages of the population growth curve (Lejeune *et al.*, 2000). Carotene content was always high, and maximum during the linear phase of growth. Results confirm that fresh *Azolla* has a good potential as a source of provitamin A.

As for fish, the digestibility of *A. filiculoides* by *Oreochromis aureus* is reported to be good (Léonard *et al.*, 1998). Partial substitution of soybean meal and maize by sun dried *A. pinnata* in isonitrogenous diets of *Clarias gariepinus*, produced negative effects on all the measured parameters (Fasakin and Balogun, 1998). Such negative results have often been obtained in aquarium conditions (Almazan *et al.*, 1986; Shiomi and Kitoh, 1987; El-Sayed, 1992). Kanangire (2001) tested three isoproteic and isocaloric ratios, containing 0, 30 and 50% *Azolla*, on *Oreochromis niloticus* and *C. gariepinus* reared in fishponds. For *O. niloticus*, growth and production parameters were good and similar in all the three treatments. For *C. gariepinus*, growth and production parameters declined with increasing quantities of *Azolla* in the diet, but remained acceptable. New data on *Azolla* composition in relation with fish feeding are provided by Shiomi and Kitoh (2001). They describe the amino-acid, ascorbic acid, dehydro-ascorbic acid, and fatty acid composition of *A. filiculoides*. A technique for intensive production of *A. imbricata* as forage in arid regions of China (Gansu province) has been described (Liu Y. *et al.*, 2000).

2.5. *Component of Integrated Farming Systems*

Azolla maintenance requires care and using it only for rice cultivation, especially where a single crop is produced per year, is not encouraging for farmers. But when *Azolla* is utilized as animal feed or in integrated farming systems, its permanent production is more attractive. Combined farming systems such as rice and *Azolla*, rice and fish or rice and duck have been traditional practices in South-East Asia. Recent research efforts have successfully developed integrated rice, fish and *Azolla*, rice, duck and *Azolla* or even rice, fish, duck and *Azolla* farming. Some of these are currently being used at the farm level.

2.5.1. *The Rice-Fish-*Azolla *System*

Mainly conducted at the *Azolla* Research Centre in Fuzhou (China) and in the Philippines (see Lejeune *et al.*, 1999), research in this field has given rise to wide application of this integrated farming system during the last few years in the Chinese provinces of Fujian, Jianxsi, Sichuan, Hunan, Hubei, Jiangsu, Liaoning and Jilin (Huang, personal communication). Liu R.Q. *et al.* (2000) have observed positive effects of the rice-fish-*Azolla* system on soil structure, chemistry, microflora and productivity. Tang *et al.* (2000) proposed a model of "high output, low input and less pollution" based on such a system, which produces 10-13 t of rice and 3-4 t of fish per hectare, with a concomitant 50% reduction in chemical fertilizer, pesticide utilization and methane emission.

2.5.2. *The Rice-Duck-*Azolla *System*

In Japan rearing ducks into paddy fields has gained popularity since the Eighties. The addition of *Azolla* to this system is more recent (Kishida, 1996; Sakamoto, 1996; Liu *et al.*, 1998; Kishida and Utsumiya, 1998a,b; Kishida and Okazaki, 1999). The promoter of this system, presently adopted by many organic farmers in Japan, recently published a comprehensive book, with detailed description of the cultural practices he has developed in his own farm, the constraints, the various interactions between the protagonists and the beneficial effects of this association (Furuno, 2001).

2.5.3. *The Rice-Fish-Duck-*Azolla *System*

A two-year (three crops) comprehensive study of ten rice-based systems has been conducted at Central Luzon State University, Philippines (Cagauan *et al.*, 1996; Cagauan, 1999; Cagauan *et al.*, 1999; Cagauan *et al.*, 2000). The systems were: rice and rice-fish cultures, rice and rice-fish cultures including herbicides and molluscicides, rice-*Azolla*, rice-duck, rice-fish-*Azolla*, rice-fish-duck, rice-duck-*Azolla* and rice-fish-duck-*Azolla*. Production of rice grain, Nile tilapia fish and mallard duck eggs were evaluated and the economics of the different farming systems were compared (cost and return analysis). Interactions between the major components of the systems and the impacts of the two locally abundant pests, the golden apple snail (*Pomacea canaliculata*), and weed, were analysed. Taken together, the data of this study have confirmed the advantage of combining rice and *Azolla*, rice and fish, rice and ducks as well as rice, fish and *Azolla* and rice, duck and *Azolla*. The newly experimented rice-fish-duck-*Azolla* system has been shown to be highly profitable under the local conditions. Indeed it gives net returns ten times higher than the conventional rice monoculture. In addition to its positive effect on production and profitability, the system also provides a friendly environment for aquatic biodiversity since it does not make use of herbicides and molluscicides (but see section 3). Ducks indeed significantly control *P. canaliculata* population, while ducks, fish and *Azolla* each contribute to weed control. The rice-fish-duck-*Azolla* system has recently been successfully adopted at the farm level in Japan (Furuno, 2001). However, the high initial investment cost of this system, essentially due to the duck component, may limit its adoption by small-scale farmers.

2.6. Selection

One important task for applied scientists is the selection of *Azolla* strains adapted to a diversity of utilizations in various environments (Lejeune *et al.*, 1999). Only limited progress has been achieved in this field. *Azolla* clones with 50% reduced phosphorus requirements have been obtained by somaclonal selection using the mutagen N-methyl-N'-nitro-N-nitrosoguanidine (Vaishampayan *et al.*, 1992; Sinha *et al.*, 1998). Vaishampayan *et al.* (1998) also obtained heat tolerant *Azolla* by the same method. Significant differences in carotene content among different *Azolla* species/strains have been observed (Lejeune *et al.*, 2000). Selection programmes aimed at obtaining improved strains for feed therefore need to include carotene content as one criterion.

Rai and Rai (1999) were able to induce salt tolerance in *A. pinnata* by stepwise transfer in culture media of increasing salinity.

2.7. Sporulation

In many areas continuous vegetative production of *Azolla* is not possible due to various factors, mainly temperature and water availability. One solution to this problem could be the conservation of *Azolla* as spores during the unfavourable season. Unfortunately, all attempts to control *Azolla* sporulation have failed until now (Marsh *et al.*, 1998). Moreover, numerous strains never sporulate. Others only sporulate during certain periods. It has recently been shown (Kar *et al.*, 1999, 2000, 2001) that in some strains (*e.g.*, *A. caroliniana*, *A. microphylla* and *A. pinnata*), sporulation during these periods is substantially stimulated by application of phosphorus or gibberellic acid.

3. *AZOLLA* AS A WEED

In certain areas of the world some *Azolla* species are considered as invading weeds. A programme aimed at the eradication of *Azolla filiculoides* by introducing *Stenopelmus rufinosus*, a North American weevil predator of *Azolla*, has been launched in South Africa (Hill, 1997). Preliminary trials in controlled conditions (Hill, 1997; Hill, 1998a) have shown that among the 26 plant species belonging to 15 families, only the three *Azolla* species (*A. filiculoides* of American origin, and the two indigenous species *A. nilotica* and *A. pinnata*) were affected, with *A. filiculoides* being the most suitable host for the weevil. A first batch of *S. rufinosus* was therefore released in 1997 (Hill, 1998b). Having then been introduced in 46 sites they established in at least 31 and completely controlled *A. filiculoides* in 20 (Hill and Cilliers, 1999; Hill 1999). Since the predator certainly ignores countries borders, reconciling the opposing views of promoters of *Azolla* development for agronomic purposes and promoters of its destruction is really difficult. Since the insect also feeds on native *A. nilotica* and *A. pinnata*, there is clearly a risk of destabilisation of the populations of these two species. It is worth mentioning that an environment-friendly herbicide for aquatic vegetation (including *Azolla*) has been developed recently (Pullen and Pullen, 1998). In Zimbabwe, where *A. filiculoides* appeared in the early 1980s, Gratwicke and Marshall (2001) studied slow streams and compared areas covered with *Azolla* to areas devoid of vegetation or with natural, submerged macrophytes. They showed that the presence of *Azolla* generally leads to a decrease in oxygen partial pressure, abundance of tadpoles and fishes, and invertebrate abundance and biodiversity. The possible negative impact of *A. filiculoides* on such small perennial streams, which constitute important refuges for small cyprinid fishes and frogs, may be considerable.

4. *AZOLLA* AND PHYTOREMEDIATION

Ecological engineering, including the use of plants for abatement in soil and water pollution, is receiving considerable attention and has resulted in the current studies in phytoremediation. In recent years, a large number of papers on *Azolla* have focused on this topic with the aim of removing excess nutrients and heavy metals from water.

4.1. Human and Agricultural Effluents

Since nitrogen can be a limiting factor for growth, the nitrogen-fixing symbiosis between *Azolla* and *Anabaena* seems a good candidate to use for the removal of phosphorus from contaminated water (Shiomi and Kitoh, 1996). Although a few reports are fairly optimistic (Costa *et al.*, 1999; Forni *et al.*, 2001), other tests show that *Azolla* has a poor capacity to remove N and P from waste waters (human effluents: Shiomi and Kitoh, 1996; Vermaat and Hanif, 1998; Costa *et al.*, 1999; pig effluents: Shiomi and Kitoh, 1996; Costa *et al.*, 1999; aquaculture wastewater: Redding *et al.*, 1997; Forni *et al.*, 2001). In comparative studies, *Azolla* was generally less efficient than other submersed, emergent or floating plants. This was mostly due to the very limited growth, or even death, of *Azolla* on the wastewater media possibly due to its sensitivity to toxic organic acids, nitrites (Kitoh *et al.*, 1993), and ammonium ions (Shiomi and Kitoh, 1996). The latter has different effects on individual partners in the *Azolla-Anabaena* association, with growth of *Azolla* and nitrogen fixation by *Anabaena* being affected independently, indicating that the tolerance of the symbiotic system could be determined separately by *Azolla* and *Anabaena* (Maejima *et al.*, 2001).

4.2. Heavy Metals

The potential for growing *Azolla* on metal-contaminated water (Al, Cd, Cu, Cr, Ni, Pb, Se, U, Zn, separately or in combination, in model solutions or effluents) to eliminate the metals, needs further studies to be convincing. Although some encouraging results are noted, like the fact that certain heavy metals can be as much as a thousand fold more concentrated in *Azolla* than in its surrounding medium (Sela *et al.*, 1989), they usually have a marked negative effect on growth (Sela *et al.*, 1989; Ayala-Silva and Al-Hamdani, 1997; Sajwan and Ornes, 1997; Wilson and Al-Hamdani, 1997; Powel *et al.*, 1998). The heavy metals mostly concentrate in roots rather than in shoots and bind to the pectin of cell walls and peptides and polyphosphate granules (Tel-Or *et al.*, 1997). The kinetics of Cd absorption by *A. pinnata* has been shown to be biphasic with a rapid initial uptake followed by a slow accumulation, possibly indicating two different mechanisms (Noraho and Gaur, 1996). Heavy metals cause a leaching of Na, K, Mg and, to a lesser extent, Ca from the roots of *Azolla* (Sela *et al.*, 1989), and may cause their roots to get detached (Powel *et al.*, 1998). Cd, Ni and Zn considerably reduce the nitrogen-fixing ability of the symbiotic system (Sela *et al.*, 1989). The bioaccumulation of metals by *Azolla* should also lead to exercise of caution if the plant is to be used as feed for animals (snails, ducks, fish, etc.) (Powel *et al.*, 1998). *A. caroliniana* has been tested for its potential to volatilise and accumulate Se, a serious pollutant in agricultural and industrial wastewaters. *Azolla* was found to be highly efficient but its low biomass compared to other aquatic plants tested, rendered it of little use (Pilon-Smits *et al.*, 1999).

An attempt was made to use *A. filiculoides* as a sensitive and quantitative biosensor for the determination of gamma-ray emitting radionuclides in surface waters. However, the results should be interpreted with care when dealing with a radionuclide for which substantial contributions from atmospheric deposition might be suspected (Walterbeek and van der Meer, 1996).

The use of dried and milled *Azolla* for the removal of metals (Au, Cd, Cr, Cu, Ni, Pb, U, Zn) from industrial or mine effluents, is more promising. It seems that all the metals share common adsorption sites and share a similar mode of adsorption. Certain elements however show a higher binding affinity: Ca interferes with the adsorption of Cd, Ni and Zn, and Cr displaces these metals (Zhao *et al.*, 1999a). Very high adsorption ratios were observed in batch experiments with the removal of up to 93 mg Pb (Sanyahumbi *et al.*, 1998) and 120 mg Cr (Zhao and Duncan, 1997a) per g dried *Azolla*. Dried *A. pinnata* adsorbed twice as much Cd as fresh ones (Noraho and Gaur, 1996). Preliminary studies using model solutions rather than effluents, showed a good efficiency of *Azolla* to bind Au (Antunes *et al.*, 2001) and Pb (Sanyahumbi *et al.*, 1998). When columns with packed *Azolla* are used for effluent treatment, an efficient metal desorption with acids followed by column regeneration allow five to ten cycles of column utilization. The metals could be efficiently recovered and this has lead to suggestions for the industrial reuse of Cr (Zhao and Duncan, 1997a; Zhao and Duncan, 1997b), Ni (Zhao and Duncan, 1998a) and Zn (Zhao and Duncan, 1998b; Zhao *et al.*, 1999b) so obtained. These methods could be interesting alternatives to conventional, and costly methods of treating metal-polluted waters that generate large volumes of toxic sludge.

5. *AZOLLA* AND TROPICAL DISEASES

Abdel Hafez *et al.* (1997) showed that in laboratory conditions there is a clear negative correlation between the density of an *A. pinnata* mat and the hatchability of eggs, growth and survival of *Biomphalaria alexandrinus*, a snail that acts as an intermediate host in schistosomiasis. The reason for this effect is not known and more systematic studies are needed to quantify the results better.

Balakrishnan and Sharma (2000) confirmed previous results concerning the possible role of *Azolla* as mosquito inhibitor (see Van Hove and Lejeune 1996). They observed that in rice fields a mat of *A. microphylla* has an inhibitory effect on ovipositing females of various mosquito species. This leads to a decrease in larval and pupal densities of Anophelines (intermediary hosts in malaria) and Culicines (intermediary host in Japanese encephalitis). It can be expected, however, that a significant reduction in mosquitoes would only be reached if most of the water surfaces of a given area were covered with *Azolla*, a situation that is probably difficult to reach.

6. "*ANABAENA AZOLLAE*" IN BIOTECHNOLOGY

Shi and Hall (1988b) suggested the design of bioreactors based on the immobilization of *A. azollae* in polyurethane and polyvinyl foams for production of ammonium, H_2, antibiotics and other compounds. A number of papers have since been published on this matter (see e.g. Uma and Kannaiyan, 1996; Kannaiyan *et al.*, 1997a,b; Rajini and Subramanian, 1997). No clear demonstration of the origin of the cyanobacteria is however provided in any of these papers even though serious doubt has often been expressed on the possibility to cultivate *A. azollae in vitro* (see e.g. Van Coppenolle *et al.*, 1995).

7. *AZOLLA* IN SPACE

Potential of *Azolla* for maintaining oxygen and carbon dioxide balances and providing food in spacecrafts and space stations has been considered as early as 1980 (Shepelev *et al.*, 1983; Shepelev *et al.*, 1984). Research in this field is still progressing (Cuello *et al.*, 1998; Chen *et al.*, 1999; Chen *et al.*, 2000).

8. PERSPECTIVES

As illustrated in figure 1 the golden age for *Azolla* research seems to be behind us. One explanation for the present disaffection is that numerous problems, basic as well as applied, have been solved. Another more important reason is the limited success, and sometimes the failures, of the considerable efforts that have been devoted to the promotion of *Azolla* utilization.

It must nevertheless be recognized that quantitative data on the actual utilization of *Azolla* at the farm level in various parts of the world, is very scarce and often confusing. Indeed, it often happens that scientists extrapolate their results obtained in laboratory or research stations to the field conditions. Although a decline in use of *Azolla* in China and Vietnam, for reasons that need not apply to other countries, is well known (Van Hove, 1989), the new successful small-scale applications have drawn less attention. Drawing up an inventory on this matter should help orienting future research and development efforts. Furthermore, improvement of *Azolla* utilization still requires a better knowledge of *Azolla* biology (Lejeune *et al.*, 1999), especially the selection of strains adapted to various environments, control of the sexual cycle, nutritive value, optimisation of *Azolla* integration in various farming systems and effect of *Azolla* on NH_3 volatilisation.

ACKNOWLEDGEMENTS

Thanks are due to Yi Bin Huang, from the Fujian Academy of Agricultural Sciences, for his friendly and efficient help in providing and analysing the recent Chinese literature and to Béatrice Lambillotte for her constant and patient updating of the *Azolla* data bank.

REFERENCES

Abdel-Hafez, A.M., Zidan, Z.H., Abdel-Megeed, M.I.,el-Emam, M.A., Ragab, F.M. and el-Deeb, F.A. (1997) Effect of the plant *Azolla pinnata* on survival, growth rate, fecundity and hatchability of egg-masses of *Biomphalaria alexandrina* snails, *Journal of the Egyptian Society of Parasitology* **27**, 825-841.

Adhya, T.K., Bharati, K., Mohanty, S.R., Ramakrishnan, B., Rao, V.R., Sethunathan, N. and Wassmann, R. (2000) Methane emission from rice fields at Cuttack, India, *Nutrient Cycling in Agroecosystems* **58**, 95-105.

Almazan, G.J, Pullin, R.S.V., Angeles, A.F., Manalo, T.A., Agbyani, R.A. and Trono, M.T.B. (1986) *Azolla pinnata* as a dietary component for Nile tilapia, *Oreochromis niloticus,* in J.L. Maclean , L.B. Dizon and L.V. Hosillos (eds), *The First Asian Fisheries Forum*, Asian Fisheries Society, Manila, Philippines, pp. 523-528.

Antunes, A.P.M., Watkins, G.M. and Duncan, J.R. (2001) Batch studies on the removal of gold (III) from aqueous solution by *Azolla filiculoides*, *Biotechnology Letters* **23**, 249-251.

Aulakh, M.S., Wassmann, R. and Rennenberg, H. (2001) Methane emissions from rice fields-quantification, mechanisms, role of management, and mitigation options, *Advances in Agronomy* **70**, 193-260.

Ayala-Silva, T. and Al-Hamdani, S. (1997) Interactive effects of polylactic acid with different aluminium concentrations on growth, pigment concentrations, and carbohydrate accumulation of *Azolla*, *American Fern Journal* **87**, 120-126.

Balakrishnan, N. and Sharma, S.K. (2000) Role of water fern, *Azolla microphylla* in controlling the larval breeding of mosquitoes in paddy fields at Coimbatore (South India), *Journal of Entomological Research* **24**, 39-42.

Bharati, K., Mohanty, S.R., Singh, D.P., Rao, V.R. and Adhya, T.K. (2000) Influence of incorporation or dual cropping of *Azolla* on methane emission from a flooded alluvial soil planted to rice in Eastern India, *Agriculture, Ecosystems and Environment* **79**, 73-83.

Braemer, P. (1927) La culture des *Azolla* au Tonkin, *Revue de Botanique Appliquée et d'Agriculture Coloniale* **7**, 816-819.

Cagauan, A.G. (1999) Production, economics and ecological effects of Nile tilapia (*Oreochromis niloticus* L.), a hybrid aquatic fern *Azolla* (*Azolla microphylla* Kaulf. x *Azolla filiculoides* Lam.) and Mallard duck (*Anas platyrhynchos* L.) in integrated lowland irrigated rice-based farming systems in the Philippines, Ph. D. thesis in applied natural sciences, Institute of Applied Natural Sciences, Catholic University of Louvain, Louvain-La-Neuve, Belgium, 404 p.

Cagauan, A.G., Branckaert, R.S.D. and Van Hove, C. (1999) Rice-duck farming in Asia : Increasing its production potentials by integration with fish and the nitrogen-fixing aquatic fern *Azolla*, in The First INFPD/FAO Electronic Conference on Family Poultry, Free communication 4, http://www.fao.org/livestock/agap/lpa/fampo1/freecom4.htm.

Cagauan, A.G., Branckaert, R.D. and Van Hove, C. (2000) Integrating fish and *Azolla* into rice-duck farming in Asia, *Naga (ICLARM quaterly)* **23**, 4-10.

Cagauan, A.G., Van Hove, C., Orden, E.A., Ramilo, N.M. and Branckaert, R.D. (1996) Preliminary results of a case study on integrated rice-fish-*Azolla* ducks production system in the Philippines, in H. Hayakawa, M. Sasaki and K. Kimura (eds.), *Integrated Systems of Animal Production in the Asian Region*, Japanese Society of Zootechnical Science and Food and Agriculture Organization, pp. 35-61

Chen, G.X., Huang, G.H., Huang, B., Yu, K.W., Wu, J. and Xu, H. (1997) Nitrous oxide and methane emissions from soil-plant systems, *Nutrient Cycling in Agroecosystems* **49**, 41-45.

Chen, M., Bian, Z., Zhang, C., Liu, H. and Chen, B. (1999) Effects of *Azolla* on the change of O_2-CO_2 concentration under controlled airtight system, *Fujian Journal of Agricultural Sciences* **14**, 56-61. (In Chinese, English abstract).

Chen, M., Liu, X. and Liu, Z. (2000) The equipment of using *Azolla* for O_2-suppementation and its test, *Space Medicine & Medical engineering* **13**, 14-18. (In Chinese, English abstract).

Cisse, M. (2001) Impact of *Azolla* on urea-N cycling in flooded rice in comparison to and in combination with fertilizer placement, application of potassium chloride (KCl) and biocides, Ph. D. thesis in agricultural science, Faculty of Agriculture, Georg-August-University of Göttingen, Germany, 124 p.

Costa, M.L., Santos, M.C. and Carrapiço, F. (1999) Biomass characterization of *Azolla filiculoides* grown in natural ecosystems and wastewater, *Hydrobiologia* **415**, 323-327.

Crevost, Ch. and Lemarié, Ch. (1917) *Catalogue des produits de l'Indochine*, Tome 1, p. 426.

Cuello, J.L., Rodriguez-Eaton, S., Stryjewski, E.C. and Sager, J.C. (1998) *Azolla-Anabaena* symbionts and microbial mat as nitrogen-fixing biocatalysts for bioregenerative space life support, *Life Support & Biosphere science* **5**, 375-388.

de Macale, M.A.R., Vlek, P.L.G., Eberhardt, U. and San Valentin, G.O. (1997) The role of *Azolla* in curbing ammonia volatilisation from flooded rice environments under Philippine conditions, MSc thesis in agriculture, Faculty of Agriculture Georg-August-University of Göttingen, Germany, 53 p.

de Macale, M.A.R., Vlek, P.L.G. and San Valentin, G.O. (2002) The role of *Azolla* cover in improving the nitrogen use efficiency of lowland rice, Proceedings of the conference on Sustaining food security and managing natural resources in Southeast Asia, 8-11 January 2002, Chiang Mai, Thailand.

Diara, H.F. and Van Hove, C. (1983) The influence of *Azolla* on rice productivity: preliminary results obtained at WARDA Richard-Toll project, Senegal, *Warda Technical Newsletter* 4, 7-8

El-Sayed, A.-F.M. (1992) Effects of substituting fish meal with *Azolla pinnata* in practical diets for fingerling and adult Nile tilapia, *Oreochromis niloticus* (L.), *Aquaculture and Fisheries Management* **23**, 167-173.

Fasakin, A.E. and Balogun, A.M. (1998) Evaluation of dried water fern (*Azolla pinnata*) as a replacer for soybean dietary components for *Clarias gariepinus* fingerlings, *Journal of Aquaculture in the Tropics* **13**, 57-64.

Feng, S.L. (1994) Effect of green *Azolla* flour as feed for rabbits, *Journal of Rabbit Breeding* **4**, 7-8. (In Chinese).

Feuillée, L. (1725) Histoire des plantes médicinales de Perou et Chily, Journal des observations physiques, mathématiques et botaniques faites sur l'ordre du roi sur les côtes orientales dans l'Amérique méridionale et dans les Indes occidentales depuis 1707-12, Griffart, Paris, vol. 3, 426 p.

Fillery, I.R.P. and Vlek, P.L.G. (1986) Reappraisal of the significance of ammonia volatilisation as an N loss mechanism in flooted rice fields, *Fertilizer Research* **9**, 79-98.

Forni, C., Chen, J., Tancioni, L. and Grilli-Caiola, M., (2001) Evaluation of the fern *Azolla* for growth, nitrogen and phosphorus removal from wastewater, *Water Research* **35**, 1592-1598.

Furuno, T. (2001) *The power of duck. Integrated rice and duck farming*, Tagari Publications, Sisters Creek, Tasmania, Australia

Ge, S.A., Xu, D.X. and Shen, Z.H. (1980) Salt tolerance of *Azolla filiculoides* and its effects on the growth of paddy in Xinwei Haitu, *Zhejiang Nongye Kexue* **1**, 17-20. (In Chinese).

Gratwicke, B. and Marshall, B.E. (2001) The impact of *Azolla filiculoides* Lam. on animal biodiversity in streams in Zimbabwe, *African Journal of Ecology* **39**, 216-218.

Hill, M.P. (1997) The potential for the biological control of the floating aquatic fern *Azolla filiculoides* Lamarck (Red water fern/Rooivaring) in South Africa, Report to the Water Research Commission by the Plant Protection Research Institute, Agricultural Research Council,WRC Report n° KV 100/97.

Hill, M.P. (1998a) Life history and laboratory host range of *Stenopelmus rufinasus*, a natural enemy for *Azolla filiculoides* in South Africa, *Biocontrol* **43**, 215-224.

Hill, M.P. (1998b) *Azolla filiculoides* the first step towards biological control, *Plant Protection News* **51**, 1-3.

Hill, M.P. (1999) Biological control of red water fern, *Azolla filiculoides* Lamarck (*Pteridophyta* : *Azollaceae*), in South Africa, *African Entomology Memoir* **1**, 119-124.

Hill, M.P. and Cilliers, C.J. (1999) *Azolla filiculoides* Lamarck (*Pteridophyta* : *Azollaceae*), its status in South Africa and control, *Hydrobiologia* **415**, 203-206.

Kanangire, C.K. (2001) Effets de l'alimentation des poissons avec *Azolla* sur la production d'un écosystème agro-piscicole en zones marécageuses au Rwanda, Ph.D.thesis, Faculté des Sciences, Facultés Universitaires Notre-Dame de la Paix, Namur, Belgium, 220 p.

Kannaiyan, S, Aruna, S.J., Kumari, S.M.P. and Hall D.O. (1997a) Immobilized cyanobacteria *A. Azollae*, a symbiont of *Azolla* as a biofertilizer for rice crops, *Journal of Applied Phycology* **7**, 1-9.

Kannaiyan, S, Aruna, S.J., Kumari, S.M.P. and Hall D. O. (1997b), Immobilized cyanobacteria as a biofertilizer for rice crops, *Journal of Applied Phycology* **9**, 167-174.

Karr, P.P., Mishra, S. and Singh D.P. (1999) Influence of gibberellic acid on the sporulation of *Azolla caroliniana*, *Azolla microphylla* and *Azolla pinnata*, *Biology and Fertility of Soils* **29**, 424-429.

Kar, P.P., Mishra, S. and Singh D.P. (2000) Variability in *Azolla* sporulation in response to phosphorus application, *Biology and Fertility of Soils* **32**, 458-462.

Kar, P.P., Mishra, S. and Singh, D.P. (2001) Influence of different phosphorus management strategies on the sporulation and growth of *Azolla*, *Experimental Agriculture* **37**, 53-64.

Khatun, A., Ali, M.A. and Dingle, J.G. (1999) Comparison of the nutritive value for laying hens of diets containing atolls (*Azolla pinnata*) based on formulation using digestible protein and digestible amino acid versus total protein and total amino acid, *Animal Feed Science and Technology* **81**, 43-56.

King, G.M. and Schnell S. (1994) Ammonium and nitrite inhibition of methane oxidation by *Methylobacter albus* BG8 and *Methylosinus trichosporum* OB3b at low methane concentrations, *Applied and Environmental Microbiology* **60**, 3508-2513.

Kishida, Y (1996) Integrated farming system of crossbred duck meat-rice production in paddy fields utilizing *Azolla*, in H. Hayakawa, M. Sasaki and K. Kimura (eds.), *Integrated Systems of Animal Production in the Asian Region*, Japanese Society of Zootechnical Science and Food and Agriculture Organization, pp.93-101.

Kishida, Y. and Okazaki, A. (1999) Integrated farming system of *Azolla-Aigamo* duck meat-rice production in paddy fields. 3. Seasonal nutritive value of aquatic fern *Azolla*, *Research Association of Sôgô-Nôgaku* **46**, 6-10. (In Japanese, English abstract).

Kishida, Y. and Utsumiya, N. (1998a) Integrated farming system of *Azolla-Aigamo* duck meat-rice production in paddy fields.1. Effects of aquatic fern *Azolla* on growth of *Aigamo* duck and rice yield, *Research Association of Sôgô-Nôgaku* **46**, 19-23. (In Japanese, English abstract).

Kishida, Y. and Utsumiya N.(1998b) Integrated farming system of *Azolla-Aigamo* duck meat-rice production in paddy fields. 2. Effects of aquatic fern *Azolla* on behavioral characteristics of *Aigamo* duck, *Research Association of Sôgô-Nôgaku* **46**, 30-35. (In Japanese, English abstract).

Kitoh, S., Shiomi, N. and Uheda, E. (1993) The growth and nitrogen fixation of *Azolla filiculoides* Lam. in polluted water, *Aquatic Botany* **46**, 129-139.

Kröck, T., Alkämper, J. and Watanabe, I. (1988) Effect of an *Azolla* cover on the conditions in floodwater, *Journal of Agronomy & Crop Science* **161**, 185-189.

Ladha, J.K., Dawe, D., Ventura, T.S., Singh, U., Ventura, W. and Watanabe, I. (2000) Long-term effects of urea and green manure on rice yields and nitrogen balance, *Soil Science Society of America Journal* **64**, 1993-2001.

Lejeune, A., Cagauan, A. and Van Hove, C. (1999) *Azolla* research and development: recent trends and priorities, *Symbiosis* **27**, 333-351.

Lejeune, A., Peng, J.L., Le Boulengé, E., Larondelle, Y. and Van Hove, C. (2000) Carotene content of *Azolla* and its variations during drying and storage treatments, *Animal Feed Science and Technology* **84**, 295-301.

Le Mer, J. and Roger, P. (2001) Production, oxidation, emission and consumption of methane by soils: A review, *European Journal of Soil Biology* **37**, 25-50.

Léonard, V., Breyne, C., Micha, J.C. and Larondelle, Y. (1998) Digestibility and transit time of *Azolla filiculoides* Lamarck in *Oreochromis aureus* (Steindachner), *Aquaculture Research* **29**, 159-165.

Liu, R.Q., Ou, X.M., Xie, L.W., Xiao, S.Y., Xia, H.A., Huang, Y.B., Wen, H.B., Chen, Y.G. and Huang, D.F. (2000) Technique and application of the *Azolla*-fish system in row planted rice fields, *Soils and Fertilizers* **1**, 38-41. (In Chinese).

Liu, X., Takayama, K., Yamashita, K., Nakanishi, Y., Manda, M. and Inanaga, J. (1998) Cultivation condition and nutritive value of *Azolla* as feed resource, *Grassland Science* **44**, 266-271. (In Japanese, English abstract).

Liu, Y., Fan, T. and Tang, X. (2000) Cultivation techniques of *Azolla* with high yield and protein, *Agricultural Research in the Arids Areas* **18**, 109-112. (In Chinese, English abstract).

Liu Z.Z. (1979) Use of *Azolla* in rice production in China, in I.Watanabe (ed.), *Nitrogen and Rice*, International Rice Research Institute, Manila, Philippines, pp. 375-394.

Liu, Z.Z. (1987) Reevaluation of *Azolla* utilization in agricultural production, in *Azolla Utilization, Proceedings of the Workshop on Azolla Use, 31 March-5 Apr 1985, Fuzhou, Fujian, China*, International Rice Research Institute, Manila, Philippines, pp. 67-76.

Lumpkin, T.A. and Plucknett, D.L. (1982) *Azolla as a green manure: use and management in crop production*, Series N°15, Westview Press, Boulder, Colorado, USA.

Ly, J. and Preston, T.R. (2001) In vitro estimates of nitrogen digestibility for pigs and water-soluble nitrogen are correlated in tropical forage feeds, *Livestock Research for Rural Development* **13**, 1-6.

Maejima, K., Kitoh, S., Uheda, E. and Shiomi, N. (2001) Response of 19 *Azolla* strains to a high concentration of ammonium ions, *Plant and Soil* **234**, 247-252.

Mandal, B., Bhattacharya, K. and Mete, P.K. (1997) Effects of *Sesbania rostrata* and *Azolla microphylla* incorporation on transformation of applied zinc and copper in lateritic rice soils with different flooding regimes, *Biology and Fertility of Soils* **24**, 394-398.

Mandal, B., Vlek, P.L.G. and Mandal, L.N. (1999) Beneficial effects of blue-green algae and *Azolla*, excluding supplying nitrogen, on wetland rice fields : a review, *Biology and Fertility of Soils* **28**, 329-342.

Marsh, B.H., Corbin, J.L. and Peters G.A. (1998) Changes in soluble amino acid and polyamine composition associated with increasing plant density and the onset of sporulation in *Azolla, Symbiosis* **24**, 315-326.

Nguyen, C.T. (1930) L'*Azolle* cultivée comme engrais vert, *Bulletin Econonomique de l'Indochine* **33**, 335-350.

Nierzwicki-Bauer, S.A. (1990) *Azolla-Anabaena* symbiosis: use in agriculture, in A.N. Rai (ed.), *Handbook of Symbiotic Cyanobacteria*, CRC Press, Boca Raton, pp.119-136.

Noraho, N. and Gaur, J.P.(1996) Cadmium adsorption and intracellular uptake by two macrophytes, *Azolla pinnata* and *Spirodela polyrhiza, Archives für Hydrobiologie* **136**, 135-144.

Pilon-Smits, E.A.H., De Souza, M.P., Hong, G., Amini, A., Bravo, R.C., Payabyab, S.T. and Terry, N. (1999), Wetlands and aquatic processes : selenium volatilization and accumulation by twenty aquatic plant species, *Journal of Environmental Quality* **28**, 1011-1018.

Plessner, O.E., Chen, Y., Shenker, M. and Tel-Or, E. (1998) Iron-enriched *Azolla* as a slow-release biofertilizer for cucumber plants grown in a hydroponic system, *Journal of Plant Nutrition* **21**, 2357-2367.

Powel, G.M., Nimmo, D.W.R., Flickinger, S.A. and Brinkman, S.F. (1998) Use of *Azolla* to assess toxicity and accumulation of metals from artificial and natural sediments containing cadmium, copper, and zinc, in E.E. Little, A.J. DeLonay and B.M. Greenberg (eds), *Environmental Toxicology and Risk Assessment: Seventh Volume,* American Society for Testing and Materials, pp. 184-199.

Pullen, E.M. and Pullen, M.D. (1998) Control of aquatic vegetation with surfactant and terpene oil, Patent classifications : Main IPC : A01N025-30 , U.S.: 504150000, Secondary IPC : A01N031-04, A01N061-02, 4 p.

Rai, A.N., Borthakur, M. and Paul, D. (1996) Symbiotic Cyanobacteria: biotechnological applications, *Journal of Scientific and Industrial Research* **55**, 742-752.

Rai, V. and Rai, A.K. (1999) Growth behaviour of *Azolla pinnata* at various salinity levels and induction of high salt tolerance, *Plant and Soil* **206**, 79-84.

Rajini, V.S. and Subramanian, G. (1997) The effect of immobilization on carbon flow through *Anabaena variabilis* and *Anabaena azollae* , *Photosynthetica* **34**, 137-139.

Redding, T., Todd, S. and Midlen, A. (1997) The treatment of aquaculture wastewaters: a botanical approach, *Journal of Environmental Management* **50**, 283-299.

Roger P.A. (1993) Les biofertilisants fixateurs d'azote en riziculture: potentialités, facteurs limitants et perspectives d'utilisation, in M. Raunet (ed.), *Bas-fonds et riziculture*, Cirad, Montpellier, France, pp. 327-348.

Sajwan, K.S. and Ornes, W.H. (1997) Potential of mosquito fern (*Azolla caroliniana* willd.) plants as a biofilter for cadmium removal from wastewater, in D.W.Tedder and F.G. Pohland (eds), *Emerging Technologies in Hazardous Waste Management*, Plenum, New york, pp. 167-177.

Sakamoto, H. (1996) Duck-paddy together growing way as a developed system in organic farming, *Geographical Journal of Nara University* **2**, 1-12. (In Japanese, English abstract).

Sanyahumbi, D., Duncan, J.R., Zhao, M. and Van Hille, R. (1998) Removal of lead from solution by the non-viable biomass of the water fern *Azolla filiculoides*, *Biotechnology Letters* **20**, 745-747.

Sela, M., Garty, J. and Tel-Or, E. (1989) The accumulation and the effect of heavy metals on the water fern *Azolla filiculoides*, *New Phytologist* **112**, 7-12.

Shepelev, Y.Y., Thyoc, N.H., Kordyum, V.A., Meleshko, G.I., Galkina, T.B. and Manko, V.G. (1983) Study of effect of weightlessness on the water fern, *Azolla*, *USSR Report: Space Biology and Aerospace Medicine* **16**, 94-96.

Shepelev, Y.Y., Thyok, N., Meleshko, G.I., Antonyan, A.A., Galkina, T.B. and Naydina V.P. (1984), Physiological and ecological characteristics of the water fern, *Azolla pinnata*, and prospects of using it in biological life-support system for man, *USSR Report: Space Biology and Aerospace Medicine* **17**, 96-101.

Shi, D.J. and Hall, D.O. (1988a) The *Azolla-Anabaena* association : historical perspective, symbiosis and energy metabolism, *Botanical Review* **54**, 353-386.

Shi, D.J. and Hall, D.O. (1988b) *Azolla* and immobilized cyanobacteria (blue-green algae) : from traditional agriculture to biotechnology, *Plants Today* **1**, 5-12.

Shiomi, N. and Kitoh, S. (1987) Nutrient absorption capacity of *Azolla* from waste water and use of *Azolla* plant as biomass, *Journal of Plant Nutrition* **10**, 1663-1670.

Shiomi, N. and Kitoh, S. (1996) The growth, nitrogen fixation and nutrient removal capacity of *Azolla* in polluted water, *Lakes & Reservoirs: Research and Management* **2**, 175-179.

Shiomi, N. and Kitoh S. (2001) Culture of *Azolla* in a pond, nutrient composition, and use as fish feed, *Soil Science and Plant Nutrition* **47**, 27-34.

Sinha, R.P., Vaishampayan, A. and Hader, D.P. (1998) Plant-cyanobacterial symbiotic somaclones as a potential bionitrogen-fertilizer for paddy agriculture: biotechnological approaches, *Microbiological Research* **153**, 297-307.

Vlek, P.L.G., Fugger, W. And Biker, U. (1992) The fate of fertilizer N under Azolla in wetland rice, in *Proc. 2nd ESA Meeting*, August 1992, Warwick, UK.

Smith, J.B. (1910) *Azolla* vs. mosquitoes, *Entomological News* **21**, 437-440.

Tang, L.F., Huang, Y.B., Weng, B.Q., Liu, Z.Z. and Liu, X.S. (2000) Sustainable agricultural model of high output, low input and less pollution in paddy field, *Scientia Agricultura Sinica* **33**, 60-66. (In Chinese, English abstract).

Tel-Or, E., Sela, M. and Ravid, S. (1997) Biofiltration of heavy metals by the aquatic fern *Azolla*, in D. Rosen, E. Tel-Or, Y. Hadar and Y. Chen (eds), *Modern Agriculture and the Environment*, Kluwer Academic Publishers, Great Britain, pp. 431-442.

Uma, D. and Kannaiyan, S. (1996) Effect of the systemic fungicide, bavistin on the nitrogen status of cyanobacteria under immobilized state in polyurethane foam, *South African Journal of Botany* **62**, 127-132.

Vaishampayan, A., Reddy, Y.R., Singh, B.D. and Singh, R.M. (1992) Reduced phosphorus requirement of a mutant *Azolla-Anabaena* symbiotic N2 fixing complex, *Journal of Experimental Botany* **43**, 851-856.

Vaishampayan, A., Dey, T., Sinha, R.P. and Häder, D.-P. (1998) Successful rice cultivation with genetically manipulated thermo-tolerant *Azolla* as a bio-N fertilizer, *Acta Hydrobiologica* **40**, 207-213.

Van Coppenolle, B., McCouch, S.R., Watanabe, I., Huang, N. and Van Hove, C. (1995) Genetic diversity and phylogeny analysis of *Anabaena azollae* based on RFLPs detected in *Azolla-Anabaena azollae* DNA complexes using *nif* gene probes, *Theoretical and Applied Genetics* **91**, 589-597.

Van Hove C. (1989) *Azolla* and its multiple uses with emphasis on Africa, Food and Agriculture Organization, Rome, Italy

Van Hove, C. and Lejeune, A. (1996) Does *Azolla* have any future in agriculture? in M. Rahman (ed.), *Biological Nitrogen Fixation Associated with Rice Production*, Kluwer Academic Press, Dordrecht, pp. 83-94.

Vermaat, J.E. and Hanif, M.K. (1998) Performance of common duckweed species (*Lemnaceae*) and the waterfern *Azolla filiculoides* on different types of waste water, *Water Research* **32**, 2569-2576.

Vlek, P.L.G. and Craswell E.T. (1981) Ammonia volatilization from flooded soils, *Fertilizer Research*, **2**, 227-245.

Wagner, G.M. (1997) *Azolla*: a review of its biology and utilization, *Botanical Review* **63**, 1-26.

Walterbeek, H.Th. and van der Meer, A.J.G.M. (1996) A sensitive and quantitative biosensing method for the determination of γ-ray emitting radionuclides in surface water, *Journal of Environmental Radioactivity* **33**, 237-254.

Wilson, G. and Al-Hamdani, S. (1997) Effects of chromium (VI) and humic substances on selected physiological responses of *Azolla caroliniana*, *American Fern Journal* **87**, 17-27.

Ying, Z., Boeckx, P., Chen, G.X. and Van Cleemput, O. (2000) Influence of *Azolla* on CH_4 emission from rice fields, *Nutrient Cycling in Agroecosystems* **58**, 321-326.

Zhao, M. and Duncan, J.R. (1997a) Batch removal of sexivalent chromium by *Azolla filiculoides*, *Biotechnology and Applied Biochemistry* **26**, 179-182.

Zhao, M. and Duncan, J.R. (1997b) Column sorption and desorption of hexavalent chromium from aqueous solution and electroplating effluent using *Azolla filiculoides*, *Ressource and Environmental Biotechnology* **2**, 51-64.

Zhao, M. and Duncan, J.R. (1998a) Removal and recovery of nickel from aqueous solution and electroplating rinse effluent using *Azolla filiculoides*, *Process Biochemistry* **33**, 249-255.

Zhao, M. and Duncan, J.R. (1998b) Bed depth service time analysis on column removal of Zr^{2+} using *Azolla filiculoides*, *Biotechnology Letters* **20**, 37-39.

Zhao, M., Duncan, J.R. and Sanyahumbi, D. (1999a) Competitive sorption of multiple heavy metals by *Azolla* filiculoides, *Resource in Environment and Biotechnology* **2**, 173-183.

Zhao, M, Duncan, J.R. and Van Hille, R.P. (1999b) Removal and recovery of zinc from solution and electroplating effluent using *Azolla filiculoides*, *Water Research* **33,** 1516-1522.

Chapter 11

CYANOBACTERIA IN SYMBIOSIS WITH CYCADS

JOSÉ-LUIS COSTA AND PETER LINDBLAD
Department of Physiological Botany, Evolutionary Biology Centre, Uppsala University, Villavägen 6, SE-752 36 Uppsala, Sweden

1. INTRODUCTION

Cycads are an ancient group of seed plants that first appeared in the Pennsylvanian period and so have existed for approximately 300 million years. Nearly 100 million years ago, cycads had a wide geographical distribution, extending from Alaska and Siberia to the Antarctic. Today they are found on every continent except Europe and Antarctica, but are restricted to small populations in the tropics and subtropics of both hemispheres. Moreover, many are facing possible extinction in nature. The approximately 160 species of cycads are distributed in three families: Cycadaceae with a single genus, *Cycas*; Stangeriaceae with two genera, *Stangeria* and *Bowenia*, and *Zamiaceae* with eight genera, *Ceratozamia*, *Chigua*, *Dioon*, *Encephalartos*, *Lepidozamia*, *Macrozamia*, *Microcycas* and *Zamia*. Cycads are mostly terrestrial and arborescent, except for at least one species which is truly epiphytic (*Zamia pseudoparasitica* Yates), and have the capacity to grow in a number of different habitats. Some species may be found as components of forests (both rainforest and seasonally dry forest) and others grow in loose strands in grasslands. Structurally, cycads have an aerial stem, which is normally columnar and woody with the exception of *Zamia pygmaea* Sims. which has a subterranean stem. The aerial stem may be branched and is covered by persistent leaf bases. In general, the vegetative shoots of cycads produce two types of leaves, scale leaves or cataphylls and foliage leaves. The leaves are born terminally and are mainly pinnate with the exception of the genus *Bowenia* that has bipinnate leaves. Reproductively, cycads are dioecious, producing either pollen-bearing or ovule-bearing modified leaves called sporophylls. The ovules, like those of true gymnosperms, are naked.

Cycads produce three types of roots: (i) a tap-root that is equivalent to the primary root system found in most types of plants (a special type of root restricted to the genus *Cycas*, which is adventitious, arises from the lower side of trunk offsets and grow downwards in close proximity to the trunk), (ii) lateral roots, and (iii) a highly

A.N. Rai, B. Bergman and U. Rasmussen (eds.), Cyanobacteria in Symbiosis, 195-205

Figure 1. Macrozamia riedlei (Fisch. ex Gaudich.) C.A. Gardn. in its natural habitat in Western Australia. (A) Male specimen with three emerging cones. (B) Uninfected coralloid root. (C) Coralloid root with established symbiotic cyanobacteria.

specialised type of lateral root usually termed 'coralloid roots' and in which symbiotic filamentous cyanobacteria may be found (Norstog and Nicholls, 1997). The present review focuses on recent advances in the understanding of symbiotic cyanobacteria in cycads. The reader is referred to earlier reviews (Lindblad, 1990; Lindblad and Bergman, 1990; Adams, 2000; Rai *et al.*, 2000) for detailed discussions regarding the overall symbiotic structures and a characterisation of the symbiotic cyanobacteria.

2. CORALLOID ROOTS

Coralloid roots have been known to exist for a long time and they have been recorded in all genera, and in all cycad species examined. The ability to form this type of root is encoded by genes in the cycad, and the roots are formed before being invaded by symbiotic cyanobacteria. Instead of the more common downward growth pattern, these roots show a marked negative geotropism and grow laterally and upward toward the surface of the soil. In seedlings of *Macrozamia*, the process of coralloid root development begins with the initiation of papillose roots called "pre-coralloids" or non-infected coralloid roots (Ahern and Staff, 1994). The first "pre-coralloid" roots are adventitious and emerge from the hypocotyl immediately under the cotyledonary petioles. Their subsequent development involves phases of maturation, cyanobacterial invasion, coralloid formation, senescence, and regeneration. At least in *Macrozamia*, coralloid roots with and without symbiotic cyanobacteria are structurally different (Figure 1). In case of infection, the cyanobacteria are present in a specific cortical layer inside the root, the cyanobacterial zone. The presence of filamentous heterocystous cyanobacteria inside the root induces irreversible modifications to the growth and development of the root. The growth in length of these roots is much retarded in comparison to the normal roots of same age, but growth in diameter increases noticeably. Furthermore, the cycad cells in the cyanobacterial zone undergo marked differentiation, elongating radially to interconnect the two adjacent cortical layers (Lindblad *et al.*, 1985a). It has been suggested that these elongated cells are specialised cells responsible for the transfer of metabolites between the partners (Lindblad *et al.*, 1985a).

3. ESTABLISHMENT OF THE SYMBIOSIS

The process of infection is still unclear. Invasion of filamentous cyanobacteria may occur at any stage of development of the root, but the precise time and location of the invasion is unpredictable. Several suggestions have been made: (a) through injured parts of the root; (b) through lenticels; (c) through the papillose, and (d) through breaks in the dermal layer (Nathanielsz and Staff, 1975a). More detailed studies are needed to establish if there is a single process of infection in cycads or if different cycad genera/families have developed different strategies. Interstingly, the cyanobiont of *Zamia furfuracea* (*Nostoc* sp. strain FUR 94201) has been isolated and grown in axenic culture before being successfully reunited with a sterile seedling of the same cycad species (Ow *et al.*, 1999).

4. SPECIFICITY AND DIVERSITY OF THE SYMBIOTIC CYANOBACTERIA

Until recently very little was known about the biodiversity and specificity of symbiotic cyanobacteria in cycads. Heterologous Southern hybridisations using cloned genes from the free-living cyanobacterium *Anabaena* sp. strain PCC 7120 demonstrated a diversity when analysing cyanobacterial DNA prepared from a large number of pooled coralloid roots collected from cycads in their natural habitat in Mexico (Lindblad *et al.*, 1989). In a more detailed study the diversity and host specificity of the cyanobionts of several cycad species (*Cycas circinalis* L., *C. rumphii* Miq., *Encephalartos lebomboensis* I. Verd., *E. villosus* Lem., and *Zamia pumila* L.) collected in Fairchild Tropical Garden (Florida, USA) were examined using the $tRNA^{Leu}$(UAA) intron sequence as a genetic marker (Costa *et al.,* 1999). Nested PCR was used to specifically amplify the intron directly from the freshly isolated symbiotic cyanobionts. The intron sequences obtained from the cycad cyanobionts showed high similarities to the corresponding sequences in the free-living strains *Nostoc* sp. PCC 73102 and *N. muscorum* as well as in several lichen and bryophyte cyanobionts, indicating their *Nostoc* identity (Costa *et al.*, 2002). Although different intron sequences were found, no variation was observed within a single coralloid root. This is consistent with infection by a single cyanobiont. However, different coralloid roots from a single *E. villosus* specimen may harbour different cyanobacteria. In addition, cyanobionts in coralloid roots of two different *Encephalartos* species were found to possess the same intron sequence, indicating that the same cyanobiont is present in two different cycad species (Costa *et al.*, 1999).

When analysing and comparing a larger set of *Nostoc* $tRNA^{Leu}$(UAA) intron sequences from free-living (data not shown) and symbiotic strains (Figure 2), no significant differences could be observed. There is a similar level of variation in the $tRNA^{Leu}$(UAA) intron sequences when comparing symbiotic cyanobacteria in cycads with other symbiotic or free-living cyanobacterial strains. A detailed analysis revealed the presence of degenerate heptanucleotid repeats in the first variable region of the intron (Costa *et al.,* 2002). Size differences between the closely related *Nostoc* sequences were found to be primarily due to different numbers of copies of the heptanucleoptide repeats and, in some cases, due to additional sequences. Based on the observed sequences, two classes could be identified: one with the consensus sequence TDNGATT (with its base pairing repeat AATYHAA) and the other with the consensus sequence NNTGAGT and its base pairing repeat AACTCHN. Slipped strand mispairing was suggested to cause the differences in number of repeats, and homologous recombination within the genome has been suggested to give rise to the additional sequences (Costa *et al.*, 2002). In Figure 3 the corresponding sequences, with identified heptanucleoptide repeats highlighted, of the first variable region of the $tRNA^{Leu}$(UAA) intron of cycad cyanobionts are shown.

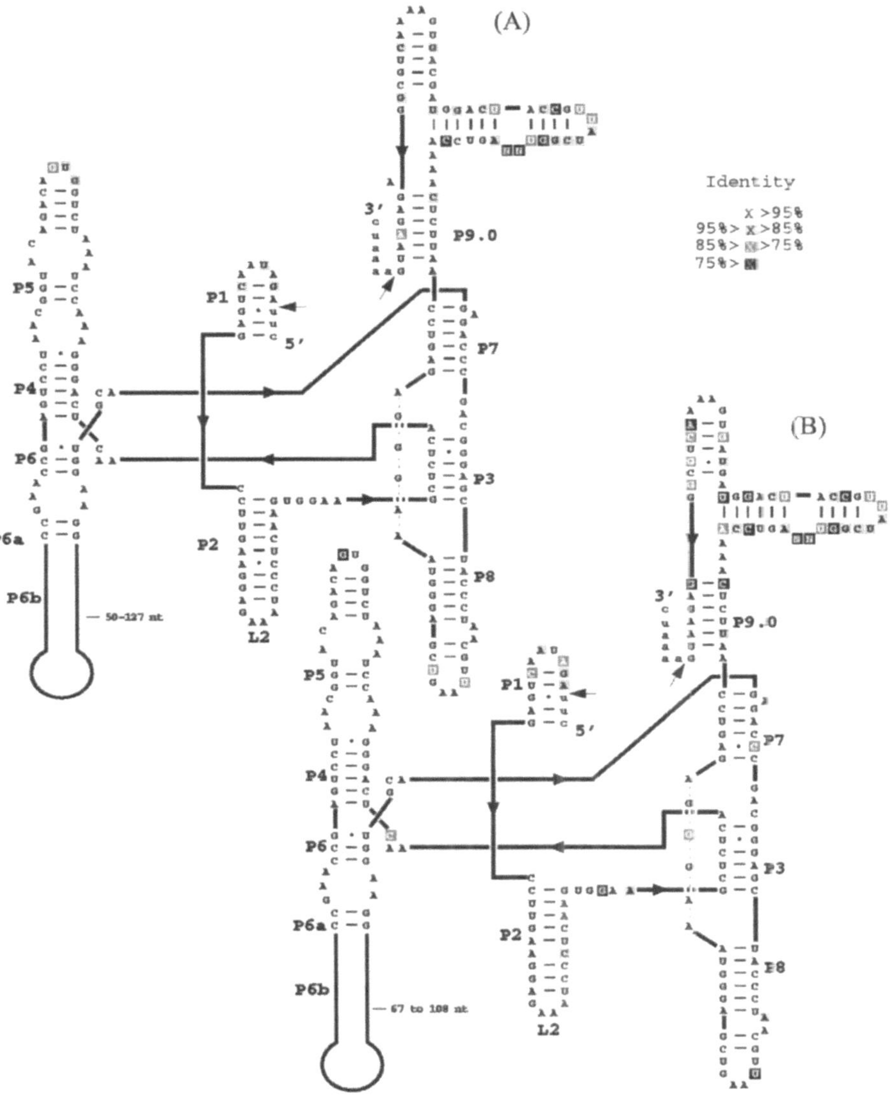

Figure 2. Variability of different positions in cyanobacterial $tRNA^{Leu}$(UAA) introns visualized in secondary structure predictions of consensus sequences. Intron sequence is shown in upper case letters and sequence of the tRNA in lower case letters. Arrows mark the intron/exon boundaries. Different degrees of shading reflect how large a fraction of the sequences supports the consensus sequence. Where alignment was not readily performed due to size variation, schematic loops showing number of nucleotides were inserted into the structures. (A) Consensus sequence of thirty-six symbiotic Nostoc *$tRNA^{Leu}$(UAA) intron sequences excluding cyanobionts of cycads (samples Nos9-17 and Nos26-54 in Costa* et al., *2002). (B) Consensus sequence of eight* Nostoc *symbionts of cycad $tRNA^{Leu}$(UAA) intron sequences (samples Nos18-25 in Costa* et al., *2002).*

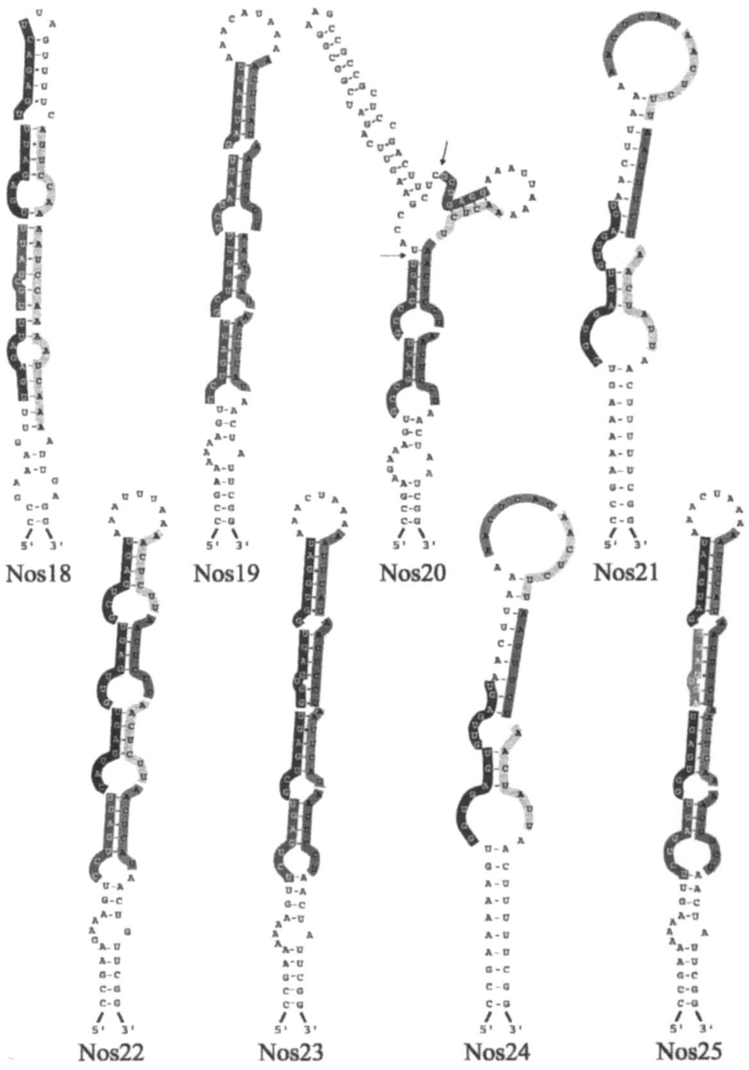

Figure 3. Representation of the hairpin formed by the first variable region of the eight tRNA^{Leu}(UAA) intron from the Nostoc *symbionts of cycads (samples Nos18-25 in Costa* et al.*, 2002). Sequences and nomenclature used are as in Costa* et al.*, 2002, and different heptanucleotide repeats are represented as blocks. Arrows indicate the positions where an additional sequence not having the heptanucleotide repeat motif has been inserted.*

Interestingly, both the sequence classes mentioned above were identified: sample 18 (the cyanobiont in coralloid roots of *Cycas circinalis*) being one and samples 19-25 (cyanobionts in several other cycad species including another *Cycas* species) being another. Note the presence of a 42 bp extra sequence in one of the two cyanobionts of *Cycas rumphii* (samples 19 and 20, respectively) (Costa *et al.*, 1999; Costa *et al.*, 2002). When examining the $tRNA^{Leu}$(UAA) intron it is clear that different *Nostoc* strains identified in the cycad coralloid roots share a high level of similarity in their sequences (Figure 2) and that the differences are mainly due to indels in the first variable region of the intron (Figure 3). In addition, these sequences show a high similarity to sequences from cyanobionts of other cyanobacterial symbioses. However, how these differences/similarities are reflected on a biological level is more difficult to ascertain. Assuming mechanisms of slipped strand mispairing and homologous recombination, together with the fact that the same features are found in other symbiotic systems like lichens and bryophytes (Costa *et al.*, 2002), it can be concluded that all symbiotically competent cyanobacteria might have once evolved from a common symbiotic ancestor. This is a very appealing concept, which should be further experimentally tested using more samples, additional symbiotic systems and other parts of the cyanobacterial genome.

5. CHARACTERISTICS OF THE SYMBIOTIC CYANOBACTERIA

5.1. The Symbiotic Filament

The filamentous heterocystous cyanobacteria within the cyanobacterial zone are located extracellularly between the elongated cycad cells and embedded in mucilage (Lindblad *et al.*, 1985a). However, some authors report an intracellular location for cyanobacteria in coralloid roots of *Cycas revoluta* Thunb. and *Macrozamia communis* L. (Nathanielsz and Staff, 1975b; Obukowicz *et al.*, 1981). Although the cyanobiont has been variously reported as *Nostoc*, *Anabaena* or sometimes *Calothrix*, the classifications were exclusively based on morphological features. However, all the molecular work detailed above is consistent with different *Nostoc* strains being the cyanobiont in cycad symbioses.

The cyanobiont differentiates into vegetative cells and heterocysts, but rarely akinetes (Lindblad *et al.*, 1985a; Lindblad *et al.*, 1985b). Structurally, the vegetative cells of the cycad cyanobionts show few modifications compared to free-living isolates. However, a high frequency of heterocysts has been described in the cyanobiont of several cycad species. Examination of successive sections of *Zamia skinneri* coralloid roots, collected from its natural habitat in Costa Rica, revealed an increased heterocyst frequency ranging from 16.7% at the growing tip to 46% in the basal (older) parts (Lindblad *et al.*, 1985b). It was also apparent that, from being mainly single heterocysts at the growing tip (86%), more and more heterocysts occur in multiples (double to quadruple) as the root becomes older.

5.2. Nitrogen Fixation And Nitrogen Metabolism

Physiologically, symbiotic cyanobacteria from coralloid roots have the ability to fix N_2. Analysis of nitrogenase activity, using the acetylene reduction assay, in sequential sections of coralloid roots clearly demonstrates a gradual decline in older parts of the coralloid root. The highest activity occurs in the growing tip of the coralloid root. Nitrogenase is localised in heterocysts (Bergman *et al.*, 1986) and appears to be confined primarily to sections of the coralloid root where single heterocysts predominate (Lindblad *et al.*, 1985b). Freshly isolated cyanobionts from coralloid roots of *Macrozamia riedlei* natural populations in Western Australia, retain nitrogenase activity even when separated from their host as long as the cells are not exposed to O_2-levels above 1% (Lindblad *et al.*, 1991). Exposure to higher levels of O_2 resulted in a dose-dependent inhibition of the nitrogenase activity. Moreover, nitrogenase activity increases considerably when the cells are exposed to light. This is because of the increased availability of ATP produced by PS-I mediated cyclic photophosphorylation. It is reasonable to suppose that the low levels of nitrogenase activity observed when assaying freshly isolated cyanobionts in darkness, reflects loss or damage of heterotrophic mechanisms that provided ATP to the cyanobiont in the intact coralloid root. Separation of the cyanobacteria from the host tissue disrupts the intercellular microenvironment and any metabolic and biochemical interactions which the cyanobiont experienced in situ in the coralloid roots.

Free-living cyanobacteria assimilate the ammonia produced by nitrogen fixation primarily via glutamine synthetase (GS) - glutamate synthase (GOGAT) enzyme system. In contrast to other cyanobacterial symbioses, the symbiotic cyanobacteria of *Cycas revoluta*, *Ceratozamia mexicana*, and *Zamia skinneri* all show *in vitro* GS activities and relative GS protein contents, similar to those found in different free-living cyanobacteria including strains originally isolated from cycads (Lindblad and Bergman, 1986). In addition, the *in vitro* activity of GOGAT in the cyanobiont of *C. revoluta* is also similar to that of free-living cyanobacteria (Lindblad *et al.*, 1987).

5.3. Transfer of Fixed Nitrogen

Using $^{15}N_2$, it has been demonstrated that the fixed nitrogen is transferred from the cyanobiont to the cycad. By analysing the xylem sap from freshly detached coralloid roots, two strategies for further processing the fixed nitrogen have been described (Table 1; Pate *et al.*, 1988). Citrulline and glutamine are the principal translocated N-solutes in *Macrozamia*, *Lepidozamia* and *Encephalartos*, all belonging to Zamiaceae. Glutamine and a smaller amount of glutamic acid, but no citrulline, are present in xylem sap of *Bowenia* (Boweniaceae) and *Cycas* (Cycadaceae). In contrast to other cyanobacterial symbioses, it seems that in cycads fixed nitrogen is translocated to the host either as a combination of citrulline and glutamine or as glutamine alone, instead of as ammonia (Table 1).

Table 1. Variation in nitrogen assimilation processes and nature of export products from cycad coralloid roots (After, Pate et al., *1988)*

Process	Cycad
Incorporation of newly fixed nitrogen into glutamine; Export of glutamine	Boweniaceae and Cycadaceae
Incorporation of newly fixed nitrogen into glutamine and citrulline; Export of a combination of glutamine and citrulline	Zamiaceae

The fixation of $^{14}CO_2$ and formation of [^{14}C]-citrulline and [^{14}C]-arginine by freshly isolated cyanobionts of *Macrozamia riedlei* (a cycad exporting the fixed nitrogen as a combination of glutamine and citrulline) and *Cycas revoluta* (a cycad exporting the fixed nitrogen as glutamine only) have been examined (Table 2). In the *Macrozamia* cyanobiont [^{14}C]-citrulline was readily synthesised, whereas in the *Cycas* symbiont [^{14}C]-citrulline was formed only when the cells were incubated together with exogenous ornithine. Similar results were obtained when determining the formation of [^{14}C]-arginine by freshly isolated cells from both *Macrozamia* and *Cycas* (Table 2). It is interesting to note the general differences between cells incubated in light and in darkness and between freshly isolated cyanobionts and corresponding free-living strains (Table 2; Lindblad *et al.,* 1991). Note the high level of [^{14}C]-citrulline formation under darkness in freshly isolated cyanobiont from *Macrozamia* (with or without the addition of external ornithine). Such citrulline production ranged from 85.7 - 544 % of that observed in free-living *Nostoc* sp. strain PCC 73102 (Table 2). As discussed above, cells freshly isolated from *Cycas revoluta,* a cycad where the fixed nitrogen is transported in the form of glutamine (Table 1), synthesise significant amount of citrulline only when incubated in the presence of exogenous ornithine. Further studies are needed to address the role of cycads in regulating the export of fixed nitrogen by cyanobionts as glutamine in some (even though the capacity to synthesise citrulline is present) and as a combination of glutamine and citrulline in others.

5.4. Carbon Metabolism

Due to their location in coralloid roots (complete darkness), the cyanobionts are expected to have a heterotrophic mode of carbon nutrition. The fixed-carbon may be provided either by the photosynthetic host (the cycad) and/or by the cyanobacteria themselves via dark CO_2 fixation. Interestingly, the symbiotic cyanobacteria retain RuBisCO, which is functional *in vivo* (Lindblad *et al.,* 1987) and they do fix CO_2 (Table 2; Lindblad *et al.,* 1991). Further investigations are needed to fully understand the fixation and further metabolism of CO_2.

Table 2. Fixation of $^{14}CO_2$ and formation of [^{14}C]-citrulline and [^{14}C]-arginine by cyanobacteria freshly isolated from coralloid roots of Macrozamia riedlei *(A; Lindblad* et al., *1991) and* Cycas revoluta *(B; Lindblad, Pate and Atkins, unpublished). The cell suspensions of freshly isolated cyanobacteria were incubated with $^{14}CO_2$ in light (90 $\mu E\ m^{-2} \cdot s^{-1}$) or darkness, with 5% O_2 or air in the gas phase, and with 0 or 10 mM (final concentration) ornithine (orn) added for 1h at 30°C before being analysed. Values given are Bq (μg chlorophyll a)$^{-1}$ h^{-1}. Values in brackets represent percentage when compared with the free-living cyanobacterium (A, Nostoc sp. strain PCC 73102; B,* Nostoc *sp. strain PCC 7422) incubated under identical conditions. ND, Not detectable (either in the freshly isolated cyanobacteria or in the corresponding free-living strain)*

(A)

Incubation conditions		Total $^{14}CO_2$ fixed	[^{14}C]-citrulline	[^{14}C]-arginine
Light	5% O_2	49.2 (2.0)	0.7 (ND)	0.2 (0.7)
Light	5% O_2 +orn	115.8 (4.0)	13.7 (35.7)	4.6 (8.0)
Light	Air	42.0 (1.9)	2.0 (ND)	0.9 (3.3)
Light	Air +orn	82.8 (2.7)	11.0 (19.1)	3.6 (5.6)
Darkness	5% O_2	12.0 (24.7)	0.6 (85.7)	ND (-)
Darkness	5% O_2 +orn	23.4 (35.5)	5.5 (220.0)	3.2 (97.0)
Darkness	Air	9.0 (23.1)	1.2 (240.0)	ND (-)
Darkness	Air +orn	13.8 (35.4)	4.9 (544.0)	1.4 (93.3)

(B)

Incubation conditions		Total $^{14}CO_2$ fixed	[^{14}C]-citrulline	[^{14}C]-arginine
Light	5% O_2	48.6 (2.3)	ND (-)	1.9 (2.7)
Light	5% O_2 +orn	66.0 (2.2)	8.9 (46.5)	4.3 (5.9)
Light	Air	24.0 (0.8)	ND (-)	0.4 (3.5)
Light	Air +orn	45.0 (1.6)	5.0 (16.7)	2.5 (5.6)
Darkness	5% O_2	12.6 (42.0)	ND (-)	ND (-)
Darkness	5% O_2 +orn	16.2 (27.0)	1.6 (77.5)	0.5 (21.8)
Darkness	Air	4.8 (16.0)	ND (-)	ND (-)
Darkness	Air +orn	10.2 (17.0)	1.3 (78.8)	0.4 (12.7)

ACKNOWLEDGEMENTS

The Swedish Natural Science Research Council/The Swedish Research Council has financially supported our research on symbiotic cyanobacteria.

REFERENCES

Adams, D.G. (2000) Symbiotic interactions, in B.A. Whitton and M. Potts (eds.), *The Ecology of Cyanobacteria*, Kluwer Academic Publishers, Dordrecht, pp. 523-561.

Ahern, C.P. and Staff, I.A. (1994) Symbiosis in cycads: The origin and development of coralloid roots in *Macrozamia communis* (Cycadaceae), *Amer. J. Bot.* **81**, 1559-1570.

Bergman, B., Lindblad, P. and Rai, A.N. (1986) Nitrogenase in free-living and symbiotic cyanobacteria: immunoelectron microscopic localization, *FEMS Microbiol. Letts.* **35**, 75-78.

Costa J.-L., Paulsrud P. and Lindblad P. (1999) Cyanobiont diversity within coralloid roots of selected cycad species, *FEMS Microbiol. Ecol.* **28**, 85-91.

Costa, J.-L., Paulsrud, P. and Lindblad, P. (2002) The cyanobacterial $tRNA^{Leu}$(UAA) intron: Evolutionary patterns in a genetic marker, *Mol. Biol. Evol.*, in press.

Lindblad, P. (1990) Nitrogen and carbon metabolism in coralloid roots of cycads, *Advances in Cycad Research I, Memoirs New York Bot. Garden* **57**, 104-113.

Lindblad, P. and Bergman, B. (1986) Glutamine synthetase: activity and localization in cyanobacteria of the cycads *Cycas revoluta* and *Zamia skinneri*, *Planta* **169**, 1-7.

Lindblad, P. and Bergman, B. (1990) The cycad - cyanobacterial symbiosis, in A.N. Rai (ed.), *Handbook of Symbiotic Cyanobacteria*, CRC Press, Boca Raton, pp. 137-159.

Lindblad, P., Bergman, B., Hofsten, A.v., Hällbom, L. and Nylund, J.E. (1985a) The cyanobacterium - *Zamia* symbiosis: an ultrastructural study, *New Phytol.* **101**, 707-716.

Lindblad, P., Hällbom, L. and Bergman, B. (1985b) The cyanobacterium - *Zamia* symbiosis: C_2H_2-reduction and heterocyst frequency, *Symbiosis* **1**, 19-28.

Lindblad, P., Rai, A.N. and Bergman, B. (1987) The *Cycas revoluta* - *Nostoc* symbiosis: enzyme activities of nitrogen and carbon metabolism in the cyanobiont, *J. Gen. Microbiol.* **133**, 1695-1699.

Lindblad, P., Haselkorn, R., Bergman, B. and Nierzwicki-Bauer, S.A. (1989) Comparison of DNA restriction fragment length polymorphisms of *Nostoc* strains in and from cycads, *Arch. Microbiol.* **152**, 20-24.

Lindblad, P., Atkins, C.A. and Pate, J.S. (1991) N_2-fixation by freshly isolated *Nostoc* from coralloid roots of the cycad *Macrozamia riedlei* (Fisch. ex Gaud.) Gardn, *Plant Physiol.* **95**, 753-759.

Nathanielsz, C.P. and Staff, I.A. (1975a) A mode of entry of blue-green algae into the apogeotropic roots or *Macrozamia communis* (L. Johnson), *Amer. J. Bot.* **62**, 232-235.

Nathanielsz, C.P. and Staff, I.A. (1975b) On the occurence of intracellular blue-green algae in cortical cells of the apogeotropic roots of *Macrozama communis* (L.Johnson), *Ann. Bot.* **39**, 363-368.

Norstog, K.J. and Nicholls, T.J. (1997) *The Biology of the Cycads*, Cornell Univ Press, Ithaca and London.

Obukowicz, M., Schaller, M. and Kennedy, G. S. (1981) Ultrastructure and phenolic histochemistry of the *Cycas revoluta* - *Anabaena* symbiosis, *New Phytol.* **87**, 751-760.

Ow, M.C., Gantar, M. and Elhai, J. (1999) Reconstruction of a cycad-cyanobacterial association, *Symbiosis* **27**, 125-134.

Pate J.S., Lindblad, P. and Atkins, C.A. (1988) Pathways of assimilation and transfer of the fixed nitrogen in coralloid roots of cycad-*Nostoc* symbioses, *Planta* **176**, 461-471.

Rai, A.N. Söderbäck, E. and Bergman, B. (2000) Cyanobacterium – plant symbioses, *New Phytol.* **147**, 449-481.

Chapter 12

THE *NOSTOC-GUNNERA* SYMBIOSIS

BIRGITTA BERGMAN
Department of Botany, Stockholm University, S-106 91 Stockholm, Sweden

1. INTRODUCTION

This chapter reviews the *Nostoc-Gunnera* symbiosis with emphasis on interactions between the symbiotic partners and aspects that are specific to *Gunnera* symbiosis as compared to other cyanobacterial and bacterial plant symbioses. For additional information on the *Nostoc-Gunnera* symbiosis, the reader is referred to the following reviews: Bonnett (1990), Osborne *et al.* (1991), Bergman *et al.* (1992a), Bergman *et al.* (1996), Bergman *et al.* (1999), Rai *et al.* (2000), Bergman and Osborne (2002).

For some reason, the herbaceous angiosperm genus *Gunnera* is the sole extant flowering plant that establishes a symbiosis with a nitrogen-fixing cyanobacterium. This is in contrast to the situation for other symbiotic eubacteria, such as *Rhizobium* and *Frankia*, which seem to have a wider host range among the angiosperms, although the hosts are still limited to a few genera. The mechanism limiting the host range among the angiosperms to only *Gunnera* family is apparently not confined to the cyanobacteria as these are able to infect several other divisions within the plant kingdom. This phenomenon is strikingly obvious from the present volume.

In keeping with the eubacterial-angiosperm symbioses, cyanobacteria in symbiosis with *Gunnera* end up intracellularly in the host plant cells but they always remain outside the host cell plasmalemma. This is in contrast to the situation in most other cyanobacterial-plant symbioses in which the cyanobacteria are located extracellularly. This specific feature makes *Gunnera* attractive for symbiotic as well as evolutionary studies (Bergman *et al.*, 1996), considering that plant chloroplasts once originated from an endosymbiotic event between a eukaryote lacking pigments and a cyanobacterium with photosynthetic competence (Douglas, 1994).

It is well documented that the ultimate outcome of the symbiosis between *Nostoc* spp. and *Gunnera* spp. is an efficient nitrogen-fixing eukaryote; i.e. a dicotyledonous angiosperm is turned into a completely nitrogen-independent organism by forming an intimate symbiosis with a prokaryote. The nitrogen-fixing capacity of the cyanobacterium is likely to be a key factor underpinning the evolution and establishment of the *Gunnera* symbiosis. Indeed, this capacity is an important trait and possibly a pre-requisite for symbiotic competence for cyanobacteria.

A.N. Rai, B. Bergman and U. Rasmussen (eds.), Cyanobacteria in Symbiosis, 207-232.

2. THE HOST PLANT – *GUNNERA*

2.1. Evolution and Morphology

The host plant *Gunnera* arose early in the history of angiosperms. The oldest pollen fossils, found in South America, are about 90 Ma old (Jarzen, 1980). The scattered global distribution of the genus *Gunnera* today also suggests antiquity (Batham, 1943). From a taxonomic point of view the genus *Gunnera* is distinctly separated from all other angiosperm genera. Also within the genus, the macromorphology varies drastically (Fig. 1). The geographical distribution of *Gunnera* is primarily within the Southern Hemisphere and a requirement for wet conditions and high rainfalls is typical for all species of the genus (see also Chapter 13).

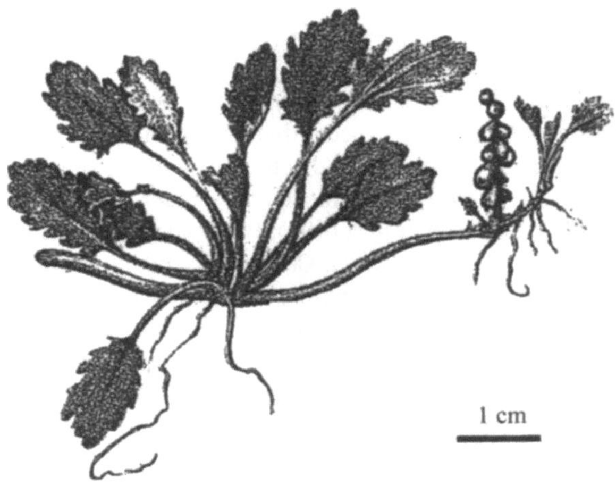

Figure 1. Gunnera dentata, *a species representing the small prostrate species with a creeping morphology and outgrowing stolons (Johnson & Brooke, 1989). This macromorphology is typical of species in New Zealand and Tasmania (see Fig. 2), and is in sharp contrast to the erect and often gigantic* Gunnera *species of Americas forming decorative long-petiolated rhubarb-like leaf rosettes of impressive heights.*

The very large (several meters high), erect and perennial *Gunnera* species are confined to Hawaii and Central and South America, while New Zealand and Tasmania are colonized by small, only a few centimeter high, matforming stoloniferous species (Fig. 1; see Bergman *et al.,* 1992a; Wanntorp, 2001). *Gunnera* spp. in South East Asia (Malaysia, Indonesia and the Philippines) and *G. perpensa* in South and East Africa represent another rather large group having more slender rhizomes.

Characteristic feature of the genus *Gunnera* are the small inconspicuous wind pollinated flowers and the endosperm rich drupes, which are probably dispersed by birds (St John, 1957; Dahlgren, 1975; Standley and Williams, 1963; see also Bergman *et al.,* 1992a). *Gunnera* may also be propagated vegetatively by discarded rhizomes and by outgrowing stolons (Fig. 1). Another peculiarity of the genus *Gunnera* is seen in their polystelic vascular system composed of numerous strands. The polystelic feature supports an aquatic ancestry of the genus. Many of these vascular strands are also seen leading to the cyanobacteria-invaded symbiotic stem tissues (Bergman *et al.,* 1992a; Söderbäck and Bergman, 1993).

2.2. Gunnera *Phylogeny*

Gunnera, with its 30-40 herbaceous species, is classified as a single genus within the family Gunneraceae. Controversy has accompanied all attempts to classify the genus and while its position among the angiopserms is not yet fully understood, recent molecular phylogeny data suggest an isolated position close to the Rosid-Caryophyllid-Asterid group (Chase *et al.,* 1993). Recent studies implicate a close relationship between *Gunnera* spp. and *Myrothamnus flabellifolius*, a morphologically different African shrub growing in dry habitats. This is surprising as all *Gunnera* spp. in contrast, prefer exceptionally wet and humid growth conditions. The close relationship to *M. flabellifolius* is inferred by using *rbcL* and the *rps16* introns as genetic tracers (Wanntorp *et al.,* 2001). The overall molecular phylogeny, based on gene sequences from 12 *Gunnera* species, representing all six subgenera and all continents, is illustrated in Fig. 2. The minute and the only annual species, *G. herteri* (Brazil and Paraguay) is a sister group, and the slender African species *G. perpensa* (Tanzania – South Africa), the first *Gunnera* species to be named (by Linneus), is also less closely related to the rest of the *Gunnera* species. The other species examined so far fall into two large and distinct clades (Fig. 2). Additional analyses using more variable gene sequences are now needed to fully resolve the phylogeny of *Gunnera* (Wanntorp, 2001). As all naturally grown *Gunnera* plants are infected by cyanobacteria it would be of great interest to search for an eventual co-evolution between infecting cyanobacteria and *Gunnera* as the receiving host.

2.3. Practical Use of Gunnera

In general, *Gunnera* plants have been of little or no practical or economic value. The very large and magnificent representatives within the family, such as *G. manicata* and *G. chilensis*, are nowadays commonly used as ornamental plants in botanical and other gardens. In some areas of Europe, having a particularly wet climate, garden escapes of *Gunnera* are currently spreading fast in the wild, such as in the Azores and the west coast of Ireland (see Chapter 13). Some *Gunnera* plants (*G. perpensa*) have traditionally been used as a herbal remedy during pregnancy in parts of Africa (Kaido *et al.,* 1997) and a pharmacological potential of *Gunnera* extracts (e.g. anthitrombin effect) has also been implicated (de Madeiros *et al.,* 2000). Additionally, the stems of *G. perpensa* are chewed upon and eaten in parts of Eastern Africa and the name *G. tinctoria* suggests a use of the plant as a coloring agent.

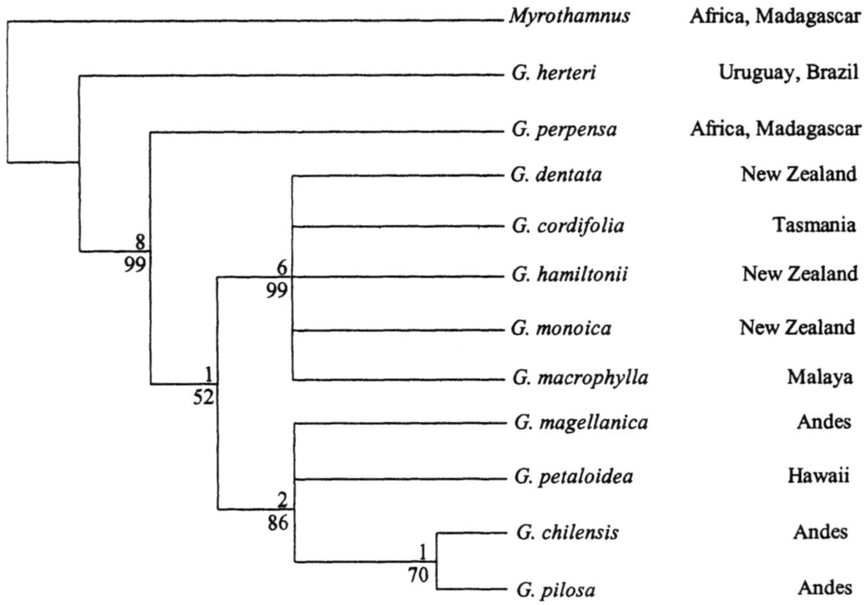

Figure 2. Phylogram of species within the genus Gunnera. *The tree combines the analyses of two genes (*rbcL *and* rps16*). After Wanntorp* et al. *(2001).*

3. THE SYMBIOSIS

It is suggested that the highly dynamic symbiosis between cyanobacteria and *Gunnera* may have arisen already ~100 Ma ago (see Chapter 16). The cyanobacterial-*Gunnera* symbiosis may therefore be the oldest angiosperm symbiosis existing today, i.e. older than angiosperm symbioses with *Rhizobium* and *Frankia*, but still younger than some of the other cyanobacterial symbioses.

3.1. Parasitism Versus Symbiosis

Common to all *Gunnera* species, irrespective of growth habitat and geographical location, is their ability to carry prokaryotic cyanbacteria as endosymbionts, a character known since 1872 (Reinke, 1872a, 1872b, 1873) and repeatedly confirmed thereafter (see Bergman *et al.,* 1992a; Wanntorp 2001; Wanntorp *et al.,* 2001). For instance, Reinke (1872b, 1873) noted that cyanobacteria were associated with all four *Gunnera* species he examined. For long, however, the exact nature of the interaction between the endosymbiont and the plant was debated (Reinke, 1872b, 1873; Jönsson, 1894; Schaede, 1951) and parasitism was often proposed. Although the symbiotic concept was introduced already in 1924 (Miehe, 1924), it took until 1969 before the interaction was defined as a true symbiosis. Silvester and Smith (1969) experimentally verified this concept by demonstrating transfer of ^{15}N from the nitrogen-fixing cyanobacterium to the plant.

3.2. Additional Microsymbionts?

There are a few reports on additional microsymbionts associated with *Gunnera* plants. However, no other prokaryote or eukaryote is known to form a true endosymbiosis with *Gunnera* plants, not even representatives of the occasionally rich microflora seen in the gland mucilage at the site of the cyanobacterial infection (Towata, 1985a). This is in keeping with most other cyanobacterial and bacterial plant symbioses. An exception is the cyanobacterial-*Azolla* symbiosis in which numerous bacteria consistently occur as additional microsymbionts among the cyanobacteria residing extracellularly in the leaf cavities (see Chapter 9).

The capacity to exclude all other potential symbionts except certain cyanobacteria points to a strict specificity between compatible cyanobacteria (*Nostoc* spp.) and *Gunnera* as a host. This in turn suggests the operation of recognition mechanisms and/or the existence of mechanisms that disguise or protect compatible cyanobacteria, but not other microorganisms, from hostile plant defense reactions (see below).

Recently however, the Hawaiian species, *G. petaloidea*, was found to also carry mycorrhizal fungi as an additional symbiont (Koske *et al.*, 1992). Whether these VAM fungi are involved in nutrient uptake supporting growth of the *Gunnera*, and thereby possibly also the endosymbiotic cyanobacterium, is not known. As only one *Gunnera* species has been examined so far, this finding needs verification. Still, this additional putative symbiont occupies a completely different site of the plant (the roots) than the cyanobacterium which is strictly confined to tissues associated intimately with the stem glands.

4. THE ENDOSYMBIOTIC *NOSTOC*

4.1. Cyanobionts

As in most other cyanobacterial plant symbioses, the endosymbionts in *Gunnera* appear to exclusively belong to the terrestrial genus *Nostoc*. This genus is probably the globally most widespread cyanobacterial taxa (Dodds *et al.*, 1995). Identification of endosymbionts in *Gunnera* has so far been based on a few morphological characters of a limited number of cyanobacterial strains (see Bergman *et al.*, 1992a). More recently, large surveys of cultured isolates from *Gunnera* spp. have been performed using DNA-fingerprinting (Rasmussen and Svenning, 1998; Nilsson *et al.*, 2000). Precise genetic identifications of natural endosymbionts to the genus level, residing in *Gunnera* or freshly isolated from *Gunnera*, are still warranted as is their phylogenetic relation to other plant-cyanobionts, whether exo- or endosymbiotic. The DNA-fingerprinting study did however prove that there is a wide diversity among the 45 cyanobacterial isolates collected worldwide from 11 *Gunnera* species (Nilsson *et al.*, 2000). Occasionally, even more than one strain was isolated from within the same *Gunnera* gland. For further discussions on cyanobacterial diversity and specificity the reader is referred to Chapter

15. Although not yet fully proven, morphological and phylogenetic analyses collectively suggest that the genus *Nostoc* by far dominates as cyanobionts in *Gunnera* (see Osborn, 1990; Bergman *et al.,* 1992a).

4.2. Reconstitution of the Symbiosis

The importance of the genus *Nostoc* as the major endosymbiont in *Gunnera* is further supported by reconstitution experiments using *G. manicata*, one of the larger species (Bonnett and Silvester, 1981; Johansson and Bergman, 1994). Only representatives circumscribed by the genus *Nostoc* showed symbiotic competence, while other even closely related cyanobacterial genera failed, including those producing hormogonia. This was the case for representatives within the genus *Anabaena* sp. (never found to be symbiotically competent), the hormogonia-forming genera *Calothrix* spp. (also symbiotically competent towards cycads and lichens, see Rai *et al.,* 2000) as well as for *Fischerella* and *Chlorogloeopsis* (hormogonia producers but with unknown symbiotic competence). This suggests that the capacity to form motile hormogonia, the "infection unit" in the *Gunnera* symbioses (Bergman *et al.,* 1996), is important but not sufficient. Furthermore, nine out of the 17 *Nostoc* species (45%) tested failed to enter *Gunnera*. Apparently, symbiotic competence is not a universal trait even within this genus. A crucial question therefore remains; which capacities are required to gain full symbiotic competence when it comes to infection of the angiosperm *Gunnera*? The fact that almost complete sequence of the symbiotically competent *Nostoc* ATCC 29133 genome is now known may be an important asset when it comes to resolving this crucial question.

It is however still an open question whether it is possible to infect any of the other *Gunnera* species with a wider array of cyanobacterial genera. Up till now, *G. manicata* is the only host species tested (Bonnett and Silvester, 1981; Johansson and Bergman, 1994). Moreover, genetic studies on freshly isolated cyanobacteria from natural stands of *Gunnera* are needed to i) verify the possible dominance of the genus *Nostoc* in nature; ii) find out if more than one cyanobacterium often occupies one *Gunnera* plant or one gland (Nilsson *et al.,* 2000), and iii) to find out whether mixtures of strains are common. Cyanobacterial strain mixtures were recently reported to occur in individual corralloid roots of cycads (Zheng *et al.,* 2002) and more than one cyanobacterium resides within individual *Azolla* plants (Braun-Howland and Nierzwicki-Bauer, 1990) and thalli of bryophytes (West and Adams, 1997). However, no such mixtures have been observed in lichens (Paulsrud *et al.,* 2000).

4.3. Morphological Plasticity

The genus *Nostoc* is characterized by being filamentous and a morphological complexity surpassed by no other prokaryote (Fig. 3). *Nostoc* easily adapts its behavior to prevailing external conditions by differentiating different cell types with various functions. The normally photosynthetic vegetative cells may for instance be converted into i) motile small celled hormogonia, ii) nitrogen-fixing thick cell-walled heterocysts and iii) perennial akinetes (spores). The cell differentiation events leading to hormogonia and heterocysts are of key-importance during establishment of the*Nostoc-Gunnera* symbioses (see below). There are no reports on differentiation of akinets in the *Gunnera* symbiosis, not even in older parts where the cyanobacterial cells appear electron-dense and partly disintegrated (Söderbäck *et al.,* 1990). This is similar to the situation in most cyanobacterial plant symbioses. Again the exception is the *Azolla* symbiosis, in which akinetes may occur (see Chapter 9). The cellular differentiation events taking place in *Nostoc,* from early infection to the establishment of the mature *Gunnera* symbiosis, are dramatic and accompany and/or govern the behavioral adaptations observed in the cyanobiont.

4.4. Surface Structures – the First Contact

Most representatives of the genus *Nostoc* are covered by thick layers of carbohydrates. Figure 3, depicts a *Nostoc* filament recently isolated (2 weeks) from stolons of the small *G. magellanica.* The vegetative cells are typically surrounded by a sphere of viscous carbohydrates, while the scattered heterocysts are mostly not.

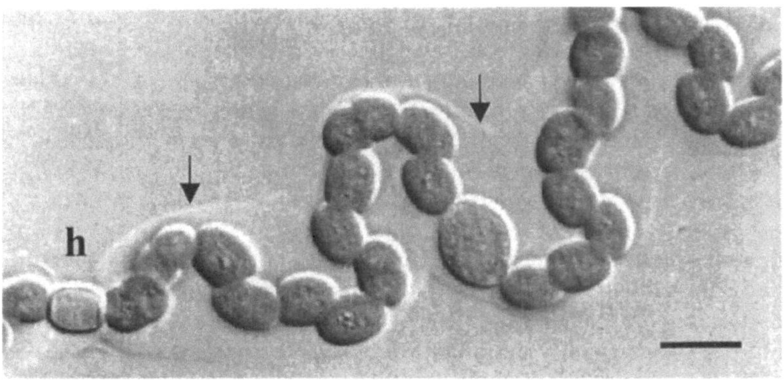

Figure 3. Nostoc *sp. recently isolated from infection sites located inthe creeping stolons of the small* G. magellanica (see Fig. 4B). *The filaments are surrounded by a pronounced carbohydrate rich mucilage (arrows) and are composed of photosynthetic vegetative cells and thick-walled nitrogen-fixing heterocysts (h). Bar: 10 µm. Photo: C.-M. Wang.*

For free-living *Nostoc* spp., external carbohydrates may have a role in protection against desiccation. *Nostoc* often covers terrestrial surface habitats regularly subjected to drought spells and the sheath having good water-holding capacity and may serve as a water reservoir. UV-protecting pigments may also be found in this external sheath (Dodds *et al.,* 1995). In the *Gunnera* symbiosis however, neither of these traits are likely to be of any importance. The host plants must provide all water and nutrients needed and the cyanobacterium ends up in darkness and is therefore in no need of protection against any UV-light. Still, sheaths may have important roles in both pathogenesis and symbiosis. It is known for instance, that carbohydrate rich surface covers may act as a disguise or camouflage to hide the true nature of pathogens and thereby avoid elicitation of host defense reactions (Krinos *et al.,* 2001). This may well be a role for the thick surface carbohydrate layer observed on *Nostoc* (Fig. 3) and possibly therefore a key feature of symbiotic competence. In addition, other surface structures are implicated to play a role during initial phases of infection in many plant-microbe associations. To reach the appropriate target host cells, mechanisms related to attraction, recognition and attachment must be operative in any bacterial-plant association (Vande Broek and Vanderleyden, 1995). Perhaps also the case in cyanobacterial-plant interactions. Fimbriae, proteinaceous long and thin "filaments", have been observed on compatible cyanobacteria (Johansson and Bergman, 1994). However, these are not yet proven to be of significance for the establishment of the *Gunnera* symbiosis. Lectins, proteins that recognizes specific carbohydrates, have also been observed in lichenized cyanobacteria (see Chapter 6; Rai *et al.,* 2000). Interestingly, it has recently been shown that the lectin composition of the invading *Nostoc* varies during the establishment of the *Nostoc-Geosiphon* symbiosis indicating a role of lectins in the infection process of this symbiosis (Schussler *et al.,* 1997). Although there are reason to believe that such mechanisms may also be operative in the *Nostoc-Gunnera* symbiosis, and possibly in all other cyanobacterial symbioses, no data are available.

5. THE *GUNNERA* GLAND – THE SUSCEPTIBLE ORGAN

5.1. Site of Infection

The stem glands ubiquitously formed on all *Gunnera* species is the sole site of infection by cyanobacteria, although the mucilage pouring out of the young non-infected glands may cover the apex and other nearby plant parts equally well. Clearly therefore, this plant structure plays an equally important role for the entrance of the microsymbiont as do root hairs for rhizobial infections of legumes. The gate of entry in *Gunnera* is however unlike those in all other cyanobacterial-plant symbiosis in which cyanobacterial infection occurs e.g. openings or pores leading into thallus cavities of liverworts and *Azolla* spp. (see Chapters 7, 9 and 10). *Gunnera* is also the only plant in which cyanobacteria occupy the stem.

The *Gunnera* organ infected by cyanobacteria shows all characteristics of plant glands (Kronstedt-Robards and Robards, 1991), both morphologically and behaviorally. For instance, *Gunnera* glands i) are composed of channels leading into each glandular tissue, ii) have cells aligning the channels that show typical transfer cell morphology and iii) continuously release a thick mucilaginous liquid (Towata, 1985a; Bergman *et al.,* 1992; Bergman and Osborne, 2002). Glands are not known to function as gates to the interior in any other known symbiosis. However, there are interesting functional similarities between infection of *Gunnera* glands by cyanobacteria and the "pollen and pistil" as well as the "sperm and egg" mechanisms used for fertilization of plants and animals. A question that still remains open is whether the only role of *Gunnera* glands is to function as a conduit for *Nostoc* into the interior of the plant. Plant glands typically act to control salt balance, to attract pollinators or to deterring herbivores or microorganisms. The bright color of the *Gunnera* gland may suggest a role in pollination of *Gunnera* or indeed in spreading of the cyanobacterium from one plant to another by insects (see also Chapter 13).

5.2. The Development of the Gland

The development of the glands in *Gunnera* is largely known but our knowledge about mechanisms underlying gland differentiation is still lacking. The glands are formed subsequent to *Gunnera* seed germination. Already at the very young cotyledon stage, and in absence of any microorganisms, distinct brightly red and structurally unique glands are seen on the slender *Gunnera* hypocotyls (Fig. 4A). These glands develop from the meristem of hypocotyl into two decussate glands in association with each new leaf primordium, just below the site of attachment on the cotyledon. At the base of each leaf, one additional gland is eventually formed (Silvester and MacNamara, 1976; Bonnett and Silvester, 1981; Bonnett 1990). The glands are therefore seen located close to the base of each new leaf petiol and occur in a very regular fashion along the stem as well as along the stolons as the plant grows. Typically therefore, several glands are formed close to the base of each new petiole (Fig. 4 C). Root primordia develop from just below these "wart-like" glandular *Gunnera* structures while stolons develop from axillary buds. Young stolons gets infected already close to the growing apex (Fig. 4B).

The *Gunnera* glands are composed of up to nine separate cell masses or papillae (Towata, 1985a; Johansson and Bergman, 1992; Osborne *et al.,* 1991; Uheda and Silvester, 2001; Bergman and Osborne, 2002). One central papilla is surrounded by other, often smaller, papillae. Cross sections of glands clearly show that the glands are penetrated by a few channels leading into the gland interior between the external papillae (Fig. 6A). A tight cell-layer also develops along the channels composed of secretory cells intervened by possibly tannin (phenolic compounds)-rich cells. The often larger secretory cells produce the mucilage, while the smaller and cytoplasm rich cells lining the channels appear electron dense after microscopic staining (Towata, 1985; Johansson and Bergman, 1992; Uheda and Silvester, 2001). Even before the papillae are exposed at the stem surface, a rich mucilage is secreted through the stomata on the covering epidermis (Schaede, 1951; Bonnett, 1990; Johansson and Bergman, 1992). Eventually the epidermis cracks open, the papillae get exposed and the glands (Fig. 4A) and the stem apex gets covered with mucilage. The whole glandular tissue continues to grow and at the bottom of the channels small groups of larger thin-walled cells are formed (Fig. 6A). These are the likely cells that get infected by cyanobacteria (Schaede, 1951; Silvester and MacNamara, 1976; Towata, 1985b; Johansson and Bergman, 1992).

5.3. Signaling

The bright red color of the gland surface (Fig. 4A) has been suggested to be due to the presence of anthocyanins and these may have several functions. If released into the mucus they may function as attractants to cyanobacteria. It is known that certain flavonoids released from roots of legumes function both as attractants of rhizobia and as

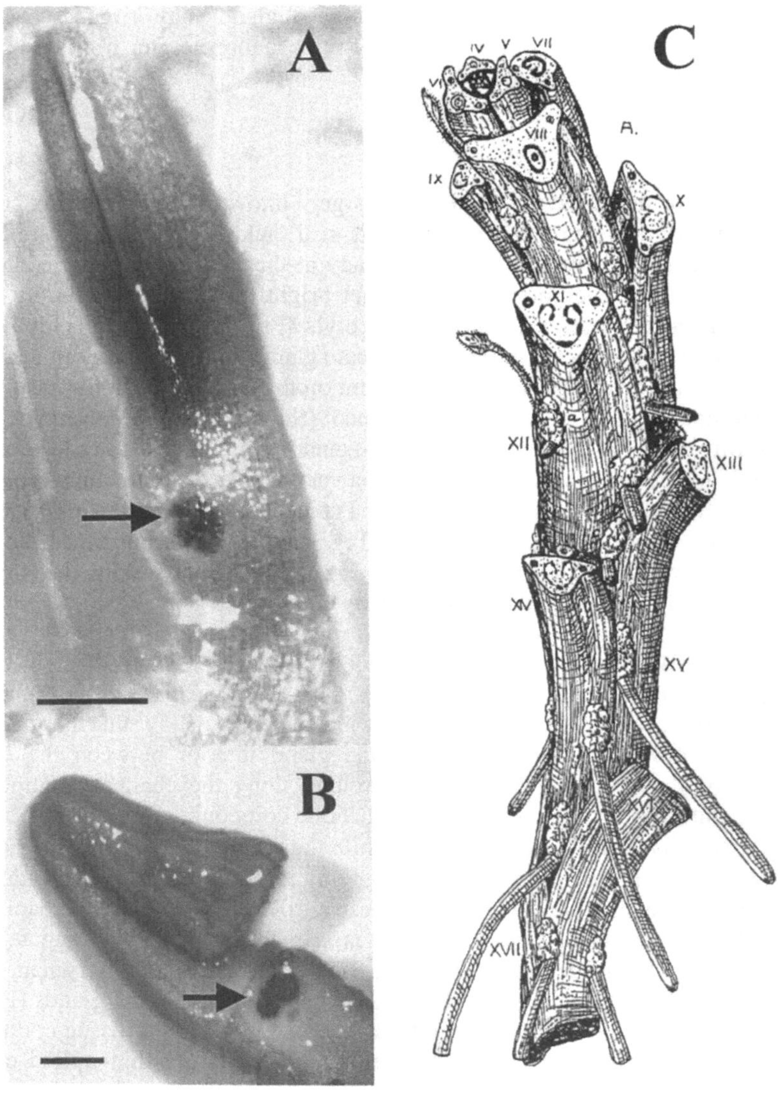

Figure 4. (A) Stem gland in Gunnera *- the site of cyanobacterial infection. A young* Gunnera *plant with a gland developing at the hypocotyl (arrow). The gland has penetrated the epidermis and secretes a viscous mucilage. Bar: ~2 mm. Photo: C. Johansson. (B) The apex of a growing* Gunnera *stolon with the infected tissue exposed (arrow) after removing the epidermis. Bar: 1 mm. Photo: C-M. Wang. (C) The location of glands on a mature* Gunnera *stem.These are "located regularly at the base of each leaf petiol, one basal and median and two lateral placed somewhat higher upon the stem". Vegetative stolons are seen developing from buds while roots develop at the base of the glands (from Baas-Becking 1947).*

signals eliciting *nod* (nodulation) gene expression. This may be the case of the red or other flavonoids released in *Gunnera*. Indeed, using rhizobia with *nod* gene::*lacZ* fusions as a reporter system, both *Gunnera* seed rise and the gland secreted mucilage have been shown to induce nod gene expression in *Rhizobium* (Rasmussen *et al.*, 1996). However, as extracts and exudates from other non-symbiotic angiosperms may also induce *nod* gene expression (Bender *et al.*, 1988), additional experiments are needed to verify the eventual role of flavonoids in *Gunnera*. There is also the possibility that the red color may function as attractants to birds or insects, which may mediate a transfer of the often sticky *Nostoc* filaments from infected to non-infected *Gunnera* glands. Alternatively, the anthocyanins may have yet other unknown functions.

6. THE "INFECTION" PROCESS

The structural aspects of the infection process in the *Nostoc-Gunnera* symbiosis have been well-described (Neumann *et al.*, 1970; Silvester and MacNamara, 1976; Towata, 1985; Söderbäck *et al.*, 1990; Johansson and Bergman, 1992; Bonnett, 1990; Uheda and Silvester, 2001). The over-whelming majority of reports deal with the symbiosis from the cyanobacterial point of view. Moreover, this is the best known and characterized infection process among the cyanobacterial-plant symbioses. The reason for this is the relative ease by which the two symbiotic partners in the *Nostoc-Gunnera* symbiosis can

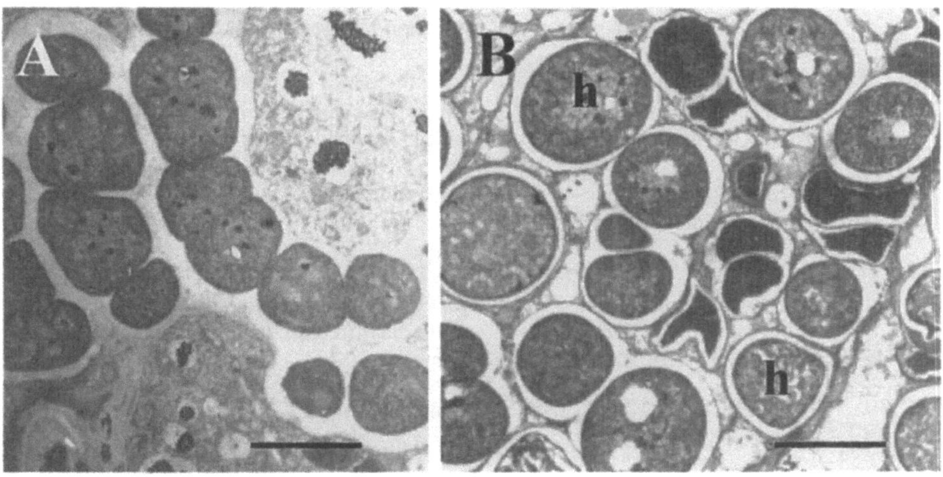

Figure 5. Transmission electron micrographs showing (A) a newly infected Gunnera *cell with the cyanobacterial filament composed of vegetative cells only (hormogonia/primordia). Bar: 5 µm. (B) A later stage when* Nostoc *has differentiate numerous heterocysts (h)and some cells are electrondense and malshaped. Bar:10 µm. Note differences in cell size/volume between the two stages. Photo: C. Johansson.*

be separated and the symbiosis reconstituted under laboratory conditions. Johansson and Bergman (1992) recognized several developmental stages using LM, TEM and SEM analyses of the infection process in *G. chilensis* and *G. manicata*. Infection embraces drastic morphological and physiological adaptations particularly in the cyanobiont. Those we know today are summarized in Table 1. A great advantage of *Nostoc*-plant symbioses, compared to other eubacterial symbioses, is the cellular plasticity of the microsymbiont which allows morphological adaptations to be easily recognized by LM

Table 1. Developmental stages identified during infection and maturation of Nostoc-Gunnera *symbiosis (after Johansson and Bergman, 1992).*

Stages	in *Gunnera*	in *Nostoc*
The infection process:		
I	gland development attraction to the plant/gland;	independent vegetative growth;
II	mucilage secreted, covers glands and apex;	accumulation at the gland surface hormogonia differentiation;
III	continued mucilage secretion through the channels;	hormogonia enter the channels; attraction & recognition;
IV	folding of cell walls lining the interior of the channels; some *Nostoc* cells sacrificed;	to the bottom of the channels; compatible cyanobacteria among folded host cell walls;
V	host cell wall dissolution hormogonia into the primordial stage;	penetration of host cells;
VI	host cell wall regeneration and closure; growth & rapid heterocyst differentiation;	*Nostoc* located intracellularly primordia differentiate into vegetative filaments;
VII	intracellular accommodation of the cyanobacterium;	spreading of the *Nostoc* infection to new compatible cortical gland cells;
At maturity:		
VIII	adaptations of the physiology; supply of C- & other nutrients to the cyanobacterium;	adaptation to a symbiotic life N_2-fixation and N-transfere to the host

and TEM. By identifying morphological adaptations in compatible *Nostoc* spp. during infection, numerous changes in gene and protein expression patterns in *Nostoc* and the underlying regulatory mechanisms may be predicted. A well-known example is the increase in heterocyst frequency as the symbiosis matures (Bergman *et al.*, 1992b).

6.1. The Two Meet – Attraction and Motility

It is perceived that the *Gunnera* plant and/or the gland attracts cyanobacteria possibly via the mucilage secreted from the gland or by a plant product(s) carried by the mucilage (Stage I-II; Table 1). Preliminary chemotactic studies using one compatible (*Nostoc* PCC 9229) and one non-compatible (*Nostoc* 268) *Nostoc* strain demonstrated no, or only a week, response towards *Gunnera* mucilage while growth was stimulated (Rasmussen *et al.*, 1994). Chemotaxis is therefore a research area in need of renewed interest, particularly since the *Gunnera* plants are always infected in nature and compatible cyanobacteria consistently end up in proximity to or on the *Gunnera* plantlets under laboratory conditions. Already three days after inoculation of the *Gunnera* plantlets with a cyanobacterial culture, all cyanobacteria gathered exclusively on the glands. This is surprising considering the acidic nature of the mucilage (Schaede, 1951, Rasmussen *et al.*, 1994) with a pH of about 4-5. This pH is unusually low for supporting and even stimulating cyanobacterial survival, yet it does (Rasmussen *et al.*, 1994).

Besides attraction, motility is required for the two partners to meet although both may occupy the same habitat. As *Gunnera* plants are sessile, cyanobacterial motility via the hormogonia is a prerequisite. As in bryophyte symbioses (see Chapter 7), the *Gunnera* host apparently releases signal molecules that induce hormogonia formation (Rasmussen *et al.*, 1994) to ensure infection. These putative low molecular weight HIF(s), hormogonia inducing factor(s), are apparently signaling molecules carried by the mucilage secreted by *Gunnera* gland (Table 1). As mentioned above, all strains known to infect *Gunnera* produce hormogonia, but not all hormogonia producing strains are competent to infect *Gunnera* glands (Johansson and Bergman, 1994). Hence, it is likely that attraction and motility are important features for the infection of *Gunnera* although mechanisms are not yet understood.

6.2. Towards the Interior

After gathering in the mucilage at the surface of the gland the hormogonia movement is directed towards the inner areas of the gland (Stage III-IV; Table 1). These are reached via the mucilage-filled channels located between the numerous papillate structures composing the outer gland structure (Fig. 6A). A few channels lead into each gland at least in theory allowing multiple sites for infection within one individual gland.

6.2.1. Signaling – from Plant to Cyanobacterium

Compatible cyanobacteria must sense attractive signal(s) from the interior of the glands in order to continue to move in that unexpected direction. This attractant could be a component of the carbohydrate mucilage *per se*, or some compound continuously released by cells in the gland and into the mucilage (Rasmussen *et al.*, 1994; Bergman *et al.*, 1996). Phenolics may have a role here. These are known to act as attractants in *Rhizobium*-legume symbioses and the red color of the *Gunnera* glands and histochemical staining of the mucilage (Schmidt, 1991) suggest the presence of phenolic compounds at the site of infection. While in the channel, signals of recognition or avoidance of plant defense reactions must also come into action in order to allow cyanobacteria to survive and compatible cyanobacteria to enter host cells. No such signals have yet been identified.

Once at the bottom of the channels, the *Nostoc* cells may need to adhere to specific target *Gunnera* cells. Adhesins are not identified although cyanobacteria synthesize several extracellular surface structures known to be involved in adhesion of cells to other organisms or surfaces. These include fimbriae, lectins and "sticky" carbohydrates (Bergman *et al.*, 1992; Johansson and Bergman, 1994; see above).

6.2.2. Signaling – from Cyanobacterium to Plant

Compatible cyanobacteria may in return influence *Gunnera* development and stimulate the formation of the symbiotic tissue, and also the subsequent plant performance. As the channels in the gland close in absence of a compatible cyanobacterium and the mature symbiotic tissue never develops, that is stage IV – VIII (Table 1) are not passed, the presence of a cyanobacterium may initiate these steps. For details see section 7 below.

6.3. The Mucilage

The importance of the gland mucilage in the *Nostoc-Gunnera* symbiosis is evident (Johansson and Bergman, 1992; Rasmussen *et al.*, 1994; Bergman *et al.*, 1996; Liaimer *et al.*, 2001). The most potent mucilage releasing glands are located close to the growing apex, while glands further down the stem produce less of the brown-yellow mucilage (Towata, 1985a). Glands that are already infected with cyanobacteria release less of the mucilage and these are soon closed by a corky outer dark-brown layer. At this stage their papillae are no longer distinguishable.

6.3.1. Composition and Function

GLC-MS analyses have shown that the mucilage is composed of carbohydrates, primarily arabinose, galactose and glucose in a 1:0.25:0.13 ratio (Rasmussen *et al.*, 1996). This indicates that the mucilage is composed of AGPs, or contains AGPs. Moreover, an antibody directed against an AGP (MAC 207) responded positively to the mucilage (Rasmussen *et al.*, 1996). AGPs are known to be of importance in cell-cell interaction events and development, such as during fertilization of plants. The pollen grains need to pass an AGP rich mucus of the transmitting tract of the style to reach the embryo. If composed of AGPs (Rasmussen *et al.*, 1994), the *Gunnera* mucilage may potentially affect cyanobacterial development and performance in symbiosis. If both partners release AGPs, a higher level may arise locally, which may lead to unexpected developmental effects. A key question is whether hormogonia release any AGPs (Stage

III; Table 1). However, additional studies are warranted to connect AGPs to any symbiotic performance in the cyanobacterium or the plant (see below).

6.3.2. Induction of Gene Expression

Recently, three genes in *Nostoc* PCC 9229 (isolated originally from *G. manicata*) have been identified using subtractive hybridization, that are specifically induced by the mucilage collected from the New Zealand species *Gunnera manicata* (Liaimer *et al.*, 2001). Six and sixteen hour treatments with the mucilage made it possible to identify three genes termed *hieA*, *hie B* and *hieC* (host induced expression) from the cDNA *Nostoc* PCC 9229 libraries constructed (Table 2). These *Nostoc* genes are expressed when subjected to the mucilage and represent, as shown by amino acid alignment, i) a putative precursor of a signaling peptide, ii) a glycoprotein and iii) a low-pH inducible gene, respectively. Although these genes suggest induction of a molecular communication between the symbionts in the *Nostoc-Gunnera* symbiosis, this needs to be further clarified.

Table 2. Effects of mucilage secreted by Gunnera *glands on cyanobacterial development and behavior (after Rasmussen* et al., *1994; Rasmussen* et al., *1996; Liaimer* et al. *2001)*

Activity	Compounds/consequence
In *Nostoc:*	
attraction	phenolics (?)
growth stimulation	carbohydrates of the mucilage?
cell differentiation	hormogonia inducing factor (HIF); a LMW protein (0.5 - 12kDa)
protein induction	a new protein (40 kDa); increased synthesis of a 65 kDa protein
gene expression	specific expression of *hieABC*
In *Rhizobium*:	
gene expression	bacterial *nod*-gene induction

6.3.3. Protein Induction

Using ^{35}S labeling, it was also shown that two proteins, a 40 kDa and a 65 kDa, are upshifted in *Nostoc* PCC 9229 subjected to mucilage treatment (Table 2) (Rasmussen *et al.,* 1996). The identity of these proteins is however not yet known, although studies in this direction are now being conducted using proteomics.

6.4. Rhizobium and Agrobacterium Gene Homologs

Another approach used to get a better insight into the eventual molecular cross-talk between *Gunnera* and *Nostoc* was to search for similarities between prokaryotic genes and proteins known to be involved in such events in the considerably better studied *Rhizobium*-legume symbiosis and the *Agrobacterium*-plant interaction. Screening showed positive hybridizations between *Nostoc* DNA and certain rhizobial genes such as *nodEF*, *nodNM* and *exoY* and the *nod*-box (Rasmussen *et al.,* 1996). The products of *nodEF* determines host-specificity, while the *nodNM* genes are supporting the synthesis of the Nod-factor released by the rhizobia in response to legume flavonoids. The *exoY* gene is involved in the synthesis of extracellular surface polysacharides while the *nod*-box has a central role in NodD-binding. However, there was no recognition between *Nostoc* DNA and the common *nodABC* genes, the regulatory gene *nodD* or the host-specific genes *nodL*. Later it was demonstrated that in spite of this finding, the *nodM* gene, although present in *Nostoc*, apparently functions only as a house keeping *glmS* gene (of which *nodM* is a second copy) encoding glucosamine. It was therefore suggested that this gene is not involved in symbiosis (Viterbo *et al.,* 1999). For instance, there was no induction of the *nodM* gene by *Gunnera* mucilage or seed rinse as would have been expected if it had a role similar to that proven for rhizobia.

Some agrobacterial genes involved in the infection process (*chvA*, *chvB* and *picA*) also hybridized to DNA from *Nostoc*, and this occurred only in compatible strains. Such hybridization was not detected with *virA* and *virG* (Rasmussen *et al.,* 1996). The significance of all these homologies suggests that potentially mechanistic similarities may exist in these highly different plant interactions. This aspect needs to be explored further.

7. PENETRATION OF HOST CELL WALLS

When infected by a compatible cyanobacterium the walls of the cells aligning the bottom of the channels start to fold, open and allow cyanobacterial penetration. Subsequently, the *Gunnera* cell walls regenerate and close up. As indicated by TEM analyses, some cyanobacterial cells are lost in this process as the cyanobacterium "squeezes" in between the *Gunnera* cells and their cell walls (Fig. 5B; Johansson and Bergman, 1992). This sequence is unique among the cyanobacterial-plant symbioses and of great general and evolutionary interest as it may represent a "blue-print" of the process that led to the genesis of chloroplasts (see Bergman *et al.,* 1996). In this process a prokaryotic cyanobacterium passes through the otherwise expelling barrier of eukaryotic plant cell walls and "infects" the plant cells (Fig. 5). In this case, a most important physiological process, N_2 fixation, is gained by the host plant.

7.1. Putative Signaling Compounds

The mechanistic details of the host cell wall penetration are largely unknown but some putative candidate compounds have been identified. Although many authors have claimed the involvment of enzymes in penetration no direct proofs are at hand. Pectolytic and cellulolytic have been recorded in glands of *Gunnera* but not in the cyanobiont, nor were cellulase activities detected in *Nostoc* isolates (see Bergman *et al.*, 1992). Two other compounds possibly involved in penetration are arabinogalactan proteins (AGPs) (see Bergman *et al.,* 1996) and the phytohormone auxin (IAA) (Sergeeva *et al.,* 2002). Both AGPs and IAA are known to influence plant cell division and development. We have recently shown that AGPs and IAA are released by symbiotically competent cyanobacteria at least when in their free-living stage. The stimulated mitotic activity in host cells aligning the gland channels and the opening of the host cells (Bergman *et al.,* 1992, Bergman *et al.,* 1996; Fig. 5) suggest that the cyanobacterium may elicit these reactions. IAA may also induce cellulase activity. It is therefore tempting to speculate that the cyanobacterium may modulate IAA levels at the site of the penetration in such a way that host cell wall loosening is induced. Prokaryotic IAA has a key-role in many other plant-bacterial interactions, both pathogenic and symbiotic (Costacurta and Vanderleyden, 1995; Glick *et al.,* 1999). The "new" IAA-concentration reached has been shown to influence plant growth and performance. Additional studies are required to verify this hypothesis. Hence, the mechanistic details of the host cell wall penetration process are largely unknown. Interestingly, extracts of the cyanobacteria-infected tissue in *Gunnera* were found to contain a substance of an "auxin" character (*Avena*-tests) while non-infected stem tissues did not (Baas Becking, 1947).

Besides in the *Gunnera* symbiosis, an intracellular penetration of the host cells by cyanobacteria seems to be a rare event among cyanobacterial symbioses. Other known examples are marine diatoms (see Chapter 1) and the fungus *Geosiphon* (see Chapter 3). Why only these three totally unrelated hosts accept intra-cellular penetration is unknown. In all these three cases, similar host cell wall penetration mechanisms must exist. The penetration process in *Geosiphon* has been examined in some detail, but events leading to the intracellular location of cyanobacteria in marine diatoms are unknown.

8. INTRACELLULAR ADJUSTMENTS

After entering the host cells and before being fully accommodated into the symbiotic life, the *Nostoc* filaments undergo yet another distinct set of morphological and physiological adjustments. The major events are the formation of the *Gunnera* symbiotic tissue (Fig. 6; termed "internal nodules" by Silvester, 1976), retarded cyanobiont growth rate, and enhanced heterocyst differentiation (Fig. 5B), nitrogen fixation and: the release of fixed-nitrogen by the cyanobiont for the benefit of the plant.

8.1. The Symbiotic Tissue

The cortical stem tissue (located below the gland surface and the cells lining the channels), which become infected with*Nostoc* filaments, is delimitated by a distinct cell layer oriented transversely to the rest of the cortical cells (Fig. 6A). This suggests that the gland, at least initially, have a precise size and a clear border to the rest of the stem cortex. It is possible however, that the cortical cells within this delimitation may also divide as infection by *Nostoc* proceeds. All the glandular cells lack chlorophyll.

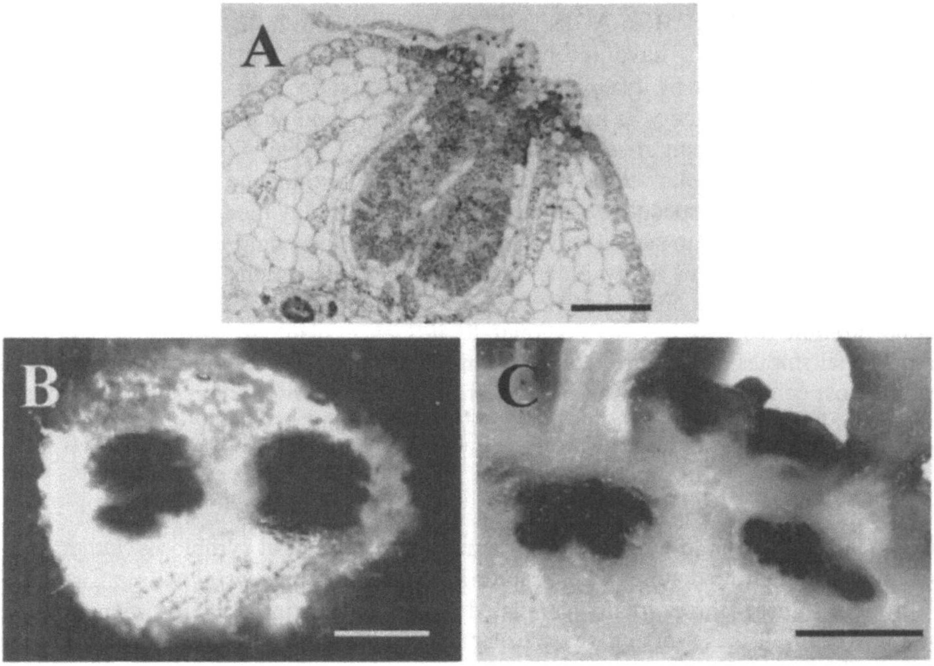

Figure 6. Cyanobacterial invaded symbiotic tissues in Gunnera.*(A) Cross sectioned* Gunnera *gland with channels leading into the interior of the gland between the papillae. The channels are aligned by a layer of more or less electron dense cells some of which get infected. The cell layer(s) delimiting the glandular tissue isalso seen. (B) Newly infected young* Gunnera *plantlet showing the cyanobacterial filled symbiotic tissuesthat develops just inside the glandswhich are located one at each side of the hypocotyl. (C) A mature* Gunnera *plant illustrating the small biomass made up of the cyanobacterial infection at this stage. Bars: 1 mm, 1 mm and 2 cm, respectively. Photos: C. Johansson.*

The *Gunnera* tissue harboring the endosymbiontic cyanobacterium is illustrated in Figure 6 (B & C). The infected tissue is taking up a large proportion in the thin hypocotyls while a small proportion in the thick fleshy rhizomes of mature plants and in particular in the large *Gunnera* species. In the small species, *Nostoc* also invades the nodes spread at regular distances along the stolons (Fig. 4B). Roots and shoots also develop from the cyanobacteria-infected stolon nodes.

The *Nostoc* filled symbiotic tissue is always delimited to only a small proportion (a few %) of the total *Gunnera* biomass and infected cells are often in close proximity to non-infected host cells (Johansson and Bergman, 1992). This suggests the existence of mechanisms that restrict the cyanobacterial growth and spreading to only specific cells inside the stem of *Gunnera*. The specific characteristics of the cells allowing infection and of the cells not allowing infection, the latter putatively with strong defense barriers, are not known.

Furthermore, it is possible that some plant metabolite(s) or the environment offered (darkness, heterotrophy, microaerobiosis etc) retard cyanobacterial growth *in planta*. This may be via an interference with the *Nostoc* cell division machinery or a retardation of cyanobacterial growth via reduced nutritional support to the cyanobacterium *in planta*, or a combination of the two. The observation that the cyanobacterial cell volume increases considerably in symbiosis compared to when free-living (Söderbäck and Bergman, 1992), suggests that cell division may be affected. An increase in cell volume is also seen in other cyanobacterial-plant symbioses (Rai *et al.*, 2000). Moreover, in *Gunnera*, the number of glands that develop is genetically predestined, and thereby the number of infection units. Infection is also restrained by the number of entrance possibilities (glands), and perhaps the subsequent spreading of the infected symbiotic tissue inwards into the *Gunnera* rhizomes and stolons is controlled by the host. The cell layer barrier is oriented in a direction perpendicular to the rest of the tissue a few cell layers down into the gland (Söderbäck, 1992; see Bergman *et al.*, 1992) which may act as a morphological barrier to the further spreading of the cyanobacterial infection. In any event, it is apparent that a strict feedback control excreted by the host plant ultimately controls the number, the size of the cyanobacteria invading and the spreading of the symbiotic tissues.

The cyanobacterial filaments eventually fill the invaded host cells, more or less completely (Fig. 5B). In *Gunnera* cells only a thin cytoplasmic rim remains along the cell walls. Although seen inside the host cell walls, the cyanobacterium is always localized outside the plant plasmalemma (Johansson and Bergman, 1992) as is the case in the other endosymbiotic cyanobacterial (*Geosiphon* and marine diatoms) and rhizobial-legume endosymbioses. The plasmalemma must greatly expand during *Nostoc* infection in order to "reach" around all *Nostoc* filaments and to exclude the prokaryote from true entry into the cell cytoplasm. Moreover, this plasmalemma must at the same time function as the interface between the symbionts and be the place where the exchange of C, N and all other metabolites released and required by the partners occurs.

8.2. Heterocyst Differentiation and Patterns

Entrance into symbiosis with *Gunnera* has a profound effect on heterocyst differentiation (compare Fig. 5 A & B). The frequency is often enhanced up to 10-fold, and considerably more so in *Gunnera* than in many other plant-symbioses (Silvester, 1976; see Bergman *et al.*, 1992b). It therefore seems likely that the plant or the conditions offered by the plant may trigger enhanced transcription of *hetR*, a key gene for the differentiation of heterocysts in free-living cyanobacteria (Buikema and Haselkorn, 1993; Adams and Duggan, 1999). Interestingly, *hetR* expression was stimulated by additions of fructose in darkness to the *Gunnera* isolate *Nostoc* PCC 9229, conditions supposed to mimic those offered in symbiosis (Wouters *et al.*, 2000). Cyanobacterial *hetR* expression profiles in intact *Gunnera* symbioses, from young to mature parts, are in progress and *hetR* expression is expected to be particularly high in newly infected stem glands. HetR was earlier suggested to respond to altered C:N ratios (Buikema and Haselkorn, 1991), and a high C:N ratio may well be the case when *Nostoc* is in transition from the non-nitrogen-fixing hormogonial stage into the multi-heterocyst carbohydrate supported stage, when heterocyst differentiation is fastest (Söderbäck *et al.*, 1990).

Yet another pronounced adaptation is the disrupted heterocyst pattern along the *Nostoc* filaments. Unlike in free-living cyanobacteria, double, triple and even quadruple heterocysts are common, in particular in mature parts of the symbiosis where heterocyst frequencies may approach 60-70% (Silvester, 1976; Söderbäck *et al.*, 1990). Expression of the *patS* gene may be modulated under such conditions. The product of this gene, a short peptide, dictates the regular distributed single-heterocyst pattern typical for *Anabaena* PCC 7120 (Yoon and Golden, 1998). PatS acts by inhibiting vegetative cells from development into heterocysts. The patterns seen in symbiosis would therefore suggest a down-regulated *patS* expression.

8.3. Nitrogen Fixation

Ever since the first evidence was presented by Silvester and Smith in 1969, nitrogen fixation by *Nostoc* living in symbiosis with *Gunnera* has repeatedly been verified (see Bonnett, 1990; Bergman *et al.*, 1992a; Rai *et al.*, 2000). Ending up in darkness, cyanobacterial N_2 fixation must depend on energy and reductant supplied by the plant (Söderbäck and Bergman, 1993). For instance, exogenously added fructose (Wouters *et al.*, 2000) and glucose (Black *et al.*, 2002) stimulate N_2 fixation activity. A positive correlation between decreased nitrogenase activities and shedding of *Gunnera* leaves has also been observed (Söderbäck and Bergman, 1993). N_2 fixation activities are positively correlated with increases in heterocyst frequencies up till about 20% heterocysts. Thereafter the correlation is negative in spite of the high heterocyst frequency seen in more mature parts of the symbiosis. This trend has been noted in *G. magellanica* (Söderbäck *et al.*, 1990) as well as in *G. monoica* (Stock and Silvester, 1994). In general, higher N_2 fixation activities are reported in symbiosis than when free-living.

8.4. Nitrogen Translocation and Assimilation

Using $^{15}N_2$ tracer it was found that nitrogen fixed in mature regions of *Gunnera* is translocated to the apical regions of the plant and that translocation is via the phloem (Stock and Silvester, 1994). The nature of the nitrogen transferred to the *Gunnera* plant has for long remained an open question, although NH_4^+ seemed to dominate in the all other cyanobacterial symbioses examined. It is however now well established that this is also the case in *Nostoc* infected *Gunnera* nodules (Silvester *et al.*, 1996). Estimates suggest that up to 90% of the nitrogen fixed by the cyanobiont is released extracellularly and primarily as asparagine (30%) and NH_3. The regulation of glutamine synthetase (the primary ammonia assimilating enzyme; GS) apparently takes place both at the level of *glnA* transcription in certain cells (lower GS quantities in heterocysts) and at GS-activity level, the latter being lowered in symbiosis compared to in the free-living stage (Söderbäck *et al.*, 1992).

The N released by the cyanobacterium is used for the benefit of the plant and *Gunnera* thus acquires a very important trait that makes it fully independent of an external nitrogen source.

9. EXCHANGE OF METABOLITES

9.1. Nutritional Support to the Endosymbiont

In order to maintain the endosymbiont physiologically active and slow growing, the plant must provide a full nutritional complement of macro- and micronutrients as well as other conditions required for survival of the normally photoautotrophic and aerobic cyanobacterium. As transfer of nutrients (besides C) from *Gunnera* to the endosymbiont has not been examined in any greater detail additional studies are required.

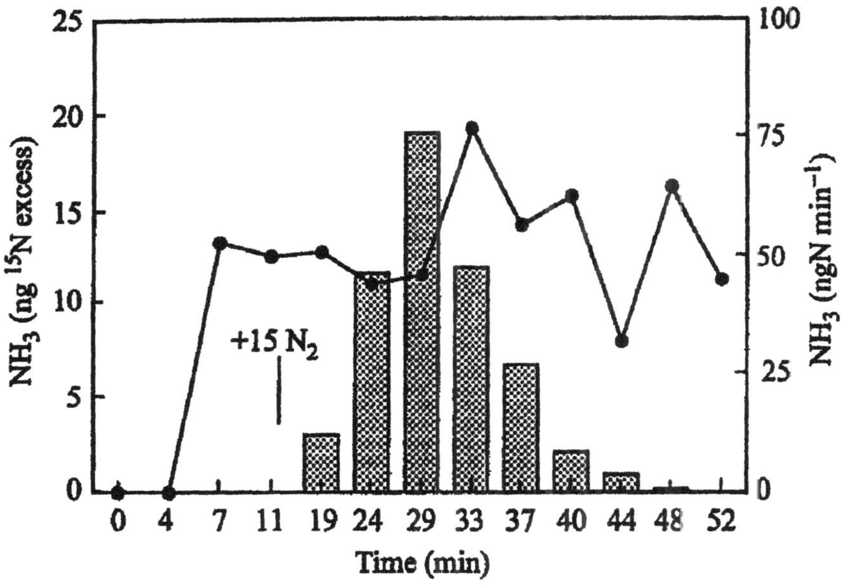

Figure 7. NH_4^+ *production and* $^{15}N_2$ *incorporated into* NH_3 *by* Nostoc *cells. The cells were freshly isolated from the internal gland nodules of* Gunnera magellanica *under the strict anaerobic conditions required to maintain activity (after Silvester et al., 1996).*

Although still responsive to light in symbiosis, and therefore photosynthetically fully competent (Silvester, 1975; Söderbäck and Bergman, 1992), they are totally dependent on *Gunnera* for a supply of carbon (Silvester, 1976) originating in *Gunnera* photosynthesis. A transfer of carbon from the plant to the endosymbiont was proposed early (Silvester, 1975) and was eventually shown for *G. magellanica* (Söderbäck and Bergman, 1993). The transfer of ^{14}C-solutes from leaves to the symbiotic tissue is likely to be facilitated through the numerous vascular strands of the petiols and rhizomes, many of which leads to the symbiotic tissues (Bathan, 1943; Bergman *et al.*, 1992a).

The chemical structure of the carbohydrate(s) transferred is still not known. The genus *Nostoc* is known to be able to live photoheterotrophically and heterotrophically on certain carbohydrates such as glucose and fructose, and this is also the case for *Gunnera* cyanobiont (Wouters *et al.*, 2000; Black *et al.*, 1999). This capacity must be held by any cyanobacterium entering into the dark interior of *Gunnera* stems.

The cyanobacterial N demands are fully covered via fixation of N_2 but nutrients such as P and Fe must be supplied to sustain for instance ATP production and nucleic acid synthesis. As cyanobacteria in symbiosis with *Gunnera* show extremely high heterocyst frequencies and high fixation rates (see Bergman *et al.*, 1992b; Rai *et al.*, 2000), Fe may be particularly in high demand being an integral part of the nitrogenase enzyme complex. The cyanobiont of *G. magellanica* contains polyphosphate granules, stores of phosphate (Söderbäck *et al.*, 1990), but other studies using TEM (Johansson and Bergman, 1994) and ESI/EELS (Jäger *et al.*, 1996) suggest that the polyphosphate number and P-contents are very low in the vegetative cells. This may suggest a low P-supply from the host plant and may perhaps, at least partly, explain the slow growth rates in symbiosis.

9.2. Nutritional Support to the Host

Like in all other nitrogen-fixing plant symbioses, a large proportion of the nitrogen fixed by the cyanobacterium is released and assimilated by the host plant. This has convincingly been shown using ^{15}N feeding experiments on cyanobacteria-infected symbiotic tissues excised from the rhizomes of *G. magellanica* under anaerobiosis (Silvester *et al.*, 1996). Up to 90 % of the nitrogen fixed is released, a considerable proportion being asparagine. Similarly, in *G. monoica*, Stock and Silvester (1994) demonstrated that the recently fixed nitrogen was translocated to the host along the stolon phloem.

Whether there are any other compounds released from the cyanobacterium *in situ*, which is subsequently taken up by the plant or which affects plant performance, is unknown. Although not yet proven, the possibility exists that growth-stimulating compounds are released by the endosymbiont while in symbiosis as discussed above. Phytohormone (IAA) release may promote plant cell division, growth and nutrient release (Glick *et al.*, 1999).

10. THE FINAL OUTCOME

The final outcome, after infection and maturity, is a fine-tuned, highly efficient and in many aspects unique endosymbiosis between a common terrestrial prokaryotic cyanobacterium and a rather odd representative (*Gunnera*) among the angiosperms present on the globe today. It is also evident from the numerous adjustments in the cyanobacterium, ranging from the molecular to the morphological and behavioral level, that competence requires highly specific characters and a high morphological and behavioral plasticity. In particular they must be able to cope with the hostile environment offered by the interior of the plant cell (Johansson and Bergman, 1992; Bergman *et al.,* 1996). The advantages for the cyanobacterium to enter into such a symbiosis is not obvious besides being served with nutrients and protected from predators. On the other hand the cyanobacterium looses in biomass production (retarded growth) and may even be prevented from reproduction and spreading.

The primary evolutionary driving force for the establishment and maintenance of this symbiosis may have been to meet the nitrogen requirements of the host plant (Box 1 in Bergman *et al.,* 1996). In this way the plant obtained a competitive ecological advantage, at least in nutrient (N) poor habitats.

11. FUTURE PERSPECTIVES

A large number of questions remain to be answered in regards to the *Nostoc-Gunnera* symbiosis from early recognition till functional maturity of the symbiosis. For instance, why only this particular angiosperm is capable of forming symbiosis with nitrogen-fixing *Nostoc*? What are the criteria needed to be fulfilled from the plant side? Likewise, we do not know what makes some *Nostoc* strains symbiotically competent, but not the other closely related cyanobacteria (even those capable of forming hormogonia including some strains of the genus *Nostoc*). It is expected that examination of the genome sequence of *Nostoc* PCC 29133 (isolated originally from a cycad) coupled to analyses of expression and protein profiling in time and space, combined with structural and behavioral studies may open new avenues to get a better understanding of this and other aspects unique to symbiosis.

ACKNOWLEDGEMENTS

The author wishes to express a sincere gratitude to Livia Wanntorp for providing the phylogenetic tree of *Gunnera* spp. Financial support was provided by the Swedish Natural Science Research Council, the Agricultural and Forest Research Council and the European Science Foundation (CYANOFIX).

REFERENCES

Adams, D.G. and Duggan, P.S. (1999) Heterocyst and akinete differentiation in cyanobacteria. *New Phytol.* **144**, 3-33.

Baas Becking, L.G.M. (1947) Notes on the endophyte of *Gunnera macrophylla*, *Biol. Jaarb.* **14**, 93-96.

Batham, E.J. (1943) Vascular anatomy of New Zealand species of *Gunnera*, *Trans. Roy. Soc. New Zealand* **73**, 209-216.

Bender, G.L., Nayudu, M., Le Strange, K.K. and Rolfe, B. (1988) The *nodD1* gene from *Rhizobium* strain NGR234 is a key determination the extension of host range to the non-legume *Parasponia*, *Mol. Plant Microbe Interact.* **1**, 259-266.

Bergman, B., Johansson, C. and Söderbäck, E. (1992a) The *Nostoc-Gunnera* symbiosis, *New Phytol.* **112**, 379-400.

Bergman, B., Rai, A.N., Johansson, C. and Söderbäck, E. (1992b) Cyanobacterial-plant symbioses, *Symbiosis* **14**, 61-81.

Bergman, B., Matveyev, A. and Rasmussen, U. (1996) Chemical signalling in cyanobacterial-plant symbioses, *Trend. Plant Sci.* **1**, 191-197.

Bergman, B., Bateman, K. and Rasmussen, U. (1999) Cyanobacteria in symbioses with plants and fungi, in Seckbach, J. (ed.), '*Origin – Evolution and Versatility of Microorganisms: Enigmatic Microorganisms and Life in Extreme Environments*', Kluwer Academic Publ., Dordrecht, pp. 613-627.

Bonnett, H.T. (1990) The *Nostoc-Gunnera* association, in Rai, A.N. (ed.), *Handbook of Symbiotic Cyanobacteria,* CRC Press, Boca Raton, pp. 161-176.

Bonnett, H.T. and Silvester, W.B. (1981) Specificity in the *Nostoc-Gunnera* endosymbiosis, *New Phytol.* **89**, 121-128.

Braun-Howland, E.B. and Nierzwicki-Bauer, S.A. (1990) *Azolla-Anabaena* symbiosis: biochemistry, physiology, ultrastructure and molecular biology, in Rai, A.N. (ed.), *Handbook of Symbiotic Cyanobacteria,* CRC Press, Boca Raton, pp. 65-117.

Buikema, W.J. and Haselkorn, R. (1991) Characterization of a gene controlling heterocyst differentiation in the cyanobacterium *Anabaena* 7120. *Genes and Develop.* **5**, 321-330.

Buikema, W.J. and Haselkorn, R. (1993) Molecular genetics of cyanobacterial development, *Ann. Rev. Plant Physiol. and Plant Mol. Biol.* **44**, 33-52.

Chase, M.W., Soltis, D.E., Olmstead, R.G., Morgan, D., Les, D.H., Mishler, B.D., Duvall, M.R., Price, R.A., Hills, H.G., Qiu, Y.-L., Kron, K.A., Rettig, J.H., Conti, E., Palmer, J.D., Manhart, J.R., Sytsma, K.J., Michaels, H.J., Kress, W.J., Karol, K.G., Clark, W.D., Hedrén, M., Gaut, B.S., Jansen, R.K., Kim, K.-J., Wimpee, C.F., Smith, J.F., Furnier, G.R., Strauss, S.H., Xiang, Q.Y., Plunkett, G.M., Soltis, P.S., Swensen, S.M., Williams, S.E., Gadek, P.A., Quinn, C.J., Eguiarte, L.E., Golenberg, E., Learn, G.H., Graham, S.W., Barrett, S.C.H., Dayanandan, S. and Albert, V.A. (1993) Phylogenetics of seed plants: an analysis of nucleotide sequences from the plastid gene *rbcL*, *Ann. Missouri. Bot. Gard.* **80**, 523-785.

Costacurta, A. and Vanderleyden, J. (1995) Accumulation of phytohormones by plant-associated bacteria, *Crit. Rev. Microbiol.* **21**, 1-18

Dahlgren, R. (1975) A system of classification of the angiosperms to be used to demonstrate the distribution of characters, *Botaniska Notiser.* **128**, 119-147.

De Meideiros, J.M.R., Macedo, M., Contanica, J.P., Mguyen, C., Cunningham, G. and Miles, D.H. (2000) Antithrombin activity of medicinal plants of the Azores, *J. Ethnopharmacol.* **72**, 157-165.

Dodds, W.K., Gudder, D.A. and Mollenhauer, D. (1995) The ecologyof *Nostoc*, *J. Phycol.* **31**, 2-18.

Douglas, S.E. (1994) Chloroplast origin and evolution, in Bryant, D.A. (ed.), *The Molecular Biology of Cyanobacteria*, Kluwer Academic Publ., Dordrecht, pp. 91-118.

Glick, B.R., Patten, C.L., Holguim, G. and Penrose, D.M. (1999) Biochemical and genetic mechanisms used by plant growth promoting bacteria, ICP, Covent Garden, London.

Jarzen, D.M. (1980) The occurrence of *Gunnera* pollen in the fossil record, *Biotrophica* **12**, 117-123.

Johansson, C. and Bergman, B. (1992) Early events during the establishment of the *Gunnera/Nostoc* symbiosis, *Planta* **188**, 403-413.

Johansson, C. and Bergman B. (1994) Reconstitution of the symbiosis of *Gunnera manicata* Linden: cyanobacterial specificity, *New Phytol.* **126**, 643-652.

Johnson, P.N. Brooke, P.A. (1989) Wetland Plants in New Zealand. DSIR Field Guide, DSIR Publishing, Wellington.

Jönsson, B. (1894) Studier öfver algparasitism hos *Gunnera* L., *Botaniska Notiser.* 1-20.

Kaido, T.L., Veale, D.J.H., Havlik, I. and Rama, D.B.K. (1997) Preliminary screening of plants used in South Africa as traditional herbal remedies during pregnancy andlabour, *J. Ethnopharmacol.* **55**, 185-191.

Koske, R.E., Gemma, J.N. and Doyle M.F. (1992) Mykorrhizal status of *Gunnera petaloidea* in Hawaii, *Pacific Science* **46**, 480-483.

Kronstedt-Robards, E. and Robards, A.W. (1991) Exocytosis in glands, in Hawes, C.R., Coleman, J.O.D. and Evans, D.E. (eds.), *Endocytosis and Exocytosis and Vesicle Traffic in Plants*, Cambridge Univ. Press, pp. 199-232.

Liaimer, A., Matveyev, A. and Bergman B. (2001) Isolation of host plant induced cDNAs from *Nostoc* sp. PCC 9229 forming symbiosis with the angiosperm *Gunnera* spp., *Symbiosis* **31**, 293-307.

Miehe, H. (1924) Entwiklungsgeschichtliche Untersuchung der Algensymbiosen bei *Gunnera macrophylla* Bl, *Flora* **72**, 211-232.

Nilsson, M., Bergman, B. and Rasmussen, U. (2000) Cyanobacterial diversity in geographically related and distant host plants of the genus *Gunnera*, *Arch. Microbiol.* **173**, 97-102.

Osborne, B., Doris, F., Cullen, A., McDonald, R., Campbell, G. and Steer, M. (1991) *Gunnera tinktoria*: an unusual nitrogen-fixing invader, *BioScience* **41**, 224-234.

Paulsrud, P., Rikkinen, J. and Lindblad, P. (2000) Spatial patterns of photobiont diversity in some Nostoc-containing lichens, *New Phytol.* **146**, 291-299.

Rai, A.N., Söderbäck, E. and Bergman, B. (2000) Cyanobacterium-plant symbiosis, *New Phytol.* **147**, 449-481.

Rasmussen, U., Johansson, C. and Bergman, B. (1994) Early communication in the *Gunnera-Nostoc* symbiosis: plant-induced cell differentiation and protein synthesis in the cyanobacterium, *Mol. Plant Microb. Interact.* **6**, 696-702.

Rasmussen, U., Johansson, C., Renglin, A., Pettersson C. and Bergman, B. (1996) A molecular characterization of the *Gunnera-Nostoc* symbiosis: comparison with *Rhizobium*- and *Agrobacterium*-plant interactions, *New Phytol.* **133**, 391-398.

Rasmussen, U. and Svenning, M. (1998) Fingerprinting of cyanobacteria based on PCR with primers derived from long and short repeated repetitive sequences, *Appl. Environ. Microbiol.* **64**, 265-272.

Reinke, J. (1872a) Über gonidienartige Bildungen in einer dicotyledonischen Pflanze, *Botanischer Zeitung.* **30**, 59.

Reinke, J. (1872b) Über die anatomischen Verhältnisseeiniger Arten von *Gunnera* L., *Göttinger Nachrichten* **6**, 100-108.

Reinke, J. (1873) Untersuchung uber die Morphologie der Vegetationsorgane von *Gunnera*, in Reinke, J. (ed.), *Morphologische Abhandlungen*, Verlag Wilhelm Engelmann, Leipzig pp. 45-123.

Schaede, R. (1951) Über die Blaualgensymbiose von *Gunnera*. *Planta* **39**, 154-179.

Schmidt, H. (1991) *Licht- und elektronenmicroskopische Untersuuchungen zum Infektionsablauf der Nostoc-Gunnera symbiose*, Dissertationes Botanica, Cramer Gebruder Geborntrāger, Berlin, Band 177.

Schussler, A., Bonfante, P., Schnepf, E., Mollenhauer, D. and Kluge, M. (1996) Characterization of the *Geosiphon pyriforme* symbiosis by affinity techniques: confocal laser scanning microscopy (CCSM) and electron microscopy, *Protoplasma* **190**, 53-67.

Schussler, A., Meyer, T., Gehrig, H. And Kluge, M. (1997) Variation of lectin binding sites in extracellular glycoconjugates during the life cycle of *Nostoc punctiforme*, a potentially endosymbiotic cyanobacterium, *Eur. J. Phycol.* **32**, 233-239.

Sergeeva, E., Liaimer, A. and Bergman, B. (2002) Evidence for biosynthesis and release of the phytohormone indole-3-acetic acid by cyanobacteria, *Planta* (in the press).

Silvester, W.B. (1975) Endophyte adaptation in *Gunnera/Nostoc* symbiosis, in P.S. Nutman (ed.), *Symbiotic Nitrogen Fixation in Plants*, Cambridge University Press, Cambridge, pp. 521-538.

Silvester, W.B. and Smith, D.R. (1969) Nitrogen fixation by *Nostoc-Gunnera* symbiosis, *Nature* **224**, 1231.

Silvester, W.B. (1975) Endophyte adaptation in the *Gunnera-Nostoc* symbiosis, in Nutman, P.S. (ed.), *Symbiotic Nitrogen Fixation in Plants*, Cambridge Univ. Press, Cambridge, pp. 521-538.

Silvester, W.B. and MacNamara, P.J. (1976) The infection process and ultrastructure of the *Gunnera-Nostoc* symbiosis, *New Phytol.* **77**, 135-141.

Silvester, W.B., Parsons, R. and Watt, P.W. (1996) Direct measurement of release and assimilation of ammonia in the *Gunnera-Nostoc* symbiosis, *New Phytol.* **132**, 617-625.

Söderbäck, E. (1992) Developmental patterns in the *Nostoc-Gunnera* symbiosis. PhD thesis, Stockholm University, Stockholm, Sweden, ISBN 91-7146-977-x.

Söderbäck, E., Lindblad, P. and Bergman, B. (1990) Developmental patterns related to nitrogen fixation in the *Nostoc-Gunnera magellanica* Lam. symbiosis, *Planta* **182**, 355-362.

Söderbäck, E. and Bergman, B. (1992) The *Nostoc-Gunnera magellanica* symbiosis: phycobiliproteins, carboxysomes and Rubisco in the microsymbiont, *Physiol. Plant.* **84**, 425-432.

Söderbäck, E. and Bergman, B. (1993) The *Nostoc-Gunnera* symbiosis: carbon fixation and translocation, *Physiol. Plant.* **89**, 125-132.

Standley, P. and Williams, L. (1963) Haloragaceae, in *Flora of Guatemala*, Fieldiana, *Botany* **24**, 564-566.

St. John, H. (1957) *Gunnera magnifica*, a new species from the Andes of Colombia, *Svensk Botanisk Tidskrift* **51**, 521-528.

Stock, P.A. and Silvester, W.B. (1994) Phloem transport of recently-fixed nitrogen in the *Gunnera-Nostoc* symbiosis, *New Phytol,* **126**, 259-266.

Towata, E.M. (1985a) Mucilage glands and cyanobacterial colonization in *Gunnera kaalensis* (Haloragaceae), *Bot. Gazette* **146**, 56-62.

Towata, E.M. (1985b) Morphometric and cytochemical ultrastructural analyses of the *Gunnera kaalensis/Nostoc* symbiosis, *Bot. Gazette* **146**, 293-301.

Uheda, E. and Silvester, W.B. (2001) The role of papillae during the infection process in the *Gunnera-Nostoc* symbiosis, *Plant Cell Physiol.* **42**, 780-783.

Viterbo, A., Matveyev, A, Rasmussen, U. and Bergman, B. (1999) Characterization of a *nodM/glmS* homologous gene in the symbiotic cyanobacterium *Nostoc* PCC 9229, *Symbiosis* **26**, 237-246.

Wanntorp, L. (2001) *Molecular phylogeny of Gunnera*, Licentiat thesis, Stockholm University, Stockholm.

Wanntorp, L., Wanntorp, H-E., Oxelman, B. and Källersjö, M. (2001) Phylogeny of *Gunnera*, *Plant Syst. Evol.* **226**, 85-107.

West, N.J. and Adams, D.G. (1997) Phenotypic and genotypic comparison of symbiotic and free-living cyanobacteria from a single field site, *Appl. Environm. Microbiol.* **63**, 4479-4484.

Wouters, J., Janson, S. and Bergman, B. (2000) The effecys of exogenous carbohydrates on nitrogen fixation and *hetR* expression in *Nostoc* PCC 9229 forming symbiosis with *Gunnera.* , *Symbiosis* **28**, 63-76.

Yoon, H.-S. and Golden, J.W. (1998) Heterocyst pattern formation controlled by a diffusible peptide, *Science* **282**, 935-938.

Zheng, W.W., Nilsson, M., Bergman, B. and Rasmussen, U. (1999) Genetic diversity and classification of cyanobacteria in different *Azolla* species by the use of PCR fingerprinting, *Theor. Appl. Genet.* **99**, 1187-1193.

Chapter 13

ECOLOGY OF THE *NOSTOC-GUNNERA* SYMBIOSIS

BRUCE A. OSBORNE AND JANET I. SPRENT*
Botany Department, University College Dublin, Belfield, Dublin 4, Ireland.
**Department of Biological Sciences, University of Dundee, Dundee, Scotland*

1. INTRODUCTION

Gunnera is the only angiosperm genus that forms a symbiotic association with cyanobacteria of the genus *Nostoc* (see Rai *et al.*, 2000). Analyses based on fossil pollen (Jarzen, 1980; Dettman and Jarzen, 1990), global_distribution (Batham, 1943; Osborne *et al.*, 1991) and molecular investigations of extant species (Drinnan *et al.*, 1994; Soltis *et al.*, 1999), indicate a genus of considerable antiquity, extending back to the Cretaceous, 90-95 million years ago. Molecular evidence also indicates that the genus forms part of an early eudicot paraphyletic group, although its affinities with other extant species are still uncertain. Rather surprisingly recent investigations associate *Gunnera* as a sister group to the *Myrothamanceaea*, a genus of xeromorphic shrubs associated with dry environments (see Wilkinson, 2000), habitats distinctly different from those occupied by *Gunnera* today (Jarzen, 1980; Osborne *et al.*, 1991).

The remarkable continuity of the genus, extending from the Cretaceous up to the present day, is all the more surprising given the major changes in climate, as well as alterations in the size of land masses and their latitudinal distribution, that have occurred over the last ~100 million years. All of these have been linked with events leading to major evolutionary modifications in plants (Willis and McElwain, 2002), suggesting the genus may have been resistant to environmental change. Recent reports of invasive populations of *G. timctoria* (Hickey and Osborne, 1998; Hickey and Osborne, 2001) are, perhaps, the first indications of an extension of the range of this genus for over 50 million years. However, we still have only a limited understanding of the ecology of this genus or its evolutionary significance.

The following review is an attempt to provide a firmer basis for further ecological investigations of this unusual genus. Given its ancient history, such information should improve our understanding of the functional significance and evolution of symbioses in general and plant-cyanobacterial symbioses in particular. It should also be useful in palaeoenvironmental reconstructions.

A.N. Rai, B. Bergman and U. Rasmussen (eds.), Cyanobacteria in Symbiosis, 233-251.

2. GLOBAL DISTRIBUTION

2.1. Biogeography

The present global distribution of the genus is largely restricted to the Southern Hemisphere (Fig. 1). The genus comprises 50-60 species in six subgenera (Table 1), distributed across South East Africa, Madagascar, Malaya, Tasmania, New Zealand, Solomon Islands, Hawaii, and the Juan Fernandes Archipelago, along both sides of the Andes from Mexico to Chile, Tierra del Fuego and the Falkland Islands. By far the greatest number and most diverse species are found in South America, ranging from gigantic herbs to small creeping stoloniferous plants (Table 1).

Essentially the species fall into three major groups based on their geographic location, a Southern African population, a combined Malaysian/Tasmanian/New Zealand population and a South American/Hawaiian population. To some extent this is consistent with the reported affinities, or lack of affinities, between different subgenera based on morphological and anatomical criteria (Wilkinson, 2000). The *Panke* subgenus is thought to be distinct from the others, whereas the *Milligania* and *Misandra* sugenera and, possibly, the *Psuedogunnera* from Malaysia share some similar features. The *Perpensum* subgenus is also thought to be distinct, possibly indicating a long separation in Africa and Madagascar. The subgenus *Ostenigunnera* is also clearly distinct from the others based on morphological and anatomical criteria (Wilkinson, 2000). This is rather surprising given its occurrence on the same landmass in Uruguay and Brazil, close to populations of the *Panke* subgenus (Fig. 1; Table 1), although a separate origin for the diversification of this species is also consistent with recent molecular evidence (Wanntorp *et al.*, 2001). Based on this information it is the possible that there were four major centres of diversification, South East Africa, Malaysia/New Zealand, Central and Andean South America and Uruguay/Brazil.

Table 1. Gunnera *subgenera, general features and biogeographic distribution. Based on Wilkinson (2000)*

Subgenus	No of Species	Features	Leaf Width	Distribution
Panke (Molina) Schindler	~50	Giant, large leafed, perennial herb; no stolons	~2m	South America, Hawai
Pseudo-Gunnera (Oersted) Schindler	1	Moderately large herbs with stolons	~70cm	Malaysia
Perpensum (Burman) Schindler	1	Intermediate size, short atems, no stolons	~25cm	South East Africa, Madagascar
Milligania (Hook. F. emend) Schindler	6	Small, creeping, mat forming, with stolons	~2.5cm	Tasmania, New Zealand
Misandra (Comm.) Schindler	3	Small, creeping, with stolons	~4cm	South America, Falkland Islands
Ostenigunnera Mattfeld	1	Small leafed, no stolons	14mm	Uruguay, South Brazil

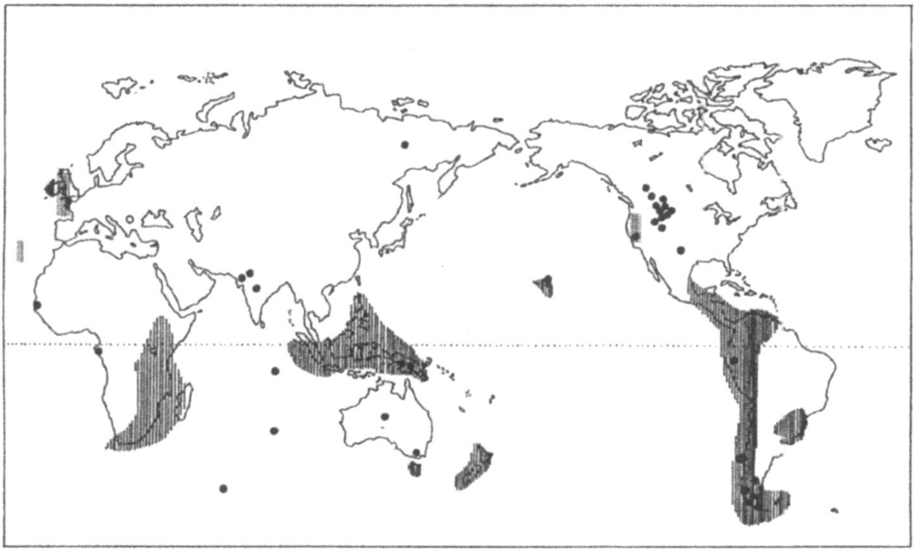

Figure 1. Global distribution of the extant Gunnera *genus together with fossil locations shown as filled circles. Updated and modified from Osborne* et al., *(1991) using new fossil records from Australia (McKensie and Kershaw, 2000) and South America (Heusser, 1995; Lequesne* et al., *1999) and revision of extant species distributions (Mora-Osejo, 1984; Hickey, 2002).*

2.2. *Climatic and Habitat Considerations*

The correlation between the distribution of the genus and areas of high rainfall/high humidity, have been well documented (Table 2; Jarzen, 1980; Osborne, 1988; Osborne et al., 1991). There is, however, some evidence of variations in rainfall between locations/countries (Table 2), perhaps indicating some differences in water requirements between species or populations of the same species or the modifying effects of high humidity. In general the smaller, creeping or stoloniferous species are more common in waterlogged areas, whereas the larger species are often associated with leached soils in regions of high precipitation (Tables 1 and 2). Invasive populations of *G. tinctoria,* as well as native species in either the northern or southern hemispheres, may have somewhat lower water requirements due, perhaps, to a lower evaporative demand at higher latitudes under more temperate conditions (Table 2).

Often an inability to survive exposure to low temperatures has been thought to be a significant factor limiting its distribution to warmer climates (Jarzen, 1980; Osborne, 1989a). However, members of the genus can be exposed to a wide range of absolute, as well as seasonal, variations in temperature (Osborne, 1989a; Bergman *et al.*, 1992). Under cool-temperate conditions, invasive or planted material of *G. tinctoria* is deciduous and can survive periods of sub-zero temperatures. In Tasmania *Gunnera* is found in subalpine habitats (Orchard, 1990) and fossil pollen has been recorded from

Southern Patagonia in the Holocence from cold tundra-like environments (Heusser, 1995). This does not mean, however, that individual species distribution may not be constrained by either low or high temperatures. There is evidence that *G. tinctoria,* for instance, may be limited in its distribution by low temperatures and/or increased seasonal variations in temperature (Osborne, 1989a). Conversely, rising temperatures in the Holocence have been proposed as an additional factor associated with the elimination of *Gunnera* from the Australian mainland (McKensie and Kershaw, 2000).

Table 2. Summary of the habitats, annual rainfall and associated plant communities in different locations. Based on a range of publications and local floras (see Jarzen,1980; Batham, 1943; Howell, 1970; van der Meijden and Caspers, 1972; Agnew, 1974; Molina, 1978; Mora-Osejo 1978; Osborne et al.*, 1991; van Wyk and Malan, 1997; Hickey, 2002). Typical rainfall data from Muller (1982), Jarzen (1980) and Hickey (2002). High altitude is above > 1,000.*

Location	Habitat	Rainfall (mm)	Plant community
Azores	Streams, gravel, sloped terrain; Low altitude	~1200	Shrubs, grasses*
Europe	Coasts, cliffs, streams, wet meadows, ditches, open woods; Low altitude	600-2000	Herbs, mostly grasses, rushes*
South East Africa	Marshes, highland rainforests, streams, cliffs; High/low altitude	1000-2000	Trees, herbs, mainly grasses
Malaysia	Montane rain forest, marshes, river banks; High altitude	2000-3500	Trees, woody shrubs
Hawai	High slopes, cliffs, leached volcanic soils; High altitude	~12.500	Herbs, shrubs
North America	Wet sloping meadows; Low altitude	800-900	Grasses, rushes*
Central America	Streamsides, road-sides, ditches, leached soils; High altitude	1500-2000	Herbs, trees, shrubs
South America	Cloud forests, cliffs, bogs, heathland, leached soils; High/low altitude	1500-2000	Trees, herbs, shrubs, tree ferns
New Zealand	Coasts, streams, bogs. Forest margins. Low altitude	800-1500	Herbs, grasses, sedges rushes, tree ferns**
Falkland Islands	Grassland, bogs, streams. Low altitude	800-1000	Herbs, mainly grasses, sedges, rushes

* Invasive populations of *G. tinctoria*
** Includes native habitats and those invaded by *G. tinctoria*

Within the constraint of a high water requirement a number of habitats can be colonised, including lake margins, streams, forest, grasslands, bogs and coastal regions, encompassing a variety of plant life forms from arborescent shrubs, trees, tree ferns, grasses and dwarf or prostrate herbs (Table 2). Most records are associated with marshes/swamps, woodland/high altitude forest, wet meadows and streams (Table 2). Often, particularly at high altitudes or on cliff faces, the plant community can be sparse, with large areas of bare, unstable soil often of volcanic origin. All these habitats would be expected to have low levels of N and P (Hickey and Osborne, 1998) and, possibly, limiting concentrations of other essential nutrients. Plants can grow either as understory species or in partial-shade, or in locations where they are fully exposed. In general, species with a sub-tropical or tropical distribution are found at high altitudes (~ 1,000m. eg. van der Meijden and Caspers, 1972; Agnew, 1974; Mora-Osejo, 1984). Typically, however, few species are found at altitudes less than 2,000m (maximum altitude ~ 4,000m) in South America where the *Panke* subgenus predominates (Mora-Osejo, 1984). For those species associated with higher northern and southern latitudes, including the invasive species *G. tinctoria*, they are found at sea level (Palkovic, 1974; Hickey, 2002).

2.3. Distribution in Relation to Other Nitrogen Fixing Symbioses

New molecular taxonomic data supports the contention that the *Gunnera* genus is widely separated from the Rosidae clade that includes all other angiosperms that have nitrogen-fixing symbionts (Hoot, Magallon and Crane, 1999). *Gunnera* species are also found principally where there are few or no nodulated legumes or actinorhizal plants. In the Falkland Islands *G. magellanica* is the only endemic nitrogen fixing plant (Moore, 1968). Here it is widely distributed, occurring from the shoreline to the highest parts of the island (J. McAdam, personal communication) and always contains cyanobacteria (Sprent, personal observation). This is not because legumes cannot grow here, as they have been successfully introduced. It is possible that the lack of suitable pollinaters has limited the spread of legumes in these areas (Sprent, unpublished). Interestingly, herbaceous nodulated legumes do grow well into the Arctic Circle, although *Gunnera* is absent from these areas. These are all members of the Papilionoideae and all nodulate and fix nitrogen (Sprent, 2001), suggesting that there is evolutionary pressure for atmospheric fixation capacity. The same arguments could be applied to *Gunnera* in its southern habitats. Why this dichotomy exists is not known.

3. PALAEOECOLOGY

3.1. Ancestral Origins

Based on similarities between the fossil pollen grains of *Tricolpites reticulatus*, which are very similar to extant *Gunnera* species, the genus is likely to have been more widespread in the past than it is today (Jarzen, 1980; Jarzen and Dettman, 1989). This is also supported by the continuity of fossil records from the Cretaceous up to the present time, on all continental landmasses. There are at least two major and related questions that can be asked on the basis of this information. Firstly, why has the genus disappeared from all its former native northern habitats and, secondly, what is the explanation for the earlier global distribution, based on the combined fossil and extant species data?

Approximately 94 million years ago global temperatures were several degrees higher and more uniform with no ice cover at the poles. This allowed many plants to colonise higher latitudes in both the northern and southern hemispheres. Importantly also a vast well-documented interior seaway had developed on the eastern side of the United States, which extended from the arctic ocean to the Gulf of Mexico (Fig. 2). This was preceded by the formation of the Mowry Sea, a consequence of a marine transgression where water from the Arctic Ocean penetrated deep into the interior of North America (see Condie and Sloan, 1997; Stanley, 1999). Clearly, therefore, there is a long history of marine incursions in this area throughout the Cretaceous. There is also evidence for another seaway close to the western seaboard of South America, providing a degree of continuity with the North America seaway *via* the Gulf of Mexico (Fig. 2). Based on the close correspondence between the fossil locations and the presence of the ancient seaways, these could have been the locations associated with the early populations of the current *Gunnera* genus in North and South America. The seaways could also have aided in the distribution of *Gunnera* seed, either through water movements or *via* birds, present since the Jurassic. Birds are known, for instance, to feed on fruits of *G. tinctoria* and thought to aid in dispersal (Hickey, 2002). The seaways are thought to have been shallow and less saline than the oceans (Condie and Sloan, 1997) and fringed by streams, swamps and marshes, habitats that are similar to those occupied by some extant *Gunnera* species, particularly the smaller members of the *Misandra* and *Malligania* subgenera (Table 1). Such environments are also thought to be characteristic of early angiosperm evolution (Willis and McElwain, 2002). Given the greater number of species currently associated with South America these seaways could have been the major focus of *Gunnera* evolution and diversification. There is also some indication of another sea corridor associated with the Asiatic coast extending from Siberia down to the Malay Peninsula. Much of southern Siberia was probably underwater, although a land bridge connected it with Alaska (Condie and Sloan, 1997). The disappearance of Asiatic populations (Fig. 1) could then have given rise to the present distribution in Malaya, New Zealand and Tasmania. This was a period when sea levels were several metres above current levels (Stanley, 1999) and shallow continental seas covered parts of Africa, southern Europe, Australia and all of New Guinea. North-western India, which at this time was attached to Madagascar, was also covered by water (Fig. 2) and the fossil finds in this region are allied more with the extant populations (*G. perpensa*) associated with Africa. It is clear that a number of these shallow seas have associated fossil *Gunnera* pollen (Fig. 1). Tidal inundation of parts of the East Coast of South America also occurred during this period and could account for a separate origin of the Uruguyan and Brazilian *G. herteri* from the other South American species. Similarly epicontinental anoxic seas were formed along the east and west coasts of Africa in areas adjacent to either fossil or extant *Gunnera* species (See Fig 1 and Stanley, 1999). If this proposal is correct, the earliest *Gunnera* relatives were associated with ancient seaways or shallow lakes that arose during a period of raised sea levels in the Cretaceous. Based on current opinion the genus is thought to have a northern Gondwana origin (Mora-Osejo, 1984), due partly to the current global distribution and because of the presence of the oldest identifiable *Gunnera* pollen from the Turonian of Peru in South America (Dettman and Jarzen, 1990). The current distribution does not, however, rule out a palaeotropical origin, consistent with the weight of evidence for many of the early angiosperms (Willis and McElwain, 2002), with subsequent radiation resulting in the colonisation of northern and southern latitudes. In both scenarios the early seaways could have provided an important migratory route for expansion of the ancestral *Gunnera* population.

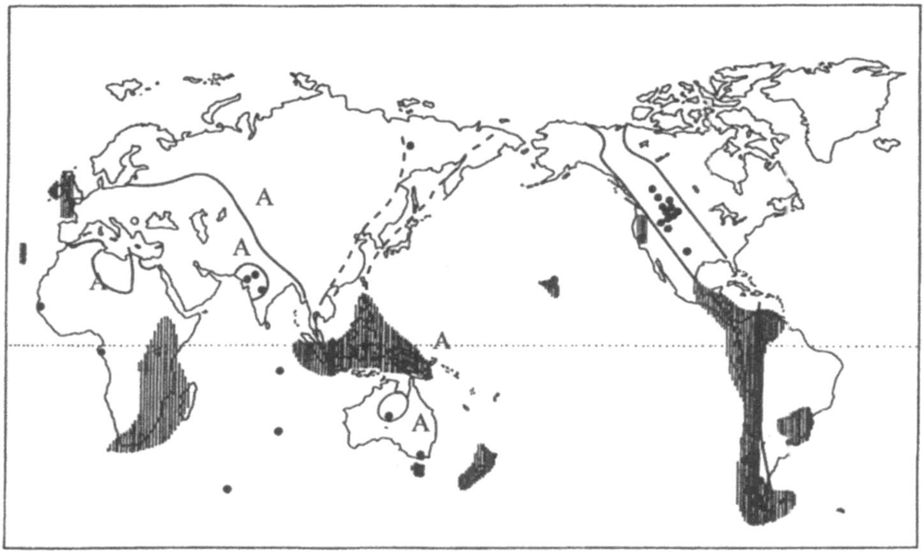

Figure 2. Documented (continuous lines) or possible (broken line) seaways and continental seas (continuous lines marked A) superimposed on fossil records and current distribution of the Gunnera *(see Fig. 1). Information on the seas and seaways is based largely on the proposed Cretaceous (94 million years ago) map of Scotese (1997), together with information from Condie and Sloan (1997) and Stanley (1999). Note that India at this time was attached to Madagascar and the positions of the fossil finds on this continent are consistent with the current distribution of* Gunnera *in South Africa and Madagascar. Australia was, with Tasmania, attached to the Antarctic continent.*

3.2. The Demise of Gunnera *from Northern Latitudes*

So why did *Gunnera* disappear from northern latitudes? This is unlikely to be due to the disappearance of the shallow seas and seaways as this genus is now more common in terrestrial habitats in the Southern Hemisphere. Fossil records also indicate that *Gunnera* was present in North America up to the middle Eocene (41-49 million years ago) long after the seaways and shallow seas had disappeared (Stanley, 1999). Interestingly the demise of *Gunnera* in North America is associated with the appearance in the fossil record of more arid vegetation types including features consistent with present day desert environments (Willis and McElwain, 2002). Two former shallow seas, with fossil records, in Australia and north-eastern India are now deserts or very arid areas. Progressive increases in aridity were a feature of many continental landmasses between the Eocene and Pliocene (Willis and McElwain, 2002). Polar deserts also developed at high latitudes in both the northern and southern hemispheres. Reduced persistence in these water-limited areas could also have been compounded by reductions in global temperature as the world cooled progressively after the Cretaceous. In North America there has been extensive development of deserts or arid lands in regions associated with the former interior seaway. In contrast desert development in South America has not been as extensive and largely confined to a narrow coastal strip in the north and to Patagonia in the south (see McGinnies *et al.*, 1968). The influence of the oceans on the climate of the southern landmasses is also greater than that of the larger northern continents and exerts a more moderating influence on the terrestrial climate. Although increased seasonality of the climate has been suggested as a reason for the demise of the genus in northern latitudes this may have been overstated in the past as extant species can survive in areas with significant annual variations in temperature and rainfall (see Osborne, 1989a).

3.3. Recent Range Modification in the Southern Hemisphere

The more recent demise of *Gunnera* from parts of Victoria in south-eastern Australia during the late Cenozoic (~5,000 years ago) has also been attributed to increases in aridity, which may have been exacerbated by man-induced burning of rainforest habitats (McKensie and Kershaw, 2000). *Gunnera* has had a particularly long history in Australia, and its only recent restriction to Tasmania reinforces the close biogeographical affinities between these two landmasses over a considerable period of geological time (McKensie and Kershaw, 2000). Fire has also been associated with recent reductions in *Gunnera* on the Juan Fernandez Islands (S. Haberle, personal communication). In contrast fire-related increases in *Gunnera* pollen in Chile (Lequesne *et al.*, 1999) and Southern Patagonia (Heusser, 1995) have been reported due, presumably, to the removal of forest cover. The extent to which fire impacted on the size of either of these populations is, however, not known.

3.4. What of the Cyanobacterial Symbiont?

Whilst the *Gunnera* host can be grown in culture in the absence of the cyanobiont *Nostoc* (Bergman *et al.*, 1992) no species has ever been found, growing in cultivation or in the wild in the absence of the cyanobacterium (see Osborne *et al.*, 1991). Additions of high concentrations of inorganic nitrogen also failed to eliminate cyanobacterial colonies from plants of *G. tinctoria* and the host was shown to have a greater dependence on the cyanobiont for supplying nitrogen (Osborne *et al.*, 1992). Effectively, therefore, this represents a more obligate nitrogen fixing symbiosis. This information, together with the ancient history of cyanobacteria, extending back to the Pre-Cambrian and the antiquity of the host, supports the contention that this is an ancient, highly evolved, partnership. The capacity to form functional associations with genetically diverse cyanobacteria (Rasmussen and Svenning, 2001) could have been an important factor in enabling the genus to colonise wet environments over a wide geographic range. However, we do not know if the earliest ancestors of extant species formed symbiotic associations with cyanobacteria or whether this capacity was gained during the post-Cretaceous period. The postulated habitats, in which *Gunnera* arose, comprising sediments of black anoxic mud, where nitrogen mineralisation would have been restricted, the development of symbiotic nitrogen fixation could have been advantageous. This could have been exacerbated by a restricted ability on the part of the host to utilise inorganic forms of nitrogen (Osborne, 1989b; Osborne *et al.,* 1992). However, nitrogen fixing associations are not common in these aquatic habitats today (Sprent and Sprent, 1990) and the symbiosis could have developed from a casual association between unusual carbohydrate-rich mucilage secreting stem glands, which serve as the sites for infection in extant species (Bergman *et al.,* 1992) and free-living cyanobacteria. At this time also atmospheric carbon dioxide levels were ~3 times the current values and this could have a significant impact on a range of microbial and plant processes which could, in turn, influence interactions between host plants and their cyanobionts (Willis and McElwain, 2002). This clearly requires further investigation.

4. ECOPHYSIOLOGICAL CONSIDERATIONS

4.1. Plant-Water Relations

Globally, water availability has been identified as one of the major factors correlated with the distribution of *Gunnera* species. This is also true at local and regional scales (Hickey, 2002), although we do not fully understand the mechanistic basis for this sensitivity to water availability. For *G. tinctoria* marked diurnal declines in photosynthesis, that are often considered to be a feature of water limitation, are found in the field even when soil and atmospheric water deficits are low (Osborne, 1988). Consequently, it has been suggested that the transport of water from root to shoot is a major limitation to the maintenance of plant water balance (Osborne, 1988). Vegetative and reproductive biomass production was also extremely sensitive to water availability, an effect that appears to be principally a consequence of the influence of water deficits on leaf expansion (Osborne *et al.,* 1991; Campbell and Osborne, 1993; Campbell, 1994). The lack of sensitivity of the stomata to major environmental variables may also contribute to poor water conservation in dry environments (Osborne, 1988; Osborne *et al.,* 1991). These characteristics appear to be independent of the symbiont as plants lacking cyanobacteria exhibit similar responses (Osborne, 1989b). We are aware of no other similar experiments on other *Gunnera* species, nor of any information on the effect of water deficits on infection, nitrogen fixation or photosynthesis.

4.2. *Plant-Temperature Relations*

Little is also known about the influence of temperature on plant performance or nitrogen fixation. Measurements of photosynthesis on *G. tinctoria*, presumably derived from somewhat warmer climates in South America, do not indicate any particular sensitivity to short term reductions in temperature (Osborne, 1989a), but there are no details of the effects of long-term exposure. Interestingly, some invasive colonies of *G. tinctoria* are probably exposed to lower temperatures than they would have experienced in their native habitats, suggesting the possibility of population differences, although this has not been explored. What is, perhaps surprising is that cultivated material of *G. tinctoria* appears to be able to survive in areas considerably outside its native or invasive range, where it is often exposed to sub-zero temperatures (Osborne *et al.,* 1991). As it is always cultivated in wet areas this suggests a moderating effect of water availability on the response to low temperatures, possibly because low temperatures reduce water uptake and transport from root to shoot (Osborne *et al.,* 1991). If this were correct colonisation of areas where plants are exposed to low temperatures would have also depended on an adequate water supply.

4.3. Plant-Nutrient Relations

Measurements of nitrogen fixation by *G. magellanica* (Silvester and Smith, 1969), and comparisons of the growth of *G. tinctoria* in response to increasing inorganic nitrogen concentrations (Osborne *et al.,* 1992), indicate that the cyanobacterium can supply all the hosts' requirements for nitrogen. Experiments also indicated that the host, in the presence or absence of the cyanobacterium, had a limited capacity to utilise inorganic forms of nitrogen *in situ,* based on *in vivo* estimates of nitrate reducatse activity (Osborne 1989b; Osborne *et al.,* 1991). This was true of both mature plants and young seedlings (Osborne, unpublished results). Whether this is related to restricted transport and/or reduced uptake/increased efflux of inorganic nitrogen is not known. Foliar applications of nitrate resulted in significant increases in the activity of this enzyme, indicating some capacity to supplement atmospherically fixed nitrogen with inorganic forms present in precipitation. The contribution of this source to the plant nitrogen budget could be significant in the high rainfall environments that these species occupy. Clearly, further experiments are required to confirm and extend these results with a wider range of species and to quantify the relative contributions of foliar absorption, uptake from the soil and atmospheric fixation to the nitrogen budget of whole plants, under a range of environmental conditions. Given that nitrogen fixation may make the largest contribution to the plant nitrogen budget consideration should also be given to the possible effects of other essential nutrients that could limit nitrogen fixation under field conditions. Of these molybdenum, iron, sulphur and phosphorus may be the more important (Brady and Weil, 1996).

4.4. The Influence of Elevated Carbon Dioxide Concentrations

An interesting question, given the likely origins of the genus in the Cretaceous during a period of high atmospheric carbon dioxide levels (see Willis and McElwain, 2002), is the extent to which these species may respond to artificial increases in the concentration of this gas. Comparative experiments to examine this using *G. tinctoria* have shown that the responses of biomass production and photosynthesis to elevated carbon dioxide were variable and dependent on the environmental conditions. For short-term exposure (weeks) to elevated carbon dioxide concentrations increases in biomass (~22%), were at the low end (see Kimball, 1983) of the range of published values (J. Hennessy and B.A. Osborne, unpublished) and associated with an increased investment in non-photosynthetic structures (roots and rhizomes). These responses are typical of a range of plant species (Berston and Bazzaz, 1996). In longer-term investigations (months) no effect of elevated carbon dioxide on plant biomass was found, although there were seasonally dependent increases in photosynthesis (Hennessy and Osborne, unpublished), indicating that acclimation occurred with longer exposures.

It has been indicated that the higher carbon dioxide concentrations associated with Cretaceous environments may have enabled some early angiosperms to expand into drier environments due to increases in water use efficiency (ratio of carbon gained to water lost *via* transpiration) (Willis and McElwain, 2002). For *G. tinctoria*, increases in water-use at elevated carbon dioxide concentrations were small and variable (Hennessy and Osborne, unpublished) and, on this basis, would not lead to major improvements in plant water balance and an increased ability to colonise drier habitats.

Together these results do not identify any major differences in the response of this species to elevated carbon dioxide despite an ancestral origin in atmospheres where the concentration was greater than it is at present. Based on the presence of a large rhizome, which provides a potentially important sink for assimilates and a reduced stomatal sensitivity to carbon dioxide, a larger response might have been predicted. Yet again, however, we need more data under different environmental conditions, as well as measurements on other species, before we can make any generalisations. There is also no information on the response of the cyanobiont, or its capacity to fix nitrogen, with elevated carbon dioxide concentrations.

4.5. *Effect of UV Exposure*

A detailed series of experiments conducted recently (see Rousseaux et al, 1999; Ballare et al., 2001) have examined the impact of stratospheric ozone depletion on ecosystems containing *G. magellanica* in Tierra del Fuego, Argentina. These habitats are particularly susceptible to enhanced UV-B levels due to significant decreases in ozone concentration over the last decade. Inhibition of leaf expansion due to exposure to UV-B was correlated with nuclear DNA damage in *G. magellanica,* suggesting it could have significant effects on plant performance and community composition in these ecosystems. The final outcome of the effect of UV at the ecosystem level is, however, unclear, due to variable responses among the species examined and because enhanced UV-B levels were found to decrease insect herbivory in *G. magellanica* (see below)

4.6. *Predation*

The genus appears to have very few predators. Whilst some grazing by sheep and cattle has been observed on naturalised populations of *G. tinctoria*, this has a minimal impact and infestation by other pests appears to be small (Hickey, 2002). Grazing by goats is, however, thought to have reduced the presence of *Gunnera* on the Juan Fernandez Archipelago (S. Haberle, personal communication). On the Azores goats and rabbits also graze introduced populations of *G. tinctoria,* and Thysanoptera and polyphagus Lepidoptera larvae may damage seedlings (Silva *et al.*, 1996).

Field experiments on *G. magellanica* growing in native ecosystems in the Tierra del Fuego National Park, Argentina, where UV-B light was removed using filters showed that this increased the susceptibility of plants to feeding by chewing insects (Rousseaux *et al.,* 1998). Again the extent to which this will influence community structure and function in these ecosystems is uncertain (Rousseaux *et al.*, 1998).

4.7. Plant-Insect Interactions and Infection

Infection of host tissue by suitable cyanobacterial strains occurs through unique and conspicuous stem glands, which produce a carbohydrate-rich mucilage (Bergman *et al.,* 1992). There is no evidence that new germinated seedlings contain *Nostoc* and each gland has to be infected with a new cyanobacterial inoculum. Rather than relying on chance infection solely *via* wind pollination (Pacheco *et al.,* 1991; Hickey, 2002) an effective vector that transfers the appropriate strain could increase the opportunity of establishing a new association. An increased reliance on atmospherically derived nitrogen would also increase the evolutionary advantages associated with such a mechanism. It has recently been suggested that a suitable vector could be a flying insect belonging to the Tipulidae or Chironomidae (J. Rikkinen, unpublished). It is well known that *Nostoc* can be colonised by chironomid larvae (see Sabater and Munoz, 2000) and such insects are common around the base of *Gunnera* plants at particular times of the year (B.A. Osborne, unpublished). Presumably the insect would be attracted to the gland due to its conspicuous red colouration, or because of the release of chemical attractants. Such a scenario recognises that the 'unique' gland structure may have originally evolved as a secretory structure or extra floral nectary and could have been involved in plant-insect interactions unrelated to its present function (J. Rikkinen, unpublished).

4.8. Gunnera *Invasions-The Second Coming?*

A number of recent accounts of invasive colonies of *G. tinctoria* have been reported in regions far outside their native range. These include records from Ireland (Osborne *et al.,* 1991; Hickey and Osborne, 1998, 2001; Hickey, 2002), the United Kingdom and France (see Hickey, 2002), New Zealand (Williams *et al.,* 2001), the Azores (Silva *et al.,* 1996) and North West America (see Howell, 1970). Dating of sediments containing pollen of *G. tinctoria* indicated that introduced plants have probably been present in some localities in Ireland for ~ 70 years (Hickey, 2002). Despite these widely scattered occurrences in different parts of the world, all the invaded habitats are associated with broadly comparable climatic zones. These are largely warm, wet temperate climates with temperatures not falling below -3°C, or cool temperate zones with woodland climates (Hickey, 2002). These climatic conditions are broadly consistent with the native range of this species in South America, although the mean and maximum temperatures are somewhat lower. Similarly, the colonies are associated with comparable habitats; wet or waterlogged infertile soils, stream sides or cliffs, generally characterised by high annual rainfall (>1000mm) and/or high humidity. In North America the annual rainfall is significantly lower (814 mm) but the plants are often subject to consistently high humidity and atmospheric water *via* fog drip (A. Dennis, personal communication). Evidence from recent records obtained for the British Isles, where *G. tinctoria* has a predominantly westerly distribution do, however, indicate an easterly expansion into areas with a significantly lower annual rainfall (Hickey, 2002). The reasons for this recent expansion into drier areas are not known. Climate warming could be a factor, as could an increased humidity. Future projections of the distribution of *G. tinctoria* based on climatic considerations suggest that it has the potential to extend its introduced range on all major landmasses, with Europe being particularly susceptible to invasion (Hickey, 2002).

Examination of the impact of invasions by *G. tintoria* in Ireland has indicated that this is not associated with increased soil nitrogen concentrations (Hickey and Osborne, 1998), despite the potential inputs from nitrogen fixation. This could be due to considerable re-cycling of nitrogen prior to canopy senescence at the end of the growing season and/or significant leaching of inorganic nitrogen in these high rainfall areas (Hickey and Osborne, 1998). An ability to retain a high proportion of nitrogen could be important in other areas, restricting the growth of competing species that have a high nitrogen requirement. Variations in nitrogen mineralisation between sites had little impact on plant performance lending further support to the proposal that these species obtain most of their nitrogen *via* N_2 fixation (Hickey and Osborne, 1998).

In areas invaded by *G. tinctoria* the diversity of species was increased, although there were alterations in community composition and an increased number of species were recorded in the colonised areas (Hickey and Osborne, 2001). Presumably this is because a greater number of microsites were available in the colonised areas. Whilst this may be unusual, as invasive species are often thought to reduce species diversity Gordon (1998) has also reported increases in native species in invaded areas. Perhaps of more importance is the fact that few seedlings survive to reproductive maturity beneath the *Gunnera* canopy, ultimately leading to a depletion of the seed bank in the absence of annual recruitment. Whether this is a feature of the seed bank associated with native populations needs to be explored.

The question arises as to why all these invasions are associated with the same species? Both *G. tinctoria* and *G. manicata* are closely related species, with similar morphological traits and habitat requirements and were widely collected and cultivated by Victorian botanists. Both have been grown extensively in estates, parks and gardens for approximately 100 years and yet there is no evidence that *G. manicata* is invasive. It could also be argued that *G. tinctoria*, given its ability to invade wet environments, should be more common in a range of habitats world-wide as it was frequently cultivated far outside its native range. Is this a consequence of genetic constraints, or is it related to particular features of the symbiosis, or the availability of suitable cyanobacterial strains? Comparisons are required of plant populations and their associated cyanobacteria to answer these questions. Whatever the reasons for the particular success of *G. tinctoria* as an invader it is clear that the genus has finally returned to high latitudes in the Northern Hemisphere after a gap of approximately 50 million years!

5. EVOLUTIONARY CONSIDERATIONS

5.1. An Evolutionary 'Dead End'?

It is tempting to assume that the formation of the *Gunnera-Nostoc* symbiosis was an evolutionary blind alley and that the introduction of a cyanobacterial-based nitrogen-fixing partnership with angiosperms was not of prolonged effectiveness in many early environments. Taxonomically the genus is now regarded as belonging to a distinct family (the Gunneraceae) with few, if any, close relatives (see Wilkinson, 2000). Based on both anatomical and physiological traits the genus is unusual among angiosperm hosts that fix atmospheric nitrogen. The symbiont is located intracellularly in relatively undifferentiated cortical tissue associated with unusual stem glands (Osborne *et al.*, 1991; Bergman *et al.*, 1992). No root or stem nodules are formed and liquid water is probably required at some stage in the infection process. In most cases the vascular architecture is anomalous (polystelic) and this, combined with relatively uncontrolled transpiration and/or limitations on root-shoot water transport (see Osborne *et al.*, 1991 and reviewed above) suggest that the genus was not on an evolutionary track that would have enabled it to survive outside wet habitats. This is further supported by the absence of any evidence for selection of species or populations that grow in significantly dry areas. If the proposed affinities between the genus *Myrothamnus* and *Gunnera* are correct this could be viewed as failed attempt at developing a successful genotype for a range of water-limited environments as *Myrothamnus* is a rather isolated group of two species, which is restricted to South Africa and Madagascar (Wilkinson, 2000).

Presumably, however, there are particular features of the host, in addition to the unusual stem glands, that are well suited to the formation of cyanobacterial associations. The question is what are they? Polystely, particularly in the larger rhizomatous species may reduce the transport pathway for supply of carbohydrate from host tissue to symbiont to support nitrogen fixation and reduce the build up of products of nitrogen fixation, as each colony should be close to a vascular bundle. However, this is not a unique feature of the genus, or of all *Gunnera* species. The presence of large starch grains in rhizome tissue (Osborne *et al.*, 1991) could also serve as a store of carbon to support N_2-fixation during periods when photosynthesis is low but, again, this is not a unique feature of this genus. Clearly what ever these factors are they remain elusive.

In other nitrogen-fixing angiosperm symbioses nutrients are supplied to the symbiont *via* the phloem and products of nitrogen fixation exported *via* the xylem (Sprent and Sprent, 1990). In *Gunnera* the phloem appears to serve both functions (Stock and Silvester, 1994). Plants also have very low inducible nitrate reductase activities *in situ* (Osborne, 1989b; Osborne *et al.*, 1992) indicating that they are less versatile than other hosts in their use of nitrogen and more reliant on their nitrogen-fixing symbionts. This could be a severe disadvantage in habitats where soil nitrate pools are high given the expected greater energetic costs associated with nitrogen fixation compared to the assimilation of inorganic nitrogen forms. This is, however, consistent with observations made in the field and in cultivation, where no host has been found that lacks the cyanobiont. Interestingly, an ability to utilise foliar-supplied inorganic forms of nitrogen (Osborne *et al,* 1992) from precipitation, could be significant in many of the high-rainfall habitats occupied by *Gunnera* species.

This view of the potential evolutionary disadvantages of the *Gunnera-Nostoc* symbiosis should, perhaps, be tempered by the fact that it has had a long and continuous existence since the Cretaceous despite major environmental change. In this sense the fact that it has survived means that it has been extremely successful through recent geological time. Presumably many other plant species have evolved and disappeared during this period. On the basis of recent fossil finds of macro-remains of *Gunnera* attributable to the subgenus *Panke*, dated at approximately 71-83 million years ago (see Wilkinson, 2000) the leaf morphology at this time may have been little different from extant forms. Willis and McElwain (2002) list 10 angiosperm genera with representatives that have persisted largely unchanged morphologically for 87-120 million years. To this could be added *Gunnera.* Clearly *Gunnera* possesses attributes that are resistant to environmental change. Of the factors identified that could confer such resistance Willis and McElwain (2002) identify hybridity, polyploidy, asexual reproduction, persistence and dormancy of propagules as important features of enduring plant species. We have little information on any of these features in relation to the survival and persistence of *Gunnera*. Certainly vegetative (asexual) reproduction does occur, although most plants are thought to establish from seed in naturalised populations (Osborne *et al.,* 1991; Hickey, 2002). Survival of seeds in naturalised habitats is, however, poor, but could be compensated by the large numbers produced annually (Hickey and Osborne, 2001). Chromosome counts on *G. manicata* (Dawson, 1983; Hanson *et al*., 2001) do not provide any evidence of polyploidy, with a diploid complement of chromosomes (2n=34). This value is consistent with estimates made on specimens of *G. magellanica* from New Zealand and *G. tinctoria* from South America (Dawson, 1983), despite their long period of apparent isolation. Rather surprisingly, however, *G. manicata* had a large genome size, whereas most ancestral angiosperms are thought to have a small genome size (Hanson *et al.,* 2001). Natural interspecific hybridisation has been reported between *G. peltata* and *G. bractaeta,* growing on Isla Alejandro Selkirk in the Juan Fernandez Archipelago (Pacheco *et al*., 1991). This may have been aided by ecological changes that have occurred recently (~4 million years), including a reduction in land area, which have brought the two species closer together and aided in hybridisation. These workers also suggest that these species probably share a common ancestor in *G. tinctoria*. More analyses of this kind examining different populations and species are warranted if we are to understand evolutionary relationships within the *Gunnera* genus.

If our proposal is correct and the early ancestral forms were largely associated with anoxic shallow seas and seaways they could have had a much more restricted, habitat-specific, distribution than their former geographic distribution appears to suggest. Increased aridity in northern latitudes resulted in a contracted Southern Hemisphere distribution. Only where waterlogged conditions were maintained or where high precipitation and/or high humidity counteracted the decreased availability of soil water did this genus survive. This was also associated with a shift from a low altitude to a predominantly high altitude distribution, based on the current predominance of the *Panke* subgenus (Table 1). On this basis the existing geographic locations of extant

Gunnera species may have continuity with populations extending back over the last 90 million years. Further analysis of Cretaceous sediments, particularly in areas known to have been associated with former continental seas, together with investigations on existing populations are required to investigate this possibility.

5.2. *What Came First?*

There has been some speculation about the features of the early *Gunnera* populations that were present in the Cretaceous (see Jarzen, 1980; Wilkinson, 2000). Where they similar to the smaller, mat-forming or stoloniferous species, belonging to the *Misandra* and *Milligania* sub-genera, or more like the larger leafed species belonging to the subgenus *Panke*? The earliest angiosperm records from approximately 120 million years ago are characterised by features such as non-palmate small leaves (2-4cm in diameter) that appear to be similar to extant plants that grow in semi-aquatic locations. These are characteristics that are consistent with an ancestor comparable to the *Misandra/Milligana* subgenus. These are the subgenera that are also regarded as the more primitive (Palkovic, 1974). Whilst the monospecific subgenus *Ostenigunnera* contains the smallest known species of *Gunnera* (*G. herteri*) with leaf widths up to 14mm, it has lobed leaves (Wilkinson, 2000), a feature that is thought to be associated with more recent Cretaceous environments. Unlike the other species it is also largely monostelic and considered more 'advanced' of all the *Gunnera* species, despite its small size. By the late Creatceous large-leafed angiosperm species were probably present (see Wilkinson, 2000). Considerable changes in angiosperm leaf size and morphology are now thought to have occurred over a relatively short period during the Cretaceous (Willis and McElwain, 2002), confounding any attempt to analyse the relatedness between early ancestral forms and extant species. Macro fossil material is required to resolve this issue.

6. CONCLUSIONS

It is clear that we still have only a limited understanding of the functional and evolutionary relationships between environmental factors and the current or past distribution of the genus *Gunnera*. What information is available is based on host responses, whilst very little is known about the influence of the cyanobacterial partner under varying environmental conditions. Most of this information is also based on one or two species and more comparative experiments are required on a wider range of species and populations. Whilst considerable emphasis has been placed on liquid water availability, atmospheric humidity *sensu stricta* could also be of ecological importance. Most species are found in super humid environments, which could counteract low soil water availability/reduced precipitation, due to a reduction in evaporative demand. We are aware of no studies, however, which have specifically examined the effects of humidity on any *Gunnera* species. Clearly, a requirement for high rainfall/high humidity environments could make the genus particularly susceptible to climate-change induced alterations in the hydrological cycle. Overall, global terrestrial precipitation and humidity has increased during the last century, although this is dependent on latitude, with increases in the Northern Hemisphere associated with decreases in the tropics and sub-tropics. However, there is considerable regional variation and future predictions are, as yet, uncertain (Folland *et al.*, 2001).

Rather surprisingly we also have little understanding of the nitrogen fixing potential of any *Gunnera* species or the extent to which this may contribute to nitrogen cycling in natural ecosystems. Not only is this information of fundamental interest, it could also be important, given the ancient ancestry of the genus, in improving our understanding of palaeoenvironmental conditions. We also need a better understanding of the ecological benefits conferred on the symbiont through the formation of this partnership. This fact is often ignored in studies on the functional significance of symbiotic systems in natural ecosystems, with most emphasis focussed on host benefits. A greater ability to exploit essential nutrients has been proposed as a factor benefiting many symbionts (Sprent and Sprent, 1990), but this requires further examination. Studies of the interrelationships between host and symbiont and the environmental limitations to infection, establishment and maintenance of a nitrogen fixing association are essential and should provide important information on the reality of establishing novel nitrogen-fixing partnerships suitable for field conditions.

ACKNOWLEDGEMENTS

We thank Betsy Hickey, Jane Hennessy and Matthew Jebb for providing information and Ian Sommerville for discussion. One of us (B. A. Osborne) thanks the support of the European Science Foundation Cyanofix Scientific Programme.

REFERENCES

Agnew, A. D. Q. (1974) *Upland Kenya Wild Flowers,* Oxford University Press, Oxford.

Ballare, C. L., Rousseaux, M. C., Searles, P. S., Zaller, J. G., Giordano, C. V., Robson, T. M., Caldwell, M. M., Sala, O. E. and Scopel, A. L. (2001) Impacts of solar ultraviolet-B radiation on terrestrial ecosystems of Tierra del Fuego (southern Argentina)-An overview of recent progress, *J. Photochem. Photobiol. B-Biology* **62,** 67-77.

Batham, E. (1943) Vascular anatomy of the New Zealand species of *Gunnera, Trans. Royal Society New Zealand* **73,** 209-216.

Bergman, B., Johansson, C. and Soderback, E. (1992) The *Nostoc-Gunnera* symbiosis, *New Phytol.* **122,** 379-400.

Berston, G. and Bazzaz, F. (1996) Belowground positive and negative feedbacks on CO_2 growth enhancement, *Plant and Soil* **187,** 119-131.

Brady, N. C. and Weil, R. R. (1996) *The Nature and Properties of Soils*, Prentice Hall, New York.

Campbell, G. J. (1994) *Water Supply, Plant Productivity and Gas Exchange Responses of Gunnera tincroria (Molina) Mirbel (Gunneraceae)*, PhD Thesis, University College Dublin.

Campbell, G. J. and Osborne, B. A. (1993) Watering regime and photosynthetic performance of *Gunnera tinctoria* (Molina) Mirbel, in M. Borghetti, J. Grace, and A. Raschii (eds.), *Water Transport in Plants Under Climatic Stress*, Cambridge University Press, Cambridge, pp. 247-255.

Condie, K. C. and Sloan, R. E. (1997) *Origin and Evolution of Earth. Principles of Historical Geology*, Prentice Hall, New York

Dawson, M. I. (1983) Chromosome numbers of three South American species of *Gunnera* (Gunneraceae), *New Zealand J. Bot.* **21,** 457-459.

Dettman, M. E. and Jarzen, D. M. (1990) The Antarctic/Australian rift valley: late Cretaceous cradle of northeastern Australasian relics? *Rev. Palaeobot. Palynol.* **65,** 141-144.

Drinnan, A. N., Crane, P. R. and Hoot, S. B. (1994) Patterns of floral evolution in the early diversification of non-mangolid dicotyledons (eudicots), *Plant Systematics Evol.* (Supplement) **8,** 93-122.

Folland. C. K., Karl, T. K., Christy, J. R., Clarke, R. A., Gruza, G. V., Jouzel, J., Mann, M. E., Oerlmans, J., Salinger, M. J. and Wang, S.-W. (2001) Observed climate variability and change, in J.T. Houghton, Y. Ding, D.J. Griggs, M. Noguer, P.J. van der Linden, X. Dai, K. Maskell, and C.A. Johnson (eds.), *Climate Change 2001; The Scientific Basis. Contribution of Working Group I to the Third Assessment Report of the Intergovernmental Panel on Climate Change*, Cambridge University Press, Cambridge, pp. 99-181.

Gordon, D. R. (1998) Effects of invasive, non-indigenous plant species on ecosystem processes: lessons from Florida, *Ecological Applications* **8,** 975-989.

Hanson, L., McMahon, K. A., Johnson, M. T. and Bennett, M. D. (2001) First nuclear DNA C-values for another 25 angiosperm families, *Ann. Bot.* **88,** 851-858.

Heusser, C. J. (1995) Three late quarternary pollen diagrams from Southern Patagonia and their palaeoecological implications, *Palaeogeography, Palaeoclimatology and Palaeoecology* **118,** 1-24.

Hickey, B. (2002) *Changes in Community Processes Associated with the Introduced and Invasive Species Gunnera tinctoria* (Molina) Mirbel, PhD Thesis, University College Dublin.

Hickey, B. and Osborne, B. A. (1998) Effect of *Gunnera tinctoria* (Molina) Mirbel on semi-natural grassland habitats in the west of Ireland, in U. Starfinger, K. Edwards, I. Kowarik and M. Williamson (eds.), *Plant Invasions: Ecological Mechanisms and Human Responses*, Backhuys Publishers, Leiden, pp. 195-208.

Hickey, B. and Osborne, B. A. (2001). Natural seed banks, seedling growth and survival in areas invaded by *Gunnera tinctoria*, in G. Brundu, J. Brock, I. Camarda, L. Child and M. Wade (eds.), *Plant Invasions: Species Ecology and Ecosystem Management*, Backhuys Publishers, Leiden, pp. 105-114.

Hoot, S. B., Magallon, S. and Crane, P. R. (1999) Phylogeny of basal eudicots based on three molecular data sets: atpB, rbcL and 18S nuclear ribosomal DNA sequences, *Ann. Missouri Bot. Garden* **86,** 1-32.

Howell, J. T. (1970) *Marin Flora: Manual of the Flowering Plants and Ferns of Marin County, California,* University of California Press, Berkeley.

Jarzen, D. M. (1980) The occurrence of *Gunnera* pollen in the fossil record, *Biotropica* **12,** 117-123.

Jarzen, D. M. and Dettman, M. E. (1989) Taxonomic revision of *Tricolpites reticulatus* Cookson ex Couper, 1953 with notes on the biogeography of *Gunnera* L, *Pollen et Spores* **31,** 97-112.

Kimball, B. A. (1983) Carbon dioxide and agricultural yield: an assemblage and analysis of 430 prior observations, *Agronomy J.* **75,** 770-782.

Lequesne, C., Villagran, C. and Villa, R. (1999) History of 'olivillo' (*Aextoxicon punctatum*) and Myrtaceae relict forests of Isla Mocha, Chile during the late Holocene, *Revista Chilena Historia Natural* **72,** 31-47. (In Spanish with English summary).

McGinnies, W. G., Goldman, B. J.and Paylore, P. (1968) *Deserts of the World*, University of Arizona Press, Tucson.

McKensie, G. M. and Kershaw, A. P. (2000) The last glacial cycle from Wyelangta, the Otway region of Victoria, Australia, Palaeogeography, *Palaeoclimatology and Palaeoecology* **155,** 177-193.

Molina, A. M. (1978) El genero *Gunnera* en la Argentina y el Uraguay (Gunneraceae). (The genus *Gunnera* (Gunneraceae) in Argentina and Uruguay), *Darwinia* **21,** 473-489.

Moore, D. M. (1968) *The Vascular Flora of the Falkland Islands*, British Antarctic Survey Scientific Reports no. 60.

Mora-Osejo, L. E. (1978) Nuevas especies de *Gunnera* L. del neotropico I. (New species of *Gunnera* in the tropics I), *Caldasia* **12,** 171-179.

Mora-Osejo, L. E. (1984) *Flora of Colombia. 3. Halagoraceae*, Instituto de Ciemcias Naturales-Museo de Historia Natural. Universidad National de Colombia, Bogota.

Muller, M. J. (1982) *Selected Climate data for A Global Set of Standards for Vegetation Science*, Dr W, Junk, The Hague.

Orchard, A. E. (1990) Gunneraceae, *Flora of Australia* **18,** 85-87.

Osborne, B. A. (1988) Photosynthetic characteristics of *Gunnera tinctoria* (Molina) Mirbel, *Photosynthetica* **22,** 168-178.

Osborne, B. A. (1989a) Effect of temperature on photosynthetic O_2-exchange and slow fluorescence characteristics of *Gunnera tinctoria* Molina (Mirbel), *Photosynthetica* **23,** 77-88.

Osborne, B. A. (1989b) Comparison of photosynthesis and productivity of *Gunnera tinctoria* Molina (Mirbel) with and without the phycobiont *Nostoc punctiforme* L, *Plant, Cell Environ.* **12,** 941-946.

Osborne, B. A., Doris, F., Cullen, A., McDonald, R., Campbell, G. and Steer, M. (1991) *Gunnera tinctoria*: an unusual nitrogen-fixing invader, *BioScience* **41,** 224-234.

Osborne, B. A., Cullen, A., Jones, P. W. and Campbell, G. J. (1992) Use of nitrogen by the *Nostoc-Gunnera tinctoria* (Molina) Mirbel symbiosis, *New Phytol.* **120,** 481-487.

Pacheco, P., Stuessy, T. F. and Crawford, D. J. (1991) Natural interspecific hybridisation in *Gunnera* (Gunneraceae) of the Juan Fernandez Islands, Chile, *Pacific Science* **45,** 389-399.

Palkovic, L. A. (1974) *The Genecology of Gunnera L. in Mexico and Central America,* PhD theses, Harvard University, Cambridge.

Rai, A. N., Soderback, E. and Bergman, B. (2000) Cyanobacterium-plant symbioses, *New Phytol.* **147,** 449-481.

Rasmussen, U. and Svenning, M. M. (2001) Characterization by genotypic methods of symbiotic *Nostoc* strains isolated from five species of *Gunnera, Arch. Microbiol.* **176,** 204-210.

Rousseaux, M. C., Ballare, C. L., Giordano, C. V., Scopel, A. L., Zima, A. M., Szwarcberg-Bracchitta, M., Searles, P., Caldwell, M. M. and Diaz, S. (1999) Ozone depletion and UVB radiation: impact on plant DNA damage in southern South America, *Proc. Natl. Acad. Sci.* **96,** 15310-15315.

Sabater, S. and Munoz, I. (2000) *Nostoc verrucosum* (Cyanobacteria) colonised by a chironomid larva in a Mediterranean stream, *J. Phycol.* **36,** 59-61.

Scotese (2001) *Atlas of Earth History. Vol. I. Paleogeography. PALEOMAP Project,* University of Texas at Arlington. (Cretaceous map and information available at www.scotese.com/cretaceo.htm).

Silva, L., Tavares, J. and Pena, A. (1996) Ecological basis for the control of *Gunnera tinctoria* on Sao Miguel Island, in *Second International Weed Control Congress,* Copenhagen, pp. 233-239.

Silvester, W. B. and Smith, D. R. (1969) Nitrogen fixation by *Gunnera-Nostoc* symbiosis, *Nature* **224,** 1321.

Soltis, P. S., Soltis, D. E. and Chase, M. W. (1999) Angiosperm phylogeny inferred from multiple genes as a tool for comparative biology, *Nature* **402,** 402-403.

Sprent, J. I. (2001) *Nodulation in Legumes,* Royal Botanic Gardens, Kew, London

Sprent, J. I. and Sprent, P. (1990) *Nitrogen-Fixing Organisms: Pure and Applied Aspects,* Chapman and Hall, London.

Stanley, S. M. (1999) *Earth System History,* Freeman, New York.

Stock, P. and Silvester, W. B. (1994) Phloem transport of recently fixed nitrogen in the *Gunnera-Nostoc* symbiosis, *New Phytol.* **126,** 259-266.

Van der Meijden, R. and Caspers, N. (1972) Haloragaceae, in C. G. G. J. van Steemis (ed.), *Flora Malesiana,* Noordhoff Publishers, Leiden, pp. 259-263.

Van Wyk, B. and Malan, S. (1997) *Field Guide to the Wild Flowers of the Highveld,* Struik, Cape Town.

Wanntorp, L., Wanntorp, H.-E., Oxelman, B. and Kallersjo, M. (2001) Phylogeny of *Gunnera, Plant Systematics Evol.* **226,** 85-107.

Wilkinson, H. P. (2000) A revision of the anatomy of Gunneraceae, *Bot. J. Linn. Soc.* **134,** 233-266.

Williams, P. A., Ogle, C. and Timmins, S. M. (2001) Chilean rhubarb (*Gunnera tinctoria*): biology, ecology and conservation impacts in New Zealand, Landcare Research Contracts Report, Lincoln, New Zealand, pp. 1-18.

Willis, K. J. and McElwain, J. C. (2002) *The Evolution of Plants,* Oxford University Press, Oxford.

Chapter 14

ARTIFICIAL CYANOBACTERIUM-PLANT SYMBIOSES

M.V. GUSEV, O.I. BAULINA, O.A. GORELOVA, E.S. LOBAKOVA, T.G. KORZHENEVSKAYA
Department of Cell Physiology and Immunology, Moscow State University, Moscow, Russia

1. INTRODUCTION

Creation of new symbioses between plants and nitrogen-fixing microorganisms, particularly cyanobacteria, are of considerable interest in view of the recent developments aimed at increasing the contribution of biological nitrogen fixation as an economical and environment-friendly nitrogen source for plants (Gusev and Koezhenevskaya, 1990; Korzhenevskaya *et al.*, 1993; Rai *et al.*, 2000b). Differrent approaches being pursued for this purpose include: (i) introducing cyanobacteria and associative microsymbiont complexes (AMC) isolated from natural symbioses, into the root system of non-symbiotrophic plants; (ii) obtaining plants with cyanobacteria and AMC in their above-ground organs; and (iii) investigating the stages of plant-cyanobacterium interactions in model systems. Obtaining new artificial symbioses and modelling the interactions between symbiotic partners implies co-cultivation of the organisms that enter into the associations. The plant partners can be isolated protoplasts, de-differentiated cells in suspension cultures, callus tissues, organs, regenerated plantlets, or whole plants. Co-cultivation can be carried out using either mixed cultures or joint cultures.

Mixed cultures represent model systems that enable us to research the developmental stages of symbiosis involving contact between partners, their spatial integration, and co-adaptation. Joint cultures are used as models for studying the communication between spatially isolated cyanobacterial and plant partners. Components of these systems interact by exchanging metabolites that diffuse in the incubation medium.

Nutrients and physio-chemical conditions for *in vitro* cultivation of plants, plant cells, and plant tissues differ from those for cyanobacteria (Gusev and Koezhenevskaya, 1990). Therefore, artificial associations are to be obtained under conditions which allow mixed cultures to grow but prevent the growth of one or both partner(s) in monoculture.

Earlier studies on the metabolic links established between cyanobacteria and plant partners during co-cultivation, have shown: (i) exchange of ^{14}C-containing compounds between them; (ii) growth of their association despite a deficiency of exogenous

A.N. Rai, B. Bergman and U. Rasmussen (eds.), Cyanobacteria in Symbiosis, 253-312.

combined nitrogen; and (iii) exchange of regulatory substances that influence various energetic and biosynthetic activities of both partners including plant morphogenesis (Gusev *et al.,* 1984; Gusev and Korzhenevskaya, 1990; Korzhenevskaya *et al.,* 1993).

This contribution reviews the results of researches on creation of experimental plant-cyanobacterium associations and the interactions between symbiotic partners at various stages of the formation and functioning of these artificial associations.

2. CREATION OF ARTIFICIAL ASSOCIATIONS OF DIAZOTROPHIC CYANOBACTERIA AND NON-SYMBIOTIC PLANTS

Diazotrophic cyanobacteria form symbioses with certain representatives of all major plant groups (Meeks, 1998; Rai *et al.,* 2000b). However, they are of little economical significance since none are crop plants. In the last decade, certain progress has been made in developing artificial associations between diazotrophic cyanobacteria and cereals. The main approaches have been:

1) Introduction of a variety of diazotrophic cyanobacteria into the plant rhizosphere and subsequent selection of promising variants;
2) Induction of *para*-nodules (*p*-nodules) in plant root, and their infection with effective strains of diazotrophs.

2.1. Introduction of Diazotrophic Cyanobacteria into the Rhizosphere of Cereals

The nature, syncyanoses form de novo in each generation of the host plant, including the primary infection of host organs and tissues by cyanobacteria. Isolation of cyanobionts from natural syncyanoses and the subsequent reconstitution of these syncyanoses demonstrated that most such symbioses are facultative (see Rai *et al.,* 2000b). Some free-living *Nostoc* strains were also competent in infecting and forming symbioses with hornworts *Anthoceros punctatus* (Enderlin and Meeks, 1983) and *Phaeroceros laevis* (West and Adams, 1997), cycad *Zamia furfuracea* (Ow *et al.,* 1999), and angiosperm *Gunnera manicata* (Bonnett and Silvester, 1981). In artificial associations of *G. manicata* and free-living *Nostoc commune* CCC1453/3, the biological nitrogen supply to the host plant (per infected gland) was higher than that in associations with symbiotic cyanobacteria (Bonnett and Silvester, 1981). Research on natural symbioses has shown promise of its applicability in developing artificial symbioses between key cereals and diazotrophic cyanobacteria, particularly *Nostoc* (Rai *et al.,* 1996). This optimism is base on the fact that all necessary growth and physiological-biochemical requirements are localized in the cells of the cyanobacteria and they may not require the expression of plant genome. Secondly, *Nostoc* spp. are capable of heterotrophic growth, can be inoculated into the soil, and fix N_2 in association with plant roots (Rai *et al.,* 1996). Besides, physiologically active metabolites secreted by cyanobacteria stimulate germination and growth of plants (Gogotov, 1988).

To date there are several reports of successful introduction of free-living cyanobacteria into the root system of cereals (Gorelova *et al.*, 1988, 1989; Lobakova *et al.,* 1989; Spiller and Gunasekaran, 1990, 1991; Tarasenko *et al.,* 1990; Gantar *et al.,*

1991a,b; Koezhenevskaya *et al.*, 1991a; Obreht *et al.*, 1993; Rai *et al.*, 1995). During co cultivation of wheat seedlings (*Triticum vulgaris*) with soil isolates of *Nostoc* (8 strains), *Anabaena* (3 strains), and *Cylindrospermum* (1 strain), adsorption of cyanobacteria on seedling roots and establishment of association (loose or tight) have been observed (Gantar *et al.*, 1991a,b). *Anabaena* strains formed only loose associations while *Nostoc* strains formed both loose and tight associations. In all cases, the cyanobacteria fixed N_2 and supported seedling growth on nitrogen-free media (Gantar *et al.*, 1991b). The growth of plants was accompanied by considerable total length of roots and nitrogen content in roots and shoots (Obreht *et al.*, 1993). In tight associations with *Nostoc* sp. 2S9B, cyanobacteria were present not only on root surfaces but also in intercellular spaces and in the cells of cortical parenchyma (Gantar *et al.*, 1991a,b). The mechanism of cyanobacterial invasion into root cells is as yet unknown. However, in addition to cyanobacteria, numerous bacteria have also been noticed in these locations and there have been suggestions that in natural symbioses, bacteria play a role in penetration and colonization of plant organs by cyanobacteria (Ozawa and Yamaguchi, 1980; Bergman *et al.*, 1993; Korzhenevskaya *et al.*, 1999b).

During prolonged co-cultivation of wheat seedlings with *Nostoc* 2S9B (Gantar *et al.*, 1991a,b), and rice seedlings (*Oryza sativa*) and rooted cuttings of alfalfa (*Medicago sativa*) with strains of *Nostoc muscorum*, the cyanobacteria also colonized the surfaces of plant stem and leaves (Baulina *et al.*, 1989, 1990; Gorelova *et al.*, 1988, 1989; Korzhenevskaya *et al.*, 1993). Cyanobacteria, aided by hormogonia formation, occupied the surface of cultivation medium, roots, stem and leaves (Baulina *et al.*, 1990; Gantar *et. al.*, 1991a; Gorelova *et al.*, 1990, 1992a). Furthermore, *Nostoc* also reached the stem tissues in *T. vulgare-Nostoc* 2S9B association (Gantar *et al.*, 1991a), and the intercellular spaces of leaves in the *M. sativa-N. muscorum* VKM 15 association (Baulina *et al.*, 1990; Gorelova *et al.*, 1990; Korzhenevskaya *et al.*, 1993).

A promising biotechnological application of diazotrophic cyanobacteria is their use in greenhouse hydroponics (Spiller and Gunasekaran, 1990, 1991). Growing wheat in a greenhouse, using transparent containers and liquid nitrogen-free media, together with *Anabaena variabilis* SA-1 mutant yielded a plant growth comparable to the growth with 3μM of nitrate in the medium. The cyanobacterium also maintained sufficient oxygen level in the medium for root respiration (Spiller and Gunasekaran, 1991).

2.2. *Introduction of Symbiotic Cyanobacteria into Rhizosphere of Non-Symbiotrophic Plants*

Cyanobacteria isolated from various natural syncyanoses could be better than free-living strains for infecting organs, tissues or whole plants of non-symbiotrophic species and forming artificial associations (Korzhenevskaya *et al.*, 1991a; Rai *et al.*, 1996). This is because of the following abilities of the cyanobionts: a) they adhere on the surface of organs and penetrate into the tissues of host plant; b) they reproduce on surface or inside inner organs of plant; c) they fix atmospheric N_2 and transfer most of the fixed nitrogen to the host plant; d) they receive and use plant products for growth and N_2-fixation.

Results of testing the ability of over 50 species of symbiotic cyanobacteria to adsorb on the surface of roots of rice seedlings confirms this assumption (Korzhenevskaya *et al.*, 1991a; Lobakova *et al.*, 1989, 2001c; Rai *et al.*, 1996; Rai *et al.*, 2001a,b). However, symbiotic origin of cyanobacteria does not guarantee effective associations (germination or rhizogenesis).

2.3. Natural Associative Complexes of Microsymbionts

Besides the dominant microsymbiont, a number of satellite microorganisms have been found in natural symbioses. Such satellite microbial components have been termed helper-bacteria. The latter apparently play significant role in formation and functioning of symbiotic systems (Smith and Douglas, 1987; Garbae and Bowen, 1989; Garbae and Duponnoids, 1990; Bergman *et al.*, 1993; Filippi *et al.*, 1995; Serrano *et al.*, 1999).

When developing artificial associations of diazotrophs with plants, survival of introduced microorganisms on the roots of inoculated plants and their competition with natural microflora is important. One of the ways to solve this problem is to co-culture microorganisms for inoculation of plants. Introducing mixed bacterial cultures improves viability of microsymbionts in the soil (Pankratova, 1987; Basan and Holgium, 1997) and on the roots of plants, where ecologic situation is considerably more stable and favorable, than in soil (Hallmann *et al.*, 1997). Besides, the level of synthesis of biologically active compounds (growth hormones, enzymes) is considerably higher in co-cultures than in corresponding monocultures (Egorov and Landau, 1982; Garbae and Bowen, 1989; Garbae and Duponnoids, 1990).

2.3.1. Associative Microsymbiont Complexes of Plant Syncyanoses

Satellite-bacteria (SB) are often isolated from natural syncyanoses along with cyanobacteria (Grobbellaar *et al.*, 1987; Fornii *et al.*, 1990; Carrapico, 1991; Lobakova *et al.*, 1992, 1996, 2001a; Serrano *et al.*, 1999). The presence of SB at all stages of development of *Azolla - Anabaena* symbiosis indicates its three-component nature (Peters and Meeks, 1989; Carrapico, 1991; Bergman *et al.*, 1993). Specific composition of SB isolated from *Azolla* fern includes not only *Arthrobacter* (Wallace and Gates, 1986; Fornii *et al.*, 1990), but also representatives of other genera: *Agrobacterium, Staphylococcus, Rhodococcus, Corynebacterium, Weeksella* (Serrano *et al.*, 1999). Participation of bacteria, and possibly fungi, in the process of host infection during the development of cycad-*Nostoc* syncyanoses has been proposed (Grushvitskii and Chavchavadze, 1978; Korzhenevskaya *et al.*, 1999b), and it may be true of the *Gunnera-Nostoc* symbiosis.

Isolation of cyanobionts from natural symbioses and their purification from satellite microflora involve methods such as γ-irradiation, UV-light, low temperature and antibiotics. Application of such methods may lead to elimination of SB, and selection of non-dominant subpopulations or mutant forms of cyanobionts (Zimmerman *et al.*, 1989). We studied associative microsymbiont complexes (AMC) isolated from *Azolla* sp., *A. pinnata* and *Cycas revoluta* using only cultural methods of microorganism isolation (Lobakova *et al.*, 2001a).

The *Azolla-Nostoc* (*Anabaena*) symbiosis is obligate and therefore, unique among natural plant syncyanoses. The AMC are transferred via the sexual stage in the development cycle of the *Azolla* (Peters and Meeks, 1989). AMC isolated from *Azolla* spp. consist of minor cyanobionts and obligate SB. The dominant cyanobiont of *Azolla* AMC is unable to grow outside the host plant or as a symbiont in association with other symbiotrophic plants (Meeks *et al.,* 1988). The AMC are characterized by obligate dependence of partners as evident from the balanced growth of partners and lack of survival of the cyanobiont in pure cultures. The cyanobacterial growth always dominates component in AMC, while SB appear on the areas of medium conditioned with metabolites of cyanobacteria or on the surface of degraded cyanobacterial colonies. The SB are always located in the vicinity to heterocyst envelopes and/or they form a network on the surface of vegetative cells and heterocysts (Fig. 1). Pure cultures of cyanobionts and SB can be obtained from AMC of *C. revoluta* coralloid roots (Lobakova *et al.,* 2001a).

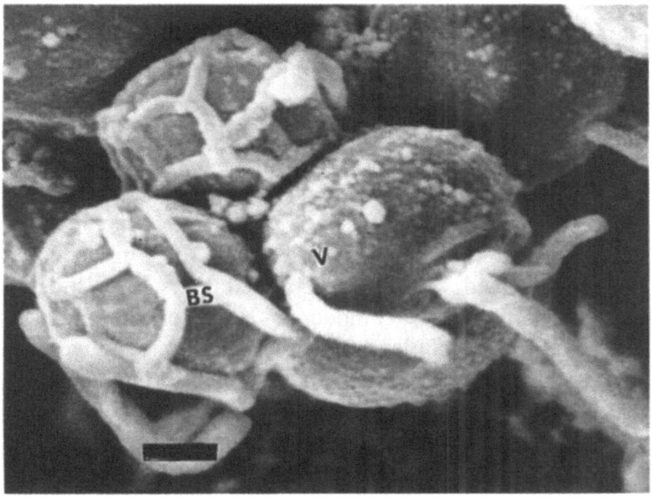

Figure 1. An AMC from A. pinnata. *H, heterocyst; V, vegetative cell; SB, Satellite bacteria. Bar, 1 μm. After Lobakova* et al. *(2001).*

2.3.2. Inoculation of Rhizosphere of Plants with AMC

Inoculation of rice seedlings, rooted tobacco cuttings (*Nicotiana tabacum* cv. Samsun and its albino mutant *P1*), or nightshade plants (*Solanum nigrum, S. dulcamara*) with AMC of *Azolla* spp., *C. revoluta,* or *Encephalortos ferox,* led to an active colonization of root, stem and leaf surfaces by cyanobacteria (Gorelova *et al.,* 1988; Lobakova *et al.,* 1989, 1992, 2001c; Korzhenevskaya *et al.,* 1991a). On plant roots, cyanobacteria were localized under root cap near the growing tip. Also, cyanobacteria were localized in cavities on the stem surface or stem ribs in the form of ascending bars, and as thin amorphous net or local microcolonies in close vicinity to guard cells of leaves. Suggestions that cyanobacterial filaments penetrate into the plant tissues through stomata are of special interest. In the tobacco callus and *A. variabilis* ATCC 29413 association, the cyanobacterium penetrated through intercellular spaces and xylem vessels into aeriferous substomatal spaces of plants and filaments were seen coming out onto the leaf surface (Gusev *et al.,* 1986; Pivovarova *et al.,* 1986; Gusev and Korzhenevskaya, 1990). One can suppose that on inoculating whole plants with AMC, microsymbionts may penetrate from leaf surface into plant tissues through stomata, and colonize intercellular spaces of various organs. On roots of such plants, *p*-nodules were formed in places where cyanobacteria were localized.

2.4. Induction of P-nodule Formation on Roots of Non-Legumes and their Colonization by Diazotrophic Microorganisms

Para-nodulation is formation of pseudonodules on roots of non-legumes, induced by abiogenic (chemical substances) or biogenic (microorganisms) agents.

2.4.1. Induction of P-nodule Formation by Abiogenic Agents of Nodulation

Abiogenic agents of nodulation (AAN) induce *p*-nodules formation on plant roots due to uncoordinated growth of dedifferentiated cells of epidermis, cortical parenchyma and pericycle (Tchan and Kennedy, 1989; Kennedy and Tchan, 1992; Glagoleva *et al.,* 1996, 1997). AAN may be: 1) plant hormones –e.g., natural auxines such as indoleacetic acid and their synthetic analogues (2.4-D, chlorambene, chlorsulphon) (Tchan and Kennedy, 1989; Nie *et al.,* 1992); or 2) enzymes which can loosen or partly destruct cell wall of plants (Al-Malah *et al.,* 1989). The frequency of *p*-nodules formation in plant roots treated with enzymic preparations is 20 times lower than in plant roots treated with hormone-like compounds. However, the practical use of AAN has limitations because AAN affect plant roots only in a limited range of concentrations and require optimal soil concentrations to be maintained throughout the plant's active growth period.

To inoculate *p*-nodules, diazotrophic microorganisms with following properties are proposed: a) ability to actively colonize *p*-nodules; b) resistance to high oxygen levels and changes in pH of the medium; c) ability to effectively fix-N_2 and share it with the host-plant (Tchan and Zeman, 1995).

Intercellular spaces of *p*-nodules of plants treated with AAN are actively colonized by diazotrophic bacteria (Tchan and Kennedy, 1989; Nie *et al.,* 1992; Tchan and Zeman, 1995; Glagoleva *et al.,* 1996, 1997; Kovalskaya *et al.,* 2001) and cyanobacteria (Gantar and Elhai, 1999). Nitrogenase activity of microorganisms in *p*-nodules is 3-6 times higher in plants treated with AAN than in non-treated plants (Tchan and Kennedy, 1989; Nie *et al.,* 1992; Gantar and Elhai, 1999; Kovalskaya *et al.,* 2001). This is apparently due to more favorable conditions for bacterial growth and nitrogen fixation inside the *p*-nodules. Microaerobic conditions inside the root improve nitrogenase functioning, and the localization of microorganisms in intercellular spaces of cortical parenchyma ensures constant flow of metabolites (apoplast transport) to diazotrophic organisms supporting their energy needs for nitrogen fixation. In addition, root metabolites of plants are known to affect the biosynthesis of bioactive compounds (Kravchenko and Borovko, 1989) and rates of nitrogen fixation (Patnaik *et al.,* 1994; Reddy *et al.,* 1997; Gantar and Elhai, 1999; Kovalskaya *et al.,* 2001) in microbes.

A comparative analysis of the ability of various diazotrophic prokaryotes to colonize *p*-nodules and to fix-N_2 indicates the following advantages of using cyanobacteria as symbionts of plants: 1) higher level of nitrogen fixation per plant; 2) optimal oxygen concentration inside *p*-nodules for nitrogen fixation by heterocyst-forming cyanobacteria (other diazotrophs need 4 times higher O_2 concentration) (Gantar and Elhai, 1999).

2.4.2. P-nodules Formation on Plant Roots by Biological Agents of Nodulation

In our view, bacterial cultures that are able to adsorb on root surfaces, and produce and excrete hormone-like compounds and/or cell wall degrading enzymes, can be used as biological agents for induction of *p*-nodules. Biological agents of nodulation (BAN) may be individual isolates or associations of microorganisms (Lobakova *et al.*, 2001b).

Adsorption of BAN on roots, penetration into cortical parenchyma, and subsequent colonization and distribution through the intercellular spaces in the entire root, induces formation of callus-like structures of *p*-nodules for the whole life-period of plant.

An example of BAN might be the AMC obtained from natural syncyanoses (Korzhenevskaya *et al.*, 1991a; Lobakova *et al.*, 1992, 2001c). Satellite-bacteria of AMC from *A. pinnata* are an association of *Arthrobacter* sp. and *Rhodococcus* sp. This association is characterized by stable composition, low level of nitrogen fixation and ability to synthesize the pectolytic enzymes pectynmethylesterase and polygalacturonase (E. Lobakova and G. Dol′nikova, unpublished results).

In associations of various plant species with AMC described earlier, increased plant cell proliferation leading to *p*-nodules formation occurred at sites where cyanobacteria were localized on plant roots (Fig. 2) (Korzhenevskaya *et al.*, 1991a; Lobakova *et al.*, 1992, 2001c). Microscopic analysis of root surface showed callus growth of de-differentiated epidermal cells and cortical parenchyma. In intercellular spaces of central part of the nodules numerous helper-bacteria, often separated from plant cells by an electron-dense layer, were found. Mass conglomerations of electron-dense globules ranging 30-90 nm in size were present in intercellular spaces of the nodules (Fig. 3). The pattern of their location indicates that bacterial cells produce them. One of the proposed functions of SB in the *Azolla-Nostoc* (*Anabaena*) syncyanose is the synthesis of the hormone indoleacetic acid (Braun-Howland and Nierzwicki-Bauer, 1990; Bergman *et al.*, 1993).

The use of the SB association (*Micrococcus* sp. and *Rhodococcus* sp.) from AMC of *C. revoluta* as BAN, also induced formation of *p*-nodules on the roots of rape seedlings (*Brassica napus*). This implies a direct role of SB in formation of *p*-nodules (Glagoleva *et al.*, 1998; Kovalskaya *et al.*, 2001).

Electron microscopy of *S. dulcamara p*-nodules in association with *Azolla* AMC indicated possibe lysis of plant cell walls by the SB. These data agree with the ability of SB to synthesize cell wall degrading enzymes even in absence of the plant partner. The ability of AMC (consisting of cyanobacteria SB) to infect various plants' organs and to induce *p*-nodules formation is apparently due to SB activity. Indeed, when whole tobacco plants (cv. Samsun or its albino mutant *P1*) are inoculated with mixed cultures of *A. variabilis* ATCC 29413 and SB from *Azolla* AMC, cyanobacteria actively reproduce and colonize only the surface of upper parts of plant and roots forming *p*-nodules. When the same plants are inoculated with pure culture of *A. variabilis*, the cyanobacterium was eliminated within a month of the inoculation.

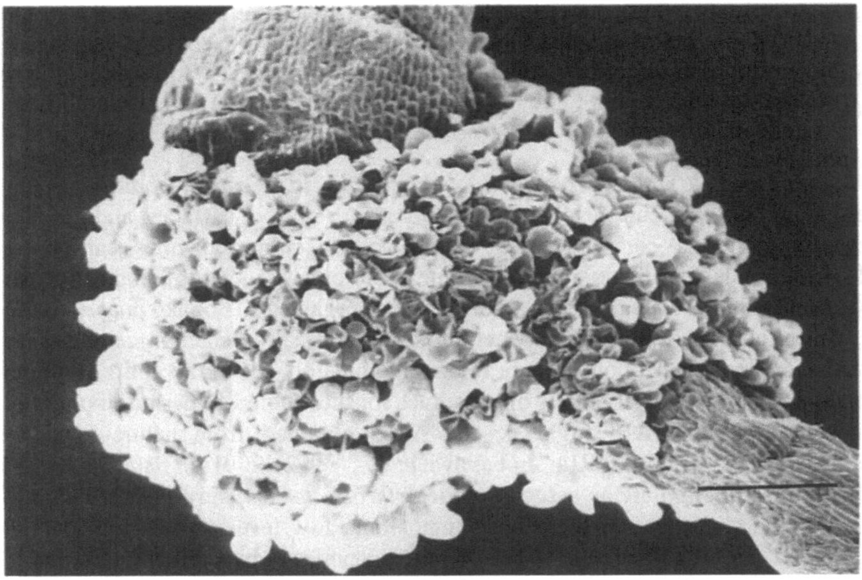

Figure 2. S. nigrum *root part showing a p-nodule. Bar 150 µm. From Lobakova* et al. *(2001c).*

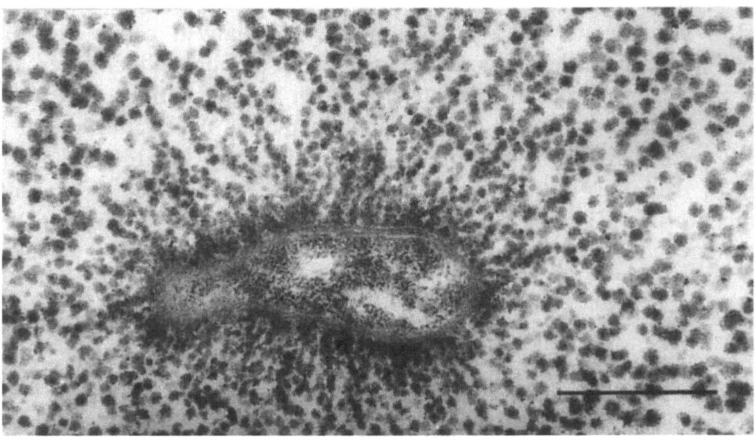

Figure 3. Electron-dense substances surrounding a satellite bacterium in the intercellular space of a p-nodule on tobacco plant root. Bar 0.5 µm. After Lobakova et al. *(2001c).*

3. THE MODELLING OF PARTNER'S INTERACTION *IN VITRO*

Thus, AMC of natural symbioses contain cyanobacteria and bacteria capable of lysing plant cell walls, producing hormone-like substances, and developing artificial diazotrophic associations.

3.1. Formation and Taxis of Cyanobacterial Hormogonia

In nature, the *de novo* formation of syncyanoses (except for the *Azolla – Anabaena* symbioses) starts from infection of host plants by cyanobacteria followed by reinfection of tissues during a plant's ontogeny. Studies on developmental aspects of natural and artificial symbioses have concluded that hormogonia formation and chemotaxis are of paramount importance for development of these symbioses (Silvester and McNamara, 1976; Bonnett and Silvester, 1981; Enderlin and Meeks, 1983; Campbell and Meeks, 1989; Baulina *et al.*, 1990; Kimura and Nakano, 1990; Meeks, 1990; Korzhenevskaya *et al.*, 1991b; Gorelova and Artamonova, 1992; Gorelova *et al.*, 1992a; Johansson and Bergman, 1992; Kerby *et al.*, 1992; Bergman *et al.*, 1993; Gantar *et al.*, 1993).

A thermolabile low molecular weight metabolite synthesized by the symbiotrophic hornwort *A. punctatus* incubated in a nitrogen-free medium, stimulates the formation of hormogonia in symbiotically compatible strains of *Nostoc* (Campbell and Meeks, 1989; Meeks, 1990). The mucilage of stem gland in *Gunnera* induces hormogonia differentiation in various *Nostoc* strains (Rasmussen *et al.*, 1994). The induction of hormogonium formation also occurs during the formation of artificial associations. For example, the wheat root exudate caused hormogonia induction in *Nostoc* 2S9B (Kerby *et al.*, 1992; Gantar *et al.*, 1993). We have shown hormogonia formation and taxis in free-living cyanobacteria *N. muscorum* CALU 304 and *N. muscorum* VKM 16, and in *Nostoc* isolated from the liverwort *Blasia pusilla* (Gorelova *et al.*, 1992b), under the influence of alfalfa (*M. sativa*) plants, alfalfa callus and the nightshade species *S. dulcamara* and *S. nigrum* (Gorelova *et al.*, 1992a).

3.1.1. Numerical Assessment of the Formation, Propagation, and Taxis of Hormogonia

In order to investigate the effects of plants and their tissues cultivated *in vitro*, on hormogonia differentiation and taxis, we developed a method for numerical assessment of the efficiency of the interaction between symbiotic partners (Korzhenevskaya *et al.*, 1994a; Gorelova *et al.*, 1995). Like the capillary-based method of Knight and Adams (1996), our method enables us to reliably assess hormogonium taxis (Gorelova *et al.*, 1996a, 1995, 1997). This method uses model systems such as joint cultures on agar media, where the plant material is spatially separated from the cyanobacteria. Hence, the partners interact via the exchange of metabolites diffusing in the agar medium. The assessment procedure includes: (i) light-optical control of the formation and propagation of hormogonia; (ii) graphic recording of changes in the propagation area and configuration of cyanobacteria on the surface of the agar medium; (iii) numerical processing of the resulting image on a Videolab-2.1A system (Uni-Export Instruments Ltd., England), and (iv) statistical evaluation of the data obtained. Following parameters are used to present the results:

$S_{\Delta j}$ and $S_{\Delta m}$ are the specific propagation areas of hormogonia (the ratio of the propagation area during the incubation period to the primary inoculum area) in the joint cultures and the monocultures, respectively.

K_S is the coefficient of propagation, whose value is used to estimate the influence of the plant partner on the propagation of hormogonia (hormogonia movement and cell division). K_S values below and above one, indicate stimulation and inhibition of hormogonia propagation, respectively. $K_S = S_{\Delta j} / S_{\Delta m}$.

K_o is the coefficient of orientation, i.e. the ratio between the areas (180° sectors) of hormogonia propagation towards the plant object (S_+) and in the opposite direction (S-). The K_o values below and above one correspond to positive and negative hormogonium taxis, respectively. $K_o = S_+/S_-$.

In our studies we used the axenic suspension cultures of free-living *N. muscorum* CALU 304 and *N. muscorum* VKM 16, *Nostoc* sp. isolated from *B. pusilla* (called *Nostoc sp. f. Blasia* hereafter), and an AMC isolated from *Azolla* (called *A. azollae* hereafter). The callus cultures of two nightshade species (*S. dulcamara* and *S. nigrum*) and alfalfa (*M. sativa*), and an axenic culture of duckweed (*Lemna minuscula*), were used as plant objects. The plant materials used can be subdivided into two groups: those that form symbiotic associations with microorganisms nature and those that do not do so. The former include *M. sativa* (a legume forming natural symbiosis with rhizobia) and *L. minuscula* (forms multicomponent epiphytic associations with various microorganisms including cyanobacteria) (Korzhenevskaya *et al.,* 1994b). *S. dulcamara* and *S. nigrum* belonged to the second group.

3.1.2. Capacity of Various Cyanobacterial Strains to Form Hormogonia

In the mono- and pure cultures we used, mass differentiation of hormogonia occurred within the first two days after transferring the cultures from liquid medium to a fresh agar medium. Hormogonia differentiation in cyanobacterial cultures from the exponential (1-2 weeks) or early and mid-stationary (3-4 weeks) growth phases was more prolific than that observed in late stationary phase culture (6 weeks and older). The strains under study did not differ significantly in the time of hormogonium formation or their frequency. However, the propagation process (i.e. the movement of hormogonia from the primary inoculum) was considerably slower in *Nostoc sp. f. Blasia* than in the other three strains, regardless of the medium composition (Table 1). Different rates of propagation over the medium surface have been noted earlier for a number of monocultures of symbiotic cyanobacteria. For instance, *Nostoc* spp. from *Gunnera arenaria* and *C. revoluta* propagated rapidly, while *Nostoc* spp. from *Anthoceros* and *Peltigera polydactyla* propagated much more slowly (Bonnett and Silvester, 1981).

Nostoc sp. f. Blasia also differed from the other cyanobacteria tested by us in its response to the culture medium. Although BG-11 medium containing combined nitrogen was somewhat more favorable to all the strains than the nitrogen-free BG-11(N^-) medium, the $S_{\Delta m}$ differences were insignificant and did not exceed 18-22% in the rapidly propagating *N. muscorum* CALU 304 and *A. azollae*. However, the $S_{\Delta m}$ value was 43-45% in *Nostoc sp. f. Blasia* (Table 1).

$S_{\Delta m}$ for *N. muscorum* CALU 304 and *A. azollae* on the L medium was nearly two-fold lower than that on the BG-11 medium. This decrease was apparently due to the influence of the acidity (pH of the L medium is 5.8). Low pH is known to inhibit hormogonia formation under red light illumination in *Nostoc* PCC 9229, a cyanobiont of *Gunnera monoika* (Rasmussen *et al.,* 1994). It also decreases the gliding rate of hormogonia in *Nostoc cycadae* from *C. revoluta* (Hirosa, 1987). Lack of N or low pH, both are equally unfavourable for hormogonia propagation in *Nostoc sp.f. Blasia*.

Table 1. Hormogonium propagation in light-incubated cyanobacterial cultures. After Gorelova et al. *(1997)*

Cyanobacteria	Inoculum age (d)	Incubation medium*	Incubation time (d)	$S_{\Delta m}$ **
Nostoc sp. f. *Blasia*	40	BG-11	5	2.35± 0.09
	40	BG-11	2	0.97± 0.11
	40	BG-11(N-)	5	1.35± 0.12
	40	BG-11(N-)	2	0.53± 0.06
	40	L	5	1.29± 0.15
N. muscorum CALU 304	40	BG-11	5	8.44± 1.11
	40	BG-11(N-)	5	6.56± 0.81
	27	BG-11(N-)	2	4.50± 1.21
	51	BG-11(N-)	2	0.80± 0.09
	40	L	5	3.32± 0.21
N. muscorum VKM 16	27	BG-11(N-)	2	4.90± 1.40
	22	BG-11(N-)	2	6.05± 0.33
	7	BG-11(N-)	2	5.58± 0.62
A. azollae	40	BG-11	5	9.87± 1.07
	40	BG-11(N-)	5	8.05± 0.89
	40	L	5	4.51± 0.50

* the following media were used: nitrogen-containing BG-11 medium (Stanier *et al.*, 1971), pH 7.0; BG-11(N-), BG-11 medium lacking $NaNO_3$; the Na^+ deficiency was compensated for by the addition of an equivalent amount of NaCl, pH 7.0; L, the medium for cultivating *L. minuscula* plants without sucrose; it contains 1 g/l KNO_3, pH 5.8.

** the means and the errors of the means.

The rate of hormogonium formation and the area of hormogonium propagation over the medium surface was greater in in dark-incubated inocula of *N. muscorum* CALU 304 (1-4 week old cultures) than in light-incubated inocula. For example, the $S_{\Delta m}$ value for a 27-day inoculum after two days of incubation in the dark and in the light was 20.38±0.37 and 4.50±1.21, respectively. However, illumination exerted no influence on inocula older than 6 weeks. Dark- and light-incubated *N. muscorum* VKM 16 showed no difference in hormogonium formation and propagation. However, intense hormogonium differentiation occurred in *Nostoc sp. f. Blasia* only under light. *Nostoc* PCC 9229 is known to form 3 times more hormogonia in the dark than in the light (Rasmussen *et al.,* 1994). In contrast, *Nostoc* CCHU 5235p isolated from *B. pusilla* forms hormogonia only in the light (Kimura and Nakano, 1990). Thus, the intensity of hormogonium formation and hormogonium propagation was not restricted to the symbiotic cyanobacteria alone but also depended on the physiological state (the culture age) of the strains used, environmental conditions (medium composition, pH, and illumination) and their interaction with the plant partners.

3.1.3. Hormogonia Formation During the Interaction of Cyanobacteria with Plants

Among the cyanobacteria tested, *Nostoc sp.f. Blasia* exhibited the lowest rate of hormogonia propagation rate in joint cultures with the callus tissues of *S. dulcamara* and *S. nigrum.* Although hormogonia differentiation occurred in the light, hormogonia did not migrate away from the inoculum site. They formed flat curls or intertwined with one another. Irrespective of the partner's age, the callus tissues of both *S. dulcamara* and *S. nigrum* had little or no effect. However, a positive influence on hormogonia differentiation in joint cultures was exerted by *M. sativa* and *L. minuscula.* Interestingly, the alfalfa callus, apart from stimulating hormogonium formation and propagation in light-incubated *Nostoc sp. f. Blasia*, also induced hormogonium differentiation in the dark. This occurred with 2- and 7-week old cyanobacterial inocula. Co-cultivating them with *L. minuscula* plants in the light resulted in 1.5- to 2-fold increase in the specific area of the propagation of *Nostoc sp. f. Blasia* hormogonia on BG-11, BG-11(N-), and L media (Table 2).

Table 2. Cyanobacterial hormogonium propagation during joint cultivation with L. minuscula plants in the light. After Gorelova et al. *(1997)*

Cyanobacteria	Incubation medium	*KS* *	*K0**
Nostoc sp. f. *Blasia*	BG-11	1.49± 0.09	0.97± 0.11
	BG-11(N-)	1.98± 0.14	1.01± 0.12
	L	2.05± 0.13	1.04± 0.17
N. muscorum CALU 304	BG-11	1.54± 0.10	1.05± 0.13
	BG-11(N-)	2.27± 0.12	0.82± 0.10**
	L	1.48± 0.09	1.03± 0.16
A. azollae	BG-11	1.26± 0.12	1.17± 0.10**
	BG11(N-)	1.36± 0.16	1.12± 0.06**
	L	1.35± 0.14	1.07± 0.15

The results were documented on the 5th day of incubation; the age of cyanobacterial inoculum was 40 days. The distance between the partners was 15 mm.
* the means and the errors of the means.
** $K_o \neq 1$ at $\alpha<0.05$.

Duckweed plants in joint cultures also caused statistically significant ($\alpha<0.05$) stimulation of hormogonia formation in other cyanobacteria (Table 2). The maximum effect occurred when *L. minuscula* interacted with *N. muscorum* CALU 304 on BG-11(N-) medium.

The stimulation of hormogonium formation and propagation by light was also observed in joint cultures of *N. muscorum* CALU and callus tissues of nightshade species (Fig. 4A). Studies using stationary phase (20-31 day old) calli yielded similar K_S values for these two species. However, the stimulatory effect of *S. nigrum* on a dark-incubated culture was virtually the same as that on a light-incubated culture, whereas the *S. dulcamara* had an inhibitory effect on dark-incubated cultures (Fig. 4B).

The pattern of differentiation in *N. muscorum* VKM 16 and *N. muscorum* CALU was depended on the plant species used in the joint culture (Fig. 4C and 4D). *S. dulcamara* callus caused almost a two-fold increase in the rate of hormogonium formation and propagation in the light. In the dark, *S. dulcamara* still exerted a stimulatory influence but it became less evident in 31-day old calli. The K_S level of *N. muscorum* VKM 16 in joint cultures with *S. nigrum* depended on the age of the plant tissue involved both in the dark and under illumination (Fig. 4C and 4D). The mean values of K_S in experiments with 16-day calli were 1.28±0.10 and 1.58±0.11 in the dark and light, respectively. The influence of the 31-day old callus was insignificant ($K_S \approx 1$).

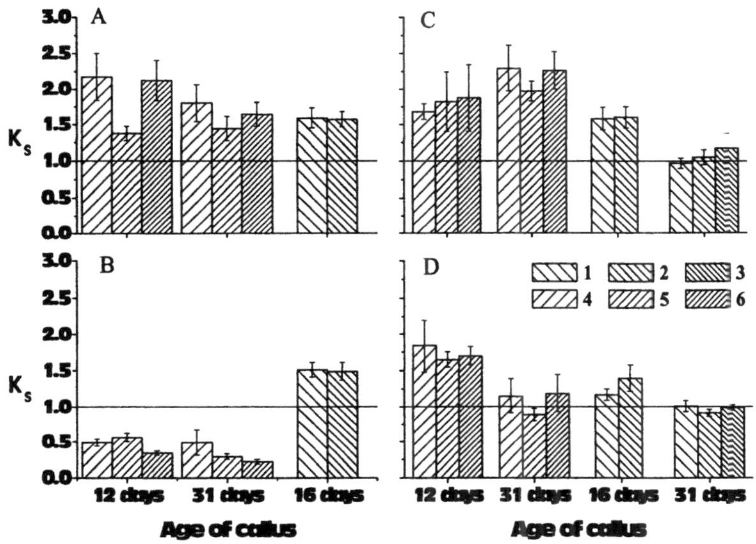

Figure 4. Influence of the S. dulcamara *(1-3) and* S. nigrum *(4-6) calli on hormogonia differentiation and propagation (K_S) in* N. muscorum *CALU 304 (A and B) and* N. muscorum *VKM 16 (C and D) in the light (A and C) and the dark (B and D). Initial distance between the partners was 10 mm (1, 4), 15 mm (2, 5), and 20 mm (3, 6). The incubation time was 2 days, and the age of the cyanobacterial inoculum was 1-4 weeks. After Gorelova* et al. *(1997).*

In addition, we determined the dependence of K_S of *N. muscorum* CALU 304 and *N. muscorum* VKM 16, on the age of *S. dulcamara* callus and the initial distance between the partners (Fig. 5). It was established that hormogonium formation and propagation were independent of the distance between the two partners (Fig. 5A and 5B) in both light- and dark-incubated joint cultures of *S. dulcamara* and *N. muscorum* VKM 16. The callus age exerted a significant influence in dark-incubated cultures only. A more intricate pattern was revealed with *S. dulcamara-N. muscorum* CALU 304 joint cultures. The influence of metabolites in the light-grown cultures depended on the distance between the partners, not on the callus age (Fig. 5C). The stimulatory effect on the hormogonium propagation had two maxima at distances of 5-9 mm and 15-20 mm between the interacting partners. K_S depended on the callus age and on the combined effect of both factors (callus age and distance) in dark-grown cultures. The weakest inhibition was caused by the plant partner at the maximum distance (15-20 mm) between the cyanobacterial inoculum and the 5-day old calli, whereas the strongest inhibition occurred at the maximum distance from 31-day calli (Fig. 5D).

We also found that *N. muscorum* VKM 16 growth is stimulated by *M. sativa*, both in suspension and callus co-cultures, and mass formation of hormogonia occurs (Gorelova, 1986; Gorelova *et al.*, 1990).

On the whole, it should be noted that plants capable of forming associations with microorganisms in nature invariably cause stimulation of hormogonium formation by various cyanobacterial strains in model systems. However, the agents that stimulate hormogonia formation are seem to have a low specificity, and their influence is not confined to the compatible partners alone.

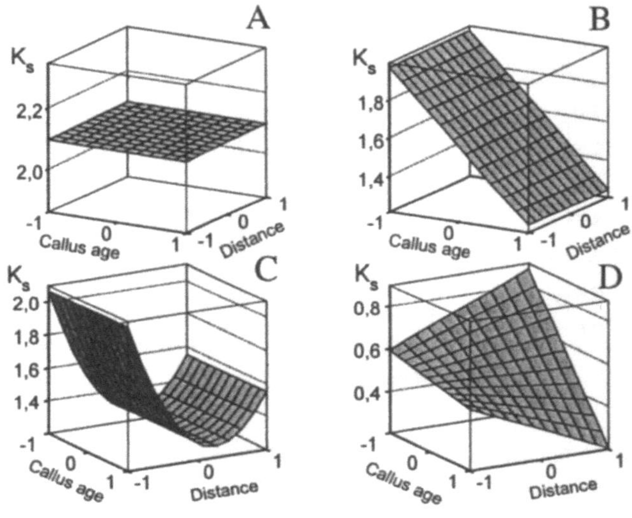

Figure 5. Dependence of the hormogonium propagation coefficient (K_S) on the callus age and distance between the partners in S. dulcamara–N. muscorum *VKM 16 (A and B) and* S. dulcamara – N. muscorum *CALU 304 (C and D) joint cultures in the light (A and C) and dark (B and D). The callus age "-1", "0", and "+1" correspond to 5, 12, and 31 days old calli, respectively, and the distance "-1", "0", and "+1" correspond to 5-9, 10-14, and 15-20 mm, respectively. The incubation time on the BG-11(N^-) medium was 2 days, and the age of the cyanobacterial inoculum was 27 days. After Gorelova* et al. *(1995).*

3.1.4. Hormogonium Taxis During Interaction with Plant Partners

Vectorial propagation of hormogonia was assessed using the orientation coefficient K_o. None of the tested plant partners induced vectorial movement of the slow propagating *Nostoc sp. f. Blasia* hormogonia (Gorelova *et al.*, 1992a, 1997). Rapidly propagating cyanobacterial strains were more sensitive to the influence of plants. For example, positive taxis of *A. azollae* towards *L. minuscula* plants (Table 2) occurred during the light-incubation of joint cultures on BG-11 and BG-11(N-) media. *N. muscorum* CALU 304 hormogonia on BG-11(N-) media exhibited positive taxis towards 5-day old *S. dulcamara* calli in the darkness (K_o= 1.70±0.21) (Gorelova *et al.*, 1995) and negative taxis during interactions with *L. minuscula* plants on the same medium in light (Table 2). An analysis of the dependence of K_o on the combined effects of the two factors involved, revealed that the distance between the partners did not influence the K_o in joint cultures of *S. dulcamara* and *N. muscorum* CALU 304, and the callus age influenced K_0 only in the dark-incubated cultures (Fig. 6). We failed to induce hormogonium taxis in the *S. dulcamara* callus-*N.muscorum* VKM 16 system either in the light or in the darkness, irrespective of the callus age and the distance between the partners (Gorelova *et al.*, 1995).

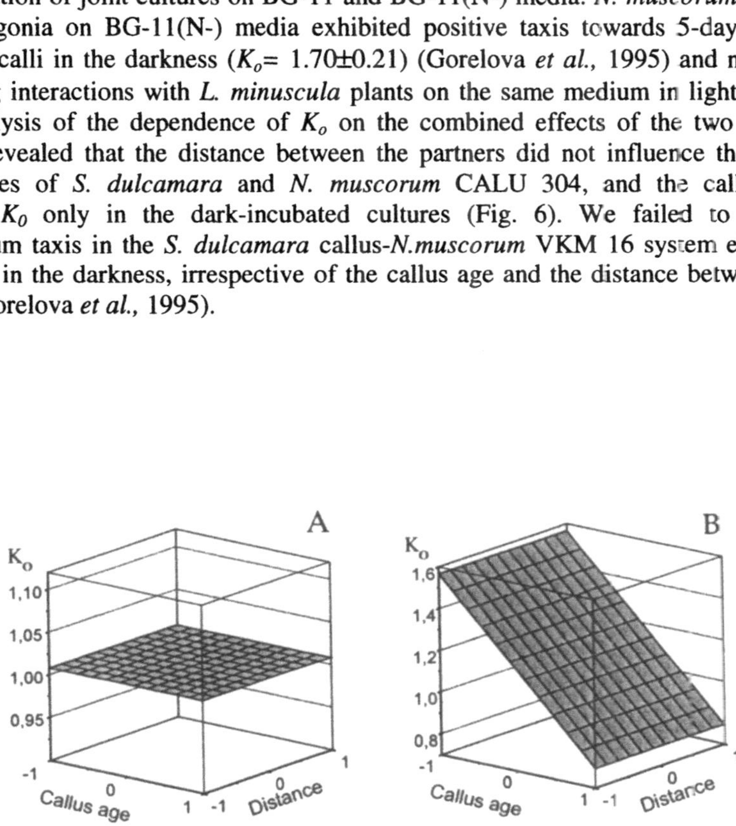

Figure 6. Dependence of the orientation coefficient (K_0) on callus age and distance between the partners in the S. dulcamara–N. muscorum *CALU 304 model system under the light (A) and darkness (B). Other details as in Fig. 5.*

Hormogonium taxis in *N. muscorum* VKN 16 was most pronounced in the mixed and joint cultures with *M. sativa.* As a result of hormogonium taxis, *N. muscorum* VKM 16 occupied the surface of alfalfa cuttings and formed intra-tissue microcolonies in the leaves (Baulina *et al.*, 1990; Gorelova *et al.*, 1990, 1992a, 1997; Korzhenevskaya *et al.*,

1993). However, cyanobacteria tended to move away from the plant tissue when the callus reached an age of 1.5 months. Although all the trichomes differentiated into hormogonia, the latter showed negative taxis.

Thus, hormogonium formation and taxis depend on the strain-specific features of the cyanobacteria and plants involved, the age of the partners, and the conditions used for establishing and incubating joint cultures (medium composition, illumination, and the initial distance between the partners). Since no correlation was found between the two criteria (K_o and K_S) used to assess the influence of the plant partners on the cyanobacteria, we conclude that the stimulation/induction or inhibition of hormogonia formation and their chemotaxis could be due to the effects of several distinct factors rather than a single one. The stimulation of hormogonia formation does not determine hormogonium taxis. Positive taxis of cyanobacteria towards the plant partners only occurred in the *M. sativa-N. muscorum* VKM 16, *S. dulcamara-N. muscorum* CALU 304, and *L. minuscula-A. azollae* systems. It was only using these partner combinations that we succeeded in obtaining stable associations with viability and balanced growth of the partners lasting several months to years (Baulina *et al.*, 1990; Gorelova *et al.*, 1990; Korzhenevskaya *et al.*, 1993, 1994a). This testifies to the adequacy of our method for numerical analysis of the formation and vectorial movement of cyanobacterial hormogonia.

3.1.5. Influence of the Plant partner on Hormogonia Differentiation and Taxis

In the absence of any direct contact between the partners in joint cultures, the K_S and K_o values are influenced by the metabolite flow between the plant tissue and the cyanobacteria. The plant partner can be considered the source of a number of compounds that can be arbitrarily classified into: (*A*) substances to be used in cyanobacterial metabolic processes ("nutrients"); (*B*) toxic substances including phytoalexins whose quantity can increase due to the stress caused by the transplantation of callus and its incubation in a non-optimal medium, or by the influence of cyanobacterial metabolites diffusing through the medium; and (*C*) signal substances, in particular those regulating the process of hormogonia formation (C_{h-} and C_{h+}) and chemotaxis (C_{c-} and C_{c+}).

It is evident from the *S. dulcamara - N. muscorum* CALU 304 model that the influence of plant metabolites can result from the combined effects of stimulators and inhibitors of cyanobacterial growth and hormogonium formation. The stimulators include at least two components. One component rapidly diffuses in the medium and is likely to be a low molecular weight compound whose gradient remains virtually unchanged within the area between the partners. This function can be performed by compound(s) *C* that induce hormogonium formation and the low molecular weight fraction of the substances of group *A* (termed hereinafter A_1). The latter group may include small peptides, amino acids, organic acids, and monosaccharides that are released into the medium as a result of the degradation of dead cells and macromolecular compounds secreted by living plant cells (Street, 1977).

The other component represents slow diffusing compounds apparently belonging to group *A* (termed hereinafter A_2). They include polysaccharides and proteins including enzymes (Street, 1977; Fowler, 1982). A_2 compounds exert a significant influence only in the immediate vicinity of the callus.

Apart from the signal substances (C_{h-}), inhibitory effect can be produced by the metabolites of group B. *S. dulcamara* produces steroid alkaloids solasodine, β-solamarine, and tomatidenol (Roddick, 1980). It is well known that steroid alkaloids are poisons or growth inhibitors for organisms ranging from bacteria to mammals. These compounds affect cell membranes by acting on the K^+ and Na^+ transports (Paseshnitchenko, 1987). Normally, the steroid compounds are located inside cells. Their release into the medium only occurs if the cell ruptures. This possibility cannot be ruled out when callus is subcultivated in an unfavorable medium. Besides, an inhibitory effect may be produced by phenol compounds including phytoalexins that are formed upon contact with specific pathogens or in response to other biotic and abiotic factors (elicitors) (Maksimova *et al.*, 1990). In a light-incubated system, substances of group B (B_L) display moderate diffusion rates, and/or their synthesis occurs only in response to cyanobacterial metabolites and, therefore, lags behind C_h and A_1 production. *N. muscorum* VKM 16 seems to be either insensitive to the spatial gradients of A_2 and B_L or it does not induce B_L production by nightshade calli. It is known that in mixed cultures, cyanobacteria can exert a variety of influences (ranging from stimulation to inhibition) on the biosynthesis of species-specific products such as sterines and steroid alkaloids (Gorelova *et al.*, 1984, 1985; Gorelova, 1986a,b; Gusev and Korzhenevskaya, 1990), sapogenins (Gusev and Korzhenevskaya, 1990; Yagodina *et al.*, 1990), indole alkaloids (Korzhenevskaya, 1990; Korzhenevskaya *et al.*, 1992) and flavonoids (anthocyanins) (Gorelova and Artamonova, 1992) by plant cells.

The pattern of the interactions between the partners in joint cultures, varies depending on whether the cultures are incubated in the dark or light. This suggests that light influences the spectrum of the plant metabolites, which affect cyanobacteria. Light is known to produce various effects on the biosynthetic activities of cultivated cells of higher plants. Light can induce or completely inhibit the synthesis of species-specific products, influence their yields, and change the ratios of differrent compounds (Nosov, 1991). Transferring dark-grown tea-plant calli to continuous light resulted in a decrease in the formation of soluble phenol compounds during the the first two subcultivations (Zagoskina and Zaprometov, 1991). Our results suggest that *S. dulcamara* calli in dark-incubated joint cultures with cyanobacteria may produce an unidentified inhibitory agent, B_d (and/or C_{h-}), which is either absent or unstable in light. This agent has a high diffusion rate that probably exceeds those of C_{h+} and A_1. The spatial gradient of B_d (C_{h-}) does not significantly change within the area between the partners; however, the B_d (C_{h-}) amount depends on the callus age. It is minimum in 5-day old calli and maximum in 31-day old calli. Thus, the effect (E) of plant metabolites on cyanobacteria can be summarised as follows:

at level "-1" (a distance of 5-9 mm) $E = \pm C_h + A_1 + A_2 - B_L - B_d$

at level "0" (a distance of 10-14 mm) $E = \pm C_h + A_1 - B_L - B_d$

at level "+1" (a distance of 15-20 mm) $E = \pm C_h + A_1 - B_d$

In joint cultures with *S. dulcamara* calli, the overall effect of plant metabolites on the growth and hormogonium formation in cyanobacteria can be evaluated using the scheme shown in Fig. 7. The overlapping of the spatial gradients of metabolites classified into different groups of compounds results in the formation of zones that differ in medium composition and, therefore, in their effects on the cyanobacteria located therein, depending on their distance from the callus.

As mentioned earlier, *N. muscorum* CALU 304 showed positive taxis in joint cultures with *S. dulcamara* under darkness, and the K_0 level did not depend on the distance between the partners. This suggests a high diffusion rate, and low molecular weight, of the chemoeffectors involved. Apart from the signal substances (C_{c+} and C_{c-}) that determine the direction of taxis, we can rule from group A_1 substances as candidates for the chemoattractant role, which are expected to be particularly important in dark-incubated systems.

Thus, the plant partners affect hormogonium formation and propagation. Longer interactions and shorter distances between the partners, result in enhancing the contribution of non-specific factors to the overall effect. Like the chemoeffectors controlling hormogonium taxis, the components that regulate (induce, stimulate, or inhibit) hormogonia formation also belong to the low molecular weight (rapidly diffusing) fraction of the metabolites produced by the plant tissue. The fact that chemoattractant activity is evident only in systems containing plants and cyanobacteria capable of establishing stable associations, suggests the existence of a selection mechanism for potential partners prior to the contact between partners and cell surface recognition.

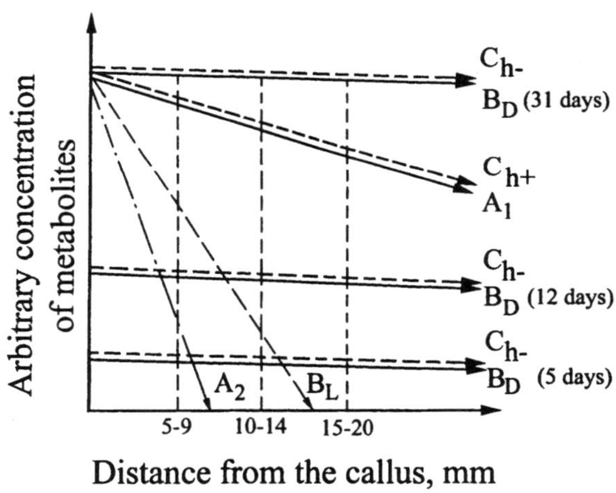

Figure 7. Schematic representation of the spatial gradients of extracellular plant metabolites. After Gorelova et al. *(1995).*

Interactions between compatible partners leads to the production of hormogonia-stimulating (-inducing) factors, hormogonium-inhibiting factors, chemoattractants, and chemorepellents by the plant partner. The activity of each of these factors depends on the age of the plant tissue and the environmental conditions. These responses enable the plant partner to direct the cyanobacterial partner to the target tissue, localize the potential infection, and set limits on its duration.

A comparison of the data obtained from artificial and from natural plant-cyanobacterium symbioses reveals that there is a close analogy between the two. A low molecular weight thermolabile hormogonuim-inducing factor (HIF) has been reported in *A. punctatus* (Campbell and Meeks, 1989; Meeks, 1990) and *Gunnera* sp. (Bergman *et al.*, 1996; Rasmussen *et al.*, 1994). Unidentified components of the exudate and aqueous extract of cycads also show the hormogonia-inducing effect (Bergman *et al.*, 1996; Ow *et al.*, 1999). These factors caused hormogonia induction in some free-living *Nostoc* strains also. In addition, a hormogonium-repressing factor (HRF) was detected in the *A. punctatus* gametophyte (Cohen and Meeks, 1997; Meeks, 1998; Meeks *et al.*, 1999). Indirect evidence of positive cyanobacterial taxis during the initiation of symbioses (Kimura and Nakano, 1990; Meeks, 1990, 1998; Rasmussen *et al.*, 1994; Bergman *et al.*, 1996) has been confirmed by demonstrating the chemoattractant activity of the exudates of N-limited *Blasia* (Knight and Adams, 1996). The chemoattractant was shown to be a thermostable, low molecular weight substance (M_r below 1,000) (Watts *et al.*, 1999). *Nostoc* sp. strain LBG1, an isolate from *P. laevis*, also exhibited positive taxis towards the exudate of wheat seedlings (Watts *et al.*, 1999).

The above similarities between artificial and natural associations suggest that plant-cyanobacterium signaling is not confined to the organisms entering into natural symbioses alone, and that it operates during the interactions between potential partners capable of establishing new associations.

3.2. Spatial Integration of the Partners

The establishment of functional artificial associations involve spatial integration of the partners into a coherent anatomical entity. Studies on a wide range of associations have revealed two complementary patterns in the formation of their structure. In one pattern, the partners form unions having close intercellular contact sites. The other pattern involves compartmentation of cyanobacteria inside (or on the surface of) the plant partner. Compartmentation is ensured by plant cell walls, membranes, and the intercellular or superficial mucous matrix whose formation accompanies the establishment of an integral system.

3.2.1. Cell Adhesion

To obtain *de novo* artificial associations in laboratory, the typically nonsymbiotrophic partners can be brought into contact *per force*, *e.g.*, by applying cyanobacteria onto a plant callus. In some cases, the partners can be brought together with the help of hormogonia (see section 3.1). The formation of mixed suspension cultures, calli, or infected plants starts with the adhesion and tight binding of some cyanobacterial cells to the plant cell wall. Cyanobacteria can attach to the plant cell wall using their surface structures, including the lipopolysaccharide outer membrane, sheaths, and extracellular mucilage containing acidic exopolysaccharides (Baulina *et al.*, 1989; Lobakova *et al.*, 1990; Korzhenevskaya *et al.*, 1993; Gorelova *et al.*, 1996b). Adhesion of cyanobacteria cannot be regarded as highly specific, since cross-reactions with irrelevant partners also occur. Washing of wheat roots or treating them with various sugars, did not affect their colonization by *Nostoc sp.* 2S9B. This is additional evidence for nonspecific binding involving extracellular polysaccharides (Gantar *et al.*, 1995). However, the location of *N. muscorum* CALU 281 and *N. muscorum* CALU 304 on the surface of rice roots was different; the cyanobacterial adhesion and growth depended on the presence of diverse carbohydrates in the medium (Gorelova *et al.*, 1989).

Based on the generally accepted views of the mutual recognition mechanisms in plants and bacteria, establishment of contacts between cell surface structures involves partner recognition using lectins. Like distant recognition, contact recognition represents a step in the sequence of sensor-signal interactions between partners that ensure specificity. It has been shown that cyanobacteria in natural symbioses contain specific sugars, pili, and in some cases lectins, which are involved in the recognition of, and adhesion to, a compatible host (Serrano *et al.*, 1999; Rai *et al.*, 2000b). The fact that cyanobacteria can firmly adhere to the surface of plant cells by means of a lipopolysaccharide, a sheath, or exopolysaccharides does not militate against the idea that an universal mechanism of lectin-receptor interaction operates in artificial associations at early stages of mutual recognition between the partners. However, nonspecific mechanisms of cell adhesion may also function in artificial symbioses since the partners may not have evolved specific mechanisms for interactions.

Spatial integration implies the adhesion of at least some of the cells involved and the concomitant formation of surface colonies. These processes can be regarded as the early stages in the formation of artificial symbioses. The subsequent spatial integration of the partners, leading to the formation of diverse anatomical structures, depends on their compatibility, the structure of the plant tissue, and the properties of individual cyanobacteria

3.2.2. Anatomical Organization

Suspension Cultures Mixed aggregates of plant cells and cyanobacteria in suspension cultures display the simplest organizational patterns. The cyanobacteria are located on the surface as well as into the interior of the plant cell aggregates. Examples include mixed aggregates of (1) *Synechococcus* sp. PCC 6301 and *N. tabacum* cv. Wisconsin-38 or *Dioscorea deltoidea* IPhR-D1 cells, (2) *A. variabilis* ATCC 29413 and *Solanum laciniatum* or *M. sativa* or *Panax ginseng* IPhR-G3 cells, (3) *Chlorogloeopsis fritschii* ATCC 27193 and *P. ginseng* IPhR-G1 or *P. ginseng* IPhR-G3 or *S. laciniatum* or *M. sativa* cells (Gorelova *et al.*, 1984, 1985; Baulina and Lobakova, 1986; Baulina *et al.*, 1989, 1994, 1995; Korzhenevskaya *et al.*, 1985, 1993; Gusev and Korzhenevskaya, 1990). In these associations, cyanobacteria are in direct contact with the plant cells which synthesize biologically active substances (Gorelova *et al.*, 1984, 1989; Lobakova *et al.*, 1984; Gorelova, 1986a,b; Gusev and Korzhenevskaya, 1990; Yagodina *et al.*, 1990; Baulina *et al.*, 1995). The cyanobacteria that do not attach to plant aggregates, get degraded and die earlier than those incorporated into the aggregates. Thus, the spatial integration contributed to the survival of cyanobacteria under the restrictive conditions imposed during the formation of artificial symbioses.

Table 3. Cyanobacteria and plants used for obtaining artificial associations using callus tissue

Cyanobacterium	Plant partner	Reference
A. variabilis CALU 458	*N. tabacum* (cv. Samsun, variegated mutant)	1
A. variabilis ATCC 29413	*N. tabacum* (cv. Samsun, variegated mutant)	2
A. variabilis ATCC 29413	*M. sativa*	3
N. muscorum VKM 16	*M. sativa*	3
N. muscorum VKM 16	*R. serpentina* K-27	
N. muscorum VKM 16	*O. sativa* (cv. Kuban-3)	
N. muscorum CALU 281	*O. sativa* (cv. Kuban-3)	4
N. muscorum CALU 304	*O. sativa* (cv. Kuban-3)	4
N. muscorum CALU 304	*S. nigrum*	
N. muscorum CALU 304	*S. dulcamara*	5
N. muscorum CALU 304	*R. serpentina* K-27	6
Nostoc sp. f. *Blasia*	*O. sativa* (cv. Kuban-3)	7
Nostoc sp. f. *Blasia*	*S. dulcamara*	
A. azollae (AMC) from *Azolla* sp.	*O. sativa* (cv. Kuban-3)	8
Nostoc sp. 2S9B	*T. aestivum* (cvs. Penawawa, Edwall, Klassic)	9
C. fritschii ATCC 27193	*R. serpentina* K-27	
C. fritschii ATCC 27193	*P. ginseng* IPhR-G3	

1) Baulina *et al* (1984); 2) Pivovarova *et al.* (1986); 3) Gorelova *et al.* (1988), Korzhenevskaya *et al.* (1993); 4) Gorelova *et al.* (1989); 5) Gorelova *et al.* (1995); 6) Gorelova (2000a,b), Gorelova and Artamonova (1992); 7) Korzhenevskaya *et al.* (1993); 8) Lobakova *et al.* (2001c); 9) Gantar (2000).

Callus Cultures. These are of two types: (i) where cyanobacteria are scattered throughout the whole thickness of the callus; and (ii) where cyanobacteria grow in the form of colonies on the surface or in the interior of the plant tissue.

The former type of anatomical organization occurs when *A. azollae* (together with SB of the AMC from the fern *Azolla sp.*) were inoculated onto the callus tissue of the albino tobacco mutant *P1* or the nightshade species *S. dulcamara* and *S. nigrum*. Even distribution of the cyanobacteria within the plant tissue resulted in the fomation of a homogeneously emerald-color mixed callus (Lobakova *et al.*, 1996, 2001c).

The latter type occurs in the mojority of associations between cyanobacteria and callus cultures (Table 3).

Among the systems listed in Table 3, a peculiar organizational pattern occurs in the *N. muscorum* CALU 304 - *Rauwolfia serpentina* mixed callus (Gorelova and Artamonova, 1992; Gorelova, 2000a). The cyanobacterial suspension that was initially applied onto the callus formed primary surface microcolonies. Their subsequent distribution on the callus surface and interior was assisted by hormogonia. The spatial integration of the partners was completed within 3-4 weeks of their growth in association. Mixed aggregates formed with a peculiar morphological pattern and these detached easily from the surrounding callus tissue. By the end of the 6th week, three cyanobacteria-containing zones became distinguishable in the aggregate (Fig. 8). *N. muscorum* CALU 304, which is located inside the aggregate, formed a central microcolony that occupies the expanded intercellular spaces and the dead *R. serpentina* cells lacking cytoplasm (zone A). The cyanobacteria in this zone represent short chains walls, while others are randomly distributed. The periphery of the central cyanobacterial colony is surrounded by numerous intertwined hormogonia and vegetative trichomes (zone B). The hormogonia and trichomes occupy the narrow intercellular spaces among densely packed *R. serpentina* cells that are smaller in size than those in the surrounding callus tissue. The surface zone inhabited by cyanobacteria is referred to as zone C.

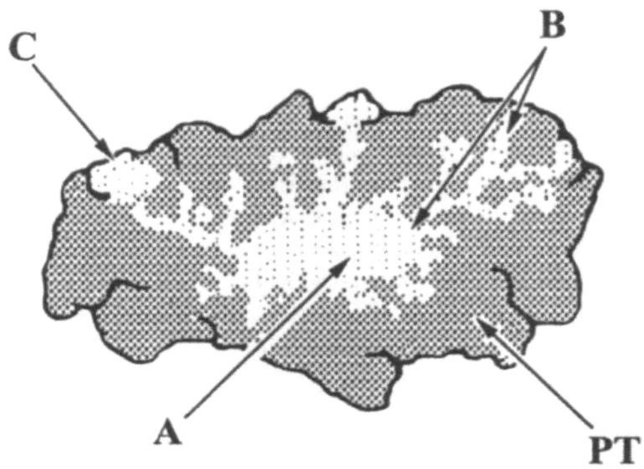

Figure 8. Organization of the mixed aggregate formed during growth of N. muscorum *CALU 304 with* R. serpentina *callus. A, B, C, zones of cyanobacterial localization; PT, plant tissue. After Gorelova (2000a).*

The growth of a mixed culture entails dynamic development of the mixed aggregates, including their aging and *de novo* formation. The surface microcolonies of *N. muscorum* CALU 304 are relocated to zone A due to a periodic local increase in the proliferative activity of plant cells surrounding them. The cyanobacteria located in the callus depth (zones A and B of the pre-existent aggregate) cease to be visible. Probably, they are eliminated or transferred (*e.g.*, as hormogonia) to a new aggregate. Subsequently, the growth of the plant tissue decelerates and the cyanobacteria reach the surface of the new aggregate (Gorelova, 2000a). The same developmental pattern characterized by periodic growth of each partner was found in other artificial associations, but not the same anatomical organization (Baulina *et al.*, 1984; Gusev and Korzhenevskaya, 1990; Lobakova *et al.*, 20001c).

In *A. Variabilis* (CALU 458 and ATCC 29413) - *N. tabacum cv. Samsun* and *N. muscorum* CALU 304 - *S. dulcamara* callus associations cultivated for a prolonged period, plant cells die in zones containing cyanobacterial microcolonies and a coherent cell wall framework forms. This framework surrounds expanded intercellular spaces. As a rule, cyanobacteria occur in modified intercellular spaces and cavities formed after the death of plant cells.

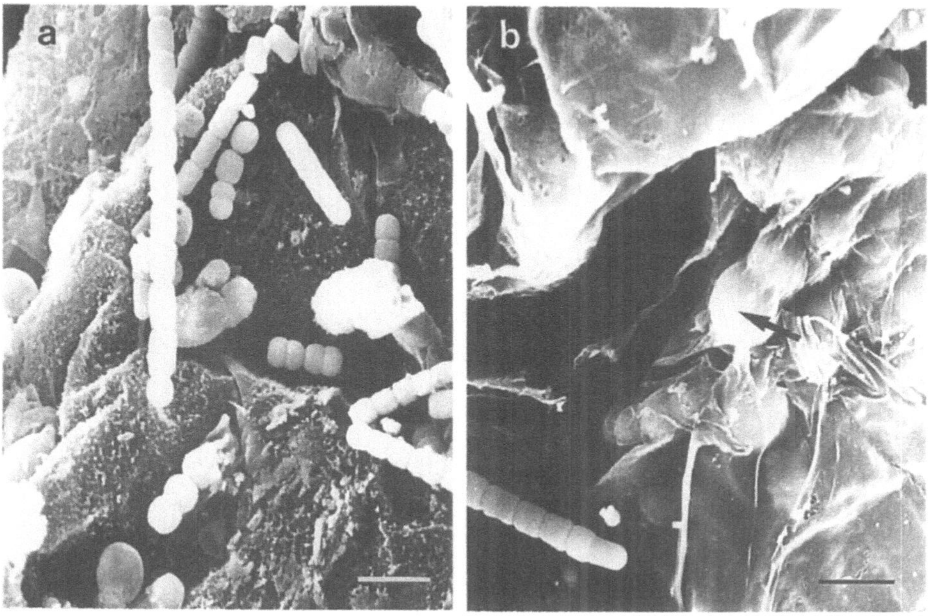

Figure 9. N. muscorum *CALU 304 in expanded intercellular space (a) and a dead plant cell (arrow, b) in zone A of mixed aggregate formed during associative growth of* N. muscorum *CALU 304 and* R. serpentina *callus (bar = 5μm). After Gorelova (2000a).*

Plants. Reconstitution of plant regenerates from mixed calli does not disrupt the spatial integration of the partners. On the contrary, it contributes to the increase in the complexity of the system's anatomical organization associated with the spreading of cyanobacteria to various plant organs and tissues. Along with the formation of intra-tissue harbors, the natural cavities in plant organs (*e.g.*, vessels and air-filled stomatal cavities) and the mucilage-coated cell zones of the root cap are colonized Pivovarova *et al.,* 1986; Baulina *et al.,* 1989; Gusev and Korzhenevskaya, 1990; Korzhenevskaya *et al.,* 1993).

Advanced spatial integration of partners, involving the compartmentation of cyanobacteria in harbors formed by plant cell wall, was also revealed in associations with whole plants. When *N. muscorum* VKM 16 was inoculated during the rooting of alfalfa cuttings on a solid medium, cyanobacterial microcolonies were visible in the leaf mesophyll as dark-green zones (Baulina *et al.,* 1990; Gorelova *et al.,* 1990; Korzhenevskaya *et al.,* 1993). Each microcolony represented a group of cyanobacterial cells with thick multilayer sheaths densely packed in harbors. The growth of microcolonies is accompanied by the expansion of intercellular spaces due to the compression of plant cells that have lost their cytoplasm. In the immediate vicinity of a harbor, degrading plant cells with numerous starch granules are found (Fig. 10).

The plant as a whole remains viable. Thus, the reorganization of the plant tissue with the formation of harbors during growth of cyanobacteria is a uniform feature of artificial associations and involves a complex multifactor process of co-adaptation of the partners.

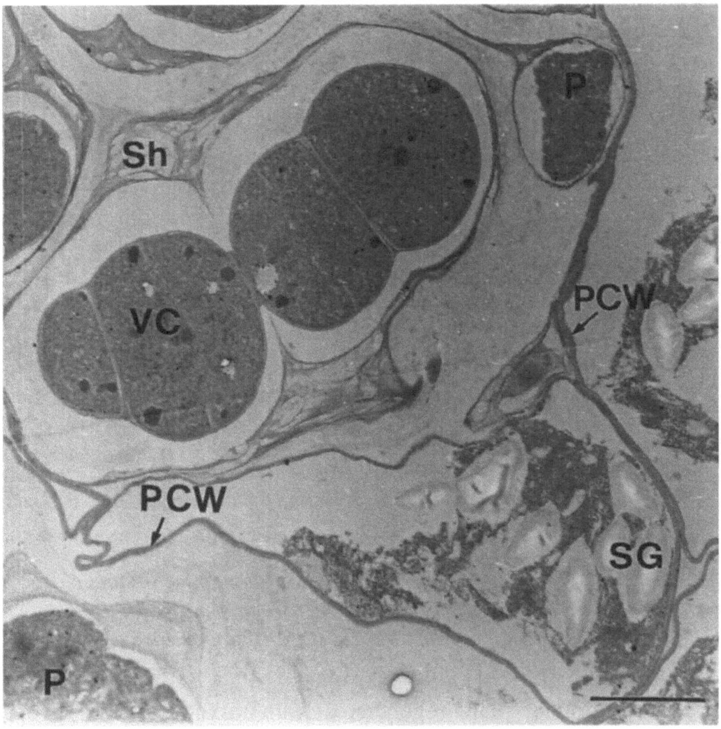

Figure 10. TEM of N. muscorum *VKM 16 inside a harbor of alfalfa leaf mesophyll. P, cyanobacterial protoplast; PCW, plant cell wall; Sh, sheath; SG, starch granules; VC, cyanobacterial vegetative cell (bar = 1 µm). After Korzhenevskaya* et al. *(1993).*

Boundary Zone Between The Partners. A manifestation of the co-adaptive modification of the partners is their involvement in the formation of the boundary zone between cyanobacterial cells *per se* and the metabolically active plant cells. Sheaths of *N. muscorum* VKM 16, in the mesophyll microcolonies of alafalfa leaves, bind together adjacent filaments belonging to distinct morphological and ultrastructural types (see section 3.4). The sheaths often merge with the surface of adjacent plant cell walls (Figs. 10, 11). As a result, the harbor is filled with a mucilaginous fibrillar matrix that, together with the surrounding plant cell wall, forms a continuous boundary-zone between the partners. It is conceivable that low molecular weight products of starch degradation diffusing *via* the boundary zone from the surrounding lysed plant cells, serve as additional substrates for mucilage production by cyanobacteria. In *N.*

muscorum CALU 304 - *R. serpentina* callus associations, only a few individual sheaths (with various morphological features) and low amounts of intercellular mucilage occur at the early stages. However, at later stages an increase in the bulk volume of the structured intercellular matrix occurs and individual sheaths become less detectable (Gorelova, 2000a).

The spatial integration, proceeding in associations at different organization levels of the plant partner, is also often accompanied by intense mucilage formation if surface growth of cyanobacteria occurs (Gusev and Korzhenevskaya, 1990; *Korzhenevskaya et al.,* 1993) . The cells' mucilaginous sheaths, amorphous mucilage, or mantle-like structures can be involved in cell adhesion and insulation. Besides, these mucilaginous structures can function as permeability barriers for plant exometabolites. It should be emphasized, however, that superficial and intra-tissue mucilage formations are not prerequisites for the establishment of stable associations. It is more characteristic of the associations involving representatives of the genera *Nostoc* and *Chlorogloeopsis.*

Mucilage deposits in the cyanobacteria-containing zones of natural symbioses (*e.g.*, the thallus cavities of *B. pussila*) can be also considered a specialized boundary zone between the partners, which performs barrier functions (Gorelova *et al.*, 1996b; Korzhenevskaya *et al.*, 1993). We also detected novel type cyanobacterial cells occurring along the entire length of the mucilaginous zone in the cortical parenchyma of cycad corraloid roots. Their ultrastructure suggests that these cyanobacterial cells specialize in excessive mucilage production (Lobakova and Baulina, 2001). Such cells are prone to destructive changes analogous to those in the mucilage-forming cells of plant root caps (Danilova and Barmicheva, 1980). The cyanobacteria involved lose their cell walls and internal structures, including the ribosomes, but retain their thylakoids.

Transport Function of the Boundary Zone. Polysaccharides in mucilaginous surface structures of cyanobacteria and the pectins of plant cell walls contain uronic acids and are capable of form gels (Baulina *et al.,* 2000a). Obviously, the gel-forming cyanobacterial exopolysaccharides and the pectins of the plant cell walls perform an important (presumably regulatory) function in the intercellular transfer of chemical substances. It has been shown that the pectin matrix is responsible for the permeability of plant cell walls (Baron-Epel *et al.,* 1988), because the gel lattice (i.e., the pores in the pectin matrix) restricts the entry of large macromolecules. The size-exclusion limit of the *Chenopodium album* cell wall was found to be ca 3.4 nm in terms of Stokes' radius (Woehleche and Ehwald, 1995). Using the method of Woehlecke and Ehwald (1995) it was shown that molecules with Stokes' radii of 1.5 to 9 nm, can freely diffuse into the mucilaginous surface structures of cyanobacterial cells in the artificial associations (Baulina *et al.,* 2000b). Production of arabinogalactan proteins (apparently involved in signalling) by the symbiotically compatible *Nostoc* PCC 9229 (Bergman *et al.,* 1996; Bateman *et al.,* 1999) stress the importance of the research on the mechanisms of their release *via* gram-negative bacterial cell walls and possible transport through the mucilaginous matrix in natural symbiotic tissues.

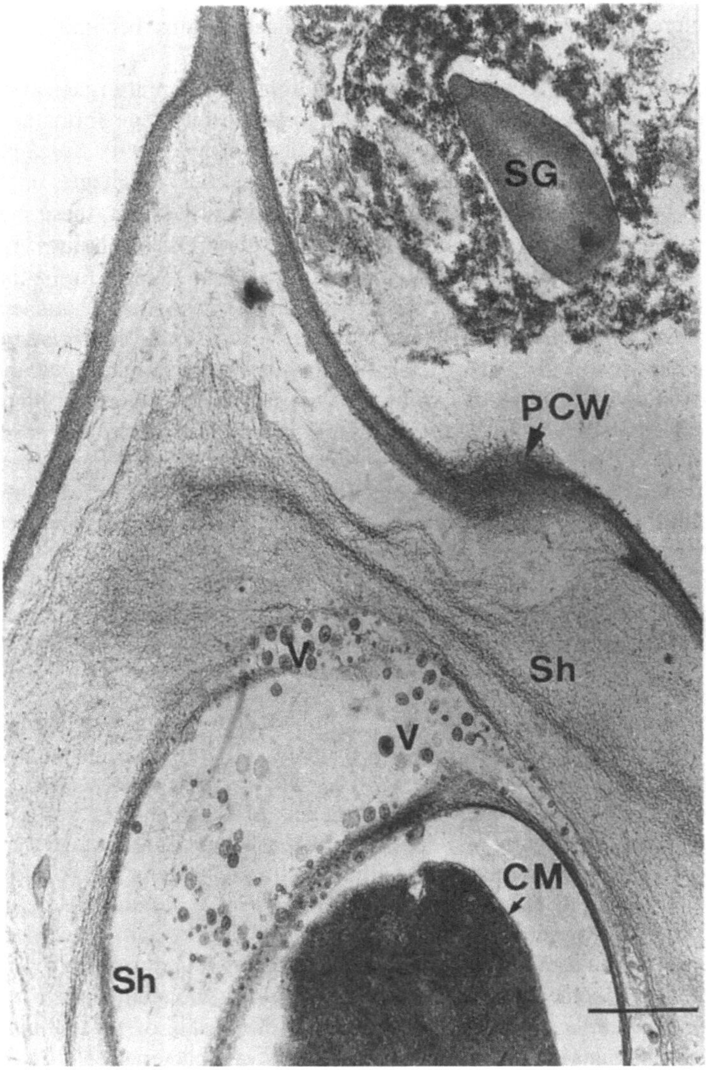

Figure 11. Protoplast of N. muscorum *VKM 16 in alfalfa leaf mesophyll. CM, cytoplasmic membrane; PCW, plant cell wall; Sh, sheath; SG, starch granules; V, vesicles (bar = 0.5μm). After Korzhenevskaya* et al. *(1993).*

The vesicle-dependent transfer of macromolecule is an alternative free diffusion across the gel matrix. Numerous vesicles with electron-dense content (Fig. 11) detach vesicles may participate in transporting sheath components and/or macromolecular exometabolites into plant cells via the boundary zone. Analogous patterns of vesicle formation by cyanobacteria was also seen in other artificial associations and natural symbioses. Importantly, vesicle exocytosis is predominantly carried out by protoplasts and spheroplasts that form part of the plant-associated cyanobacterial population (see section 3.4). The suggestion that vesicle-dependent transport occurs in natural symbioses had been also made in earlier works (Peveling, 1971).

3.2.3. Characterization of AMC-Plant Callus Tissue Interactions

P-nodules were shown to form on the roots of plants in artificial associations of rice, nightshade (*S. dulcamara* and *S. nigrum*) or tobacco (cv. Samsun and albino mutant *P1*) plants with AMC isolated from *Azolla* sp. and *A. pinnata* (see section 2.3.1). Cyanobionts of AMC also colonized the aerial plant organs. The patterns of distribution of AMC in the *p*-nodules of all tested plant species were similar: cyanobacteria inhabited the intercellular spaces in the peripheral parts of *p*-nodules, and actively dividing SB cells settled in their center. Thus, stable microsymbiont complexes dissociated in the presence of de-differentiated plant tissues.

The AMC-callus culture interaction was characterized by a specific effect of the plant tissues on the growth of AMC and its components. In mixed culture containing a callus, AMC components separated and experienced differential selective pressures. Depending on the species/strain of the plant callus involved, the mixed culture displayed (i) predominant growth of one AMC component or (ii) an oscillatory pattern of development of all AMC components (Fig.12).

As we stated above (see section 3.2.2.), AMC cyanobacteria dispersed throughout the whole thickness of the plant tissue in associations containing the two nightshade species or tobacco *P1* callus cultures. This resulted in the formation of a homogeneous emerald-green mixed callus. SB growth was not visible. However, solitary SB cells in contact with the surface of cyanobacterial cells were detected during scanning. No growth of AMC components occurred on the cultivation medium away from the plant tissues.

After inoculating a rice callus with AMC, cyanobacteria grew during the first 20 days in fissures on the callus surface. The cyanobacteria formed opaque, intensely colored grainy colonies. Cyanobacteria did not enter the interior of the callus, and their colonies were easy to separate from the callus surface. The solitary, originally opaque, cyanobacterial colonies gradually became glossy, because large amounts of transparent mucilage formed on their surface. Intense mucilage formation was accompanied by gradual "bleaching" of the colonies. Subsequently, orange opaque colonies formed on the mucilage surface. The pattern of AMC growth in combination with rice callus tissue was analogous to that in monoculture on solidified mineral medium. In both systems, each of the AMC component (Lobakova *et al.*, 2001a) grew alternately. In this association, intense SB growth also occurred on the medium surface in the immediate vicinity of the rice callus.

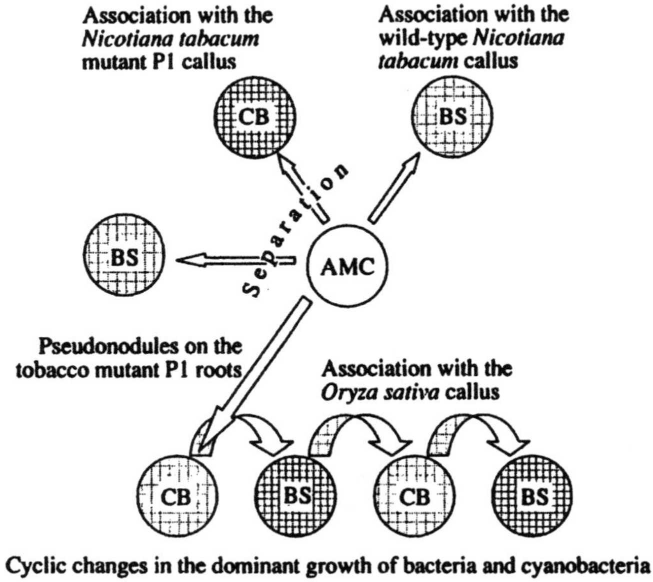

Figure 12. Interaction pattern of AMC and its components with plant tissues. CB, cyanobacteria; BS, satellite bacteria. After Lobakova et al. *(2001c).*

Inoculating AMC onto the callus tissue of tobacco (*N. tabacum* cv. Samsun) failed to yield a mixed subculturable callus. Growth of separate cyanobacterial colonies occurred only during the first 14 days of mixed cultivation. Cyanobacteria "bleached" and disappeared subsequently. On day 18—20 of the callus' growth, smooth glossy SB colonies appeared on its surface. Intense SB growth, both on the callus surface and in the cultivation medium, resulted in rapid degradation of the callus.

Different patterns of tobacco, nightshade, and rice colonization by AMC cyanobacteria are possibly related to the morphological peculiarities of the plant tissues, the cell size, and intercellular spaces. Our studies also revealed differences between the interactions of the AMC components with *N. tabacum* cv. Samsun and albino mutant *P1* of the tobacco callus culture. These cultures were similar in their appearance and morphology, but the patterns of the AMC-plant tissue interactions were quite different. It is known that not only breeds, or strains but even the differrent cells obtained from a single plant may prove to be quite dissimilar (Butenko, 1999). The spreading of an AMC in plant tissue cultures is conditional on the morplogical and metabolic characteristics of the callus cultures and, therefore, strain- (not merely species-) specific.

The differences between the patterns of growth (balanced growth; oscillatory pattern; predominance of one component) of AMC cyanobacteria and SB in mixed cultures with plant calli can be used for dissociating stable multicomponent natural symbiotic complexes and for selection of their components. Long-term subcultivation of associative calli (tobacco mutant *P1*or nightshade species in combination with AMC) provides researchers with a promising experimental system, enabling them to maintain the symbiotic competence of microbionts under laboratory conditions.

3.3. Regulation of Nitrogen Assimilation in Cyanobacteria: Influence of Plant Partners

Cyanobacterial partners provide fixed-N to their hosts, and undergo morpho-physiological modifications in natural as well as in artifical symbioses (Lee and Joseph, 1988; Meeks *et al.,* 1988, 1999; Meeks, 1998; Gusev and Korzhenevskaya, 1990; Rai, 1990; Korzhenevskaya *et al.,* 1993, 1999a; Bergman *et al.,* 1996; Rai *et al.,* 2000b). Such modifications include increased heterocyst frequency, change in activities of key enzymes of nitrogen metabolism (*e.g.*, nitrogenase and glutamine synthase), and release of ammonia or organic nitrogen. The reserve carbon and nitrogen containing cytoplasmic inclusions (glycogen and cyanophycin granules) may serve as indicators of the physiological state of cyanobacteria and the probable direction of cell differentiation.

3.3.1. Accumulation of Reserve Polymers by Cyanobacterial Cells

Glycogen granules (α–granules, or polyglucose bodies) represents the main carbohydrate reserves in cyanobacteria. Increased accumulation of α–granules occurs under photoheterotrophic growth conditions (Baulina *et al.,* 1981) and nitrogen starvation (Gromov, 1986; Jensen, 1993). Cyanophycin (multi-L-arginyl-poly-L-aspartic acid) is a unique reserve polymer, found only in cyanobacteria, which along with phycobiliproteins plays a role of endogenous source of combined nitrogen. Sometimes it also serves as a source for carbon and energy. Its biosynthesis does not require ribosomes and is catalyzed by cyanophycin synthetase. Degradation of cyanophycin is performed by the sequential action of cyanophycinase and isoaspartyl dipeptidase (Simon, 1973, 1976; Richter *et al.,* 1999; Berg *et al.,* 2000). The peptide is deposited in the cytoplasm in the form of cyanophycin or structured granules (CG). Usually, the content of CG is 0.8-3.0% of cell volume, but it may reach up to 18-25% under some growth conditions (Jensen, 1993). Cyanophycin is known not to dissolve in water and organic solvents. It is resistant to a wide range of endoproteases and exopeptidases. We have shown previously that upon degradation of cells the CG are preserved in the debris of destroyed cell structures. The amount of cyanophycin in dead cells remains equal to that in live cells of the same cyanobacterial population (Gorelova, 2001a).

The dynamics of reserve polymers accumulation in cyanobacteria during interaction with plant partners is of special interest. As in natural symbioses, heterocyst differentiation and nitrogenase activity also increase in artificial associations Gusev *et al.,* 1986; Korzhenevskaya *et al.*, 1989, 1993, 1999a; Gusev and Korzhenevskaya, 1990; Gorelova and artamonova, 1992; Gorelova, 1998). We calculated the amount and size of CG and computed the volume of cyanophycin in individual vegetative cells of *N. muscorum* CALU 304 (Gorelova, 1998; Korzhenevskaya *et al.,* 1999a; Gorelova and Korzhenevskaya, 2001a). The pure cyanobacterial cultures grown on BG - 11 medium with nitrogen (Stanier *et al.,* 1971) and nitrogen-free Allen-Arnon medium (Allen and Arnon, 1955) were studied, along with cells grown in mono- and mixed cultures with calli of *R. serpentina* and *S. dulcamara* on plant tissue culture media (R and MS_m medium, respectively) that contained 5 times more nitrogen (mainly as nitrate) than BG-11. Duration of mixed cultivation was 10 weeks with *R. serpentina* and 99 weeks with *S. dulcamara*. In both cultures, partner's interaction was followed by their spatial and metabolic integration leading to the formation of stable model associations. The detailed conditions for cultivation in mixed cultures, and some of the structural and physiological traits of cyanobacteria in such cultures, have been described by us earlier (Gorelova and Artamonova, 1992; Gorelova *et al.*, 1995, 1999, 2001b; Gorelova, 1998, 2000a; Korzhenevskaya *et al.*, 1999a).

Table 4. Cyanophycin granules and glycogen accumulation by vegetative cells of N. muscorum *CALU 304 in pure and monocultures*

Variant		Cyanophycin granules		Glycogen
Culture, medium	Incubation time, weeks	Average number per section	Diameter, nm[a]	
Pure culture,				
BG-11 medium	3	1.00	269±14	
AA medium	4	0.56	225±47	+
Monoculture,				
R medium	2	3.63	598±27	
	8	3.50	583±23	
Monoculture,				
MS_m medium	11	2.57	697±36	

[a] mean value and error of mean

Accumulation of Reserve Polymers in N. muscorum CALU 304 in Pure- and Mono-cultures. Pure cultures of *N. muscorum* CALU 304 generally consists of cell chains and smaller amount of standalone cells at early stationary phase. Incubation on nitrogen-free AA medium leads to heterocyst differentiation (4-7% of the total cells). Cellular inclusions are scarce in heterocysts. When cultivated on nitrogen-containing BG-11 medium, heterocyst differentiation does not occur. Vegetative cells contain CG (73-460 nm) but there is little or no glycogen reserves. In contrast, glycogen reserves are always present in cells grown on nitrogen-free medium, but CG are fewer and smaller (Table 4). Thus, the composition of the incubation medium affects the intracellular balance of carbon and nitrogen and thereby the corresponding reserve polymers in *N. muscorum* CALU 304 cells. This is also true for other cyanobacteria (Gromov, 1986; Dembinska and Allen, 1988; Jensen, 1993).

When transferred from liquid BG-11 medium onto the agar media for cultivating *R. serpentina* and *S. dulcamara* calli, *N. muscorum* CALU 304 remained in the form of amorphous filmy aggregates without any change in size for 2-2.5 months and subsequently died. The cyanobacterium was mainly represented by short chains of 3 to 5 cells and stand alone cells. The ultrastructure of the vegetative cells was similar to that of stationary phase cultures on BG-11 medium. The main difference was comparatively high amount CG and little or no glycogen granules. On average the number of CG per cell increased 2.6-3.6 times, while the diameter of CG increased 2.2-2.6 (Table 4).

Carbon containing inclusions that could be found were lipids (β–granules) and round granules localized mostly in cell's central part. The latter resembled poly-β–hydroxybutyrate (PHB) granules by their size, shape, electron density and limiting monolayer membrane. However, they contain fine-grained electron-dense impregnations implying heterogeneity of their chemical composition.

Media for cultivation of plant tissues contain 5 times more mineral nitrogen than BG-11 medium, and 10 to 35 times more sucrose, than required for photoheterotrophic growth of *Nostoc* cyanobionts (Vagnoli *et al.,* 1992). Increased CG content, lack of α–granules, and modest amounts of other carbon containing inclusions (PHB-like and β–granules) indicate a shift in intracellular N:C balance towards nitrogen. Such a shift may be caused by consumption of the excess exogenous nitrogen by the cyanobacterium but its inability to metabolize high concentrations of exogenous sucrose.

Accumulation of Reserve Polymers by N. muscorum CALU 304 in Mixed Caltues. In mixed cultures with *R. serpentina* callus, as in monocultures, cyanobacteria consumed exogenous N and accumulated cyanophycin (Table 5) during early stages of growth. After 2 weeks of incubation, the volume of the CG increased by 28.5 and 16.2 times in cyanobacterial cells localized on the callus surface and those away from the plant tissue, respectively (Fig. 13). By the end of sixth week of associated growth, the ultrastructure of cyanobacteria changed in a way that indicated a switch to photoheterotrophism. Mass deposits of glycogen and numerous carboxysomes appeared in vegetative cells, but PHB-like and β–granules were practically absent. The increased glycogen accumulation apparently indicated that *N. muscorum* CALU 304 is well-supplied with carbon. The amount and volume of CG decreased approximately four-fold but remained higher than those in vegetative cells of pure cultures. Continued cultivation of the association led to further decrease in carbon containing inclusions and cyanophycin (Table 5 and Fig. 13).

Table 5. Cyanophycin granules and glycogen accumulation by vegetative cells of N. muscorum *CALU 304 in mixed cultures*

Variant		Cyanophycin granules		Glycogen
Partner	Incubation time (weeks)	Average number per section	Diameter[e], nm	
R. serpentina	2	5.89	453±17	-
	6_a	1.21	526±36	+
	6_b	1.71	398±22	+
	8	0.50	367±27	+/-
R. serpentina (out of contact)	2	4.17	421±25	+/-
	10	0.83	286±23	+
S. dulcamara	11	1.69	368±24	+/-
	45_c	0.46	323±16	+/-
	45_d	0.26	138±24	+/-
	99	0.67	233±27	+

[a] young surface colony of mixed aggregate
[b] mature surface colony of mixed aggregate
[c] on callus surface
[d] inside the callus
[e] mean value and error of mean

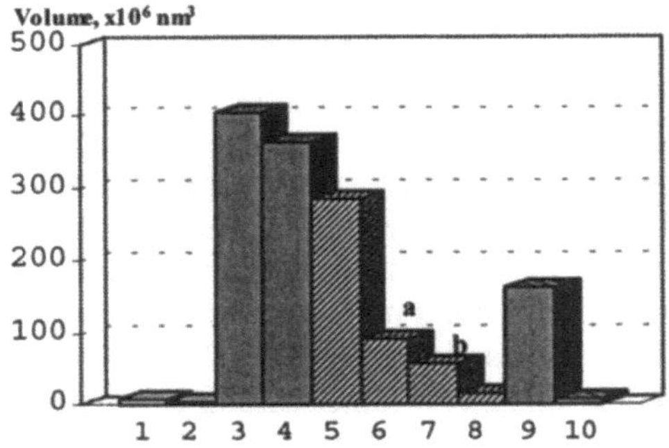

Figure 13. Cyanophycin volume per vegetative cell of N. muscorum *CALU 304: 1, 2, at the stationary growth phase in pure culture (cultivation time: 3 and 4 weeks on the BG-11 and AA medium, respectively); 3, 4, in monoculture on the R medium (cultivation time: 2 and 8 weeks, respectively); 5-8, in mixed culture containing* R. serpentina, *the partners were spatially integrated (cultivation time: 2, 6, and 8 weeks, respectively); 9, 10, in the same mixed culture but out of contact with the plant tissue (2 and 10 weeks, respectively); "a" and "b", see Table 5. After Korzhenevskaya* et al. *(1999a).*

Cyanophycin degradation was substantially slower in the cyanobacterium, especially on late cultivation stages, grown in mixed cultures with *S. dulcamara* callus (Gorelova and Korzhenevskaya, 2001a). The cyanobacterial cells contained of CG of similar sizes (approximately, 0.01μm^3) after 8-10 weeks in association with *R. serpentina*, and 10 months of growing in association with *S. dulcamara* without subcultivation (Table 5 and Fig. 14).

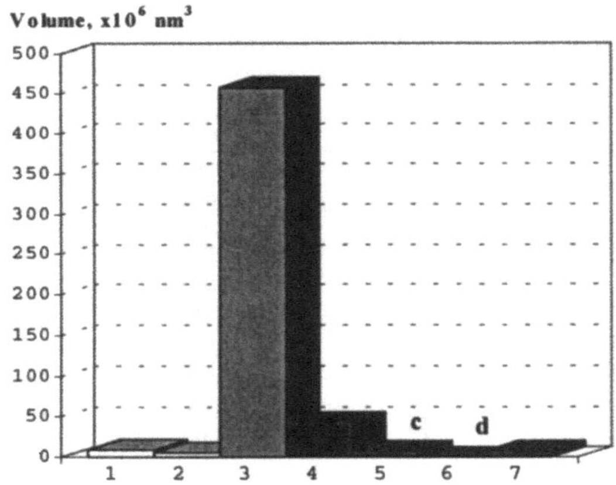

Figure 14. Cyanophycin volume per vegetative cell of N. muscorum *CALU 304: 1, 2, at the stationary growth phase in pure culture (cultivation time: 3 and 4 weeks on the BG-11 and AA medium, respectively); 3, in monoculture on the MS_m medium (cultivation time: 11 weeks); 4-7, in mixed culture containing* S. dulcamara, *the partners were spatially integrated (cultivation time: 11, 45, and 99 weeks, respectively); "c" and "d", see Table 5.*

3.3.2. Effect of Plant Partner on Nitrogen Metabolism of the Cyanobacterium

Since *N. muscorum* CALU 304 has similar characteristics when grown in monoculture either on medium R for *R. serpentina* callus or medium MS_m for *S. dulcamara* callus growth, the difference in accumulation of reserve polymers and cell differentiation in mixed cultures must be due to the influence of plant partner. In this connection, the following facts attract attention:

1. During early stages (2 weeks) of cultivation, cyanobacteria localized on the surface of *R. serpentina* calli got their nutrients through plant tissue and contained 76% more cyanophycin than cyanobacteria growing on the medium surface without direct contact with the callus (Table 5 and Fig. 13). The increase in cyanophycin content occurred due to increased *de novo* synthesis because of the increased availablity of combined nitrogen (Dembinska and Allen, 1988). Thus, *R. serpentina* calli, when their growth is not nitrogen-limited, stimulate the uptake of exogenous combined nitrogen and its accumulation in the form of cyanophycin by *N. muscorum* CALU 304 (Gorelova, 1998, 2001a; Korzhenevskaya *et al.*, 1999a).
2. Heterocyst differentiation occurred (Table 6) in mixed culture with *R. serpentina* after 5-6 weeks of associated growth (Gorelova and Artamonova, 1992; Gorelova, 1998; Korzhenevskaya *et al.*, 1999a). Vegetative cells of *N. muscorum* CALU 304 showed features of heterotrophy, and contained 10-15 and 6-9 times more cyanophycin than pure cultures grown on BG-11 and AA medium, respectively. Heterocyst frequency was 2% to 5% and nitrogenase activity was apparent only in mixed cultures grown in the light, although more heterocysts formed in the dark (Gorelova, 1998). The specific activity of nitrogenase in mixed aggregates was lower than in natural symbioses (Lindblad and Bergman, 1986, 1990; Söderbäck *et al.*, 1990) and in artificial associations of *A. variabilis* ATCC 29413 with tobacco regenerates (Gusev *et al.*, 1986). However, all the above mentioned natural and artificial symbioses, and cyanobacteria isolated from them, were grown under N-free conditions. In contrast, the mixed culture of *R. serpentina* and *N. muscorum* CALU 304 was grown in presence of high levels of combined-N. In mixed culture with *S. dulcamara*, heterocysts differentiated only when cyanophycin levels decreased by 50% below their level in pure culture of *N. muscorum* CALU 304 on AA medium (Gorelova and Korzhenevskaya, 2001a).

Table 6. Differentiation of heterocysts and nitrogenase activity of N. muscorum *CALU 304*

Culture Medium	Heterocyst Frequency, %		Nitrogenase Specific Activity $\mu mol\ C_2H_4\ h^{-1}\ mg\ chl\ a^{-1}$	
	Light	Dark	Light	Dark
[a] Pure, BG-11 medium	0	0	0	0
[a] Monoculture, R medium	0	0	0	0
[a] Mixed culture with *R. serpentina* (in aggregate), R medium	3.2-3.5	4.7-5.0	2.07±0.44	0
[b] Mixed culture with *R. serpentina* (out of contact), R medium	2.0-1.5	-	1.27±0.26	-

[a] 45 day of incubation
[b] 58 day of incubation

3. Preliminary analysis of the dynamics of cyanophycin expenditure, prior to the heterocyst differentiation, in *N. muscorum* CALU 304 cells shows that during the interactions with *S. dulcamara* and *R. serpentina* the alteration in cyanophycin volume *(V)* can be described by the single equation: $V=a\cdot t^{-1}$ (where *t* is incubation time and *a* is coefficient for both systems). Induction of differentiation of heterocysts and nitrogenase expression coincides with a two-fold increase in specific rate of cyanophycin degradation [O.A. Gorelova and S.Yu. Kleimenov, unpublished] and decrease in carbon reserves (Korzhenevskaya *et al.*, 1999a).

Thus, it seems that the usual system of nitrogen assimilation in *N. muscorum* CALU 304 is altered in the presence of *R. serpentina* callus. Apparently, the plant tissue not only affects the activity of highly specific enzymes of cyanophycin synthesis and degradation, but possibly exerts overall control on nitrogen metabolism of *N. muscorum* CALU 304 and replaces cyanobacterial intra-filamentous nitrogen deprivation sensing-signaling pathway with its own extracellular or inter-partner pathway in the initiation cascade for heterocyst differentiation, maturation and spacing. Formation of multiple contiguous heterocysts is consistent with this.

The possibility of existence of a symbiotic regulatory system activated by products of host plants to control nitrogen metabolism of microsymbionts has been suggested in case of lichens (Rai *et al.*, 1980, 2000b) and natural symbioses of cyanobacteria with plants (Meeks, 1990, 1998; Campbell and Meeks, 1998; Bergman *et al.*, 1996; Meeks *et al.*, 1999). Our data fully conform to a working model of the initiation cascade for heterocyst differentiation in symbiosis proposed by Meeks *et al.* (1999). This model presumes the possibility of symbiosis specific signal influence upon the activity of *ntcA* gene. The *ntcA* gene activity affects not only the early phase in heterocyst differentiation, but also the later phases of heterocyst maturation and *nif*HDK expression (Flores *et al.*, 1999; Meeks *et al.*, 1999; Wolk *et al.*, 1999). Moreover, the enhanced assimilation of exogenous combined nitrogen by *N. muscorum* CALU 304 during early stages of interaction with *R. serpentina*, can also be explained by the influence of plant partner on *ntcA* expression or the activity of its product. Tthe NtcA-dependent activation of the expression of cyanobacterial nitrogen-assimilation genes also includes the regulation of nitrate assimilation (Flores *et al.*, 1999; Meeks *et al.*, 1999).

3.4. Structural Modification of Associated Cyanobacteria

The establishment of artificial associations and their functioning are accompanied by reciprocal co-adaptation of the partners. Because cyanobacteria (prokaryotes) exhibit comparatively higher plasticity, it is they that undergo structural-functional changes, which can be very substantial in some cases. Cyanobacteria use diverse options with respect to cell differentiation: they may form hormogonia, heterocysts, or, less frequently, akinetes. Besides, some vegetative cells undergo morphological changes involving cell wall modifications. There is good reason to suggest that these changes play an important role in the ontogeny of stable associations. For example, pronounced cell heteromorphism occurs in the microsymbionts of long-lasting associations between cyanobacteria and plants. In addition, the transition of cyanobacteria to the symbiotic state in most natural symbioses is accompanied by chenges in their morphology: filaments tend to disappear, single cells become common, cell size increases, cell shape and division patterns change, and the cell wall becomes thinner (Rai, 1990; Meeks, 1998; Rai *et al.*, 2000b). More drastic changes in cell surface structures that may result in cell wall reduction were found in the cyanobionts of the lichen *P. polydactyla* (Peat, 1968), *G. kaalensis* (Towata, 1985), the liverwort *B. pusilla* (Gorelova *et al.*, 1992, 1996b; Korzhenevskaya *et al.*, 1993; Baulina *et al.*, 2000a), and various species of cycads (Grilli-Caiola, 1980; Grobbellaar *et al.*, 1988; Baulina *et al.*, 2000a; Lobakova *et al.*, 2000; Lobakova and Baulina, 2001). In addition, a decrease in the growth rate is characteristic of morphologically changed cyanobacteria in natural symbioses. Nevertheless, no data have yet been presented on the mechanisms involved in control of the microsymbiont growth. The factors responsible for cell heteromorphism and its functional role(s) also remain obscure.

The experiments using artificial associations are expected to improve our understanding of the factors, mechanisms, and biological implications of the structural variability of cyanobacterial cells during their interaction with plant tissues.

3.4.1. Heteromorphic Cell Forms

The data accumulated up to now concerning the morphology of cyanobacteria in artificial associations have made it possible to compile a general list of the morphological cell types classified by us into the heteromorphic group:

(1) deviant cell size (relative to the species-specific size parameters), but retaining rigid cell walls; very small or gigantic forms may occur;

(2) anomalous division patterns;

(3) irregular shapes with incomplete cell division;

(4) defective cell wall (the peptidoglycan layer considerably disrupted or reduced). Both very small and gigantic forms may occur, including spheroplasts and protoplasts. Protoplasts and spheroplasts will be referred to as the forms with reduced cell walls (FRCW) (Fig. 15).

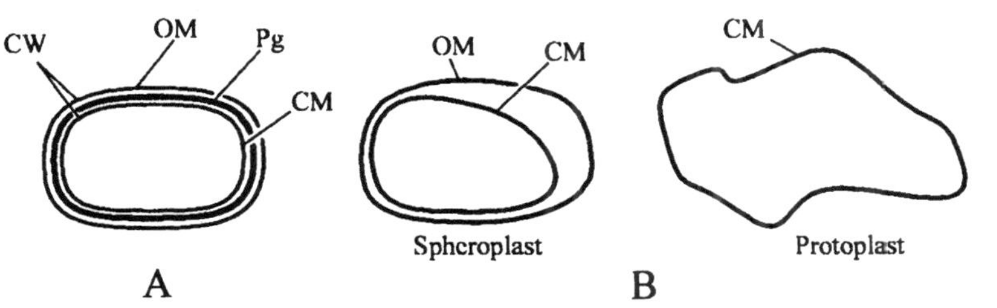

Figure 15. Scheme of the cell surface organization of intact vegetative cell (A) and the forms with reduced cell walls (B). CW, cell wall; OM, outer membrane; Pg, peptidoglycan; CM cytoplasmic membrane.

3.4.2. Heteromorphism in Associated Cyanobacteria

In the mixed suspension cultures of *Synechococcus* sp. PCC 6301 and the tobacco cells (*N. tabacum cv.* Winconsin-38) (Baulina *et al.*, 1988, 1994) the cyanobacterial cells become elongated, particularly the cells of the biotin-dependent mutant of *Synechococcus*. Several nucleoid zones are seen in elongated cells, suggesting unbalanced growth and division processes in them. The cell surfaces become uneven and wavy. The peptidoglycan layer also becomes uneven in thickness (Fig. 16). Long-term maintenance of the mixed culture was impossible. This could be due to the low structural and metabolic plasiticity of *Synechococcus*. This obilgate phototroph is incapable of cell differentiation and displays the most conservative metabolism type in contrast to other cyanobacteria (it is incapable of photoheterotrophic or chemoheterotrophic growth and does not fix nitrogen). However, unicellular cyanobacteria do occur in natural symbioses (Schenk, 1992; Rai *et al.*, 2000b). Nevertheless, a majority of the cyanobacteria that form symbiotic systems with plants belong to the genus *Nostoc*.

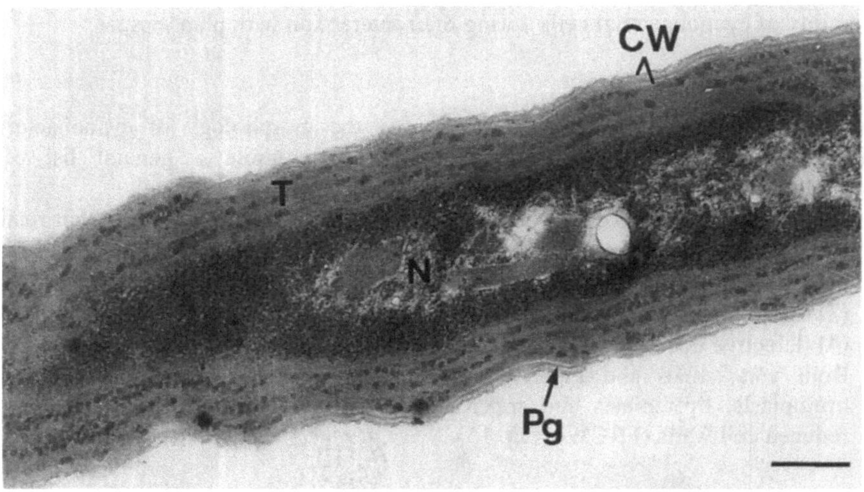

Figure 16. The ultrastructure of Synechococcus *sp. PCC 6301 (mutant strain bio-2 N1142) inside an aggregate of tobacco cells in the mixed suspension culture. CW, cell wall; N, nucleoid; Pg, peptidoglycan; T, thylakoid (bar = 0.2μm). After Baulina* et al *(1994).*

Morphological and ultrastructural changes were observed during the long-term cultivation of the *N. muscorum* VKM 16 in association with alfalfa. Development of the association included the stages of callusogenesis induction in the presence of cyanobacteria, subcultivation of mixed callus, organogenesis, and plant regeneration (Gorelova *et al.*, 1988, 1990; Baulina *et al.*, 1990; Korzhenevskaya *et al.*, 1993). In addition to the ordinary vegetative cells, all the heteromorphic cell forms listed above occurred in the population. The frequency of forms with defective cell wall (FDCW) varied depending on the stage of growth and population development. Protoplasts dominated among the FDCW at some of the stages. Cyanobacterial heteromorphism also manifested itself in monocultures, but the ratio between various morphological types and the time of their formation differed as compared with associative systems. Destructive processes resulting in monoculture death prevailed in the population.

We also used *A. variabilis* ATCC 29413 in similar experiments. Although heteromorphism did occur in the population of this cyanobacterium during its interaction with the alfalfa tissue, no protoplasts/spheroplasts formed (Gorelova *et al.*, 1988). Some of the cells increased in size and appeared like akinetes of *A. cf. flos-aque* (Fig. 17; Stulp and Stam, 1985). Moreover, the ultrathin sections of such modified cells demonstrated their organizational difference from typical cyanobacterial akinetes. In particular, they lacked their characteristic envelopes and granules of reserve polymers. Thus, the structural variabiliy of different cyanobacterial species in associations with a plant species can differ.

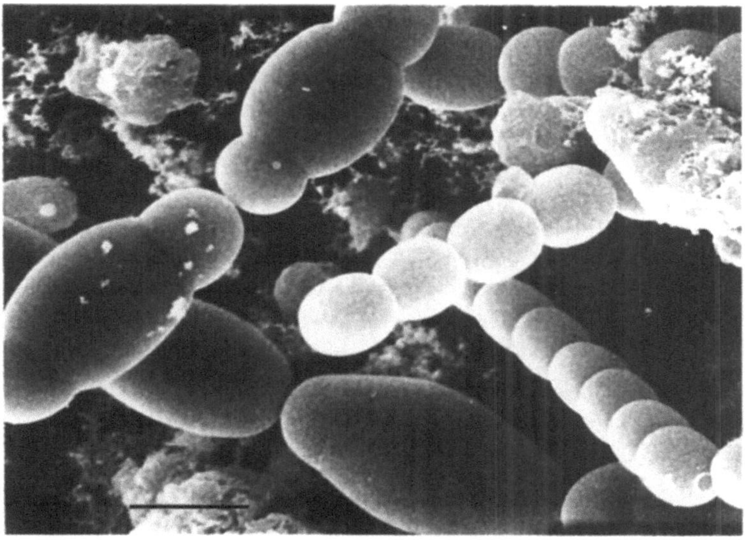

Figure 17. Cell heteromorphism in A.variabilis *ATCC 29413 grown in mixed culture with alfalfa callus (bar = 1 μm).*

The occurrence of short filaments, unicellular and anomalously dividing cells and FDCW including protoplasts is a characteristic feature of *N. muscorum* VKM 16 micro-colonies in the mesophyll harbors of alfalfa leaves (see section 3.2.2) (Figs. 10,11) (Baulina *et al.*, 1990; Gorelova *et al.*, 1990; Korzhenevskaya *et al.*, 1993). In addition, heterocysts were detected during cultivation on nitrogen-free media.

All the above heteromorphic forms also appeared during the growth of the *N. muscorum* CALU 304-*R. serpentina* mixed cultures, even in cyanobacterial populations on the agar medium which were not in direct contact with the plant callus (Gorelova,

2000a,b). They are particularly abundant and multifarious in the central microcolony (zone A, see section 3.2.2) at the later stages of growth. A peculiar group comprises gigantic cell forms (GCF) including gigantic cells (GC) with rigid cell walls (Fig. 18), GCF with a modified peptidoglycan structure, and gigantic spheroplasts (Fig. 19) (Gorelova and Korzhenevskaya, 2000a, 2001b; Gorelova, 2001b). GCF invariably appeared after 6-7 weeks of mixed cultivation. In contrast to other heteromorphic forms, GCF formed only in mixed cultures (Gorelova, 2000a). The GC is almost spherical and has an average size of 14.3±0.8 μm x 12.8±0.7 μm, although larger cells with a diameter of 16.5-24.5 μm were also detected. Gigantic spheroplasts often have an irregular amoeboid shape, and their size somewhat exceeds that of GC. However, the similarity of the internal ultrastructure and the occurrence of intermediate GCF with varying degrees of integrity of the murein layer suggest possible formation of protoplasts from GC or interconversion of these two forms.

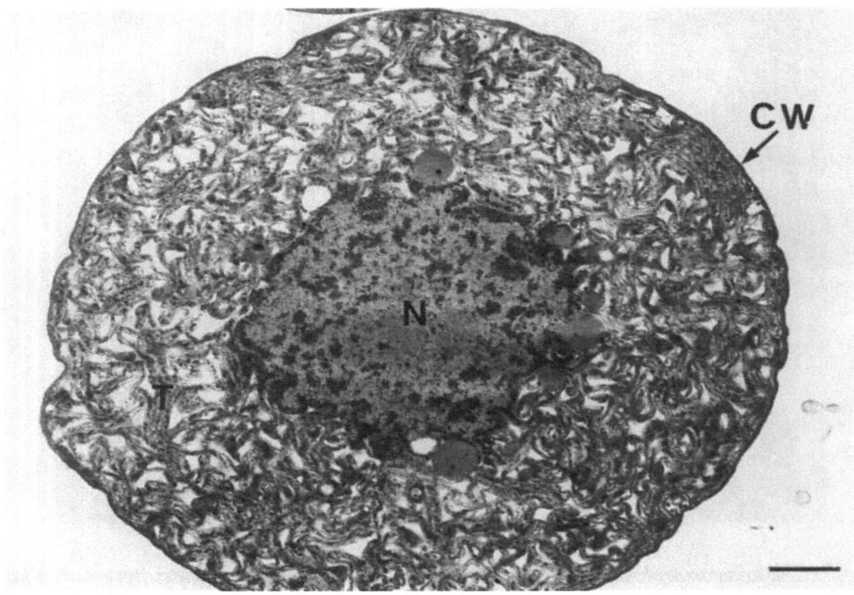

Figure 18. Gigantic cell of N. muscorum *CALU 304 in mixed culture with* R. serpentina. *CW, cell wall; N, nucleoid; T, thylakoid (bar = 1 μm). After Gorelova and Korzhenevskaya (2001b).*

Cytoplasmic membranes of GCF form deep invaginations and occasionally lomosome-like structures, leading to an increase in surface area that would promote their interactions with the environment. The intra-cytoplasmic membrane system of GCF includes a large number of curved thylakoids that form a "lacy" network or are arranged in concentric circles and parallel rows. The intra-thylakoid spaces are expanded, indicating low photosynthetic capacity.

The central GCF part is occupied by several nucleoid zones or a large single zone with sharp boundaries, and ribosome agglomerations that are predominantly located on the periphery. The nucleoid is several times larger than that in ordinary cells. Apparently, cytokinesis lags behind replication processes in GCF. The latter probably contain "excess" hereditary substances that can be potentially sufficient for the formation of several cells. The fact that large agglomerations of ribosomes and polysome figures simultaneously occur in a cell seems to indicate an enhancement in the processes of transcription and translation.

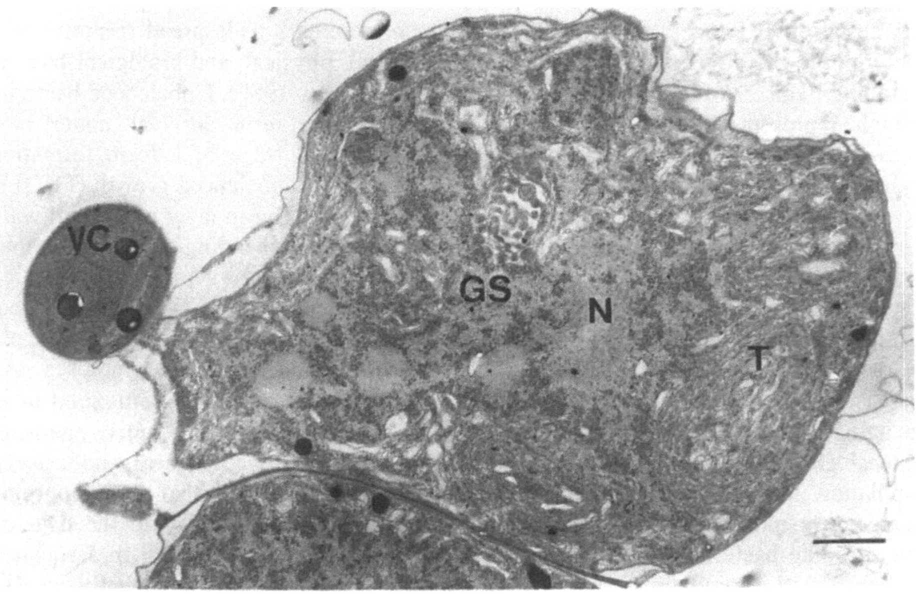

Figure 19. Gigantic spheroplast (GS) in contact with a vegetative cell (VC) of N. muscorum *CALU 304 in mixed culture with* R. serpentina. *N, nucleoid; T, thylakoid (bar = 1 µm). After Gorelova and Korzhenevskaya, 2001b).*

Various GCF are also formed in *N. muscorum* CALU 304 during mixed culture with *S. dulcamara* calli. Evidence of the existence of gigantic *Nostoc* cells in aging vacuolated cells of *G. manicata* was presented in a work by Johansson (1994). It was suggested that cell division and growth processes can revert to normal in GC.

3.4.3. Heteromorphism as a Manifestation of the Unbalanced Growth

Most researchers dealing with natural cyanobacterial symbioses, interpret cells with extraordinary shapes as degenerate because their number increases as symbiotic tissues age and they exhibit structural peculiarities (Ducket *et al.*, 1977; Obukowicz *et al.*, 1981; Towata, 1985; Söderbäck *et al.*, 1990; Johansson and Bergman, 1992; Johansson, 1994). *Nostoc* cells of peculiar shapes (with curved cell walls) prevail at the later stages of infection in *G. magellanica* stem glands (Söderbäck *et al.*, 1990). These cells have an electron-dense cytoplasm with numerous densely packed membranes, but they lack reserve granules. However, Towata (1985) found a large number of osmiophilic irregular protoplast-like *Nostoc* cells in the apical parts (young symbiotic tissues) of *G. kaalensis*. The emergence of curved cyanobacterial cells in narrow intercellular spaces at the initial stages of infection was also observed in *G. chilensis* (Ducket *et al.*, 1977). We detected heteromorphic cell forms including FDCW in growing and functionally active cyanobiont populations of various natural and artificial associations (Korzhenevskaya *et al.*, 1993; Gorelova *et al.*, 1996b; Baulina *et al.*, 2000a; Gorelova, 2000a, 2001a; Lobakova *et al.*, 2000).

In our opinion, the above heteromorphic forms in cyanobiont population are a manifestation of unbalanced growth. The latter represents a widespread response of a bacterium to the effects of a wide variety of chemical, physical, and biological factors, including inducers of L-transformation (Prozorovsky *et al.*, 1987). Unbalanced bacterial growth represents a unique adaptation aimed at short-term survival under new environmental conditions; it can also represent an initial stage of L-form formation under certain circumstances. The formation of forms with unbalanced growth (FUG) is always accompanied by structural disruption of the peptidoglycan layer of the cell wall. However, the pattern and extent of these changes may vary depending on the deleterious factor and the cell's physiological state. FUG are characterized by anomalous division resulting in the formation of both viable and unviable cells (*e.g.*, mini-cells that do not contain the complete genome). As discussed above, all these characteristics also apply to morphologically modified cyanobionts in natural and artificial symbioses. An increase in the number of heteromorphic cell forms during prolonged cultivation or in the aging parts of symbiotic plant organs does not militate against the adaptive character of these changes, because it is related to the prolonged existence of the cyanobacterial population under changed conditions. There is also evidence that heteromorphic changes are induced by the plant partner (see 3.4.6). In an analogy to the data on heterotrophic bacteria (Prozorovsky *et al.*, 1987), the fact that the FUG in associated cyanobacterial populations contain structurally intact and functionally active spheroplasts and protoplasts (see section 3.4.4) testifies to a high probability of the transition from the stage of unbalanced growth to the L-form formation.

3.4.4. Forms with Reduced Cell Wall

In addition to other heteromorphic forms, spheroplasts and protoplasts also occur in associations (Table 7). Protoplasts and spheroplasts are characterized by variable shapes, including an ameboid habit (Figs. 11, 20, 21). As a rule, small membrane vesicles detach from their surface. In some systems, the outer membrane of spheroplasts is located very close to their CM. The CM appears to be intact. No significant differences between the internal organization of FRCW and that of intact vegetative cells were found, except that reserve granules were scanty. Similar protoplast/spheroplast have been found in the cyanobionts of cycads (Fig. 22; Baulina, 2000a; Lobakova *et al.*, 2000; Lobakova and Baulina, 2001) and the liverwort *B. pusilla* (Fig. 23; Gorelova *et al.*, 1992b, 1996b; Korzhenevskaya *et al.*, 1993; Baulina *et al.*, 2000a) . We suggest that protoplast-vegetative cell interconversions could take place in the liverwort symbioses. This is consistent with the fact that the internal structure of the protoplasts and vegetative cells is similar (Baulina and Gorelova, 1996; Gorelova et al., 1996b). Such protoplasts are typical components of the *Nostoc* population in initial subcultures of the cyanobiont when isolated from *B. pusilla* and in stored inactive cultures.

The protoplasts and spheroplasts described above are likely to be metabolically active. Spheroplasts obtained by short-term lysozyme treatment of cyanobacterial cells retained their capacity for bioenergetic and biosynthetic processes on the level of intact vegetative cells (Semenova *et al.*, 1982).

Table 7. Artificial associations yielding cyanobacterial FRCW

Cyanobacterium	Plant Partner
N. muscorum VKM 16	*M. sativa*
A. variabilis CALU 458	*N. tabacum* cv. Samsun, variegated mutant
C. fritschii ATCC 27193	*P. ginseng* IPhR-G1
C. fritschii ATCC 27193	*S. laciniatum*
N. muscorum CALU 304	*S. dulcamara*
N. muscorum CALU 304	*R. serpentina* K-27

3.4.5. L-form Generation in Associated Cyanobacteria

It is possible to induce L-transformation in cyanobacteria under laboratory conditions (Gusev *et al.*, 1981, 1983). Inoculation of spheroplast suspensions of *A. variabilis* CALU 458 and *C. fritschii* ATCC 27193 (obtained using lysozyme) on solid media resulted in formation of colonies resembling the L-colonies of chemoheterotrophic bacteria. The formation of such colonies also occurred in subcultures. The symptoms of L-transformation, including ultrastructural changes, occurred most persistently in *C. fritschii*. All attempts to induce L-transformation in the obligate phototroph *Synechococcus* sp. PCC 6301 were unsuccessful.

The presence of L-forms of the cyanobiont in artificial associations and natural symbioses is difficult to prove unambigiously, because of the relatively small size of samples amenable to assessment by transmission electron microscopy. In order to conclude that L-transformation actually occurs, it is necessary to identify supplementary symptoms, in addition to detecting the ultrastructurally intact and apparently metabolically active spheroplasts and protoplasts. These symptoms are: gigantic FRCW, elementary bodies, and amorphous agglomerations of cell debris. We failed to detect FRCW in any of the *A. variabilis* ATCC 29413-containing associations tested. However, microcolonies of this cyanobacterium whose morphology, growth pattern, and ultrastructure raise the possibilty of L-transformation, were detected on the medium surface, in the vicinity of the callus, when *A. variabilis* was grown in mixed culture with tobacco (Baulina *et al.*, 1988; O.I. Baulina and L.V. Pivovarova, unpublished results).

Small cells (about 0.5 nm in diameter) are frequently seen (under scanning electron microscope) in populations of *A. variabilis* CALU 458 located in the harbors of *N. tabacum* cv. Samsun (variegated mutant) callus subcultured for 2.5 years (Baulina et al., 1984). These small cells occasionally detach from larger cells by budding. However, in ultrathin sections, they cannot be identified as elementary bodies devoid of cell wall or as anomalous mini-cells. Nevertheless, amorphous agglomerations (Fig. 20) and solitary gigantic cells were detected in *A. variabilis* CALU 458 populations, in addition to the protoplasts and small cell types.

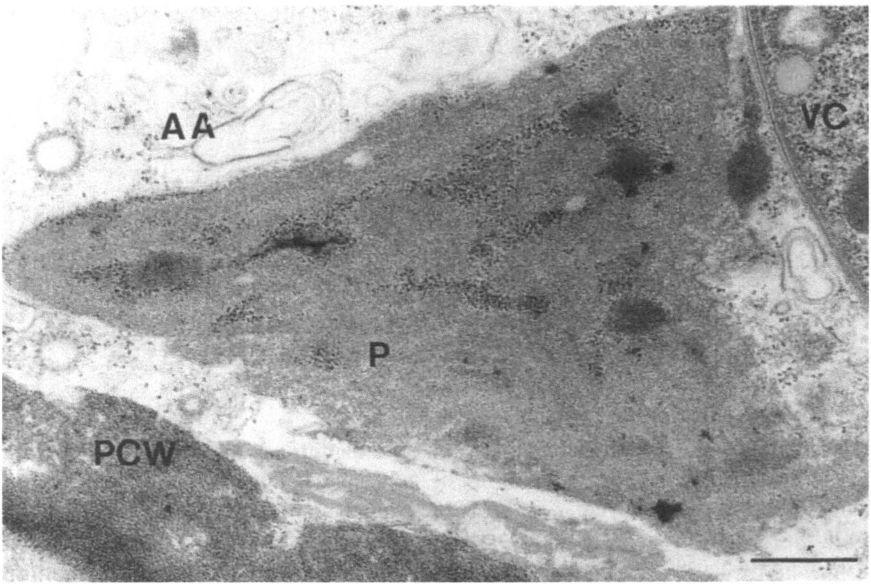

Figure 20. Part of a microcolony of A. variabilis *CALU 458 inside a harbor of the mixed culture with* N. tabacum *(cv. Samsun, variegated mutant) callus. AA, amorphous agglomerations; P, protoplasts; PCW, plant cell wall; VC, vegetative cell (bar = 0.5μm).*

All cell forms and amorphous agglomerations characteristic of the colonies of L-forms also occured in *N. muscorum* CALU 304 mixed cultures with *R. serpentina* and *S. dulcamara* (Gorelova, 2000a, 2001a).

In order to confirm the L-transformation, it is necessary to demonstrate that protoplasts can reproduce by dividing or forming elementary bodies. The latter are regarded as minimum structural units and presumably main reproductive elements of L-forms (Prozorovsky *et al.*, 1981; Dominique, 1995). The capacity of L-forms to reproduce in the absense of cell walls, provides for long-term existence of modified populations that retain species-specific traits while in a host organism, despite changes in some metabolic properties, the surface antigens, and the macromolecular organization of CM (Prozorovsky *et al.*, 1981; Nishiyama and Yamaguchi, 1990; Dominique, 1995; Hoischen *et al.*, 1997). It is difficult to reliably establish protoplast budding or equivalent division in compact intratissue colonies, although this seems a distinct possibility.

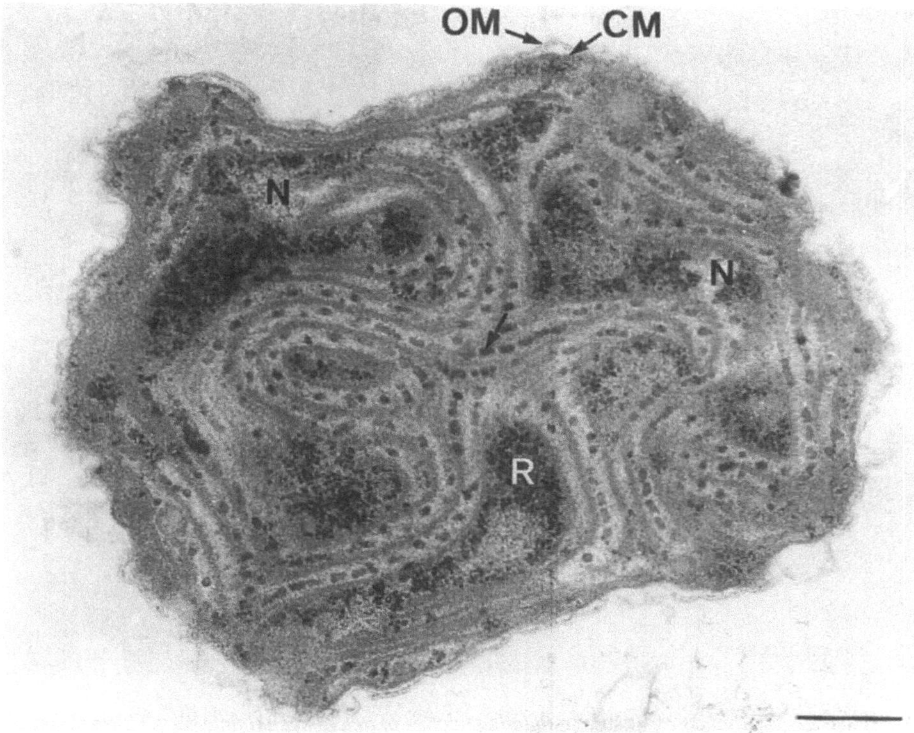

Figure 21. N. muscorum *CALU 304 spheroplast in mixed culture with* R. serpentina. *CM, cytoplasmic membrane; N, nucleoid; OM, outer membrane; R, ribosome; T, thylakoid; α, glycogen granule. (bar = 0.25μm). After Gorelova (2001a).*

The spheroplast-like GCF of *N. muscorum* CALU 304 that form during the mixed culture with *R. serpentina* calli display intracellular vesicles. A part of them presumably, represent elementary bodies (Gorelova and Korzhenevskaya, 2000a, 2001b; Gorelova, 2001b). The intrathylakoid space expands at the sites where the photosynthetic membranes are in contact with the nucleoid. The thylakoid membrane invaginates into these protrusions, forming round vesicles (Figs. 19, 24). Solitary vesicles or their agglomerations are seen in ultrathin sections. They contain easily distinguishable dense granules and thin fibrils, analogous to the structural elements of the nucleoid zone. They probably represent ribosomes and DNA strands. The thylakoid membrane can pinch off, incorporating ribosomes and DNA molecules into the resulting vesicle. This is consistent with the evidence that a part of the ribosomes and the nucleoid in cyanobacteria, in contrast to other bacteria, attach to the protoplasmic side of the photosynthetic membranes instead of the CM. Based on these observations, involvement of the intracellular membrane systems of cyanobacteria in DNA replication and segregation of daughter chromosomes has been suggested based on these data (Pinevich, 1997).

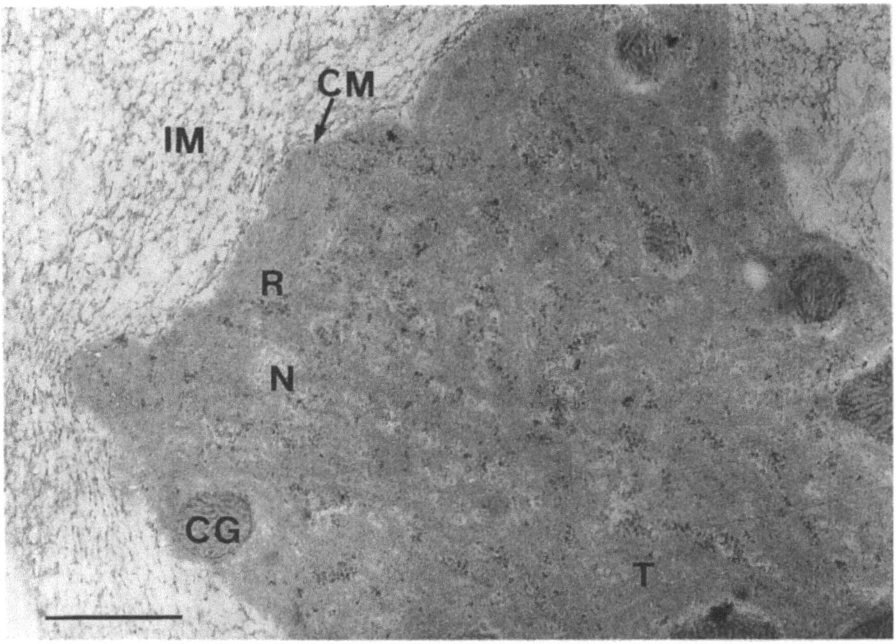

Figure 22. Protoplast of Nostoc *in mucilage zones of active cyanobacteria in* C. revoluta *coralloid roots. CG, cyanophycin granule; CM, cytoplasmic membrane; IM, intercellular matrix; N, nucleoid; R, ribosomes; T, thylakoid (bar = 0.5µm).*

The vesicle size varies between 80 and 1060 nm. Since the cyanobacterial genome is significantly larger than that of other bacteria, it is evident that only large vesicles can contain the complete gene set of parental cells. In our opinion, the vesicles described above are similar to the elementary bodies. They form inside spheroplast-like GCF and enter the intercellular matrix after their destructon. In addition to vesicles, small cells (0.9-1.7 µm) exhibiting binary division and protoplasts with under developed thylakoids occur in this system (Gorelova and Korzhenevskaya, 2001b). Thus, the research on artificial associations has identified a novel symbiotic phenomenon of cyanobionts' L-transformation and its implication for cyanobacterium-plant symbioses. The results obtained from natural and artificial symbioses are consistent with the general concepts on the significance of the L-transformation process as a unique adaptation stratagy under new environmental conditions.

In contrast to the typical culture conditions, host tissues are distinguished by a number of unique physical and chemical properties (microaerobiosis, low illumination, low pH, and space constraints) (Meeks, 1998; Rai *et al.*, 2000b) and by the metabolic interactions between the symbiotic partners. Presumably, the morphological changes in cyanobacteria, resulting from the transition of vegetative cells to the stages of unbalanced growth and L-transformation during the ontogeny of the symbiosis, are due to the influence of the host organism as a unique habitat. Cyanobacterial L-forms represent a special cell type (Meeks *et al.*, 1999) formed under the influence of the plant partner (Gorelova and Korzhenevskaya, 2001b).

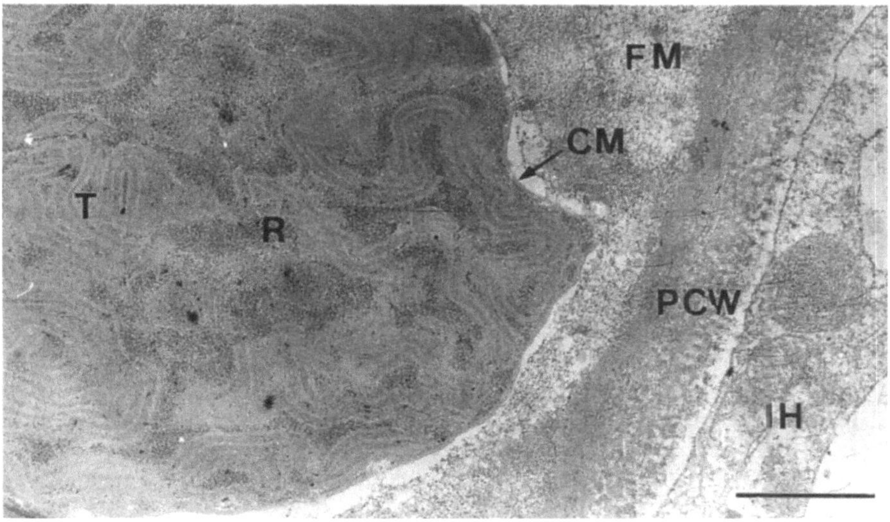

Figure 23. Protoplasts of symbiotic Nostoc *sp. in the cavity of a* B. pusilla *thallus. CM, cytoplasmic membrane; FM, fibrillar matrix; IH, intracavity hair of liverwort; PCW, plant cell wall; R, ribosomes; T, thylakoid (bar = 0.5µm). After Korzhenevskaya* et al *(1993).*

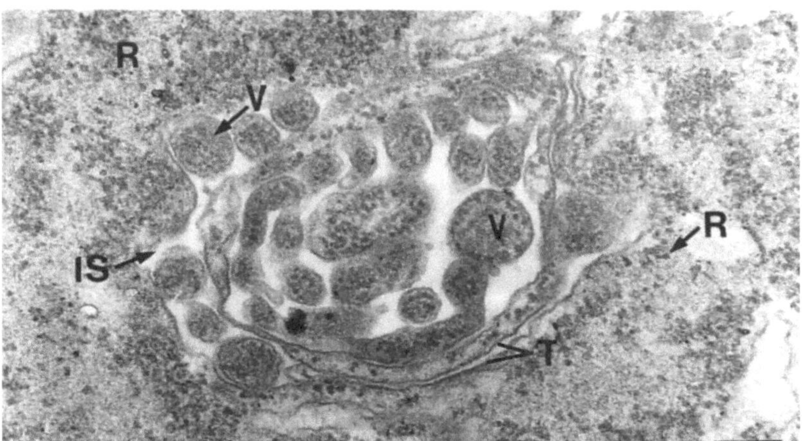

Figure 24. Part of a N. muscorum *gigantic spheroplast from Fig. 19. IS, intrathylakoid space; R, ribosomes; T, thylakoid; V, vesicles (bar, 0.25µm). After Gorelova and Korzhenevskaya (2001b).*

3.4.6. Heteromorphism Induction by the Plant Partner

3.4.6. Heteromorphism Induction by the Plant Partner

The plant partner's influence on the cell differentiation strategy of the cyanobiont has been described above with the examples of hormogonia induction (see section 3.1) and heterocyst differentiation (see section 3.3). In artificial associations, yielded heteromorphic changes in cyanobacterial cells are also due to the influence of the plant partner. For instance, alfalfa callus induces the formation of heteromorphic, akinete-like (see section 3.4.2) cell types in *A. variabilis* ATCC 29413. Heteromorphic changes also occurred in *Synechococcus* sp. 6301 grown in mixed suspension with tobacco cells (see section 1.2). Moreover, the peptidoglycan layer of the *Synechococcus* cell wall lost its rigidity only in the presence of tobacco cells, indicating the induction of this process by the plant partner (Baulina *et al.*, 1994).

The number and diversity of heteromorphic cell forms is enhanced in the populations of *N. muscorum* CALU 304 grown with *S. dulcamara* calli. This is also true for a mixed culture of *N. muscorum* CALU 304 with *R. serpentina* callus (Table 8). In absence of the plant partner, cultivation media *per se* failed to promote FDCW formation or maintain their viability (Gorelova, 2001a). As we mentioned above, GCF only form in the presence of a plant tissue. Cell heteromorphism increases with an increase in the duration of the cyanobacterium-plant partner interaction. Because enhanced heteromorphism does not occur in monoculture, and because heteromorphism manifests itself in mixed cultures irrespective of the direct contact with *R. serpentina* tissues, the factors inducing heteromorphic changes in *N. muscorum* are likely to be of plant origin and diffusible in agar medium (Gorelova, 2000a).

Table 8. Detection of heteromorphic cell forms in N. muscorum *CALU 304 populations in pure cultures, monocultures, and in mixed cultures with* R. serpentina

Variant		Heteromorphic Cell Forms			FDCW % frequency
Culture	Incubation time weeks	GCF	Minicells	FDCW	
Pure culture					
BG-11 medium	3	-	+	+	3-9
AA medium	4	-	+/-	+	3-9
Monoculture	2	-	+	+	<2
	8	-	-	-	0
Mixed culture [1]					
S	2	-	+	+	3-9
A	6	+/-	+	+	3-9
B	6	-	+/-	-	0
C	6*	-	+/-	-	0
C	6**	-	+/-	+	<2
A	8	+	+	+	≥10
C	8**	+	+	+	≥10
Mixed culture (out of contact)	2	-	+	+	<2
	10	+	+	+	≥10

[1]Data on the heteromorphic cell forms in microcolonies (i) on the primary surface (S) of microcolony; and (ii) in the zones A, B, and C of the mixed aggregate (see text for more details); * and **, the cyanobacteria from the young and mature surface colonies, respectively, of the mixed aggregate.

Mechanisms of Heteromorphic Changes. A decrease in the thickness of the peptidoglycan layer and formation of pores therein are characteristic of all cyanobacterial species in the artificial associations obtained by us. The peptidoglycan damage can in principle result from impairment of certain stages of its biosynthesis or from the direct effects of endo- or exogenous lytic enzymes. The effects of exogenous lytic enzymes can be exemplified by simultaneous local lysis of the outer membrane and peptidoglycan layer observed in *C. fritschii* ATCC 27193 in a stable mixed suspension culture with ginseng cells (strain IPhR-G1) (Baulina *et al.*, 1995).

The variable pattern of ultrastructual changes in the peptidoglycan of heteromorphic cell types was demonstrated in a *N. muscorum* CALU 304 - *R. serpentina* callus association (Gorelova and Baulina, 2000; Gorelova, 2001a). The peptidoglycan layer loses its rigidity and becomes uneven (patch-like bulges and zones with reduced thickness occur). Occasionally, the peptidoglycan layer appears to be perforated or a loose, thin fibrillar network in the expanded periplasm. This layer is partly or completely lacking in spheroplasts and protoplasts. A part of the cell population (mostly actively growing hormogonia) sustains damage and exhibits characteristic symptoms of autolysis. As we stated above, longer incubation of *N. muscorum* CALU 304 with *R. serpentina* callus results in an enhancement of FDCW numbers. These facts suggest a plant partner-induced activation of cyanobacterial autolytic enzymes. This causes, lysis of the peptidoglycan as well as the impairment of its synthesis. Indeed we observed (i) a change in the shape of vegetative cells (barrel-like elongated cells become spherical or discoid shaped) and (ii) incompletely divided cells, in which the apposition of the newly synthesized septal peptidoglycan is oriented from the periphery to the cell's center as well as towards the outer membrane where it forms folds (Fig. 25). Peripheral deposits of peptidoglycan characterized by an opposite orientation relative to the septum, also occured in the *N. muscorum* VKM 16 - alfalfa, *N. muscorum* CALU 304 - *S. dulcamara*, and *A. variabilis* CALU 485 - tobacco systems.

The balanced operation of multienzyme complexes including both biosynthetic and lytic enzymes (transpeptidases, transglycosidases, endopeptidases, etc.), is a prerequisite for peptidoglycan biosynthesis during the bacterial cell cycle that leads to cell elongation and septation (Begg *et al.,* 1990; Bramhill, 1997). Our data suggest an imbalance between the lytic and biosynthetic components of enzyme complexes and point to a shift in the ratio of PBP2 and PBP3 activities (Gorelova, 2001a). These proteins ensure the correct pattern of peptidoglycan assemblage during cell elongation and septum formation.

The decoupling of cell growth from cell division, during the formation of gigantic cells, can be due to disrupted synthesis of the septal peptidoglycan and impaired functioning of ring FtsZ (Gorelova and Baulina, 2000; Gorelova, 2001a) that causes cell constriction at the division initiation site (Zhang *et al.,* 1995; Doherty and Adams, 1999). The fact that constriction is asymmetrical, or does not occur at all in some cells at the site of septum synthesis, supports this suggestion. It was proposed earlier that the *ftsZ* gene (or its homologue) can serve as a target for the signal mechanism involved in controlling cell division of the cyanobiont by the host plant (Bergman *et al.,* 1996). The results of our studies give us grounds for the suggestion that plant products directly affect protein FtsZ (Gorelova, 2001a). *R. serpentina*, an indole alkaloid producer, retains its capacity for synthesizing the species-specific products when growing in mixed culture with cyanobacteria (Gorelova and Artamonova, 1992; Korzhenevskaya *et al.*, 1992). Indole alkaloids are known to prevent the assemblage of tubulin dimers into microtubules in eukaryotes (Nadezhdina *et al.*, 1995). Based on the high homology of the amino acid sequences of tubulin and protein FtsZ, and the similarity of their biochemical properties, it seems likely that the indole alkaloids of *R. serpentina* callus disrupt FtsZ polymerization. Disrupting FtsZ assemblage results in changing the invagination shape and, therefore, the morphology of the septum formed upon cell division (Bramhill, 1997). Another important point about the initiation of cell division concerns the location of the initiation site of septum formation and the position of the division plane.

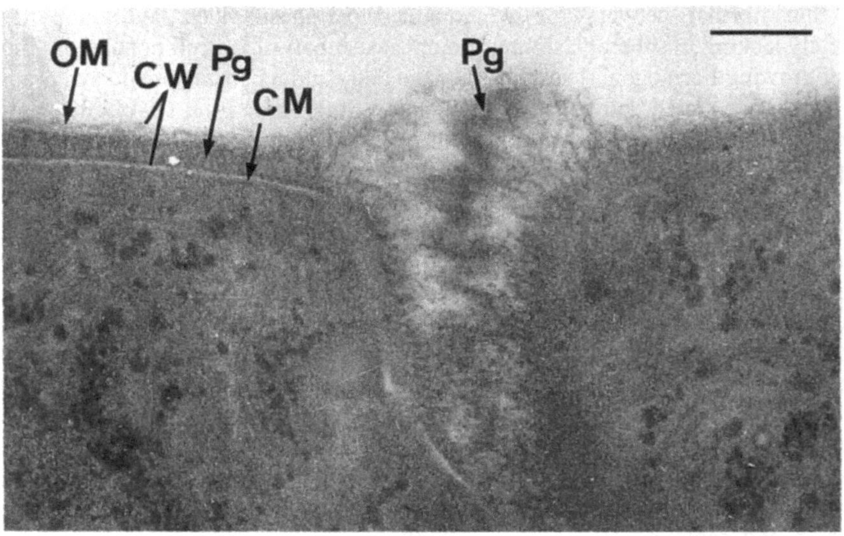

Figure 25. A portion of N. muscorum *CALU 304 dividing cell with the septal peptidoglycan layer that is synthesized toward the outer membrane. CM, cytoplasmic membrane; CW, cell wall; OM, outer membrane; Pg, peptidoglycan (bar = 0,25μm). After Gorelova (2001a).*

The representatives of the order Nostocales are characterized by division in single plane, which is perpendicular to the filament axis (Hensyl, 1989). Division in additional planes are common in many artificial associations (Baulina *et al.*, 1984; Gorelova *et al.*, 1999, Korzhenevskaya *et al.*, 1993). This can be considered a supplementary criterion of disrupted cell division occurring during the interaction between symbiotic partners.

Thus, the plant partner seems to cause heteromorphic changes in associated cyanobacteria by lysing peptidoglycan, disrupting its biosynthesis, and impairing cell division.

3.5. Communication between Spatially Separated Cyanobacteria and Plant Partners in Model Systems

The formation of natural or artificial symbioses of cyanobacteria with higher plants results in morphological and physiological changes in cyanobacteria. Much of the current research in the field of symbiology is concerned with the relevant phenotypic Traits, including the studies on the activities of key enzymes invoved in metabolic processes, and the identification of the genes whose expression determines the changes occurring in cyanobacteria. It has been proposed that the interactions that lead to a stable symbiotic association appear to be mostly unidirectional, from plant to cyanobacterium (Meeks, 1998). Multicomponent sensing-signaling systems have been assumed to regulate cyanobacterial behaviors (Bergman *et al.,* 1996; Meeks, 1998; Meeks et al., 1999; Rai *et al.,* 2000b)].

Bearing in mind these ideas, we made an analysis of plant tissue-induced modifications of cyanobacteria (see sections 3.1, 3.3, and 3.4.) in the absence of direct physical contacts between the partners. Joint cultures, where the partners are spatially separated and metabolite exchange can only proceed by diffusion through the medium, were employed as model systems. We also compared cyanobacterial populations that developed at different locations in mixed cultures on solid media (at the surface or in the depth of the plant tissue i. e., with the partners spatially integrated, and at the medium surface away from the plant tissue). This research revealed that the plant partner may be the source of exocellular agents (EA) which diffuse into the medium and affect the cyanobacterial partner without the need for a direct contact between the partners (Korzhenevskaya *et al.,* 1985, 1999a; Gorelova *et al.,* 2000; Gorelova and Korzhenevskaya, 2000b; Gorelova, 2000a, 2001a). EA influence the growth and viability of cyanobacterial populations as a whole. The stimulatory, inhibitory, or inductive effects of EA on cyanobacterial activities manifest themselves in relation to at least three processes investigated by us: (i) hormogonium formation and taxis; (ii) regulation of nitrogen assimilation; and (iii) heteromorphic changes (Table 9).

The effects of EA vary depending on the combinations of partners used, their age, the distance between them, the time of interaction, and the incubation conditions (medium composition and illumination regimen). EA differ in their degree of specificity. Their effects may temporarily coincide or lag behind the analogous effects observed during interactions between spatially integrated plant tissues and cyanobacteria. For instance, heterocyst differentiation in *N. muscorum* CALU 304 colonies on the medium surface out of contact with the *R. serpentina* tissue occurred 3-4 weeks later than in cyanobacteria growing in mixed aggregates (Gorelova, 1998; Korzhenevskaya *et al.,* 1993). The heterocyst frequency and the level of nitrogenase activity were approximately 40% lower than in mixed aggregates (see section 3.3.2, Table 6). Nevertheless, *N. muscorum* CALU 304 vegetative cells did not suffer from a lack of combined nitrogen and they contained more cyanophycin than gegetative cells in pure culture grown on AA medium. The cyanobacteria concomitantly accumulated numerous glycogen granules. Apparently, the time and intensity of the effects of plant EA can be conditional on the size of EA molecules, apart from the influence of the concentration decrease caused by EA dilution by the medium. The size of EA molecules determines (i) the diffusion rate of EA in an agarized medium and (ii) the permeability of cyanobacterial cell surface structures (sheaths, capsules, cell walls, and cytoplasmic membranes) for them. In addition, cyanobacteria obviously differ in their sensitivity threshold and competence to EA.

The chemical nature of EA has not yet been discovered. However, it seems likely that they include, apart from signal molecules triggering the cascade systems of various differentiation pathways in cyanobacterial cells, compounds "directly" influencing the partner's metabolism (enzyme activities and FtsZ operation) without affecting the gene expression. It should also be emphasized that cyanobacteria *per se* can influence the biosynthetic activities of the plant partner, changing the spectrum and yield of the products it synthesizes (Gorelova *et al.*, 1984, 1985; Gorelova, 1986a,b; Gusev and Korzhenevskaya, 1990; Korzhenevskaya, 1990; Yagodina *et al.*, 1990; Gorelova and Artamonova, 1992; Korzhenevskaya *et al.*, 1992).

4. CONCLUSION

Studies conducted during the last decade have significantly improved our knowledge of the differrent stages in formation and function of artificial associations involving cyanobacteria and furnished new data on the involvement of satellite bacteria in this process. The strenuous efforts to create artificial associations between important agricultural plants and nitrogen-fixing microorganisms have yielded promising results particularly for introducing heterocyst-forming cyanobacteria into the plant rhizosphere. Important progress has also been made in the induction of root paranodule and colonization of such para-nodules by microsymbionts.

It was demonstrated earlier that modification of the partners in artificial associations are analogous to those occurring in natural symbioses (Gusev and Korzhenevskaya, 1990; Korzhenevskaya *et al.*, 1989,1993). The data presented in this contribution demonstrate that use of model systems to investigate symbiont interactions in mixed and joint cultures can provide valuable information on the developmental stages and metabolic activities of cyanobacterial-plant symbioses. It seems likely that natural associations also employ the factors that regulate microsymbiont behavior in model systems. Therefore the following aspects should be examined in natural symbioses:

(i) the influence of the age of the plant tissue on the formation and taxis of cyanobacterial hormogonia;

(ii) the influence of the plant tissue on peptidoglycan metabolism and cytokinesis of cyanobacteria;

(iii) the induction by the plant partner of heteromorphic changes in cyanobacteria that are linked to the transition of vegetative cells to the phases of unbalanced growth and L - transformation.

Successful attempts to create artificial associations together with the recent data on interactions between partners suggest that plants and cyanobacteria have genetic potential to form new symbiotic systems, in addition to those found in nature.

Table 9. Changes in the structural and physiological characteristics of cyanobacteria induced by plant exocellular agents

Cyanobacterial Response to the EA of the Partner	Potential Significance for Symbiogenesis
Induction and/or stimulation of hormogonium differentiation	Formation of infective units, an increase in the probability of en counters between the partners, their spatial approach, and the rein fection of macrosymbiont tissues during the ontogeny
Inhibition of hormogonium differentiation	Preventing infection during the interaction of incompetent partners and/or under conditions disfavoring the formation of an efficient association, "anchoring" the symbiont (rendering it immotile), and changing cell differentiation programs
Positive hormogonium taxis	Attraction of the potential microsymbiont to the site used for spatial integration; preliminary selection of compatible partners prior to their contact-dependent recognition
Negative hormogonium taxis	Repulsion of incompatible partners; reinfection of the macrosym biont during its ontogeny by cyanobacteria if they are banished by aging tissues
Changes in the dynamics of cyanophycin synthesis and consumption	Plasticity with respect to supplying the system *in toto* with combined nitrogen: cyanophycin storage while an exogenous source is available and cyanophycin mobilization when combined-N is exhausted
Heterocyst differentiation	Supplying the whole system with combined nitrogen under the macrosymbiont's control
Nitrogenase expression	Ditto
Cytokinesis retardation; dis balance of growth and cell di vision	Decrease in the growth rate of the cyanobiont; balanced growth of the partners under the macrosymbiont's control
Increase in the frequency of asymmetrical and anomalous cell divisions	Ditto
Faulty FtsZ operation	Ditto
Changes in the peptidoglycan metabolism	Ditto
L-form generation	The cyanobiont's adaptation (changes in the antigen properties and the reproduction mode) enabling it to persist in the macrosymbiont

ACKNOWLEDGEMENTS

We are very greatfull to our co-workers Dr. I.B. Yagodina, Dr. L.V. Pivovarova, Dr. S.Yu. Kleimenov, Mrs. G.A. Dol`nikova, Dr. A.V. Oleskin, and Dr. A.V. Kitashov for their suggestions and help in preparation of this manuscript.

Some of the work presented in this chapter was supported by grants nos.94-04-12583, 97-04-48363, 00-04-48708, and 00-15-97-911 from the Russian Foundation for Basic Research, and the International Science Foundation, grant no. J6J100.

REFERENCES

Allen, M.B. and Arnon, D.I. (1955) Studies on nitrogen-fixing *Anabaena cylindrica* Lemm., *Plant. Physiol.* **30**, 366-372.

Al-Malah,M.K., Davey, M.R., and Cocking E.C. (1989) Formation of nodular structures on rice seedling by rhizobia, *J. Experimental Bot.*, **40**, 473-478.

Baron-Epel, O., Gharyal, P.K., and Schindler, M. (1988) Pectins as mediator of wall porosity in soybean cells, *Planta*, **175**, 389-395.

Bashan, Y. and Holgiun, G. (1997) *Azospirillium*-plant relationships: environmental and physiological advances (1990-1996), *Can. J. Microbiol.* **43**, 103-121.

Bateman, K., Rasmussen, U., and Bergman, B. (1999) A putative arabinogalactan protein is secreted by prokaryotic cyanobacteria, in *Book of Abstr. of 9th Int. Congress Molecular Plant-Microbe Iinteractions*, Amsterdam, p. 206.

Baulina, O.I. and Lobakova, E.S. (1986) Morphology and ultrastructure of ginseng-cyanobacterium association in suspension cultures, in R.G. Butenko (ed.), *Plant Cell Culture & Biotechnology*, Nauka, Moscow, pp. 252-259 (in Russian).

Baulina, O.I. and Gorelova, O.A. (1996) Crystal-like inclusions in the symbiotic cyanobacterium *Nostoc sp.*, *Microbiology (*Translated of *Mikrobiologiya)* **65**, 621-623.

Baulina, O.I., Mineyeva, L.A., Suleymanova, Sh.S., and Gusev, M.V. (1981) Cell ultrastructure of blue-green algae under the conditions of photogeterotrophic cultivation, *Mikrobiologiya* **50**, 523-527 (in Russian).

Baulina, O.I., Agaphadorova, M.N., Korzhenevskaya, T.G., Gusev, M.V., and Butenko R.G. (1984) Cyanobacteria in artificial association with tobacco callus culture, *Mikrobiologiya* **53**, 997-1001 (in Russian).

Baulina, O.I., Yagodina, I.B., Pivovarova, L.V., and Agahfadorova, M.N. (1988) Heteromorphism of cyanobacteria during their interaction with tobacco cells and tissues in artificial associations, in *Abstr. of Int. Conf. Biology of Cultivated Cells and Biotechnology*, Novosibirsk, p. 381 (in Russian).

Baulina, O.I., Lobakova, E.S., Pivovarova, L.V., Skripnikov, A.Yu., Gorelova, O.A., Korzhenevskaya, T.G., and Gusev, M.V. (1989) Morphological and physiological characterization of artificial syncyanoses, in A.A. Baev and Ya.I. Buryanov (eds.), *Molecular and Genetic Mechanisms of Microorganism-Plant Iinteractions,* Scientific Center of Biological Research of the Academy of Sciences of the USSR in Pushchino, Pushchino, Moscow, pp.193-198 (in Russian).

Baulina, O.I., Gorelova, O.A., and Korzhenevskaya, T.G. (1990) Organization of cell surface structure in the sites of cyanobacterium localization in alfalfa tissues, in *Abstr. III All-Union Conf. Biosynthesis of Cellulose and Other Ccompounds of Cell Wall* ("Cellulose-90"), Kazan, p.5 (in Russian).

Baulina, O.I., Yagodina, I.B., Korzhenevskaya, T.G., and Gusev, M.V. (1994) Morphology and ultrastructure of the cyanobacterium *Synechococcus elongatus* growing in associations with plant cells, *Mikrobiologiya* **63**, 643-656 (in Russian).

Baulina, O.I., Lobakova, E.S., Korzhenevskaya, T.G., Butenko, R.G., and Gusev, M.V. (1995) Ultrastructure of ginseng cells and the cyanobacteria *Chlorogloeopsis fritschii* in the association cultivated in the dark, *Moscow University Biological Sciences Bulletin* (Translated of *Vestnik Moskovskogo Universiteta, Biologiya*) **50**, 1-11.

Baulina, O.I., Gorelova, O.A., Lobakova, E.S., Gusev, M.V., and Korzhenevskaya T.G. (2000a) Cyanobacteria with a reduced cell wall in natural and model plant symbioses, in H.C. Weber, S. Imhof, and D. Zeuske (eds.), *Third Int. Congress on Symbiosis, Progr., Abstr. and Papers*, Marburg, p. 31.

Baulina, O.I., Titel, C., Malai, O.B., and Gorelova, O.A. (2000b) Research on the permeability of mucilaginous surface structures of cyanobacteria to neutral hydrophilic macromolecules, in N.N. Kolotiliva (ed.), *Problems of Ecology & Physiology of Microorganisms,* Dialog-MGU, Moscow, p. 39 (in Russian).

Begg, K.J., Takasuga, A., Edwards, D.H., Dewar, S.J., Spratt, B.G., Adachi, H., Ohta, T., Matsuzawa, H., and Donachie, W.D. (1990) The balance between different peptidoglycan precursors whether *Escherichia coli* cells will elongate or divide, *Bacteriology* **172**, 6697-6703.

Berg, H., Ziegler, K., Piotukh, K., Baier, K., Lockau, W., and Volkmer-Engert, R. (2000) Biosynthesis of the cyanobacterial reserve polymer multi-L-arginyl-poly-L-aspartic acid (cyanophycin). Mechanism of the cyanophycin synthethase reaction studied with synthetic primers, *Eur. J. Biochem.* **267**, 5561-5570.

Bergman, B., Matveyev, A., and Rasmussen U. (1996) Chemical signaling in cyanobacterial-plant symbioses, *Trends in Plant Science* **1**, 191-197.

Bergman, B., Rai, A.N., Johansson C., and Söderbäck, E. (1993) Cyanobacterial-plant symbioses, *Symbiosis* **14**, 61-81.

Bonnett, H.T. and Silvester, W.B. (1981) Specificity in the *Gunnera-Nostoc* endosymbiosis, *New Phytol.* **89**, 121-128.

Bramhill, D. (1997) Bacterial cell division, *Ann. Rev. Cell Dev. Biol.* **13**, 395-424.

Braun-Howland, E. and Nierzwicki-Bauer, S.A. (1990) *Azolla-Anabaena* symbiosis: biochemistry, ultrastructure, and molecular biology, in A.N. Rai (ed.), *Handbook of Symbiotic Cyanobacteria*, CRC Press, Boca Raton, Florida, pp. 65-117.

Butenko, R.G. (1999) *Biology of higher plant cells in vitro and biotechnology of their base,* FBK-Press, Moscow (in Russian).

Campbell, E.L. and Meeks, J.C. (1989) Characteristics of hormogonia formation by symbiotic *Nostoc* spp. in responce to the presence of *Anthoceros punctatus* or its extracellular products, *Appl.& Envir. Microbiol.* **55**, 125-131.

Campbell, E.L. and Meeks, J.C. (1992) Evidence for plant-mediated regulation of nitrogenase expression in the *Anthoceros-Nostoc* symbiotic association, *J. Gen. Microbiol.* **138**, 473-480.

Carrapiço, F. (1991) Are bacteria the third partner of the *Azolla-Anabaena* symbiosis, *Plant and Soil* **137**, 157-160.

Cohen, M.F. and Meeks J.C. (1997) A hormogonium regulating locus, *hrmUA*, of the cyanobacterium *Nostoc punctiforme* strain ATCC 29133 and its response to an extract of a symbiotic plant partner *Anthoceros punctatus*, *Molecular Plant-Microbe Interactions* **10**, 280-289.

Danilova, M.Ph. and Barmicheva, E.M. (1980) Root cap, in M.Ph. Danilova and G.M.Kozubov (eds.), *Atlas of Plant Tissue Ultrastructure*, Karelia Publishing House, Petrozavodsk, pp. 331-346 (in Russian).

Dembinska, M.E. and Allen, M.M. (1988) Cyanophycin granule size variation in *Aphanocapsa*, *J. Gen. Microbiol.* **134**, 295-298.

Doherty, H.M. and Adams, D.G. (1999) The organization and control of cell division genes expressed during differentiation in cyanobacteria, A. Peschek, W. Loffelhardt, and G. Shmetterer (eds.), *The Phototrophic Prokaryotes*, Kluwer Acadamic/Plenum Publishers, New York, pp. 453-461.

Dominque, G.J. (1995) Electron dense cytoplasmic particles and chronic infection – a bacterial pleomorphy hypothesis, *Endocytobiosis & Cell Res.* **11**, 19-40.

Duckett, J.G., Prasad, A.K.S.K., Davies, D.A., and Walker, S. A (1977) Cytological analysis of the *Nostoc-Bryophyte* relationship, *New Phytol.* **79**, 349-362.

Egorov, N.S. and Landau, N.S. (1982) Biologically active substances of gramnegative bacteria, *Prikladnaya Biochim. i Mikrobiol.* **18**, 835-849 (in Russian).

Enderlin, C.S. and Meeks J.C. (1983) Pure culture and reconstitution of the *Anthoceros-Nostoc* symbiotic association, *Planta* **158**, 157-165.

Filippi, C., Bagnoli, G., and Giovannetti, M. (1995) Bacteria associated to arbutoid mycorrhizae in *Arbutus unedo* L., *Symbiosis* **18**, 57-69

Flores, E., Muro-Pastor, A.M., and Herrero, A. (1999) Cyanobacterial nitrogen assimilation genes and NtcA-dependent control of gene expression, in G.A. Peschek, W. Loffelhardt, and G. Schmetterer (eds.), *The Phototrophic Prokaryotes*, Kluwer Academic/Plenum Publishers, New York, pp. 463-477.

Fornii, C., Gentili, S., Van Hove, C., and Grilli Caiola M. (1990) Isolation and characterization of the bacteria living in the sporocarps of *Azolla filiculoides* Lam., *Ann. Microbiology* **40**, 235-243.

Fowler, M.W. (1982) Substrate utilisation by plant cell cultures, *J. Chem. Tech. & Biotech.* **32**, 338-346.

Gantar M. (2000) Co-caltivation of N_2-fixing cyanobacterium *Nostoc* sp strain 2S9B and wheat callus, *Symbiosis* **29**, 1-19.

Gantar, M. and Elhai, J. (1999) Colonization of wheat *para* – nodules by the N_2 – fixing cyanobacterium *Nostoc* sp. strain 2S9B, *New Phytol.* **141**, 373-379.

Gantar, M., Kerby, N.W., and Rowell, P. (1991a) Colonization of wheat (*Triticum vulgare* L.) by N_2 – fixing cyanobacteria: II. An ultrastructural study, *New Phytol.* **118**, 485-492.

Gantar, M., Kerby, N.W., Rowell, P., and Obrent, Z. (1991b) Colonization of wheat (*Triticum vulgare* L.) by N_2 – fixing cyanobacteria: I. A survey of soil cyanobacterial isolates forming associations with roots, *New Phytol.* **118**, 477-483.

Gantar, M., Kerby, N.W., and Rowell, P. (1993) Colonization of wheat (*Triticum vulgare* L.) by N_2 – fixing cyanobacteria: III The role of a hormogonia – promoting factor, *New Phytol.* **124**, 505-513.

Gantar, M., Rowell, P., Kerby, N.W., and Sutherland, I.W. (1995) Role of extracellular polysaccharide in the colonization of wheat (*Triticum vulgare* L.) roots by N_2-fixing cyanobacteria, *Biology & Fertility of S oils* **19**, 41-48.

Garbae, J. and Bowen, C.D. (1989) Stimulation of ectomycorrhizal infection of *Pinus radiata* by some microorganisms associated with the mantle of ectomycorrhizas, *New Phytol.* **112**, 383-388.

Garbae, J. and Duponnoids, R. (1990) Specificity and function of mycorrhization helper bacteria (MHB) associated with the *Pseudotsuga menziesii – Laccaria laccata*, in *Abst. of Int. Symbiosis Congress*, Jerusalem, p. 50.

Glagoleva, O.B., Koval'skaya, N.Yu., and Umarov, M.M. (1996) Endosymbiosis formation between nitrogen-fixing bacteria *Pseudomonas caryophylli* and rape root cells, *Endocytobiosis & Cell Research*, **11**, 147-158.

Glagoleva, O.B., Koval'skaya, N.Yu., Kireev, I.I., Lobakova, E.S., and Umarov, M.M. (1997) *Para*-nodulation of rape upon inoculation with nitrogen-fixing rhizosphere bacteria, *Microbiology (Translated of Mikrobiologiya)* **66**, 455-460.

Glagoleva, O.B., Lobakova, E.S., Koval'skaya, N.Yu., Korzhenevskaya, T.G., and Umarov, M.M. (1998) Endosymbiosis formation between rhizosphere nitrogen-fixing bacteria and rape roots, *Docladi Akademii Nauk SSSR* **362**, 283-285 (in Russian).

Gogotov, I.N. (1988) The possibility of using nitrogen-fixing phototrophic bacteria in biotechnology, in I.N. Gogotov (ed.), *Phototrophic Microorganisms*, Scientific Center of Biological Research of the Academy of Sciences of the USSR in Pushchino, Pushchino, pp.95-107 (in Russian).

Gorelova, O.A. (1986a) Growth and biosynthetic activity of higher plant cells in mixed cultures with nitrogen-fixing cyanobacteria, *Cand. Sci. (Biol.) Dissertation*, Moscow State Univ., Moscow (in Russian).

Gorelova, O.A. (1986b) Biosynthesis and accumulation of steroid compounds by *Solanum laciniatum* cells under conditions of mixed culturing with cyanobacteria, in R.G. Butenko (ed.), *Plant Cell Culture and Biotechnology*, Nauka, Moscow, pp. 261-263 (in Russian).

Gorelova, O.A. (1998) Modification of the cyanobacterium *Nostoc muscorum* CALU 304 in a mixed culture with *Rauwolfia* callus, in *Problems of Modern Mycology, Algology and Phytopatology (Proceedings of the Int. Conference)*, Moscow State Univ. – ID "Muravei", Moscow, pp. 322-323 (in Russian).

Gorelova O.A. (2000a) Spatial integration of the partners and heteromorphism of the cyanobacterium *Nostoc muscorum* CALU 304 in a mixed culture with the *Rauwolfia* tissue, *Microbiology (Translated of Mikrobiologiya)* **69**, 471-479.

Gorelova, O.A. (2000b) Inducible heteromorphism of the cyanobacterium *Nostoc muscorum* CALU 304 during its interaction with *Rauwolfia serpentina tissue*, in N.N. Kolotiliva (ed.), *Problems of ecology and physiology of microorganisms*, Dialog-MGU, Moscow, p. 51 (in Russian

Gorelova O.A. (2001a) Surface ultrastructure of the heteromorphic cells of *Nostoc muscorum* CALU 304 in a mixed culture with the *Rauwolfia* callus tissue, *Microbiology (Translated from Mikrobiologiya)* **70**, 285-294.

Gorelova, O.A., (2001b) Formation of elementary body-like vesicles by heteromorphic cells of *Nostoc muscorum* CALU 304 during its interaction with a plant tissue, in *Abstr. All-Russia Conf. Agricultural Microbiology in XIX-XXI Centuries*, St. Petersburg, p. 49 (in Russian)..

Gorelova, O.A. and Artamonova, K.V. (1992) A new morphological structure in the association of a cyanobacterium with plant tissue, in *Abstr. 14th Conf. on Electron Microscopy (Biology and Medicine)*, Moscow, p. 192 (in Russian).

Gorelova, O.A. and Baulina, O.I. (2000) Ultrastructure of the cell surface of heteromorphic cells of *Nostoc muscorum* CALU 304 in mixed culture with *Rauwolfia serpentina* tissue, in N.N. Kolotiliva (ed.), *Problems of Ecology and Physiology of microorganisms*, Dialog-MGU, Moscow, p. 52 (in Russian).

Gorelova, O.A. and Korzhenevskaya, T.G. (2000a) Gigantic cell forms of cyanobacteria resulting from their interaction with cultured *Rauwolfia* tissue, in N.N. Kolotilova and A.I. Netrusov (eds.), *Autotrophic Microorganisms*, MAKS Press, Moscow, pp. 56-57 (in Russian).

Gorelova, O.A. and Korzhenevskaya, T.G. (2000b) Communication of spatially separated cyanobacteria and plant tissues in model systems, in N.N. Kolotilova (ed.), *The Problems of Ecology and Physiology of Microorganisms*, Dialog-MGU, Moscow, p. 53 (in Russian).

Gorelova, O.A. and Korzhenevskaya, T.G. (2001a) Cyanophycin accumulation by cyanobacteria during the interaction with plant tissues in model associations, in *Abstr. All-Russia Conf. Agricultural Microbiology in XIX-XXI centuries*, St. Petersburg, pp. 49-50 (in Russian).

Gorelova, O.A. and Korzhenevskaya, T.G. (2001b) Formation of gigantic and ultramicro forms of *Nostoc muscorum* CALU 304 during the interaction with culturing *Rauwolfia* tissues, *Microbiology (Translated of Mikrobiologiya)*, in press.

Gorelova, O.A., _e_abek, J., Korzhenevskaya, T.G., Butenko, R.G., and Gusev, M.V. (1984) The growth and biosynthetic activity of *Solanum laciniatum* cells in mixed cultures with nitrogen-fixing cyanobacteria, *Docladi Akademii Nauk SSSR* **279**, 253-256 (in Russian).

Gorelova, O.A., _e_abek, J., Korzhenevskaya, T.G., Butenko, R.G., and Gusev, M.V. (1985) Influence of the cyanobacterium *Chlorogloeopsis fritschii* on growth and biosynthetic activity of *Solanum laciniatum* cells in mixed culture, *Fisiologiya Rastenii* **32**, 1158-1165 (in Russian).

Gorelova, O.A., Baulina, O.I., Lobakova, E.S., and Korzhenevskaya, T.G. (1988) Heteromorphism of cynobacteria in artificial associations with rice and alfalfa plants, in *Abstr. 13th Conf. on Electron Microscopy (Biology and Medicine)*, Moscow, p. 241.

Gorelova, O.A., Lobakova, E.S., Luk, K., Baulina, O.I., and Korzhenevskaya, T.G. (1989) Cyanobacteria of the *Nostoc* genus in artificial associations with rice callus and plants, *in Abst. YIII Conf. Spore-Forming Plants of Middle Asia and Kazachstan*, Fan Publ. Co, Tashkent, 51 (in Russian).

Gorelova, O.A., Baulina, O.I., and Korzhenevskaya, T.G. (1990) Ultrastructure of cyanobacteria developing in alfalfa leaf tissue upon infection, in *Abstr. IV Republican Conf. on Electron Microscopy (Electron Microscopy and Modern Technology)*, Kishineu, pp. 50-51 (in Russian).

Gorelova, O.A., Artamonova, K.V., and Korzhenevskaya, T.G. (1992a) Hormogonia formation in plant-cyanobacteria artificial associations, in *Abstr. 8th East. Eur. Symp. on Biological Nitrogen Fixation "Nitrogenfix-92"*, Saratov, PD 3.

Gorelova, O.A., Shchelmanova, A.G., Baulina, O.I., and Korzhenevskaya, T.G. (1992b) Ultrastructure of the population of the symbiotic cyanobacterium *Nostoc* sp., in *Abstr. 14th Conf. on Electron Microscopy (Biology and Medicine)*, Moscow, p. 49 (in Russian).

Gorelova, O.A., Korzhenevskaya, T.G., and Gusev, M.V. (1995) Formation and oriented propagation of cyanobacterial hormogonia in model systems with higher plant tissues, *Moscow University Biological Sciences Bulletin (Translated from Vestnik Moskovskogo Universiteta, Bioloigiya)* **50**, 17-25.

Gorelova, O.A., Baulina, O.I., Gusev, M.V., and Korzhenevskaya, T.G. (1996a) Formation of hormogonia and their taxis during the interaction of cyanobacteria and plants, in *Abstr. Conf. Autotrophic Microorganisms*, Moscow, p. 30 (in Russian).

Gorelova, O.A., Baulina, O.I., Shchelmanova, A.G., Korzhenevskaya, T.G. and Gusev, M.V. (1996b) Heteromorphism of the cyanobacterium *Nostoc* sp., a microsymbiont of the *Blasia pusilla* moss, *Microbiology (Translated of Mikrobiologiya)* **65**, 719-726.

Gorelova, O.A., Baulina, O.I., Korzhenevskaya, T.G. and Gusev, M.V. (1997) Formation of hormogonia and their taxis during the interaction of cyanobacteria and plants, *Microbiology (Translated of Mikrobiologiya)* **66**, 669-675.

Gorelova, O.A., Baulina, O.I., and Korzhenevskaya, T.G. (1999) Peculiarities of cell division in *Nostoc muscorum* CALU 304 in monoculture and associations with plant tissues, *Microbiology (Translated of Mikrobiologiya)* **68**, 460-464.

Gorelova, O.A., Korzhenevskaya, T.G., Baulina, O.I., and Gusev, M.V. (2000) Communication of spatially separated cyanobacteria and plant tissues in the model systems, in H.D. Weber, S. Imhof, and D. Zeuske (eds.), *Third International Congress on Symbiosis, Progr., Abstr. and Papers*, Marburg, pp. 79.

Grilli Caiola, M. (1980) On the phycobionts of the cycad coralloid roots, *New Phytol.* **85**, 537-544.

Grobbelaar, N., Joubert, L., Chang, D., and Coetzee, J. (1988) *In situ* EM studies of the endophytes of cycad (*Encephalartos spp.*) coralloid roots, in H. Bothe, F.J. de Bruijn and W.E. Newton (eds.), *Nitrogen Fixation: Hundred Years After*, Gustav Fischer, Stuttgart, p. 209.

Grobbellaar, N., Scott, W.E., and Marshall, J. (1987) The identification of the coralloid root endophytes of the southern african cycads and ability of the isolates to fix dinitrogen, *S. Afr. J. Bot.* **85**, 111-113.

Gromov, B.V. (1986) Inclusions, in B.V. Gromov (ed.), *The Functional Structure of Cyanobacteria*, Izdatelstvo Leningradskogo Universiteta, Leningrad, pp. 108-115 (in Russian).

Grushvitskii, I.V. and Chavchavadze, E.S. (1978) The class *Cycadopsida*, in A.A.Phedorov (ed.), *The Life of Plants* **4**, Izdatelstvo Prosveshenie, Moscow, pp. 268-295 (in Russian).

Gusev, M.V. and Korzhenevskaya, T.G. (1990) Artificial associations, in A.N. Rai (ed.), *Handbook of Symbiotic Cyanobacteria*, CRC Press, Boca Raton, Florida, pp. 173-230.

Gusev, M.V., Semenova, L.R., Levina, G.A., and Mineyeva, L.A. (1981) Study of conditions of L-transformation in cyanobacteria, *Mikrobiologiya* **50**, 114-121.

Gusev, M.V., Baulina, O.I., Semenova, L.R., Mineyeva, L.A., and Kaz, L.N. (1983) Electron-microscopic study of the cyanobacterium *Chlorogloea fritschii* L-like colonies, , 629-633 (in Russian).

Gusev, M.V., Butenko, R.G., and Korzhenevskaya, T.G. (1984) Cyanobacteria in association with cultivated cells of higher plants, in V.P.Skulachev (ed.), *Soviet Scientific Reviews, Sect. D* **4**, *Physicochemical Biology Reviews*, Harwood Acad. Publ., N.Y., pp. 1-40.

Gusev, M.V., Korzhenevskaya, T.G., Pivovarova, L.V., Baulina O.I., and Butenko, R.G. (1986) Introduction of a nitrogen-fixing cyanobacterium into tobacco shoot regenerates, *Planta* **167**, 1-8.

Hallmann, J., Quadt-Hallmenn, A., Mahafee, W.F., and Kloepper, J.W. (1997) Bacterial endophytes in agricultural crops, *Can. J. of Microbiology* **43**, 895-914.

Hensyl, W.R. (1989) *Bergey's Manual of Systematic Bacteriology*, Williams & Wilkins, London, (p.1780).

90. Hirosa, M. (1987) Energy supply system for the gliding movement of gormogonia of the cyanobacterium *Nostoc cycadae*, *Plant Cell Physiol.* **28**, 587-597.

Hoischen, C., Gura, K., Luge, C., and Gumpert, J. (1997) Lipid and fatty acid composition on cytoplasmic membranes from *Streptomyces hydroscopicus* and its stable protoplast-type L-form, *J. Bacteriol.* **179**, 3430-3436.

Jensen, T.E. (1993) Cyanobacterial ultrastructure, in T. Berner (ed.), *Ultrastructure of Microalgae*, CRC Press, Boca Raton, Ann Arbor, London, Tokyo, pp. 7-50.

Johansson, C. (1994) Establishment of the *Gunnera-Nostoc* symbiosis, *Doctoral Dissertation*, Stockholm University, Stockholm.

Johansson, C. and Bergman, B. (1992) Early events during the establishment of *Gunnera-Nostoc* symbiosis, *Planta* **188**, 403-413.

Kennedy, I.R. and Tchan, Y.T. (1992) Biological nitrogen fixation in non-leguminous field crops: resent advances, *Plant & soil* **141**, 93-118.

Kerby, N., Rowell, P., Gantar, M., and Obreht, Z. (1992) Novel nitrogen-fixing associations between cereals and cyanobacteria, in *Abstr. 8th East. Eur. Symp. on Biological Nitrogen Fixation "Nitrogenfix-92"*, Saratov. PD 2.

Kimura, G. and Nakano, T. (1990) Reconstitution of the *Blasia-Nostoc* sybiotic association under axenic conditions, *Nova Hedwigia* **50**, 191-200.

Knight, C.D. and Adams, D.G. (1996) A method for studying chemotaxis in nitrogen fixing cyanobacterium-plant symbioses, *Physiological and Molecular Plant Pathology* **49**, 73-77.

Korzhenevskaya, T.G. (1990) Experimental symbiology (applied to plant syncyanoses), *Doctoral Dissertation*, Moscow State Univ., Moscow (in Russian).

Korzhenevskaya, T.G., Gorelova, O.A., Gusev, M.V., and Butenko, R.G. (1985) Co-cultivation of higher plant cells and nitrogen-fixing cyanobacteria, *Fisiologiya Rastenii* **32**, 88-96 (in Russian).

Korzhenevskaya, T.G., Gorelova, O.A., Skripnikov, A.Yu., Pivovarova, L.V., Lobakova, E.S., Baulina, O.I., Butenko, R.G., and Gusev, M.V. (1989) Interaction of nitrogen-fixing cyanobacteria with plant cells and tissues in artificial associations, in A.A. Baev, Ya.I. Buryanov (eds.), *Molecular and Genetic Mechanisms of Microorganisms-Plants Interaction*, Scientific Center of Biological Research of the Academy of Sciences of the USSR in Pushchino, Pushchino, Moscow, pp. 189-193 (in Russian).

Korzhenevskaya, T.G., Baulina O.I., Gorelova, O.A., and Lobakova, E.S. (1991a) Functional and morphological co-adaptations in experimental syncyanoses, in *Abstr. of Int. Symbiosis Congress*, Jerusalem, p. 54.

Korzhenevskaya, T.G., Lobakova, E.S., Gorelova, O.A., and Gusev, M.V. (1991b) Reconstruction of symbioses (lichen and plant syncyanoses as example), in N.S. Golubkova (ed.), *Main Problems in Experimental Lichenology in the USSR*, USSR Academy of Sciences, Komarov Botanical Institute, Leningrad, pp. 21-29 (in Russian).

Korzhenevskaya, T.G., Baulina, O.I., Gorelova, O.A., Lobakova, E.S., Shchelmanova, A.G., Burtseva, O.I., Artamonova, K.B., Kleimenov, S.Yu., and Gusev, M.V. (1992), Development of fundamental and methodical grounds for symbiogenesis, in *Abstr. II All-Union Conf. on the Gene and Cell engineering*, Moscow, pp. 74-75 (in Russian).

Korzhenevskaya, T.G., Baulina, O.I., Gorelova, O.A., Lobakova, E.S., Butenko, R.G., and Gusev, M.V. (1993) Artificial syncyanoses: the potential for modeling and analysis of natural symbioses, *Symbiosis* **15**, 77-103.

Korzhenevskaya, T.G., Gusev, M.V., and Ehwald, R. (1994a) Associations of *Lemnaceae* plants with microorganisms, *Vestnik Moskovskogo Universiteta, Biologia* no. **2**, 15-20 (in Russian).

Korzhenevskaya, T.G., Gorelova, O.A., Lobakova, E.S., Baulina, O.I., Polesskii, A.V., and Gusev, M.V. (1994b) Development of fundamental and methodical grounds for symbiogenesis, *Mol. Genet., Mikrobiol.& Virusol.* no. **4**, 20-21 (in Russian).

Korzhenevskaya, T.G., Gorelova, O.A., Baulina, O.I., and Gusev, M.V. (1999a) Accumulation of reserve polymers by *Nostoc muscorum* CALU 304 cells grown in mixed culture with plant tissues, *Microbiology (Translated of Mikrobiologiya)* **68**, 158-163.

Korzhenevskaya, T.G., Lobakova, E.S., Dol'nikova, G.A., and Gusev, M.V. (1999b) Topography of microsymbionts in apogeotropic roots of the cycads *Cycas revoluta* Thunb. and *Encephalartos horridus* (Jacq.) Lehm., *Microbiology (Translated of Mikrobiologiya)* **68**, 111-116.

Koval'skaya, N.Yu., Lobakova, E.S., and Umarov, M.M. (2001, in press) The formation of artificial nitrogen-fixing symbioses with the rape (*Brassica napus* var. *napus*) plants in nonsterile soil, *Microbiology (Translated of Mikrobiologiya)* **70**.

Kravchenko, L.V., and Borovko, A.V. (1989), Selection of microorganisms and plants in order to creating an association with increased auxine production, in *Abstr. All-Union Conf. Microorgamisms as stimulators and inhibitors of plants and animals growth,* Tashkent, 105 (in Russian).

Lee, K.Y., and Joseph, C.M. (1988) Glutamine synthetase specific activity and protein concentration in symbiotic *Anabaena* associated with *Azolla caroliana*, *Antonie van Leeuwenhoek* **54**, 345-355.

Lindblad, P. and Bergman, B. (1986) Glutamine syntetase: activity and localization in cyanobacteria of the cycads *Cycas revoluta* and *Zamia skinneri*, *Planta* **169**, 1-7.

Lindblad, P. and Bergman, B. (1990) The cycad-cyanobacterial symbiosis, in A.N. Rai (ed.), *Handbook of Symbiotic Cyanobacteria*, CRC Press, Boca Raton, Florida, pp. 137-159.

Lobakova, E.S. and Baulina, O.I. (2001) Morphology and ultrastructure of symbiotic cyanobacteria in apogeotropic roots of dormant cycad plants, in *Abstr. All-Russia Conf. Agricultural Microbiology in XIX-XXI centuries,* St. Petersburg, pp. 60-61 (in Russian).

Lobakova, E.S., Korzhenevskaya, T.G., Gusev, M.V., and Butenko, R.G., (1984) Synthesis of biologically active compounds by ginseng cells association with the cyanobacterium *Chlorogloea fritschii, Docladi Akademii Nauk SSSR,* **276,** 1017-1018 (in Russian).

Lobakova, E.S., Gorelova, O.A., and Korzhenevskaya, T.G (1989) N_2-fixing cyanobacteria as stimulators and inhibitors of plants growth, in *Abstr. All-Union Conf. Microorgamisms as Stimulators and Inhibitors of Plants and Animals Growth,* Tashkent, 51 (in Russian).

Lobakova, E.S., Baulina, O.I., and Korzhenevskaya, T.G. (1990) Distinct patterns of adhesion of *Chlorogloeopsis fritshii* cyanobacterium on ginseng cells in mixed culture, in *Abstr. III All-Union Conf. Biosynthesis of Cellulose and Other Compounds of Cell Wall* ("Cellulose-90"), Kazan, p.19 (in Russian).

Lobakova, E.S., Shchelmanova, A.G., Baulina, O.I., and Korzhenevskaya, T.G. (1992) Electron microscopic study of the microsymbiotic complex from *Azolla* sp. in an artificial association with tobacco callus, in *Abstr. 14th Conf. on Electron Microscopy (Biology and Medicine)*, Moscow, p. 57 (in Russian).

Lobakova, E.S., Dol'nikova, G.A., Korzhenevskaya, T.G., and Gusev, M.V. (1996) Cyanobacterial-bacterial complexes of fern *Azolla sp.* in artificial associations with tobacco callus and plants, in *Abstr. Conf. Autotrophic Microorganisms*, Moscow, p. 44 (in Russian).

Lobakova, E.S., Zagoskina, N.V., Troshina, U.V., Baulina, O.I., and Korzhenevskaya, T.G. (2000) Structure of coralloid roots and the synthesis of phenol compounds therein: a study with different cycad species, in H.C. Weber, S. Imhof, and D. Zeuske (eds.), *Third International Congress on Symbiosis, Progr., Abstr. and Papers*, Marburg, p. 122.

Lobakova, E.S., Dol'nikova, G.A., and Korzhenevskaya, T.G. (2001a) Cyanobacterial-bacterial complexes in plant syncyanoses, *Microbiology (Translated of Mikrobiologiya)* **70**, 111-116.

Lobakova, E.S., Koval'skaya, N.Yu., and Umarov, M.M. (2001b) New approach in creating artificial associations of plants with nitrogen-fixing bacteria, in *Abstr. All-Russia Conf. Agricultural Microbiology in XIX-XXI centuries*, St. Petersburg, pp.61-62 (in Russian).

Lobakova, E.S., Shchelmanova, A.G., Korzhenevskaya, T.G., and Gusev, M.V. (2001c) Infection of plants and plant tissue cultures with cyanobacteria-bacteria complexes, *Microbiology (Translated of Mikrobiologiya)* **70,** 299-305.

Maksimova, N.I., Merzlyak, M.N., and Gusev, M.V. (1990) Cell and tissue cultures in studying pathogen – host plant relationships, *Nauchye Doklady Vysschei Shkoly, Biologicheskie Nauki* no. **2**, 6-21 (in Russian).

Meeks, J.C. (1990) Cyanobacterial-bryophyte associations, in A.N. Rai (ed.), *Handbook of Symbiotic Cyanobacteria*, CRC Press, Boca Raton, Florida, pp. 43-63.

Meeks, J.C. (1998) Symbiosis between nitrogen-fixing cyanobacteria and plant, *BioScience* **48**, 266-276.

Meeks, J.C., Joseph, C.M., and Haselkorn, R. (1988) Organization of the *nif* genes in cyanobactheria in symbiotic association with *Azolla* and *Anthoceros*, *Arch. Microbiol.* **150**, 61-71.

Meeks, J.C., Campbell, E., Hagen, K., Hanson, T., Hitzeman, N., and Wong F. (1999) Developmental alternatives of symbiotic *Nostoc punctiforme* in response to its plant partner *Anthoceros punctatus*, in G.A. Peschek, W. Loffelhardt, and G. Schmetterer (eds.), *The Phototrophic Prokaryotes*, Kluwer Academic/Plenum Publishers, New York, pp. 665-678.

Nadezhdina, N.S., Kashina, A.S., and Severin, F.F. (1995) Microtubule proteins, in V.T Ivanova and V.M.Lipkin, (eds.), *Proteins and Peptides*, V.1, Nauka Publ. Co, Moscow.

Nie, Y.F., Vesk, M., Kennedy, I.A., Sriskandarajah, S., Lane, E., and Tchan, Y.T. (1992) Structure of 2,4-D induced *para*-nodules with *Rhizobium* on wheat, *Phytochem.* **11**, 67-73.

Nishiyama, Y., and Yamaguchi, H. (1990) Morphological detection of filipin-sterol complexes in the cytoplasmic membrane of staphylococcal L-form, *Microbiol. Immunol.* **34,** 25-34.

Nosov, A.M. (1991) Control of secondary metabolite biosynthesis in suspension cell cultures, in R.G. Butenko (ed.), *Biology of Cultivated Cells and Biotechnology of Plants*, Nauka, Moscow, pp. 5-20 (in Russia)

Obreht, Z., Kerby, N.W., Gantar, M., and Rowell, P. (1993) Effect of root-associated N_2-cyanobacteria on the growth and nitrogen content on wheat (*Triticum vulgare* L.) seedlings, *Biology and Fertility of Soils* **149,** 68-72.

Obukowicz, M., Schaller, M., and Kennedy, G.S. (1981) Ultrastructure and phenolic histochemistry of the *Cycas revoluta-Anabaena* symbiosis, *New Phytol.* **87**, 751-759.

Ow, M.C., Gantar, M., and Elhai, J. (1999) Reconstitution of a cycad-cyanobacterial association, *Symbiosis* **27**, 125-134.

Ozawa, T. and Yamaguchi, M. (1980) Increase in cellulase activity in cultured soybean cells caused by *Rhizobium japonicum*, *Plant Cell Physiology* **21**, 331-337.

Pankratova, E.M. (1987) Participation of cyanobacteria in nitrogen turnover in soil and it`s fertility, A.A. Imshentsky (ed.), *Uspechi Mikrobiologii* **21**, Nauka, Moscow, 181-212 (in Russian).

Paseshnitchenko, V.A. (1987) Biosynthesis and biological activity of plant terpenoids and steroids, *Itogi Nauki i Techniki, Ser. Biol. Chimiya* **25**, 1-196 (in Russian).

Patnaik, G.K., Kanungo, P.K., and Rao, V. (1994) Interaction of 2,4-dichlorophenoxyacetic acid (2,4-D) with nitrogen fixing bacterial populations and nitrogen fixation associated with rice, *Microbiological Research* **149**, 291-295.

Peat, A. (1968) Fine structure of the vegetative thallus of the lichen *Peltigera polydactyla*, *Arch. Mikrobiol.* **61**, 212-222.

Peters, G.A. and Meeks, J.C. (1989) The *Azolla – Anabaena* symbiosis: basis biology, *Ann. Rev. Plant Physiol. & Plant Mol. Biol.* **40**, 193-210.

Peveling, E. (1973) Vesicles in the phycobiont sheath as possible transfer structures between the symbionts in the lichen *Lichina pygmaea, New Phytol.* **72**, 343-346.

Pinevich, A.V. (1997) Intracytoplasmic membrane structures in bacteria, *Endocytobiosis & Cell Res.***12**, 9-40.

Pivovarova, L.V., Korzhenevskaya, T.G., Butenko, R.G., and Gusev, M.V. (1986) Localization of cyanobacteria growing in association with callus culture and with regenerated plants of tobacco, *Fisiologiya Rastenii* **33**, 74-81 (in Russian).

Prozorovsky, S.V., Katz, L.N., and Kagan, G.Ya. (1981) *L-Forms of Bacteria (mechanism of formation, structure, role in pathology)*, Meditcina, Moscow (in Russian).

Prozorovsky, S.V., Zigangirova, N.A., Konstantinova, N.D., and Katz, L.N. (1987) The phenomenon of unbalanced growth in bacteria, *Zhurnal Mikrobiologii, Epidemiologii i Immunologii* **12**, 94-102 (in Russion).

Rai, A.N. (ed.) (1990) *Handbook of Symbiotic Cyanobacteria*, CRC Press, Boca Raton, Florida.

Rai, A.N., Rowell, P., and Stewart, W.D.P. (1980) NH_4^+ assimilation and nitrogenase regulation in the lichen *Peltigera aphtosa* Willd., *New Phytol.* **85**, 545-555.

Rai, A.N., Borthakur, M., and Paul, D. (1996) Symbiotic cyanobacteria: biotechnological applications, *J. of Scientific & Industrial Research* **55**, 742-752.

Rai, A.N., Malin, C., and Bergman B. (2000a) Creation of new nitrogen-fixing association, in *Abst. of Summer School and Workshop on "Cyanobacterial Symbioses, their Evolution and the Creation of New Cyanobacterial Symbioses"*, Ballyvaughan, Ireland, 88.

Rai, A.N., Söderbäck, E., and Bergman, B. (2000b) Cyanobacterium-plant symbioses, *New Phytol.* **147**, 449-481.

Rasmussen, U., Johansson, C., and Bergman, B. (1994) Early communication in the *Gunnera-Nostoc* symbiosis: plant-induced cell differentiation and protein synthesis in the cyanobacterium, *Mol. Plant-Microbe Interact.* **7**, 696-702.
Reddy, P.M., Ladha, J.K., So, R.B., Hernandes, R.J., Ramos, M.S., Angeles, O.R., Dazzo, F.B., and Debruijn, F.J. (1997) Rhizobial communication with rice roots – induction of phenotypic changes, mode of invasion and extent of colonization, *Plant & Soil* **194**, 81-98.
Richter, R., Hejazi, M., Kraft, R., Ziegler, K., and Lockau, W. (1999) Cyanophycinase, a peptidase degrading the cyanobacterial reserve material multi-L-arginyl-poly-L-aspartic acid (cyanophycin). Molecular cloning of the gene of *Synechocystis* sp. PCC 6803, expression in *Escherichia coli*, and biochemical characterization of the purified enzyme, *Eur. J. Biochem.* **263**, 163-169.
Roddick, J.G. (1980) Steroidal alkaloids, in E.A. Bell and B.V. Charlwood (eds.), *Secondary Plant Products, Encycl. Plant Physiol., New Ser.* **8**, Springer-Verlag, Berlin, Heidelberg, New York, pp.174-176.
Schenk, H.E.A. (1992) Cyanobacterial symbioses, in A.Ballows, H.G. Truper, M. Dworkin, W. Harder, and K-H. Schleifer (eds.), *The Prokaryotes,* 2nd edition, Springer-Verlag, N.Y., pp.3819-3854
Semenova, L.R., Mineyeva, L.A., and Gusev, M.V. (1982) The effect of osmotic stabilizing agent on the formation of spheroplasts in cyanobacteria and their photosynthetic activity, *Mikrobiologia* **51**, 259-266 (in Russian).
Serrano, R., Carrapiço, F., and Vidal, R. (1999) The presence of lectins in bacteria associated with the *Azolla-Anabaena* symbiosis, *Symbiosis* **27**, 169-178.
Silvester, W.B. and McNamara, P.J. (1976) The infection process and ultrastructure of the *Gunnera-Nostoc* symbiosis, *New Phytol.* **77**, 135-141.
Simon, R.D. (1973) The effect of chloramphenicol on the production of cyanophycin granule polypeptide in the blue-green alga *Anabaena cylindrica, Arch. Microbiol.* **92**, 115-122.
Simon, R.D. (1976) The biosynthesis of multi-L-arginyl-poly(L-aspartic acid) in the filamentous cyanobacterium *Anabaena cylindrica, Biochim. Biophys. Acta* **422**, 407-418.
Smith D.C. and Douglas, A.E. (1987) *The Biology of Symbiosis*, Edward Arnold, London.
Söderbäck, E., Lindblad, P., and Bergman, B. (1990) Developmental patterns related to nitrogen fixation in the *Nostoc-Gunnera magellanica* Lam. Symbiosis, *Planta* 182, 355-362.
Spiller, H. and Gunasekaran, M. (1990) Ammonia – excreting mutant strain of cyanobacteraun *Anabaena variabilis* support growth of wheat, *Appl. Microbiol. & Biotechnol* **33**, 477-480.
Spiller, H. and Gunasekaran, M. (1991) Simultaneous oxygen production and nitrogenase activity of an ammonium-exreting mutant of the cyanobacterium *Anabaena variabilis* in co-culture with wheat, *Appl. Microbiol. & Biotechnol.* **35**, 798-804.
Stanier, R.Y., Kunisava, R., Mandell M., and Cohen-Bazire, G. (1971) Purification and properties of unicellular blue-green algae (order *Chroococcales*), *Bact. Revs.* 35, 171-205.
Street, H.E., ed. (1977) *Plant tissue and cell culture, Botanical Monographs.* **11**, Blackwell Scientific Publications, Oxford, London, Edinburgh, Melbourne.
Stulp, B.K. and Stam, W.T. (1985) Taxonomy of the genus *Anabaena* (*Cyanophyceae*) based on morphological and genotypic criteria, *Arch. Hydrobiol. Suppl.* **71**, 257-268
Tarasenko, V.A., Nguen Van Hoa, and Tarnavskii, E.B (1990) Investigation of prospects components of artificial associations wheat plants with cyanobacteria, in *Abstr. IV Republican Conf. on Electron Microscopy (Electron Microscopy and Modern Technology)*, Kishineu, pp. 109-110 (in Russian).
Tchan, Y.T. and Kennedy I.R. (1989) Possible N_2-fixing root nodules induced in nonlegumens, *Agric. Sci.* **2**, 57-59.
Tchan, Y.T. and Zeman, M.M. (1995) N_2-fixation (C_2H_2 reduction) in 2,4-dichloro-phenoxacetic acid (2,4-D) treated wheat inoculated with free-living diazotrophs, *Soil boil. & biochem.* **27**, 453-457.
Towata, E.M. (1985) Morphometric and cytochemical ultrastructural analyses of the *Gunnera kaalensis-Nostoc* symbiosis, *Bot. Gaz.* **146**, 293-301.
Vagnoli, L., Margheri, M.C., Allotta, G., and Materassi, R. (1992) Morphological and physiological properties of symbiotic cyanobacteria, *New Phytologist* **120**, 243-249.
Wallace, W.H. and Gates, J.E. (1986) Identification of eubacteria isolated from leaf cavities of four species of N-fixing *Azolla* fern as *Arthrobacter* conn and dimmick, *Appl. Enuiron. Microbiol. 52*, 425-429.
Watts, S., Knight, C.D., and Adams, D.G. (1999) Characterization of plant exudates inducing chemotaxis in nitrogen-fixing cyanobacteria, in G.A. Peschek., W. Loffelhardt, and G. Schmetterer (eds.), *The Phototrophic Prokaryotes*, Kluwer Academic/Plenum Publishers, New York, pp.679-684.
West, N.J. and Adams, D.C. (1997) Phenotypic and genotypic comparison of symbiotic and free-living cyanobacteria from single field site, *App. Environ. Mocrobiol.* **63**, 4479-4484.

Woehleche, H. and Ehwald, R. (1995) Characterization of size-permeation limits of cell walls and porous separation materials by high performance size-exclusion chromatography, *J. Chromatography*, **708**, 263-271.

Wolk, C.P., Zhu, J., and Kong, R. (1999) Genetic analysis of heterocyst formation, in G.A. Peschek, W. Loffelhardt, and G. Schmetterer (eds.), *The Phototrophic Prokaryotes*, Kluwer Academic/Plenum Publishers, New York, pp. 509-515.

Yagodina, I.B., Gorskaya, N.V., Korzhenevskaya, T.G., Gusev, M.V., and Butenko, R.G. (1990) Associative interactions of *Dioscorea* and cyanobacteria cells in mixed culture, *Izv. Akad. Nauk SSSR Ser. Biol.* no. **3**, 329-337 (in Russian).

Zagoskina, N.V. and Zaprometov, M.N. (1991) Tissue culture of tea plants: some aspects of the phenolic compound formation, in R.G. Butenko (ed.), *Biology of Cultivated Cells and Biotechnology of plants*, Nauka, Moscow, pp. 32-35 (in Russia)

Zhang, C.C., Huguenin, S., and Friry, A. (1995) Analysis of genes encoding the cell division protein FtsZ and a glutathione synthetase homologue in the cyanobacterium *Anabaena sp.* PCC 7120, *Res. Microbiol.* **146**, 445-455.

Zimmerman, W.J., Rosen, B.H., and Lumpkin, T.A. (1989) Enzymatic, lectin and morphological characterization and classification if presumptive cyanobionts from *Azolla* Lam., *New Phytol.* **113**, 497-503.

Chapter 15

CYANOBACTERIAL DIVERSITY AND SPECIFICITY IN PLANT SYMBIOSES

ULLA RASMUSSEN AND MALIN NILSSON
Department of Botany, Stockholm University, 10691 Stockholm, Sweden

1. INTRODUCTION

An important aspect in the understanding of the onset and maintenance of cyanobacterial symbioses, is to investigate and gain information on the diversity of the involved cyanobacteria and the specificity of the interaction. Cyanobacteria, predominantly belonging to the genus *Nostoc*, have the ability to form symbioses with a range of taxonomically different hosts, to infect and colonize different tissues and organs in the host, and to be inter- or intra-cellularly located in the host cells. This is a unique phenomenon among nitrogen-fixing symbioses and naturally raises questions related to diversity and specificity, and calls for a clear definition and description of the symbiont. In this Chapter we are addressing these questions and give a general statement of our current knowledge of diversity and specificity.

2. CYANOBACTERIA IN SYMBIOSES

Cyanobacteria are a large group of microorganisms, of which only a few genera have symbiotic capacity. To be symbiotically competent, an extreme adaptability of the cyanobacteria is required. They must be capable of adapting to the conditions offered by the host and alter their metabolism to be in a mutualistic state, exchanging metabolites with the host. Another prerequisite of symbiotic cyanobacteria, which form symbioses with plants, is their capacity to differentiate motile filaments, hormogonia, which constitute the "infection unit". Although numerous cyanobacteria produce motile hormogonia, the capacity to enter into symbioses with plants is restricted to a few genera (Table 1). In addition, the cyanobacteria must exist as free-living in the close vicinity of the potential host. *Nostoc* is the most dominating genus in terrestrial symbiotic systems, forming symbioses with fungi, bryophytes, *Azolla*, cycads, and *Gunnera*.

A.N. Rai, B. Bergman and U. Rasmussen (eds.), Cyanobacteria in Symbiosis, 313-328.

Table 1. Examples of cynobacterial genera involved in symbioses

Cyanobacterial genera	**Symbiotic partner**
Aphanocapsa	Marine sponges
Anabaena	*Azolla*, Cycads, Mosses
Calothrix	Lichens, Liverworts and hornworts, Cycads, Diatoms
Chlorogloeopsis	Liverworts and hornworts
Cyanelles	Glaucophytes, *Paulinella*
Cyanothece	Diatoms
Fischerella	Lichens
Gloeocapsa	Lichens
Gloeothece	Lichens
Hyella	Lichens
Nostoc	Lichens, Bryophytes, Cycads, *Gunnera*
Oscillatoria	Marine sponges, Mosses
Phormidium	Marine sponges, Mosses
Pleurocapsa minor	Bacteria
Scytonema	Lichens
Synechocystis	Lichens, Marine sponges
Richelia	Diatoms

Nostoc is the only cyanobacterial genus found in *Gunnera*, and is predominating in cycads, lichens and bryophytes. In contrast, several different cyanobacterial genera have been recorded in lichen symbioses. Unicellular (*Gloeocapsa, Gloeothece, Synechocystis* and *Hyella*), filamentous heterocystous (*Nostoc, Calothrix,* and *Scytonema*), and heterocystous branched-filamentous (*Fischerella*) forms have been detected in these associations (Rai, 1990). Even though *Nostoc* is predominating in cycads, suggestions and a few investigations have indicated the possible presence of *Anabaena* and *Calothrix* (Obukowicz *et al.*, 1981; Zhu, 1982; Grobbelaar *et al.*, 1987). Also in bryophytes, investigations have shown the presence of *Calothrix* and *Chlorogloeopsis* (West and Adams, 1997). In the water fern *Azolla*, the symbiotic cyanobacteria have been identified as *Anabaena* (Moore, 1969; Lumpkin and Pluknett, 1980). However, new data on the taxonomic affiliation strongly indicate that the symbiont belongs to the genus *Nostoc* (Meeks *et al.*, 1988; Plazinski *et al.*, 1990). The preference of *Nostoc* as the symbiotic partner in terrestrial systems could be explained by the fact that *Nostoc* is dominating among cyanobacteria in terrestrial habitats and has an extremely widespread geographic distribution (Vincent, 1988).

In aquatic symbiotic systems, cyanobacteria are found in association with diatoms, sponges or dinoflagellates. Diatoms have been found to associate with the cyanobacteria *Rhichelia intracellularis* and *Calothrix rhizosolenia* (Ostenfeld and Schmidt, 1901). More recent studies have also suggested association with the cyanobacteria *Cyanothece*

(Carpenter and Janson, 2000). Marine sponges form symbioses with four cyanobacterial genera: *Aphanocapsa, Synechocystis, Oscillatoria* and *Phormidium* (Adams, 2000).

3. METHODS USED FOR IDENTIFICATION OF THE SYMBIONTS

Identification of a symbiont is often problematic because 1) the morphological and physiological behavior might change in the symbiotic conditions compared to isolated cultures or free-living stage of the organisms, 2) when culturing a symbiotic organism, a selection cannot be excluded, or in the extreme, it might not be cultivable. In symbiotic cyanobacteria both factors have to be considered. The change in morphology is well documented for the genus *Nostoc*, which exhibits an extreme flexibility influenced by environmental conditions, including those imposed by the hosts (Potts, 2000; Dodds and Gudder, 1995). The latter is also a well-known fact of the cyanobacteria in *Azolla* (Plazinski *et al.*, 1990; Tang *et al.*, 1990). Just a few decades ago identification was mostly based on morphological/physiological observations of isolated strains or by direct microscopic examination. Based on these criteria, an identification and affiliation to a certain genus can be ensured, but below that taxonomic level morphological identifications display too low resolution. The development and use of different molecular methods have created new possibilities for identification in general, and for studying the diversity and specificity of symbiotic cyanobacteria in particular (Wilmotte, 1994). The different approaches used in the study of symbiotic cyanobacteria are given in Table 2. Based on the results from the individual studies, our knowledge has expanded in respect to the genetic diversity of the symbionts within particular host genera. However, it is important to notice that different genetic methods have been used in the separate investigations. The results are therefore difficult to compare since the varying methods might have different degrees of genetic resolution. Thus, one should be careful when comparing the individual symbioses in respect to the degree of diversity and specificity, at least with the current information.

4. DIVERSITY OF THE SYMBIONTS IN THE INDIVIDUAL SYMBIOSES

With the exception of the *Azolla*, and possibly lichen symbioses, each new offspring of the cyanobacterial hosts is infected *de novo*. This imposes a natural limitation in the possible diversity of infecting cyanobacterial strains in *Azolla* and lichens and could in theory suggest high diversity in the others. Below is a summary of our current knowledge in regards to the cyanobacterial diversity in each symbiosis (see also Fig. 1).

4.1. Azolla

Among the cyanobacterial-plant symbioses, the one with the water fern *Azolla* is unique in the sense that the symbiont never leaves the host. There is continuity of the symbiosis throughout the reproductive cycle of the fern (Peters and Meeks, 1986). As a consequence there seems to be no, or very little genetic diversity of the symbiont within a particular species of *Azolla* (Zheng *et al.*, 1999). Moreover, phylogenetic analyses of the symbiont from the different host species strongly support the hypothesis of co-

Table 2. General overview of the different methods used for cyanobacterial diversity and specificity investigations

Host	Method	Result	Source
Lichens	Southern hybridization with *nif* probes	Difference between fresh and cultured isolates	Kardish *et al.*, 1990
	Physiological/morphological	Classification at genus level	Rai, 1990
	Sequence analysis ($tRNA^{Leu}$ (UAA) intron)	Limited number of *Nostoc* strains, high specificity, one strain/one host species	Paulsrud and Lindblad, 1998; Paulsrud *et al.*, 1998, 2000, 2001
Bryophytes	Reconstitution experiments	Several different *Nostoc* strains can associate with the host, also strains from other symbioses	Ridgeway, 1967; Rodgers and Stewart, 1977; Enderlin and Meeks, 1983
	PCR fingerprinting and ITS-RFLP	Diversity among the cavities in a thallus - no diversity within a cavity	West and Adams, 1997
	Sequence analysis ($tRNA^{Leu}$ (UAA) intron)	Diversity within thalli, one strain per cavity	Costa *et al.*, 2001
Azolla	Morphological	One cyanobacterial species- *Anabaena azollae*	Moore, 1969; Lumpkin and Plucknet, 1980
	Immunological	Differences between symbiotic and free-living isolates	Gates *et al.*, 1980; Ladha and Watanabe, 1982; Arad *et al.*, 1985; McCowen *et al.*, 1987;
		4 subgroups of *Anabaena azollae*	Lui *et al.*, 1989
	Southern hybridization	Genetic variation of *Anabaena* in different hosts	Plazinski *et al.*, 1988, 1990
	RFLP using *nif* probes	Cyanobacteria from host sections *Euazolla* and *Rhizosperma* have different evolutionary lines	Franche and Cohen-Bazire, 1987; van Coppenolle *et al.*, 1995

Host	Method	Result	Source
	Fatty acid profiles	Confirmation of differences between cyano-bacteria from host sections *Euazolla* and *Rhizosperma*	Caudales *et al.*,1995
	PCR fingerprinting	Taxonomy of the cyanobacteria follow that of the host	van Coppenolle *et al.*, 1993
		Identification of *Anabaena azollae* strains from 4 *Azolla* species	Eskew *et al.*, 1993
		One cyanobacterial strain/ one *Azolla* species, taxonomy of cyanobactria follow that of the host	Zheng *et al.*, 1999
Cycad	Morphological	Three major types of *Nostoc* with preferences for specific cyacad groups. Symbiont classified as *Nostoc* in most cases	Grilli Caiola, 1980 Grobbelaar *et al.*, 1987 Lindblad and Bergman, 1990
	RFLP	More than one *Nostoc* strain can associate with a particular plant	Lindblad *et al.*, 1989
	Zymogram	Confirmation of morphological differences	Zimmerman and Rosen, 1992
	DNA methylation pattern, total DNA restriction pattern	Extensive diversity among cyanobacterial strains within the same host species	Lotti *et al.*, 1996
	Sequence analysis ($tRNA^{Leu}$ (UAA) intron)	No cyanobacterial diversity within one coralloid root, more than one strain can associate with the plant as a whole	Costa *et al.*, 1999
	Reconstitution experiments	Reconstitution possible with isolates from cycads and with a soil bacteria	Ow *et al.*, 1999

Host	Method	Result	Source
	PCR fingerprinting	Numerous *Nostoc* strains in one coralloid root	Zheng *et al.*, In Press
Gunnera	Reconstitution experiments	Reconstitution possible, indicates low specificity, selection seemed to occur at recognition or entrance stage	Silvester and McNamara, 1976; Johansson and Bergman 1994
	RFLP	Different cyanobacteria existed in hosts in a local area and in hosts from two different geographical locations	Zimmerman and Bergman, 1990
	PCR fingerprinting	Genetic similarities in *Nostoc* strains from different hosts, as well as genetic differences among cyanobacteria from the same host	Rasmussen and Svenning, 1998
		Diversity within and between different host species. More than one strain in a gland. Restricted spreading of related strains in the world	Nilsson *et al.*, 2000
	16S rRNA and ITS-RFLP, DGGE of *hetR*	High diversity among symbiotic isolates in the 16S rRNA and ITS region	Rasmussen and Svenning, 2001
Diatoms	Sequence analysis of *hetR*	High specificity was observed. Host specificity not dependent on geographic location	Janson *et al.*, 1999

evolution between the host and the cyanobacteria as suggested previously (van Coppenolle *et al.*, 1993; Caudales *et al.*, 1995). All attempts to maintain the symbionts of *Azolla* on artificial medium have failed. However, the morphological and physiological studies, which have been performed on the isolates, indicate that the

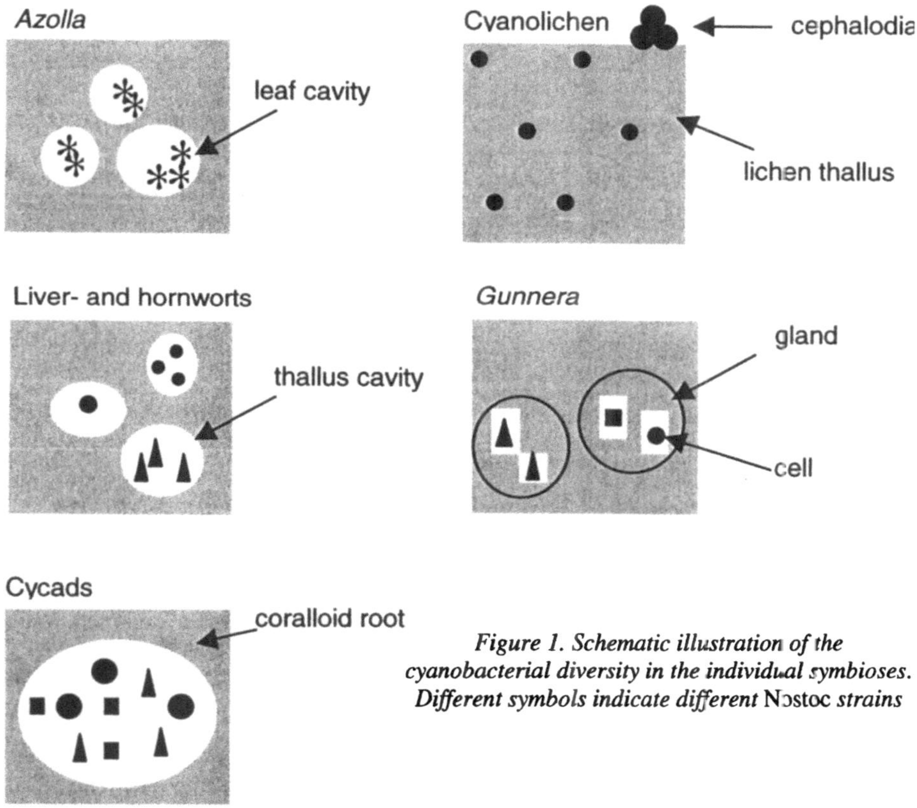

Figure 1. Schematic illustration of the cyanobacterial diversity in the individual symbioses. Different symbols indicate different Nostoc *strains*

symbionts are specific to the fern (Vaginoli *et al.*, 1992). The taxonomic status of the cyanobiont has been changing with time between the genus *Nostoc* and the genus *Anabaena*. However, more and more evidence, both morphological (e.g., its ability to form hormogonia) and genetic, indicate that the cyanobiont belongs to the genus *Nostoc* (Meeks *et al.*, 1988; Plazinski *et al.*, 1990). Several attempts have been made to classify the cyanobacteria inhabiting the different *Azolla* species by using serological methods, fatty acid profiling, RFLP analysis and PCR fingerprinting (Franke and Cohen-Bazire, 1987; Plazinski *et al.*, 1988; Liu *et al.*, 1989; van Coppenolle *et al.*, 1993, 95; Caudales *et al.*, 1990, 1995; Zheng *et al.*, 1999). The study by Zheng *et al.* (1999), using PCR fingerprinting is so far the only one where direct analysis of the cyanobacterial filaments has been performed. It is also the first study where the symbionts from all species of *Azolla* have been analyzed in the same study. A different genetic fingerprint pattern was generated from the cyanobacteria isolated from each individual *Azolla* species, indicating a high specificity. Furthermore, the results indicate that the phylogeny of the symbionts follows the taxonomy of the host *Azolla* and supports the hypothesis of co-evolution between *Azolla* and the cyanobacteria. (see also Chapter 9).

4.2. Bryophytes (Liverworts and Hornworts)

Diversity and identification of cyanobacteria in this symbiosis was investigated by using morphological and physiological criteria in early studies by Duckett *et al.* (1977). These investigations suggested low cyanobacterial diversity, even in plants collected from widely separated locations. However, in the last few years different molecular techniques have been used to study the cyanobacterial diversity (West and Adams, 1997; Costa *et al.*, 2001). By PCR fingerprinting using arbitrary primers and PCR amplification of the 16S-23S rRNA internal transcribed spacer (16S-23S ITS), a large number of genetically different *Nostoc* strains could be identified in the hornwort *Phaeoceros* sp. (West and Adams, 1997). Furthermore, different cavities in the thallus were colonized by different strains of *Nostoc*, indicating low specificity in the association. The same conclusions were drawn from the study of Costa *et al.* (2001). Using sequence analysis of the $tRNA^{Leu}$ (UAA) intron, a high diversity of *Nostoc* strains infecting the hornwort *Antoceros fusiformis* and the liverwort *Blasia pusilla* could be displayed. However, only one *Nostoc* strains could be detected in a single cavity. In both studies, the cyanobacterial samples were taken directly from field populations of bryophytes at different locations. In addition, free-living cyanobacteria isolated from the vicinity of the host were included in the study by West and Adams (1997). None of the 40 free-living strains were identical to the symbionts although most of them show symbiotic competence when tested by reconstitution experiments in the laboratory. Moreover, a very localized distribution of a strain was demonstrated. The same *Nostoc* strain was never found in hosts at two separate locations (West and Adams, 1997). This is in contrast to the observations by Costa *et al.* (2001), who could demonstrate that the same strain could be identified at locations separated by at least 200m. (see also Chapter 7).

4.3. Cycads

The symbionts of cycads have historically been identified as *Anabaena cycadae* (Spratt, 1911; Chaudhuri and Akhtar, 1931) or *Nostoc cycadae* (Watanabe, and Kiyohara, 1963). In recent years, the symbionts isolated were classified as *Nostoc* spp. (Grilli-Caiola, 1980; Bonnett and Silvester, 1981; Lindblad and Bergman, 1990). Although cyanobacteria identified as *Calothrix* sp. have been isolated from the cycad *Encephalartos* (Grobbelaar *et al.*, 1986) the main symbiont in all cycad species is *Nostoc*. Detailed morphological characterization of the symbiont could verify phenotypic differences between the cyanobacteria from different host species, and even within a single plant. Thus, a low specificity of the cyanobacteria towards the host was claimed (Grilli-Caiola, 1980; Grobbelaar *et al.*, 1987). This was further confirmed by the work of Zimmerman and Rosen (1992), using electrophoretic protein profiles and zymograms. Using genetic methods, a large genetic diversity among cyanobacterial isolates from different cycad species and genera, as well as between individual coralloid roots in a single plant, was demonstrated (Lindblad *et al.*, 1989; Lotti *et al.*, 1996). Genetically different *Nostoc* strains could be found in different coralloid roots of a single cycad plant, while only one cyanobacterial strain was observed in a single coralloid root (Costa *et. al.*, 1999). However, a recent study by Zheng *et al.* (2002) has clearly demonstrated that many genetically different *Nostoc* strains were present in a single coralloid root, suggesting a low specificity even within the symbiotic unit. This investigation was performed on indigenous *C. revoluta*, which might be of importance for explaining the observed high cyanobacterial diversity in the roots. In previous studies the cyanobacteria have been collected from plants grown in botanical gardens or greenhouses, where the "local" cyanobacterial flora in the soil might not represent the one found in a natural habitat. This new finding is extremely interesting in respect to specificity, which in cycads must be considered as being the lowest among all the cyanobacterial-plant symbioses. It also stresses the importance of using natural field samples when examining cyanobactrial diversity in plant symbioses. (see also Chapter 11).

4.4. Gunnera

In *Gunnera*, the cyanobacteria are located intracellularly in individual glands located on the stem. Morphological observations of individual isolates from *Gunnera* have clearly shown that a high diversity exists between isolates collected from naturally grown *Gunnera* species (Bergman *et al.*, 1992). The different morphological feature of *Nostoc* strains isolated from different *Gunnera* species are illustrated in Figure 2. The high morphological diversity could be confirmed by molecular analysis such as RFLP analysis and PCR fingerprinting (Zimmerman and Bergman, 1990; Rasmussen and Svenning, 1998; Nilsson *et al.*, 2000). PCR fingerprinting of cyanobacterial isolates from different naturally grown *Gunnera* species have revealed a high genetic diversity among the isolates, both within and between different host species. Furthermore, it could be demonstrated that different *Gunnera* species can be infected by the same *Nostoc* strain (Rasmussen and Svenning, 1998; Nilsson *et al.*, 2000). Moreover, within an individual plant a high genetic diversity of the cyanobacteria, collected from different glands has been documented (Nilsson *et al.*, 2000). In addition, there are indications that a single gland on the stem can be infected by different strains (Nilsson *et al.*, 2000). In this context one should keep in mind that cyanobacteria are located intracellularly in this symbiosis. It still remains to be investigated whether one cell can be infected by multiple cyanobacteria. Analyses of cyanobacteria collected from *Gunnera* sp. grown worldwide demonstrated a local geographic distribution of the cyanobacteria. However, within a country, or even within a local area, the diversity could be extensive (Nilsson *et al.*, 2000; see also Chapter 12).

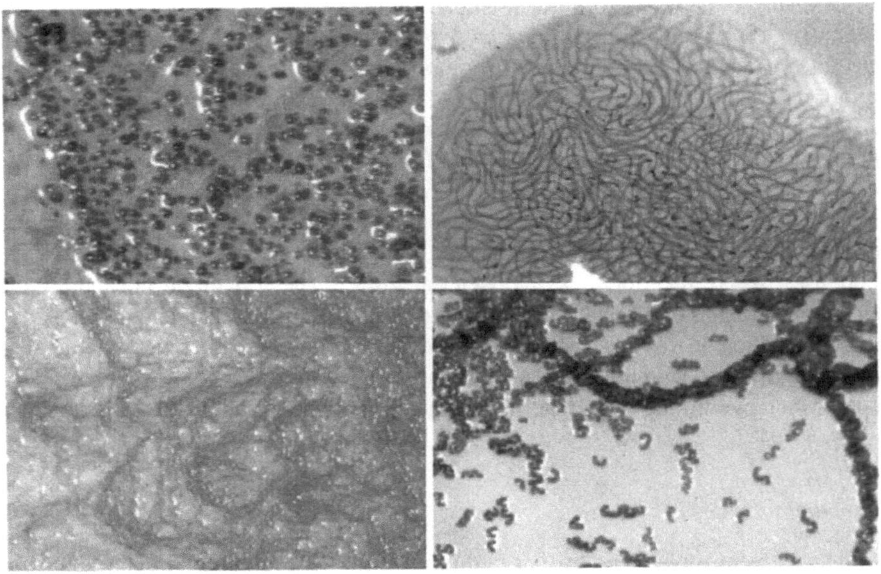

Figure 2. Different morphology among Nostoc *isolates from field-grown* Gunnera *spp. (Photo, E. Söderbäck).*

4.5. Cyanolichens

The cyanolichens differ from the other cyanobactrial symbioses in many respects for example in terms of biological existence (no organism without its symbiotic partner). Furthermore, the vegetative transfer of the symbiont, and the physiological location of the cyanobacteria are distinct characters. The differences are also reflected in the diversity and specificity of the cyanobacterial partner. Although the predominating symbiont is *Nostoc*, many other cyanobacterial genera have been identified in lichens (Table 1). In studies where the diversity of the cyanobiont has been investigated, a surprisingly low diversity was recorded (Paulsrud and Lindblad, 1998; Paulsrud *et al.*, 1998; Paulsrud *et al.*, 2000; Paulsrud *et al.*, 2001). Studies on the lichen *Peltigera* sp. using sequencing of the $tRNA^{Leu}$ (UAA) intron, could demonstrate, that only one *Nostoc* strain is present in the lichen indicating that the thallus is colonized only once or that there is a high degree of specificity. Even in the case where the lichens were collected from geographically distinct areas, or even different countries, the same *Nostoc* strain was found in all the samples. This high specificity could further be demonstrated by field studies, where colonization by *Nostoc* strains isolated from other lichen species was tried. None of the introduced strains did successfully colonize the lichen (Paulsrud *et al.*, 2001). The high specificity seen in this symbiosis could be a result of the vegetative transfer of the symbionts to each new generation. The open physiological location of the cyanobacteria within the thallus could, however, suggest a possibility of infection by multiple strains. However, since the methods used indicate that this is not the case, a strict unknown mechanism of specificity is active (see also Chapter 4).

4.6. Diatoms and Marine Sponges

The cyanobacterial diversity in diatoms is low, reflecting a high specificity. When investigating different diatom specimen, the same cyanobacterial strain was found, even in geographic distant areas (Janson *et al.*, 1999). The low diversity often seen in marine habitats could be a result of the high demands of adaptation required of cyanobacteria to withstand the harsh conditions experienced in the oceans. The diversity and specificity within marine sponges has not been investigated so at the present stage (see also Chapter 1 and 2).

5. SPECIFICITY

The unique capacity of the genus *Nostoc*, to form symbioses with representatives from all divisions within the plant kingdom, to be located in different tissues and even to be intra- or extracellular in different hosts, raises the question about the specificity in the interactions.

Specificity has been studied, for example, by reconstitution experiments. The ability to grow *Gunnera*, cycads and bryophytes, as well as cyanobacteria under sterile conditions in the laboratory has been the key to success in such experiments (Enderlin and Meeks, 1983; Ridgeway, 1967; Rodgers and Steward, 1977; Enderlin and Meeks, 1983; Johansson and Bergman, 1994; Ow *et al.*, 1999). Re-infection showed that *Nostoc* strains isolated from one host could infect another host, even one from a different host taxa. This indicate a very low host specificity and an enormous flexibility and adaptation of the cyanobacteria, since one strain must have the capacity to infect and be located in different symbiotic tissue and even to be located intra- or extracellularly.

The capacity of infection is there, but does it actually occur under natural conditions? Maybe most hosts, which are not bound to a specific symbiont strain through co-evolution, have the ability to associate with a wide range of symbionts in the artificial conditions that occur in a laboratory. These hosts may however, in their natural conditions, display a much higher apparent specificity due to many factors such as the local environment or the availability of suitable strains (see section 5.1). In the *Bradyrhizobium-Amphicarpaea bracteata* interaction it has been shown that symbiont effectiveness is dependent on host genotype. This was concluded through reconstitutions between the host and symbiotic bacterial strains collected from widely separated areas, where the most effective symbionts were strains collected from the host plant origin or from genetically related plants of the host (Wilkinson *et al.*, 1997). This further supports that reconstitutions are possible between potential but not natural partners, and that the highest yield of the association occurs with the natural symbionts best suited for that particular plant at that particular location.

Diversity and specificity in symbiosis are two phenomena closely linked together, high diversity-low specificity and vice versa. Based on the results of the investigations of cyanobacterial diversity in each particular symbiosis, we can propose the following: high specificity in the *Azolla*, lichen and diatom symbioses and low specificity in the bryophyte, cycads and *Gunnera* symbioses (West and Adams, 1997; Paulsrud and Lindblad, 1998; Paulsrud *et al.*, 1998; Costa *et al.*, 1999; Janson *et al.*, 1999; Zheng *et al.*, 1999, 2002; Nilsson *et al.*, 2000). However, it is problematic to talk about specificity in the interactions as long as we do not have any well-defined criteria to be used in grouping the organisms below the genus level. So far, most of the symbiotically competent *Nostoc* strains are referred to as the species *Nostoc punctiforme*, type strain PCC 73102. However, the identification of the species is exclusively based on morphological criteria, cell size and the formation of a *punctiforme* stage (sheath-bound filaments with terminal heterocysts) in the life cycle when growing on artificial medium. The lack of any adequate species concept makes it difficult, or impossible to determine the specificity. Strictly speaking we can only conclude that a plant is infected by *Nostoc* and that genetically different *Nostoc* spp. are present in the tissue. If any specificity exists towards different taxonomical hosts it cannot be determined with the data existing today. However, a few studies have indicated that symbiotic *Nostoc* strains do not constitute a homogenous group. Based on DNA-DNA hybridization between different cyanobacteria including two symbiotic strains PCC 73102 and PCC 7422 (originally isolated from *Cycas* sp), a very low level of genetic relatedness between the two strains (56% relative binding) was observed (Lachance, 1981). In addition, RFLP analysis of the conserved 16S rRNA region of isolates from different *Gunnera* species shows a high genetic diversity, further indicating a heterogeneity of symbiotic *Nostoc* strains (Rasmussen and Svenning, 2001).

5.1. What Affects Diversity and Specificity?

The local community of cyanobacteria in the vicinity of the potential host is one of the features that might be of importance for the diversity seen in the established symbioses. The natural flora might be determined by a lot of factors such as humidity, temperature, pH, etc. However, our knowledge concerning the natural community in the soil is rudimentary. Only a few studies have been done on naturally grown hosts where the free-living cyanobacteria, isolated form the vicinity of the host, have been included. In field studies the cyanobacteria from the hornwort *Phaeoceros* were compared with free-living cyanobacterial strains (West and Adams, 1997). Surprisingly, identical strains were never found both in symbiosis and free-living, although symbiotic competence could be demonstrated for most of the free-living strains by reconstitution experiments in the laboratory. The interpretations of these results are difficult and can at present only be speculations. It could be that the natural community of the free-living cyanobacteria has changed in time due to environmental factors since the infection of the host took place. West and Adams (1997) could also demonstrate that the local flora is changing, even within short distance, where free-living isolates from different sampling sites were never identical. This change might consequently also change the diversity observed in the host.

If many of the free-living strains are symbiotically competent why do they not infect? Many unknown factors in the microenvironment might be of crucial importance.

Significant is also the ability of the cyanobacteria to respond to chemical signals secreted from the host, inducing the differentiation into motile hormogonia. Hormogonia inducing factors has been shown to be secreted from *Gunnera* and *Anthoceros* (Campbell and Meeks, 1989; Rasmussen *et al.*, 1994; Knight and Adams, 1996). There have also been indications that the non-hosts wheat, corn, bean, sugar beat and rice can induce hormogonia (Gantar *et al.*, 1993; Svircev *et al.*, 1997). The attraction of the hormogonia to the host is still an open question but it is interesting to note that roots of wheat and rice do attract *Nostoc*. In laboratory experiments where rice was inoculated with a mixture of different symbiotic *Nostoc* strains, it was demonstrated that some kind of selection towards specific *Nostoc* strains does occur (Nilsson, unpublished data). However, the success of a particular strain could be due to factors such as fast growth, a high hormogonia forming ability, or inhibition of neighboring strains. Most likely, the answer lies in a combination of all these factors and probably numerous additional ones.

6. CONCLUDING REMARKS AND FURTHER PERSPECTIVES

Our knowledge in regards to the diversity of symbiotic cyanobacteria has expanded tremendously in recent years due to the introduction of molecular methods. Different methods have, however, been used in each symbiotic system making comparisons in a general sense almost impossible. An important future desire would therefore be to use comparable methods to be able to draw more valid conclusions on specificity and diversity. Furthermore, since the definition of species and strain is somewhat unclear at the present state, the formation of guidelines and criteria for such definition is crucial for any future investigations. Another interesting, but so far rather uninvestigated area, is the marine environment, where only a few cyanobacterial symbioses have been found, but undoubtedly many more await to be discovered. Many questions have to be addressed in understanding the specificity of symbiosis, for example what are the criteria for being symbiotically competent, concerning both cyanobacteria and host, and what are the factors affecting this event. Another way to gain insight into general specificity mechanisms is through the investigation of the capacity of symbiotic cyanobacteria to form association with non-host plant partners, such as agricultural important plants. Special focus on taxonomic definition of the cyanobacteria, natural micro-flora in the soils, the influence of environmental and geographic differences on specificity are other interesting and important aspects for future understanding of the symbiotic associations of cyanobacteria.

ACKNOWLEDGEMENTS

Financial support for our studies on cyanobacterial diversity has been obtained from the Carl Tryggers Foundation and SIDA/SAREC.

REFERENCES

Adams, D.G. (2000) Symbiotic interactions, in B.A. Whitton and M. Potts (eds.), *The Ecology of Cyanobacteria*, Kluwer Academic Publishers, Dodrecht, pp. 523-561.

Arad, H., Keysari, A., Tel-Or, E., and Kobiler, D. (1985) A comparison between cell antigens in different isolates of *Anabaena azollae*, *Symbiosis* **1**, 195-204.

Bergman, B., Johansson, C. and Söderbäck, E. (1992) Transly Review No. 42. The *Nostoc-Gunnera* symbiosis, *New Phytol.* **122**, 370-400.

Bonnett, H.T., and Silvester, W. (1981) Specificity in the *Gunnera-Noctoc* endosymbioses, *Symbiosis* **14**, 61-81.

Carpenter, E.J. and Janson, S. (2000) Intracellular cyanobactreial symbionts in the marine diatom *Climacodium frauenfeldianum*, *J. Phycol.* **36**, 540-544.

Caudales, R., Wells, J.M., Antoine, A.D. and Butterfield, J.E. (1995) Fatty acid composition of symbiotic cyanobacteria from different host plant (*Azolla*) species: evidence for coevolution of host and symbiont, *Int. J. Syst. Bacteriol.* **45**, 364-370.

Chaudhuri, H. and Akhtar, A.R. (1931) The coral-like toots of *Cycas revolute*, *Cycas circinalis* and *Zamia floridana* and the alge inhabiting them, *J. Indian Bot. Soc.* **10**, 43-59.

Costa, J.L., Paulsrud, P. and Lindblad, P. (1999) Cyanobiont diversity within coralloid roots of selected cycad species, *FEMS Microbiol. Ecol.* **28**, 85-91.

Costa, J.L., Paulsrud, P., Rikkinen, J. and Lindblad, P. (2001) Genetic diversity of *Nostoc* symbionts endophytically associated with two bryophyte species, *Appl. Environ. Microbiol.* **67**, 4393-4396.

Campbell, L.E. and Meeks, J.C. (1989) Characteristics of hormogonia formation by symbiotic Nostoc spp. In response to the presence of *Anthoceros punctatus* or its extracellular products, *Appl. Environ. Microbiol.* **55**, 125-131.

Dodds, K., Gudder; D.A. and Mollenhauer, D. (1995) Thr ecology of *Nostoc*, *J. Phycol.* **31**, 2-18.

Duckett, J.G., Prasad, A.K.S.K., Davids, D.A. and Walker, S. (1977) A cytological analysis of the *Nostoc*-bryophyte relationship, *New Phytol.* **79**, 349-362.

Eskew, D.L. Caetano-Anollés, G., Bassam, B.J. and Gresshoff, P.M. (1993) DNA amplification fingerprinting of the *Azolla-Anabaena* symbiosis, *Plant. Mol. Biol.* **21**, 363-373.

Enderlin, C.S. and Meeks, J.C. (1983) Pure culture and reconstitution of the *Anthoceros-Nostoc* symbiotic association, *Planta* **158**, 157-165.

Franche, C. and Cohen-Bazire, G. (1987) Evolutionary divergence in the *nif*HDK gene region among nine symbiotic *Anabaena azollae* and between heterocystous cyanobacteria, *Symbiosis* **3**, 159-178.

Gantar, M., Kerby, N.W. and Rowell, P. (1993) Colonization of wheat (*Triticum vulgare* L.) by N2-fixing cyanobacteria:III The role of a hormogonia-promiting factor, *New Phytol.* **124**, 505-513.

Gates, J.E., Ficher, R.W., Goggin, T.W. and Aztolan, N.I. (1980) Antigenic difference between *Anabaena azolla* fresh from the *Azolla* fern leaf cavity and free-living cyanobacteria, *Arc. Microbiol.* **128**, 126-129.

Grilli-Caiola, M. (1980) On the phycobionts of the cycad corroloid roots, *New Phytol.* **85**, 537-544.

Grobbelaarn, N., Hattingh, W. and Marshall, J. (1996) The occurrence of coralloid roots on the South African species of the Cycadales and their ability to fix nitrogen symbiotically , *S. Afr. J. Bot.* **52**, 467-471.

Grobbelaar, N., Scott, W.E., Hattingh, W. and Marshall, J. (1987) The identification of the coralloid root endophytes of the southern African cycads and the ability of the isolates to fix dinitrogen, *S. Afr. J. Bot.* **53**, 111-118.

Janson, S., Wouters, J., Bergman, B. and Carpenter, E.J. (1999) Host specificity in the *Richelia*-diatom symbiosis revealed by *hetR* gene sequence analysis, *Environ. Microbiol.* **1**(5), 431-438.

Johansson, C. and Bergman, B. (1994) reconstitution of the symbiosis of *Gunnera manicata* Linden: cyanobactrial specificity, *New Phytol.* **126**, 643-652.

Kardish, N.D., Rotem Abarbanell, D., Zilbberstein, A. and Galun, M. (1990) Comparison between the symbiotic *Nostoc* of the lichen *Nephroma laevigatum* Ach. And its cultured isolated *Nostoc* by recombinant DNA, *Symbiosis* **8**, 135-146.

Knight, C.D. and Adams, D.G. (1996) A method for studying chemotaxis in nitrogen-fixing cyanobacterium-plant symbioses, *Physiol. Molec. Plant Pathol.* **49**, 73-77.

Lachance, M-A. (1981) Genetic relatedness of heterocystous cyanobacteria by deoxyribonucleic acid-deoxyribonucleic acid reassociation, *Int. J. Syst. Bacteriol.* **31**, 139-147.

Ladha, J.K. and Watanabe, I. (1982) Antigenic similarity among *Anabaena azollae* separated from different species of *Azolla*, *Biochem. Biophys. Res. Commun.* **109**, 675-682.

Lindblad, P. Haselkorn, R., Bergman, B. and Nierzwicki-Bauer, S.A. (1989) Comparison of DNA restriction fragment length polymorphism of *Nostoc* strains in and from cycads, *Arch. Microbiol.* **152**, 20-24.
Lindblad, P. and Bergman, B. (1990) The cycad-cyanobacterial symbiosis, in A.N. Rai, (ed), *Handbook of Symbiotic Cyanobacteria*, CRC Press Boca Raton, Florida, pp. 137-159.
Lotti, F., Giovannetti, L., Margheri, M.C., Ventura, S. and Materassi, R. (1996) Diversity of DNA methylation pattern and total DNA restriction pattern in symbiotic *Nostoc*, *World J. Microb. Biot.* **12**, 38-42.
Lui, C.C., Cheng, Y., Tong, L.F., Zheng, Q., Song, T.Y., Cheng, M., Li, Y.Y. and Lin, T.L. (1989) Studies on preparation of monoclonal antibody against *Anabaena azollae*, *J. Fujian Acad. Agr. Sci.* **32**, 562-569.
Lumpkin, T.A. and Plucknett, D.L. (1980) *Azolla*: botany, physiology and use as green manure, *Econ. Bot.* **34**, 111-153.
McCowen, S.M., MacArtur, L. and Gates, J.E. (1987) *Azolla* fern lectins that specifically recognize endosymbiotic cyanobacteria, *Current Microbiol.* **14**, 329-333.
Meeks, J.C., Joseph, C.M. and Haselkorn, R. (1988) Organization of the *nif* genes in cyanobactreia in symbiotic association with *Azolla* and *Anthoceros*, *Arch. Microbiol.* **150**, 61-71.
Moore, A. W (1969) *Azolla*: biology and agronomic significance, *Bot. Rev.* **35**, 17-34.
Nilsson, M., Bergman, B. and Rasmussen, U. (2000) Cyanobacterial diversity in geographically related and distant host plants of the genus *Gunnera*, *Arch. Microbiol.* **173**, 97-102.
Obukowicz, M., Schaller, M. and Kennedy, G.S. (1981) Ultrastructure and phenolic histochemistry of the *Cycas revolute-Anabaena* symbioses, *New Phytol* **87**, 751-754.
Ostenfeld, C. H. and Schmidt. J, (1901) Plankton fra det Röde hav og Adenbugten, *Vidensk. Meddel. Naturh. Forening i Kbhvn*., pp.141-182.
Ow, M.C., Gantar, M. and Elhai, J. (1999) Reconstitution of a cycad-cyanobacterial association, *Symbiosis* **27**, 125-134.
Paulsrud, P. and Lindblad, P. (1998) Sepuence variation of the tRNA Leu intron as a marker for genetic diversity and specifity of symbiotic cyanobacteria in some lichens. *Appl. Environ. Microbiol.* **64**, 310-315.
Paulsrud, P., Rikkinen, J. and Lindblad. P. (1998) Cyanobiont specificity in some *Nostoc*-containing lichens and in a *Peltigera aphthosa* photosymbiodem, *New Phytol.* **139**, 517-524.
Paulsrud, P., Rikkinen, J. and Lindblad. P. (2000) Spatial patterns of photobiolt diversity in some *Nostoc*-containing lichens, *New Phytol.* **146**, 291-299.
Paulsrud, P., Rikkinen, J. and Lindblad. P. (2001) Field investigations on cyanobacterial specificity in *Peltigera aphthosa*, *New Plytol.* **152**. 117-123.
Peters, G.A. and Meeks, J.A. (1989) The *Azolla-Anabaena* symbiosis: basic biology, *Ann. Rew. Plant Physiol. and Plant Mol. Biol.* 40, 193-210.
Plazinski, J., Franche, C., Liu, C.C., Lin, T., Shaw, W., Gunning, B.E.S. and Rolf, B.G. (1988) Taxonomic status of *Anabaena azollae*: an overview, *Plant Soil* **108**, 185-190.
Plazinski, J., Zheng, Q., Taylor, R., Croft, L., Rolfe, B.G. and Gunning, B.E.S. (1990) DNA probes show genetic variation in cyanobacterial symbionts of the *Azolla* fern and a closer relationship to free-living *Nostoc* strains than to *Anabaena* strains. *Appl. Environ. Microbiol.* **56**, 1263-1270.
Potts, M. (2000) *Nostoc*, in B.A. Whitton and M. Potts (eds.), *The Ecology of Cyanobacteria*, Kluwer Academic Publishers, Dodrecht, pp. 523-561.
Rai, A.N. (1990) Cyanobacteria-fungal symbioses: the cyanolichens, in A.N. Rai (ed) *Handbook of Symbiotic Cyanobacteria*, CRC Press, Boca Raton, Florida, pp. 9-41.
Rasmussen, U, Johansson, C. and Bergman, B. (1994) Early communication in the *Gunnare-Nostoc* symbiosis: Plant-induced cell differentiation and protein synthesis in the cyanobacterium, *Appl. Environ. Microbiol.* **7**, 696-702.
Rasmussen, U. and Svenning, M.M. (1998) Fingerprinting of cyanobacteria based on PCR with primenrs derived from short and long tandemly repeated repetitive sequences, *Appl. Environ. Microbiol.* **64**, 265-272.
Rasmussen, U. and Svenning, M.M. (2001) Characterization by genotypic methods of symbiotic *Nostoc* strains isolated from five species of *Gunnera*, *Arc. Microbiol.* **176**, 204-210.
Ridgway, J.E. (1967) The biotic relationship of *Anthoceros* and *Phaeoceros* to certain cyanobacteria, *Ann. Mo. Bot. Gard.* **54**, 95.
Rodgers, G.A. and Stewart, W.D.P. (1977) The cyanophyte-hehatic symbiosis I. Morphology and physiology, *New Phytol.* **78**, 441-458.
Silvester; W.B. and Mc Namara, P.J. (1976) The infection process and ultrastructure of the *Gunnera-Nostoc* symbiosis, *New Phytol.* **77**, 135-141.

Spratt, E.R. (1911) Some observations on the life-history of *Anabaena cycadeae*, *Ann. Bot.* **25**, 369-382.

Svircev, Z., Tamas, I., Nenin, P. and Drobac, A. (1997) Co-cultivation of N2-fixing cyanobactreia and some agriculturally important plants in liquid and sand cultures, *Appl. Soil Ecol.* **6**, 301-308.

Tang, L.F., Watanabe, I. and Liu, C.C. (1990) Limited multiplication of symbiotic cyanobacteria of *Azolla* ssp. On artificial media, *Appl. Environ. Microbiol.* **56**, 3623-3626.

van Coppenolle, B., Watanabe, I., van Hove, C. and McCouch, S.R. (1993) Genetic diversity and phylogeny analysis of *Azolla* based on DNA amplification by arbirtary primers, *Genome* **36**, 686-693.

van Coppenolle, B., McCouch, S.R., Watanabe, I., Huang, N. and van Hove, C. (1995) Genetic diversity and phylogeny of *Anabaena azollae* based on RFLPs detected in *Azolla-Anabaena azollae* DNA comlex using *nif* probes, *Theor. Appl. Genet.* **91**, 589-597.

Vagnoli, L., Margheri, M.C., Allotta, G. and Materassi, R. (1992) Morphological and physiological properties of symbiotic cyanobacteria, *New Phytiol.* **120**, 234-249.

Vincent, W.F. (1988) *Microbial ecosystems of Antarctica*. Cambridge University Press, Cambridge.

Watanabe, A. and Kiyohara, T. (1963) Symbiotic blue-green algae of lichens, liverworts and cycads, in *Studies on Microalgae and Photosynthetic Bacteria*,Jap. Soc. Plant Phys. (ed) University of Tokyo Press, pp. 186-196.

West, N.J. and Adams, D.G. (1997) Phenotypic and genotypic comparison of symbiotic and free-living cyanobacteria from a single field site, *Appl. Environ. Microbiol.* **63**, 4479-4484.

Wilkinson, H. H., Spoerke, J.M. and Parker, M.A. (1997) Divergence in symbiotic compatibility in a legume-*Bradyrhizobium* mutualism, *Evolution* **50**, 1470-1477.

Wilmotte, A., (1994) Molecular evolution and taxonomy of the cyanobacteria in D.A. Bryant, (ed) *The molecular biology of cyanobacteria*. Kluwer Academic Publisher. Printed in the Netherlands, pp. 1-25.

Zheng, W.W., Nilsson, M., Bergman, B. and Rasmussen, U. (1999) Genetic diversity and classification of cyanobacteria in different *Azolla* species by the use of PCR fingerprinting, *Theor. Appl. Genet.* 99, 1187-1193.

Zheng, W.W., Song, T., Bao, X., Bergman, B. and Rasmussen, U. (in press) High cyanobacterial diversity in coralloid roots of cycads revealed by PCR fingerprinting, *FEMS Microbiol. Ecol.*

Zhu, C. (1982) Fine structure of blue-green algae and the cells lined along the endophyte cavity in the coralloid roots of *Cycas*, *Acta Bot. Sin.* **24**, 109-112.

Zimmerman, W.J. and Bergman, B. (1990) The *Gunnera* symbiosis: DNA restriction fragment length polymorphism and protein comparison of *Nostoc* symbionts, *Microb.Ecol.* **19**, 291-302.

Zimmerman, W.J. and Rosen, B.H. (1992) Cyanobiont diversity within and among cycads of one field site, *Can. J. Microbiol.* **38**, 1324-1328.

Chapter 16

EVOLUTION OF CYANOBACTERIAL SYMBIOSES

J. A. RAVEN
Division of Environmental and Applied Biology
Biological Sciences Institute, School of Life Sciences
University of Dundee, Dundee DD1 4HN, UK

1. INTRODUCTION

The other chapters in this book show the great diversity of extant symbioses involving potentially free-living cyanobacteria and a wide variety of other organisms. In this contribution the evolution of these symbioses is considered, with emphasis on the molecular phylogeny of the cyanobacteria and their symbiotic associates and the number of independent origins of the symbioses. The chapter also considers the timing of the origin of cyanobacteria, of organisms symbiotic with cyanobacteria, and of the symbiotic associations, with consideration of the possible evolutionary significance of the symbioses in the context of the environments, which have occurred during and since their origins (see Raven, 2002b).

Also considered is the most widespread and evolutionarily and biogeochemically important cyanobacterial symbiosis, i.e. the symbiosis which gave rise to all plastids. This symbiosis involved genetic integration such that the genome of the original cyanobacterium has largely been entirely lost, or transferred to the past nuclear genome, so that there is (currently) no possibility of reconstituting a free-living cyanobacterium from a plastid (see Raven, 2002b).

2. CYANOBACTERIA AND THE ORIGIN OF PLASTIDS

Cyanobacteria may have evolved as long as 3.45 billion years ago if fossil evidence from that time is reliable (Golubic and Lee, 1999). They certainly occurred from billion years ago as judged from molecular fossils (Brocks *et al.*, 1999; Summons *et al.*, 1999). The origin of eukaryotic cell has been dated at 2.7 billion years ago from molecular fossil data (Brocks *et al.*, 1999) with the body fossil *Grypanea* from 2.1 billion years ago. *Grypanea* is the earliest known organism, which is probably a eukaryote, and is most likely to be an alga, i.e. a photosynthetic eukaryote not at the level of organisation of the embryophytes or higher plants. The earliest known fossil and a eukaryotic alga that can be referred not just to an extent algae Division (Rhodophyta) or Class

A.N. Rai, B. Bergman and U. Rasmussen (eds.), Cyanobacteria in Symbiosis, 329-346.

(Bangiophyceae) but very probably an extent family (Bangiaceae) is *Bangiomorpha pubescens* (Butterfield, 2000) from 1.2 billion years ago. Molecular phylogenetic evidence from plastids, mitochondria and past genomes suggests that a single endosymbiotic event involving a cyanobacterium and a eukaryotic, mitochondria-containing cell gave rise to all of the plastid-containing (Bodyl, 1997) organisms (LaRoche *et al.*, 1996; Bhattacharya and Medlin, 1995, 1998; S.E. Douglas, 1998; Tanaka *et al.*, 1998; Tomitani *et al.*, 1999; Gupta, 2000; Mareira *et al.*, 2000).

However, some molecular genetic (the largest subunit of RNA polymerase II) evidence favours a divergence of the past (eukaryotic) nucleus of red algae from before the origin of green algae and plants, animals and fungi (Stiller and Hall, 1997, 1998; Stiller *et al.*, 2001).

The diversity of plastids in eukaryotes other than those belonging to the Chlorophyta (and Embryophyta), Glaucocystophyta and Rhodophyta is explicable in terms of secondary endosymbiosis. Green algae cells symbiotic with two different chemoorganotrophic eukaryotes gave rise to the chlorarachniophytes and photosynthetic euglenoids. Red algal cells symbiotic with (probably) four different chemoorganotrophic eukaryotes to yield the cryptophytes, haptophytes, heterokontophytes and (probably) dinophytes. It is important to note that, accepting the arguments of Stiller and Hall (1997, 1998) and Stiller *et al.* (2001) the monophyly of plastids can be maintained if the red algae are secondarily endosymbiotic. This would require the loss of the 'extra' membranes (additional to the two normal envelope membranes) which otherwise characterise plastids originating by endosymbiosis (Stiller and Hall, 1997).

Passing from the early phase of the primary symbiosis involved a chemoorganotrophic past and a symbiotic cyanobacterium which was still capable of an independent existence to the current state with plastids having a very restricted genome, there must have been a loss of genes from the cyanobacterium. Plastid genomes have fewer genes than do cyanobacteria. Some of the missing genes have been transferred to the host nucleus, and others have been lost entirely from the symbiosis (see Allen and Raven, 1996). Clearly all of the genes needed for plastid functioning are retained by photosynthetic eukaryotes, with a majority of them in the nucleus.

In addition to the genes in the past nucleus that are clearly needed for plastid synthesis, replication and function, molecular phylogenetic analyses have shown that a significant number of other genes were denied from cyanobacteria precursor of plastids (Rujan and Martin, 2001). From a comparison of a fraction of the (known) complete nuclear genome (25,000 genes) with the complete genome of the cyanobacterium *Synechocystis* (3,600 genes), it has been estimated that between 1.6% and 9.2% of *Arabidopsis* nuclear genes were derived from the cyanobacterial precursor of the plastids.

In some cases the cyanobacterial genes replaced pre-existing genes derived from the archean which gave rise to the eukaryotic cell and, to a smaller extent, from the proteobacteria ancestor of mitochondria. Thus, the 12 enzymes of the photosynthetic carbon reduction cycle in eukaryotes are encoded by nuclear genes of a mixture of cyanobacterial and proteobacteria origin (Martin and Scharrenberger, 1997). We note that both large and small subunits of RUBISCO are plastid-encoded in red algae and in the algae derived from them by secondary endosymbiosis, and the large subunit of RUBISCO is plastid–encoded and the small subunit is nucleus-encoded in green algae and algae derived from them by endosymbiosis and in higher plants. In this case the RUBISCO of green algae and their derivatives is cyanobacterial in origin, while the RUBISCO of red algae and the algae derived from them is proteobacterial. The Form II nuclear encoded RUBISCO of peridinin-containing dinoflagellates is also of proteobacterial origin. Of the other genes, the cyanobacterial sugar bisphosphate-1-phosphatase acts on fructose-1, 6-bisphosphate and a sedoheptulose-1,7-bisphosphate. The plastids of eukaryotes have separate phosphatases for the two sugar bisphosphates. Miyagawa *et al.* (2001) showed that incorporating the cyanobacterial bifunctional phosphatase in to the higher plant increased the rate of growth and photosynthesis under the conditions tested. Such results are of interest when considering the evolutionary selective pressures relating to gene replacement.

In addition to replacing some genes, the cyanobacteria contributed new metabolic capacities to these eukaryotes (in addition to photosynthesis). One probable example is cellulose synthase (Noble *et al.*, 2001), although as yet there seems to be no molecular genetic evidence.

What determines the extent of transfer of genes from the cyanobacterial (plastid) genome to the past nuclear genome as opposed to their loss from the symbiosis or retention in the plastid genome? Clearly the genes required for the functioning of plastids must be retained by the symbiosis, at least for organisms which have lost the capacity for chemoorganotrophic growth.

The question of what determines the transfers of cyanobacterial genes from the plastid genome to the nuclear genome (and the corresponding problem of the extent of transfer of proteobacterial genes from the mitochondrial genome to the nucleus) has been addressed by several authors (e.g. Allen and Raven, 1996; Martin and Scharrenberger, 1997; Race *et al.,* 1999; Zerges, 2000; Selosse *et al.,* 2001). It has been suggested that the possibility of damage to DNA in energy–transducing organelles is increased as a result of the high rate of production of active oxygen species (Allen and Raven, 1996, Palmer *et al.*, 2000). This exacerbates the effect of Muller's Ratchet, i.e. the accumulation of deleterious mutations in asexually propagated genomes such as those of (almost all) organelles. These two factors have been considered as favouring transfer of genes to the nucleus (Allen and Raven, 1996). Once some genes have been transferred to the nucleus the gene-specific resource cost of retaining the organelle genome increases, in that the full replication, transcription and translation apparently has to be retained even if only one protein-encoding gene remains in the organelles genome. This constitutes another possible reason for moving all genes out of the organelle genome. Selosse *et al.* (2001) further suggest that selection for rapid replication could be a general cause of organelle genome reduction.

With all of these factors which might favour the transfer of all organelle genes to the nucleus, what factors might favour retention of genes in organelles? One argument is that some proteins which function in the plastids could not be transferred into the plastids if their genes were in the nucleus and the proteins were synthesised in the cytosol. However, this argument cannot apply to the large subunit of RUBISCO since this protein can be transported into plastids in natural (dinoflagellates) and experimental (genetically modified tobacco) conditions (see Allen and Raven, 1996). J.F. Allen has suggested that retention of genes in the organelles related to the control of transcription and translation of the genes under the control of organelle signals (Allen, 1993; Allen and Raven, 1996; Pfannschmidt *et al.,* 1999, 2001). Martin and Schnarrenberger (1997) have suggested that organelle-encoded gene products might be toxic when expressed in compartments other than those of the plastid.

The analysis of the genes for enzymes of the photosynthetic carbon reduction cycle by Martin and Schnarrenberger (1997) also falsifies the 'product-specific corollary', a hypothesis which predicts that the products of nuclear-encoded genes are only targeted to the organelles from which the genes were derived. Thus, some genes for the photosynthetic carbon reduction cycle enzymes of eukaryotes were derived from the mitochondrial ancestor, so that the genes have evolved plastid targeting regardless of whether they were derived from the plastid precursor on the mitochondrial precursor (Martin and Schnarrenberger, 1997).

Whatever the reason(s) for the retention of genes in organelles, it is unlikely to be a 'frozen historical accident'. One argument is that there is a continuing transfer of genes from organelle genomes to the nuclear genome, at least in cases (flowering plant plastids and mitochondria), where there is no bar to gene transfer imposed by changes in the genetic code in the organelle relative to the 'universal code' (e.g. the mitochondria of metazoans) (Adams *et al.*, 2000; Palmer *et al.*, 2000; Adams *et al.,* 2001a,b; Miller *et al.*, 2001). Despite this continuing transfer of genes from organelles to nuclei there is a core of 45 plastid genes which are common to all photosynthetic eukaryotes examined, i.e. a glaucocystophyte, a rhodophyte, a diatom, a euglenophyte and five higher land plants (Martin *et al.,* 1998). Such an irreducible minimum of genes retained in organelles is consistent with the hypothesis of Allen (see Allen, 1993; Allen and Raven, 1996). Furthermore independent parallel gene transfer from the plastid to the nucleus in multiple lineages exceed phylogenetically unique transfers by more than 4:1.

Perhaps more central to the theme of this article is the nature of the cyanobacteria involved in the primary endosymbiosis, and the nature of the relationship between the cyanobacterium and the host. The plastid today is primarily involved in the photochemical generation of NADPH and ATP with H_2O as the electron donor, assimilation of inorganic C, NO_2^-, NH_4^+ and SO_4^{2-} (the cytosolic NO_3^- reductase was probably derived from the cyanobacterial ancestor of plastids) and in aspects of lipid synthesis. Accordingly, the most economical hypothesis is that the role of the symbiotic cyanobacterium in the pre-plastid stage of the symbiosis was to bring these genes for these processes to the symbiosis. Of course the mitochondrial precursor could have contributed some of these attributes (with the exception of the two-photosystem thylakoid machinery), a possibility that is supported by the occurrence of nitrate and sulphate assimilation processes in many fungi; the eukaryotic ancestors of the fungi were apparently never photosynthetic. There are, of course, precedents for CO_2 assimilation and, presumably, NO_3^- and SO_4^{2-} assimilation in extant symbioses between cyanobacteria and chemoorganotrophic eukaryotes (Table 1).

Table 1. Associations involving cyanobacteria as symbionts. The list excludes fully integrated plant symbioses with only about 10% of genes needed for plastid function retained in plastid genomes. Based on Table 5.2 of Smith and Douglas (1987), Tables 1 and 2 of Rai et al. *(2000) and Table 1 of Raven (2002b), with additional data from Carpenter and Janson (2000), Gordon* et al. *(1994), Larkum* et al. *(1987), Larsen* et al. *(1992), Lucas (1991), Sara* et al. *(1998), Wanntorp* et al. *(2001), Wilkinson (1999, 1992).*

Hosts	Symbionts	Growth forms	Location	N_2 fixation	Photosynthesis by cyanobiont *in situ*
Bacillariophyceae (Diatoms) (P)					
Marine: *Hemiaulus, Rhizosolenia* and 5 other genera (l)	*Richelia intracellularis;*	f	i	+	?
Freshwater: *Rhopalodia, Epithemia, Denticulata* (l)	*Synechococcus* sp.?	u	i	?+	?
Dinophyta (N,P)		u, f	i, e	o	o
Porifera (N): ~38 marine genera (l)	*Aphanocapsa, Phormidium; Synechococcus?*	u; f	e, i, i+e	o, +	+
Echiuroid worms (N)					
Ikedosoma gogoshimense (l)	?			o	o(-)
Bonnellia fuliginosa (l)	?			o	o(-)
Ascidians (N): Didemnid ascidians (l)	*Prochloron* (+ *Oscillatoria* sp)	u	e	?	+
Zygomycotina (N): (Protolichen)	*Nostoc*	f	i	+	+
Geosiphon pyriforme (l)	*Chroococcidiopsis,*	u	e	(o)	(+)
Winfrenatia reticulata (d)	*Gloeocapsa*				
Ascomycotina (N) (Lichens): ~40 genera (l) (+ Chlorophyta (P)) (- Chlorophyta (N))	*Gloeothece, Gloeocapsa, Synchocystis, Hyella*	u	e	o	+
	Calothrix, Fischerella, Nostoc, Scytonema	f	e	+	+
Basidiomycotina (N) (Lichens) At least 5 genera (l)	*Scytonema*	f	e	?+	+
Liverworts (P) *Blasia, Cavicularia* (l)	*Nostoc*	f	e	o, +	?
Hornworts (P) *Anthoceros, Dendroceros, Notothylas, Phaeoceros* (l)	*Nostoc*	f	e	+	?
Mosses (P) *Sphagnum* (l)	*Nostoc, ? Anabaena*	f	i	+, o	?
Pteridophytes (P) *Azolla* (l) All extant spp have cyanobionts	*Anabaena azollae* (really a *Nostoc* sp?)	f	e	+	(+)
Gymnosperms (P): Cycadales (l), 150 known spp, cyanobionts in all spp examined.	*Nostoc* (*Calothrix*?)	f	e(i)	+	-
Angiosperms (P): ~50 *Gunnera* spp (l), cyanobionts in all spp examined.	*Nostoc*	f	i	+	-

d = extinct, i.e. fossil (only); e = extracellular; f = filamentous; i = intracellular; l = living, i.e. extant; N = nonphotosynthetic; P = photosynthetic; u = unicellular; o = not tested; ? = not clear; - = absent; + = proven

The other metabolic innovation supplied to some extant symbioses by cyanobacteria is diazotrophy (Table 1). While symbioses of diazatrophic cyanobacteria with photosynthetic eukaryotes typically involve no, or minimal, photosynthesis by the cyanobacteria, symbioses of N_2–fixing cyanobacteria with chemoorganotrophic eukaryotes involve the supply of all of the organic C and combined N by the cyanobiont (Table 1; Rai *et al.,* 2000). These extant symbioses in which the cyanobacteria contribute both photosynthesis and diazotrophy to the symbiosis raise the possibility that the cyanobacterial plastid ancestor was diazotrophic as well as photosynthetic. Evidence on the occurrence of diazotrophy in the nearest living cyanobacterial relative of the plastid ancestor is equivocal (Palenik and Haselkorn, 1992; Urbach *et al.,* 1992; Young, 1992; Neilsson *et al,* 1995; Honda *et al.,* 1999; Turner *et al.,* 2001).

Even if the cyanobacterial plastid ancestor was diazotrophic, there are a number of possible reasons as to why diazotrophy was not retained. One is incompatibility of diazotrophy with the oxygenic photosynthesis in plastids within a eukaryote, although potentially free-living intracellular and extracellular cyanobacteria symbiotic with eukaryotic O_2-evolvers tolerate similar conditions to those in plastids (Table 1). In a number of cases heterocysts are involved; this does not seem possible in plastids as we know them today.

Another possibility (Douglas, 1996) is that the diazatrophic proto-organelles could not get into the germ line and be vertically transmitted. This clearly cannot apply to unicellular hosts (diatoms dinoflagellates) or to any N_2 fixation in plastids which are transmitted through meristems, including reproductive meristems, in higher plants. Even if there was genetic differentiation of plastids between N_2-fixing leucoplasts in non-green tissues and non N_2-fixing chloroplasts after the initial symbiosis and genetic integration, both plastid types could be vertically transmitted. There is also the possibility that a single kind of proplastid could differentiate into a 'diazoplast' as well as chloroplast, chromoplast, leucoplast and amyloplast. As with photosynthetic (chloroplast), attractant or warning (chromoplast) and storage (amyloplast) functions, many of the genes required for 'diazoplast' function could have been transferred to the host nucleus.

A further possible reason for the absence of a 'diazoplast' relates to the general genetic homogeneity of organelles within eukaryotes. This contrasts with the genetic diversity of symbiotic diazotrophs in the natural environment, e.g. *Nostoc* from among cyanobacteria, as well as rhizobia and *Frankia.* In the natural habitat, the host can associate with different diazotrophs and during each host generation. This ability to 'mix and match' symbionts as a function of environmental conditions is, of course, dependent on the availability of the relevant symbiont genotypes from the environment. There is certainly evidence from cnidarian symbioses with the dinoflagellate *Symbiodinium* spp. for the occurrence of different genotypes of *Symbiodinium* in a given cnidarian as a function of the habitat (Rowan and Knowlton, 1995; cf. Bythell *et al.*, 1997; Goodwin *et al.,* 2001; Toller *et al.,* 2001).

Another aspect of the evaluation of genetically integrated organelles from cyanobacterial symbioses is the environmental conditions, which might favour the evolution of the initial symbiosis. Clearly photosynthetic symbionts can alleviate a shortage of organic C for the early chemoorganotrophic eukaryotes. It is likely that the evolution of eukaryotes with their endomembrane system, which permits endocytosis phagotrophic nutrition, would have greatly increased the capacity for chemoorganotrophs to consume photolithotrophs. Before phagotrophy the recycling of organic matter from photolithotrophy would have involved 'abiotic' death (e.g. of planktonic obligate photolithotrophs after they had sunk out of the photic zone), viral parasitism and lysis, and parasitism by bacteria or archeans. The eukaryotic chemoorganotrophs would have extended the possibilities for parasites (and parasitoids) and, especially, phagotrophs to attack photolithotrophs (Raven, 1997, 1998). This would have caused a greater pressure of parasites on the photolithotrophs, and introduced phagotrophs and thus grazing. These innovations in biotic removal of photolithotrophs would have provided niches for better defended photolithotrophs, including eukaryotes as well as cyanobacteria.

3. CYANOBACTERIA AS DIAZOTROPHIC SYMBIONTS WITH PHOTOSYNTHETICALLY COMPETENT DIATOMS

The marine planktonic centric diatoms *Hemiaulus* (all species) and *Rhizosolenia* (some species) contain filaments of the heterocystous cyanobiont *Richellia* (Heinboke, 1986; Villareal 1991, 1994; Carpenter *et al.*, 1999; Janson *et al.*, 1999). *Richellia* contributes combined N to the symbiosis, and probably performs little or no photosynthesis. Vertical migration of non-symbiotic *Rhizosolenia* species (like that of *Ethmodiscus*) in oligotrophic stratified waters brings NO_3^- from the thermocline into the more higher illuminated near surface waters. This comprises an alternative to N_2 fixation *via Richellia* in increasing combined N availability to the diatoms, but presumably depends, like N_2 fixation by the free-living non-heterocystous filamentous cyanobacterium *Trichodesmium*, on aeolian Fe input and possibly a higher P availability (Sañudo-Wilhemy *et al.*, 2001). It is not clear whether *Hemiaulus* and *Richellia* containing *Rhizosolenia* undergo vertical migrations; such migrations are also possible for the gas-vacuolate *Trichodesmium*. The marine diatom *Climacodium frauenfeldianum* has unicellular cyanobacterial symbionts (Carpenter and Janson, 2000).

The (predominantly) freshwater benthic pennate diatoms *Denticula*, *Epithemia* and *Rhopalodia* contain (presumed) unicellular cyanobacteria which are thought to account for the observed N_2 fixation in *Rhopalodia* (Kies, 1992). These organisms occupy similar habitats to N_2-fixing benthic freeliving cyanobacteria and (in coastal marine waters) the cyanobacterial lichen *Lichina* with the heterocystous *Calothrix* as its diazotrophic symbiont (Raven *et al.*, 1990).

Molecular phylogenetic evidence calibrated by the fossil record suggests that the diatoms evolved some 250 million years ago at the beginning of the Triassic (Kooistra and Medlin, 1996; Medlin *et al.*, 1997; cf. Scheiber *et al.*, 2000). The molecular genetic and fossil evidence shows that the marine centric diatoms evolved earlier than the freshwater centric diatoms. The fossil record gives no evidence of the occurrence of cyanobacterial symbionts.

4. CYANOBACTERIA AS SOLE SYMBIONTS WITH NON-PHOTOSYNTHETIC AND NON-DIAZOTROPHIC EUKARYOTES

In cases where cyanobacteria are the sole photosynthetically eukaryotes a major metabolic contribution of the cyanobacteria to the symbiosis is photosynthate (Table 1). Examples here are heterotrophic dinoflagellates poriferans, ascidians, *Geosiphon*, cyanobacterial ascolichens without green algal symbionts and basidiolichens (Table 1). There seems to be no convincing evidence available to the occurrence of photosynthesis with transfer of photosynthate to the eukaryote in the case of symbiosis of cyanobacteria with dinoflagellates and echiuroids (Table 1).

For some of these symbioses of non-photosynthetic eukaryotes with cyanobacteria as their sole photobiont the cyanobacterial component is a diazotroph, so combined N is also contributed to the symbioses (*Geosiphon*, cyanolichens; Table 1).

In considering the selective forces acting on the symbioses of cyanobacteria with non-photosynthetic, non-diazotrophic symbionts it is helpful to know the timing of the origin of the symbioses and hence the environmental conditions when the symbioses evolved. The occurrence in the fossil record of the eukaryotic components is quite well known for some taxa (Porifera, zygo- asco- and basidio-mycete fungi). Fossil evidence for the Porifera goes back to the upper Neoproterozoic ~ 600 Ma ago, while that for zygo-, asco- and basidio-mycetes goes back to the Ordovician (? Cambrian), Ordovician and Carboniferous, respectively (Taylor and Taylor, 1994; Printzen and Lumbsch 2000). Alas, the evidence that they were symbiotic with cyanobacteria is much less frequent; as will be seen. This is a recurrent theme in our consideration of the evolution of cyanobacterial symbioses. By the fortunate accident of preservation in the Rhynie Chert, the most convincing fossil evidence for a cyanolichen is from the Lower Devonian (~400 million years ago), but represents a combination of symbionts (presumably non-diazotrophic cyanobacteria with a zygomycete) which does not occur today (Taylor *et al.,* 1997). The combination in *Geosiphon* of a zygomycete with the endosymbiotic, diazotrophic, photosynthetic cyanobacterium *Nostoc* does not have a known fossil record, but its' molecular phylogenetic relationship to the Glomales (the arbuscular mycorrhizal zygomycetes) has recently been established (Redecker *et al.,* 2000).

Molecular phylogenetic attempts to date the origin of higher taxa can be applied to all extant organisms, but suffer from problems of extrapolating the rate of nucleotide substitution unless a fossil-based calibration is possible. Such molecular evidence suggests a much earlier origin of Porifera and Fungi (>800 Ma ago) (Redecker *et al.,* 2000). Both the fossil and the molecular phylogenetic evidence as to the time of origin of the sponges and fungi put them clearly after the crisis in the abiotic (electrical storm) production of combined N during the transition of the atmosphere to an oxygenated state (Navarro-Gonzalez *et al.,* 2001).

This does not, of course, mean that the supply of combined N from abiotic sources was abundant relative to biological requirements for combined N in the face of denitrification losses of combined N, and a need for biological N_2 fixation would have been evident. As for carbon dioxide availability, the Earth presumably went through episodes of low CO_2 paralleling the low temperatures of the Lower and Upper

Neoproterozoic (Raven *et al.*, 2002). However, the absolute level of CO_2 required to establish very large-scale glaciations depends on the disposition of continents and hence latitudinal heat transfer as well as of the increase through geological time of the radiant energy output. There was certainly a low CO_2 episode in the Carboniferous.

The low CO_2 episodes could have imposed a limitation on photosynthesis by submersed marine benthic photosynthetic organisms with the diffusion resistance imposed by the diffusion boundary layer. There would also have been a CO_2 supply limitation for any waterlogged poikilohydric organisms on land. The restriction on CO_2 supply to a benthic cyanobacterial mat in the ocean could have been partially relieved by endosymbiosis in sponges with their flagellar ventilation processes related to their prior sole dependence on aerobic, phagotrophic nutrition (Raven, 1993, 1997; Cheshire *et al.*, 1995, 1997). Similar considerations could apply to didemnid ascidians where there is also a ciliary ventilation system (Pandy and Royce, 1992). For the marine intertidal lichen *Lichina* with the diazotrophic *Calothrix* as its' photobiont, the ascomycete component may have a role in facilitating CO_2 transfer to the cyanobacterium (Raven *et al.*, 1990). For terrestrial cyanolichens the hydrophobic proteins (hydrophobins) of the fungal partner's cell wall can restrict waterlogging of lichen thalli and thus facilitate CO_2 supply from the bulk atmosphere in the gas phase to the cyanobacteria relative to the situation in terrestrial cyanobacterial mats (Raven, 1986, 1996, 2002a; Palmqvist, 2000; Honneger, 1998). However, the extent to which this mechanism maintains the (potential) intercellular gas spaces of terrestrial lichens clear of liquid water during and just after rainfall, dew or flooding events requires further work. It is certain that none of the lichen symbioses, including those involving cyanobacteria, have achieved the homoiohydric state, so cyanobacteria and their associates have not achieved the state found in their tracheophytes with their fully integrated (plastid) cyanobionts (Raven, 1986, 2002a).

5. CYANOBACTERIA AS SYMBIONTS WITH NON-PHOTOSYNTHETIC AND NON-DIAZOTROPHIC EUKARYOTES IN THE PRESENCE OF ANOTHER PHOTOBIONT

The major group of organisms in this category are those ascolichens which have both cyanobacterial and chlorophyte symbionts (Table 1). In this case the photosynthesis in the symbiosis is almost all carried out by the green algal photobionts, with the cyanobacteria mainly acting in N_2 fixation. The mainly diazotrophic role of cyanobacteria in a photolithotrophic symbiosis in which they are exposed to light but in which the cyanobacteria have a very small role in photosynthesis in terms of the symbiosis as a whole resembles that in those symbioses of cyanobacteria with embryophytes in which the cyanobacteria are illuminated (i.e. all except cycads and *Gunnera* where the cyanobacteria do not have the chance to photosynthesise because they are not exposed to light) (Table 1; Rai *et al.*, 2000).

The role of the photosynthetic apparatus of the cyanobacteria in such symbioses probably relates largely to diazotrophy and to the growth and maintenance of the cyanobacteria. The resource costs of symbiotic cyanobacterial N_2 fixation relative to that by other symbiotic diazotrophs (rhizobia, *Frankia*) will be considered in the next section in the context of diazotrophy in illuminated and unilluminated cyanobacteria symbiotic with embryophytes (Table 1).

The timing of the evolution of these ascolichen symbioses relative to that of lichens as a whole can only be discovered by examining molecular phylogentic analyses at the genus and species levels, since lichens with cyanobacterial and chlorophyte symbionts can occur in the same genus as those with only one kind of photobiont.

6. CYANOBACTERIA AS DIAZOTROPHIC SYMBIONTS WITH PHOTOSYNTHETICALLY COMPETENT EMBRYOPHYTES

Table 1 shows symbioses in which the cyanobacteria are exposed to light (those with liverworts, hornworts, mosses and *Azolla*) and those in which the cyanobacteria are associated with underground structures and are permanently in darkness (cycads and *Gunnera*). Clearly the underground cyanobionts of the cycads and *Gunnera* cannot photosynthesise *in hospice*, yet they retain significant amounts of photosynthetic structure and function (Rai *et al.*, 2000). For the symbionts in the embryophytes at the bryophyte and tracheophyte grade of organisation where the cyanobionts are exposed to light the complete photosynthetic machinery occurs in all cases that have been examined, although the cyanobionts only make a minor contribution to photosynthesis by the symbiosis (Rai *et al.*, 2000).

There is no direct fossil evidence for the occurrence of cyanobionts in embryophytes, even for Triassic cycads from the Antarctic where root structure is so well preserved that arbuscular mycorrhizal are very clearly identifiable (Taylor and Taylor, 1994). From the times at which the earliest fossils of the eukaryotic component of the symbioses have been first identified in the fossil record. For the bryophytes this means an origin 480, or even 510, million years ago, based on fossil spores. The branching order of the liverworts and hornworts is still a contentious matter (Kendrick and Crane, 1997; Qiu *et al.*, 1998; Renzaglia *et al.*, 2000). However, it is clear that the mosses were the last branching of the bryophytes before the polysporangiophytes (extant and extinct tracheophytes, and extinct tracheophyte- like plants whose endohydric conducting system is not true xylem).

Although the polysporangiophytes, including tracheophytes at the pteridophyte grade of organisation, are known in the fossil record from ~ 420 million years ago, the only fern genus known to be symbiotic with cyanobacteria is *Azolla*, with fossils from Lower Cretaceous deposits up to 120 million years old. The timing of the earliest fossils of *Azolla* is consistent with the molecular phylogenetic evidence which shows *Azolla* to be a relatively late-branching member of the leptosporangiate ferns.

Among the tracheophytes with cyanobacterial symbionts, the earliest known cycads are from the Triassic (up to 250 million years ago), while *Gunnera* goes back to at least 70 million years ago on the basis of fossilized pollen grains assigned to the form genus *Tricolpites* (Taylor and Taylor, 1994; Wanntorp *et al.*, 2001).

These diazotrophic symbioses are poikilohydric in the case of the bryophytes; although the tracheophytic *Azolla* spp. have the structures associated with homoiohydric it is likely that these organisms are not functionally homoiohydric; interfacial aquatic vascular plants frequently have non-functional stomata (Raven, 2002). *Gunnera* spp. are also restricted to wet habitats and have a restricted range of stomatal responses to environmental conditions (Osborne *et al.*, 1991). This leaves the cycads as the only examples of diazotrophic cyanobacterial symbioses, which are functionally homoiohydric in non-wetland habitats. The cycads were significantly more abundant and diverse in the mesozoic than they are today. The extant vascular plants with diazotrophic symbionts which live in relatively dry conditions are all flowering plants symbiotic with rhizobia and *Frankia*. These symbioses could, from a mixture of fossil and molecular phylogenetic data, have originated in the Upper Cretaceous.

There is minimal taxonomic overlap; even at the level as high as that of classes of embryophytes, between organisms with *Frankia* or rhizobia and those with cyanobacteria as their diazotrophic symbionts; *Gunnera* is the only angiosperm with cyanobionts, and there are no non-angiosperm embryophytes with rhizobia or *Frankia*. However, as we have seen in the context of water relations, these is some ecological overlap, e.g. of cycads and legumes in N_2 fixation in fire-prone environments where combined N is lost in the fires. Here it is possible to compare the resource costs of N_2 fixation in cyanobacterial and non-cyanobacterial symbioses and provide evidence as to which diazotrophic symbiont might be selectively favoured in a given habitat.

N_2 fixation is known to have a high Fe requirement (see Raven 1988, 1990; Sañudo-Wilhelmy *et al.*, 2001). This is caused by the very high Fe content per nitrogenase – nitrogenase reductase complex, and the low specific reaction rate of the complex, regardless of whether the enzyme is of the most common Mo type, the V type, or the 'no metal other than Fe' type. Additional Fe requirements result from the involvement of Fe in leghaemoglobin which is always involved in rhizobial symbioses, and sometimes, in actinorhizas as a means of supplying adequate O_2 to aerobic diazotrophic symbionts while maintaining a very steady state O_2 concentration at the site of N_2 fixation and thus minimizing O_2 damage to nitrogenase (Thumfort *et al.*, 1994, 1999, 2000). Diazotrophy by cyanobacteria symbiotic with embryophytes (and fungi and diatoms) involves heterocystous organisms in which O_2 protection is provided by the wall lipid layers in the heterocyst wall, analogous to the walls of *Frankia* vesicles of actinorhizas. The absence of leghaemoglobin from diazotrophic symbiosis involving cyanobacteria means a lower Fe demand for diazotrophy than in rhizobial, and many actinorhizal, symbiosis.

The potential for a lower Fe cost of diazotrophy in symbioses involving photosynthetic organisms and cyanobacteria has to be offset against the occurrence of some or all of the photosynthetic apparatus; this is the case where the cyanobionts are never exposed to light as well as where the cyanobionts are illuminated and hence could use photosynthetic reactions in N_2 fixation and, to a lesser extent, CO_2 fixation.

The cases in which the cyanobionts are not illuminated mean that Fe associated with photosynthetic reactions, which are not common to respiratory reactions has no obvious role in diazotrophy. Thus the two (or three) Fe associated with photosystem II and the 12 Fe associated with I (in F_X, F_B and F_A) are not common to respiratory reactions and have no obvious role in diazotrophy. The cytochrome b_6 – f complex (common to linear and cyclic electron transport processes and to respiration) with five Fe per complex and the NAD(P)H – PQ oxidoreductase (common to some variants of cyclic electron flow and to respiration) with 8-18 Fe per complex (Raven *et al.*, 2002) have direct roles in diazotrophy in vegetative cells of cyanobacteria. The continued expression of the photochemical reaction centres in symbiotic cyanobacteria even when they never receive light *in hospice* appears to constitute a Fe sink with no catalytic role for the Fe in diazotrophic or other non-photosynthetic reactions. While 'photorhizobia' in stem modules of such legumes as *Aeschynomene* and *Sesbania* express their photoreaction centre, this is much less iron costly (one Fe per reaction centre) than photoreaction one, or even protraction two, of cyanobacteria, and furthermore, there is evidence that it is used in N_2 fixation in the light. More quantitative modelling, taking into account the quantity of Fe in leghaemoglobin in *Parasponia* legumes and many actinorhizal plants and in the photochemical reactions (cyanobacteria) relative to that in the nitrogenase complex is needed, as are measurements of Fe in rhizobia and *Frankia* plus that in leghaemoglobin and of Fe in naturally un-illuminated symbiotic cyanobacteria.

7. MULTIPLE INDEPENDENT ORIGINS AND LOSSES OF CYANOBACTERIAL SYMBIONTS IN DIFFERENT HIGHER TAXA

For the higher taxa of eukaryotic pasts in Table 1 there are three cases in which there was a single origin of cyanobacterial symbiosis with no loses. These cases are the order Cycadales and the genera *Azolla* and *Gunnera*. For *Gunnera* the cyanobiont is always *Nostoc*, while for *Azolla* the unique symbiotic genus could be *Nostoc* rather than *Anabaena* (Zheng *et al.*, 1999). The cycads generally sometimes have *Calothrix*.

The situation is more complex in the bryophytes. The three higher taxa at this grade of organisation (liveworts, hornworts, mosses) presumably acquired their symbionts independently. Only two genera of liveworts have cyanobionts; these are closely related and so probably reflect a single symbiotic event (Schuster, 1979).

In the hornworts the symbiotic state is common but not universal; cyanobionts are absent from *Megaceras* and (which also lacks pyrenoids and a carbon concentrating mechanism) *Sphagnum* is the only moss with cyanobacterial associates.

Molecular phylogenetic analyses of lichens have shown that at least five independent origins of the lichenized state occurred in the ascomycetes and basidiomycetes (Gargas *et al.*, 1995). Not all of these symbioses involve cyanobacteria as the microbiont, but all of the basidiomycete microbionts are cyanobacteria so it is clear that cyanobacterial lichenization of basidiomycetes was polyphyletic. Since only a relatively small subset of ascolichens were analysed by Gargas *et al.* (1995) it is likely that cyanobacterial ascolichens were also polyphyletic. There has also been a very significant loss of lichenization in the ascomycetes, with major chemoorganotrophic lineages derived from symbiotic lichen ancestors (Lutzoni, 1997; Lutzoni *et al.*, 2001). This work shows that losses of lichenization have been more frequent than lichenization events (Lutzoni *et al.*, 2001). Lutzoni and Pagel (1997) point out that fungi which are symbiotic with cyanobacteria and chlorophytes (and with liverworts; see Read *et al.*, 2000) show increased rates of nucleotide substitutions in nuclear ribosomal DNA relative to that in non-symbiotic fungi.

Among the zygomycetes the only two cyanobacterial symbioses are very likely to be of independent origin. *Geosiphon pyriforme* is a zygomycete chemoorganotrophic symbiont which was derived from an arbuscular mycorrhizal symbiont of embryophytes

(Redecker *et al.*, 2000). The zygomycete ancestors of the Lower Devonian *Winfrenatia reticulata* are not well understood (Taylor *et al.*, 1997).

8. VERTICAL TRANSMISSION OF CYANOBACTERIAL SYMBIONTS AND SPECIFICITY OF SYMBIONTS.

The cyanobacterial symbionts of diatoms are transmitted during vegetative cell divisions, and presumably also *via* sexual reproduction. It is also known that structurally intact cyanobacteria occur in the oocytes and nurse cells of several sponges (Smith and Douglas, 1987; Usher *et al.*, 2001) and *via* a plant rake and algal pouch in larvae of the didemnid ascidian *Diplosoma similis* (Hirose, 2000) and assexual reproduction of lichens *via* soredia distributes fungus with photobiont. *Azolla* transmits its symbionts *via* the megasporocaps and megaspores. On other cases the symbionts are known to be acquired a new in each generation (cycads, *Gunnera*, liverworts, hornworts).

A.E. Douglas (1998) has argued that horizontal transmission of symbiotic microorganisms is related to three factors. One of these is selection pressure on the microbial symbionts to exploit the host, leading to a reduced benefit of the symbiosis to the host. A second factor is the variation with environmental conditions in the amount of host benefit from different genotypes in the natural environment. The third factor is the spatial or temporal variability in the availability of free-living symbionts, which can infect the host. The second and third factors would counter the selection pressure on hosts to specialise; i.e. to form symbioses only with highly effective symbionts, at least under the most commonly encountered environmental conditions. These two latter factors would be expected to counter the selection pressure to specialise in symbionts which confer the greatest most benefit (under a given set of conditions). A.E. Douglas (1998) argues that vertical transmission is a key means of allowing specialisation by the host on effective symbioses.

9. CONCLUSIONS AND PROSPECTS

Much remains to be done in investigating the evolution of cyanobacterial symbioses. Molecular phylogenetic studies will increase our understanding of the relationships of the cyanobacteria and their plastid descendants, and of the relationships of the hosts. Combined with the fossil record, such molecular studies will improve our understanding of when the host taxa and the ancestors of the symbiotic bacteria evolved. More difficult is determination of when the symbioses themselves evolved, but at least a 'not before' date can be found and the numbers of gains and losses of symbionts can be determined (Gargas *et al.*, 1995; Lutzoni *et al.*, 2001; Saldarriaga *et al.*, 2001). Such studies can be related to our increasing understanding of the environmental conditions at particular times in the past, and of the communities and ecosystems at those times. The use of these different kinds of knowledge can help us to understand the possible evolutionary selective factors which favoured cyanobacterial symbioses. Such studies must be related to other knowledge, e.g. that of the role of symbioses in increasing metabolic potential of the host (Douglas, 1996, 1998) and the known increase in the rate of evolution on transition to mutualism (Lutzoni and Pagel, 1997). New molecular and ecological techniques can help to show the advantages of vertical and of horizontal transmission of symbionts, and the origin of specificity.

REFERENCES

Adams, K.L., Daley, D.O., Qiu, Y.L., Wheland, J. and Palmer, J.D. (2000) Repeated, recent and diverse transfers of a mitochondrial gene to the nucleus in flowering plants, *Nature* **408**, 354-357.

Adams, K.L., Ong, H.C. and Palmer, J.D. (2001a) Mitochondrial gene transfer in pieces; fission of the ribosomal protein gene rp1z and partial or complete gene transfer to the nucleus, *Mol. Biol. Evol.* **18**, 2289–2297.

Adams, K.L., Rosenbluth, M., Qiu, Y.L. and Palmer, J.D. (2001b) Multiple losses and transfers to the nucleus of two mitochondrial succinate dehydrogenase genes during angiosperm evolution, *Genetics* **158**, 1289-1300.

Allen, J.F. (1993) Control of gene-expression by redox potential and the requirement for chloroplast and mitochondrial genomes, *J. Theoret. Biol.* **165**, 609-631.

Allen, J.F. and Raven, J.A. (1996) Free-radical-induced mutation *vs* redox regulation: costs and benefits of genes in organelles, *J. Mol. Evol.* **42**, 482-492.

Bhattacharya, D. and Medlin, L. (1995) The phylogeny of plastids: a review based on comparisons of small subunit ribosomal RNA coding regions, *J. Phycol.* **31**, 489-498.

Bhattacharya, D. and Medlin, L. (1998) Algal phylogeny and the origin of land plants, *Plant Physiol.* **31**, 489-498.

Bodyl, A. (1997) Mechanism of protein targetting in the chlorachniophyte plastids and the evolution of complex plastids – a hypothesis, *Botanica Acta* **110**, 395-400.

Brocks, J.J., Logan, G.A., Buick, R. and Summons, R.E. (1999) Archean molecular fossils and the early rise of eukaryotes, *Science* **285**, 1033-1036.

Butterfield, N.J. (2000) *Bangiomorpha pubescens* n.gen, n. sp: implications for the evolution of sex, multicellularity, and the Mesoproterozoic/Neoproterozoic radiation of eukaryotes, *Palaeobiology* **26**, 386-404.

Bythell, J.C., Douglas, A.E., Sharp, V.A., Searle, J.B. and Brown, B.E. (1997) Algal genotypes and photoacclimatory responses of the symbiotic alga *Symbiodinium* in natural populations of the sea anemone *Anemonia viridis*, *Proc. Royal Soc. London Series B – Biological Sciences* **264**, 1277-1282.

Carpenter, E.J., Montoya, J.P., Burns, J., Mulholland, M.R., Subramaniam, A. and Capone, D.G. 1999) Extensive bloom of a N_2-fixing diatom/cyanobacterial association in the tropical Atlantic Ocean, *Marine Ecol.–Progress Series* **185**, 273-283.

Carpenter, E.J. and Janson, S. (2000) Intracellular cyanobacterial symbionts in the marine diatom *Climacodium frauenfeldianum* (Bacillariophyceae), *J. Phycol.* **36**, 540-544.

Cheshire, A.C., Wilkinson, C.R., Seddon, S. and Westphalen, G. (1997) Bathymetric and seasonal changes in photosynthesis and respiration of the phototrophic sponge *Phyllosporongia lamellasa* in comparison with respiration by the heterotrophic sponge *Iarthella basta* on Davies Reef, Great Barrier Reef, *Marine and Freshwater Reseach* **48**, 589-599.

Cheshire, A.C., Butler, A.J., Westphalen, G., Rowland, B., Stevenson, J. and Wilkinson, C.R. (1995) Preliminary study of the distribution and photophysiology of the temperature phototrophic sponge *Cymbastella* sp. from South Australia, *Marine and Freshwater Research* **48**, 1211-1216.

Douglas, A.E. (1996) Symbiotic *Interactions*, Oxford Scientific Publications, University Press, Oxford

Douglas, A.E. (1998) Host benefit and the evolution of specialisation in symbiosis, *Heredity* **81**, 599-603.

Douglas, S.E. (1998) Plastid evolution: origins, diversity trends, *Current Opinions in Genetics and Development* **8**, 655-661.

Gargas, A., Depriest, P.T., Grube, M. and Tehler, A. (1995) Multiple origins of lichen symbioses in fungi suggested by SSU rDNA phylogeny, *Science* **268**, 1492-1495.

Golubic, S. and Lee, S.J. (1999) Early cyanobacterial fossil record: preservation, palaeoenviornments and identification, *Europ. J. Ecol.* **34**, 339-348.

Goodwin, M.S., Whitehead, L.F. and Douglas, A.E. (2001) Symbiotic dinoflagellates in marine Cnidaria: diversity and function, *Hydrobiologia* **461**, 79-82.

Gordon, N., Angel, D.L., Neori, A., Kress, N. and Kimor, B. (1994) Heterotrophic dinoflagellates with symbiotic cyanobacteria and nitrogen limitation in the Gulf of Aqaba, *Marine Ecol.–Progress Series* **107**, 83-88.

Gupta, R.S. (2000). The natural evolutionary relationships among prokaryotes, *Critic. Rev. Microbiol.* **26**, 111-131.

Heinboke, J.F. (1986) Occurrence of *Richellia intracellularis* (Cyanophyta) within the diatoms *Hemiaulus haukii* and *Hemiaulus membranaceus* off Hawaii, *J. Phycol.* **22**, 399-403.

Hirose, P. (2000) Plant rake and algal pouch of the larvae in the tropical ascidian *Diplosoma similis* : an adaptation for vertical transmission of photosynthetic symbionts, *Zoological Science* **17**, 233-240.

Honda, D., Yokota, A. and Sugiyama, J. (1999) Detection of seven major evolutionary lineages in cyanobacteria based in the 16S rRNA gene sequence analysis with new sequences of five marine *Synechococcus* strains, *J. Mol. Evol.* **48**, 723-739.

Honneger, R. (1998) The lichen symbiosis – What is so spectacular about it? *Lichenologist* **30**, 193-212.

Janson, S., Wouters, J., Bergman, B. and Carpenter, E.J. (1999) Host specificity in the *Richelia* – diatom symbiosis by hetR gene sequency analysis, *Environ. Microbiol.* **1**, 431-438.

Kenrick, P and Crane, P.R. (1997) *The Origin and Early Diversification of Land Plants. A Cladistic Study*, Smithsonian Institution Press, Washington.

Kies, L. (1992) Glaucocystophyceae and other protists harbouring procaryotic endocytobionts, in W. reisser (ed.), *Algae and Symbiose,* Biopress Limited, Oxford, pp. 353-377.

Kooistra, W.H.C.F. and Medlin, L.K. (1996) Evolution of the diatoms (Bacillariophyta). Reconstruction of their age from small subunit rRNA coding regions and the fossil record, *Mol. Phylogen. Evol.* **6**, 391-407.

Larkum, A.W.D., Cox, G.C., Hiller, R.G., Parry, D.L. and Dibbaywan, T.P. (1987) Filamentous cyanophytes containing phycourobilins in symbiosis with sponges and an ascidian of coral reefs, *Mar. Biol.* **95**, 1-13.

LaRoche, J., Van der Stacy, G.W.M., Partensky, F., Ducret, A., Aebersold, R., Li, R., Golden, S.S., Hiller, R.G., Wrench, P.M., Larkum, A.W.D. and Green, B.R. (1996) Independent evolution of the prochlorophyte and green plant chlorophyil *a/b* light-harvesting proteins, *Proc. Natl. Acad. Sci. USA* **93**, 15246-15248.

Larsen, J. (1992) Endocytobiotic consortia with dinoflagellate hosts, in W. Reisser ed.), *Algae and Symbioses*, Biopress Limited, Bristol, pp. 427-442.

Lucas, I.A.N. (1991) Symbients of the tropical Dinophysiales (Dinophycaeae), *Ophelia* **33,** 213-224.

Lutzoni, F.M. (1997) Phylogeny of lichen – and non-lichen-forming amphalorid mushrooms and the utility of testing for combinability among multiple data sets, *Systematic Biol.* **46**, 373-406.

Lutzoni, F. and Pagel, M. (1997) Accelerated evolution as a consequence of transitions to mutualism, *Proc. Natl. Acad. Sci. USA* **94**, 11422-11427.

Lutzoni, F., Pagel, M. and Reeb, V. (2001) Major fungal lineages are derived from lichen symbiotic ancestors, *Nature* **411**, 937-940.

Mareira, D., Le Guyaden, H. and Philippe, H. (2000) The origin of red algae and the evolution of chloroplasts, *Nature* **405**, 69-72.

Martin, W. and Schrarrenberger, C. (1997) The evolution of the calvin cycle from prokaryatic to eukaryotic chromosomes: a case study of functional redundancy in ancient pathways through endosymbiosis, *Current Genetics* **72**, 1-18.

Martin,W., Stoebe, B., Goremykin, V., Hansmann, S., Hagesara, M. and Kowallik, K.V. (1998) Gene transfer to the nucleus and the evolution of chloroplasts, *Nature* **393**, 162-165.

Medlin, L.K., Kooistra, W.H.C.F., Gersande, R., Sims, P.A. and Wellbrock, U. (1997) Is the origin of the diatoms related to the end-Permean moss extinction? *Nova Hedwigia* **65**, 1-11.

Miller, R.S., Olmstead, R.G., Adams, K.L., Palmer, J.D., Lao, N.T., Higgie, L., Kavanaugh, T.A., Hibbend, J.M., Giray, J.L., Marden, C.W., Calie, P.J., Jermiin, L.S. and Wolfe, K.H. (2001) Many parallel losses of infA from chloroplast DNA during angiospem evolution from multiple independent transfers to the nucleus, *Plant Cell* **13**, 645-658.

Miyagawa, Y., Tami, M. and Shigeoka, S. (2001) Overexpression of a cyanobacterial fructose-1,6 - /sedoheptulose -1,7 – bisphosphatase in tobacco enhances photosynthesis and growth., *Nature Biotechnol.* **19**, 965-969.

Navarro-Gonzalez, R., Mckay, C.P. and Mvondo, D.N. (2001) A possible nitrogen crisis for Archaean life due to reduced nitrogen fixation by lightning. *Nature* **412**, 61-64

Neilssen, B., Vandepeer, Y., Wilmotte, A. and Deivachter, R. (1995) An early origin of plastids within the cyanobacterial divergence is suggested by evolutionary trees based on complete 16S ribosomal- RNA sequences, *Mol. Biol. Evol.* **12**, 1160-1173.

Noble, D.R., Romanowicz, D.K. and Brown, R.M. (2001) Cellulose in cyanobacteria. Origin of vascular plant cellulase synthase? *Plant Physiol.* **127**, 529 – 542.

Osborne, B.A., Doris, F., Cullen, A. McDonald, R., Campbell, G. and Steen, M. (1991).*Gunnera tintoria* – an unusal nitrogen fixing invader – this water loving species may offer insight into the development of terrestrial plants, *Bioscience* **41**, 224-234.

Palenik, B. and Haselkorn, R. (1992) Multiple evolutionary origins of prochloraphytes, the chlorophyll - b - certaining prokaryotes, *Nature* **355**, 265-267.

Palmer, J.D., Adams, K.L., Cho, Y.R., Parkinson, C.L., Qiu, U.L. and Song, K.M. (2000) Dynamic evolution of plant mitochondrial genomes: mobile genes and introns and highly variable mutation rates, *Proc. Natl. Acad. Sci. USA* **97**, 6960– 6966.

Palmquist, K. (2000) Carbon economy in lichens, *New Phytol.* 148, 11-36.

Pandy, R.L. and Royce, C.L. (1992) Ascidians with algal symbionts, in W Reisser (ed.), *Algae and Symbioses,* Biopress Limited, Bristol, pp 215-230.

Pfannschmidt, T., Nilsson, A. and Allen, J.F. (1999) Photosynthetic control of chloroplast gene expression, *Nature* **397**, 625-628.

Pfannschmidt, T., Allen, J.F. and Oelmullen, R. (2001) Principles of redox control in photosynthetic gene expression, *Physiol. Plant.* **112**, 1-9.

Printzen, C. and Lumbsch, H.T. (2000) Molecular evidence for the diversification of extant lichens in the late Cretaceous and Tertiary, *Mol. Phylogen. Evol.* **17**, 379-387.

Qiu, Y.L., Cho, Y., Cox, J.C. and Palmer, J.D. (1998) The gain of three mitochondrial introns identifies liverworts as the earliest land plants, *Nature* **394**, 671-674.

Race, H.C., Hermann, R.G. and Martin, W. (1999) Why have organelles retained genomes? *Trends Genetics* **15**, 364-370.

Rai, A.N., Söderbäck, E. and Bergman, B. (2000) Cyanobacterium–plant symbioses, *New Phytol.* **147**, 449–481.

Raven, J.A. (1986) Evolution of plant life forms, in T Gwnish (ed.), *On the Economy of Plant Form and Function*, Cambridge University Press, Cambridge, pp 421-492.

Raven, J.A. (1988).The iron and molybdenum use efficiences of plant growth with different energy, carbon and nitrogen sources, *New Phytologist* **109**, 279-287

Raven, J.A. (1990). Predictions of Mn and Fe use efficiences of phototrophic growth as growth and C assimilation pathway. *New Phytol.* **116**, 1-18.

Raven, J.A. (1993) Energy and nutrient acquisition by autotrophic symbioses and their asymbiotic ancesters, *Symbiosis* **14**, 33-60.

Raven, J.A. (1996) Into the voids: The distribution, function, development and maintenance of gas spaces in plants, *Ann. Bot.* **78**, 137-142.

Raven, J.A. (1997) Phagotrophy in phototrophs, *Limnol. Oceanograph.* **42**, 198-205.

Raven, J.A. (1998).Small is beautiful. The picophytoplankton, *Functional Ecol.* **12**, 503-513.

Raven, J.A. (2002a). Selective pressures in stomatal evolution. *New Phytologist* **153**, 371-386

Raven, J.A., Johnston, A.M., Handley, L.L. and McInroy, S.G. (1990) Transport and assimilation of inorganic carbon by *Lichina pygmaea* under emersed and submersed conditions, *New Phytol.* **114**, 407- 417.

Raven, J.A. (2002b) The evolution of cyanobacterial symbioses, *Biol. Environ. Proc. Royal Irish Academ.*, in press.

Raven, J.A., Johnston, A.M., Kübler, J.E., Korb, R., McInroy, S.G., Handley, L.L., Scrimgeour, C.M., Walker, D.I., Beardall, J., Clayton, M.N., Vanderklift, M., Fredrikson, S. and Dunton, K.H. (2002) Algae in the Cold, *Ann. Bot.*, in revision.

Read, D.J., Duckette, J.G., Francis, R., Lignore, R. and Russell, A. (2000) Symbiotic fungal associations in 'lower' land plants, *Philos.l Trans. Royal Soc. London B* **355**, 815-831.

Redecker, D., Morton, J.B. and Bruns, T.D. (2000) Ancestral lineages of arbuscular mycorrhizal fungi (Glomales), *Mol. Phylogen. Evol.* **14**, 276-284.

Renzaglia, K.S., Duff, R.J., Nickrent, D.L. and Garbary, D.J. (2000) Vegetative and reproductive innovations in early land plants: implications for a unified phylogeny, *Philos. Trans. Royal Soc. London B* **355**, 769-793.

Rowan, R. and Knowlton, N. (1995) Intraspecific diversity and ecological zonation in coral-algae symbioses, *Proc. Natl. Acad. Sci. USA* **92**, 2850-2853.

Rujan, T. and Martin,W. (2001) How many genes in *Anabidopsis* come from cyanobacteria? An estimate from 386 protein phylogenes, *Trends Genetics* **17**, 113-120.

Saldarriaga, J.F., Taylor, F.J.R., Keeling, P.J. and Cavalier-Smith, T. (2001) Dinoflagellate nuclear SSU rRNA phylogeny suggests multiple plastid losses and replacements, *J. Mol. Evol.* **53**, 204-213.

Sara, M., Bavestrello, G., Cattaneo-Vietti, R. and Cernano, C. (1998) Endosymbiosis in sponges: relevance for epigenesis and evolution, *Symbiosis* **25**, 57-70.

Sañudo-Wilhelm, S., Kustka, A.B., Gobler, C.J., Hutchins, D.A., Yang, M., Lwiza, K., Burns, J., Capore, D.G., Raven, J.A. and Carpenter, E.J. (2001) Phosphorus limitation of nitrogen fixation by *Trichodesmium* in the Central Atlantic Ocean, *Nature* **411**, 66-69.

Scheiber, J., Krinsley, D and Rieipati, L (2000) Diagenetic origin of quantz mudstone and implications for silica cycling, *Nature* **406**, 981-985.

Schuster, R.M. (1979) The phylogeny of the Hepaticae, G.C.S. Clark and J.G. Duckett (eds.), *Bryophyte Systematics*, Academic Press, London and New York, pp 41-82.

Selosse, M.A., Albert, B.R. and Gatelle, B. (2001) Reducing the genome size of organelles favours gene transfer to the nucleus *Trends Ecol. Evol.* **16**, 135-141.

Smith, D.C. and Douglas, A.E. (1987) *The Biology of Symbiosis*, Edward Arnold, London.

Stiller, J.W. and Hall, B.D. (1997) The origin of red algae: implications for plastid evolution, *Proc. Natl. Acad. USA* **94**, 4220-4525.

Stiller, J.W. and Hall, D.B. (1998) Sequences of the largest subunit of RNA polymerase II from two red algae and their implications for rhodophyte evolution, *J. Phycol.* **34**, 857-864.

Stiller, J.W., Riley, J. and Hall, B.D. (2001) Are red algae plants? A critical evolution of these key molecular data sets, *J. Mol. Evol.* **52**, 527-539.

Summons, R.E., Jahnke, L.L., Hope, J.M. and Logan, G.A. (1999) 2-Methylhaponoids as biomarkers for cyanobacterial oxygenic photosynthesis, *Nature* **400**, 554-557.

Tanaka, A., Ito, H., Tanaka, R., Tanaka, N.K., and Yoshida, K. and Okada, K. (1998) Chlorophyll *a* oxygenase (CAO) is involved in chlorophyll *b* formation from chlorophyll *a*, *Proc. NatlAcad. Sci. USA* **95**, 12719-12723.

Taylor, T.N. and Taylor, E.C. (1994) *The Biology and Evolution of Fossil Evidence*, Prentice-Hall, Englewood Cliffs.

Taylor, T.N., Hass, H. and Kerp, H. (1997) A cyanolichen from the Lower Devonian Rhynie Chert, *Amer. J. Bot.* **84**, 992-1004.

Thumfort, P.P., Atkins, C.A. and Ayzell, D.B. (1994) A re-evaluation of the role of the infected cell in the control of O_2 diffusion in legume modules, *Plant Physiol.* **105**, 1321-1333.

Thumfort, P.P., Layzell, D.B. and Atkins, C.A. (1999) Diffusion and reaction of oxygen in the central tissue of ureide-producing legumes model, *Plant Cell Enviorn.* **22**, 1351-1363.

Thumfort, P.P., Layzell, D.B. and Atkins, C.A. (2000) A simplified approach for modelling diffusion into cells, *J. theoret. Biol.* **204**, 47-65.

Toller, W.W., Rowan, R. and Knowlton, N (2001) Zooxanthellae of the *Montastraea annularis* complex: patterns of distribution of four taxa of *Symbiodinium* on different reefs and across depths, *Biol. Bull.* **201**, 348-359.

Tomitani, A., Okada, K., Miyashita, H., Matthijs, H.C.P., Ohne, T. and Tanka, A. (1999) Chlorophyll *b* and phycolylins in the common ancestor of cyanobacteria and chlorplasts, *Nature* **400**, 159-162.

Turner, S., Huang, T.C. and Chaw, S.M. (2001) Molecular phylogeny of nitrogen–fixing unicellular cyanobacteria, *Bot. Bull. Acad. Sinica* **42**, 181-186.

Urbach, E., Robertson, D.L. and Chisholm, S.W. (1992) Multiple evolutionary origins of prochlorophytes within the cyanobacterial radiation, *Nature* **355**, 267-270.

Usher, K.M., Kuo, J., Fromert, J. and Sutton, D.C. (2001) Vertical transmission of cyanobacterial symbionts in the marine sponge *Chondrilla australiensis* (Desmospongiae), *Hydrobiologia* **461**, 15-23.

Villareal, T.A. (1991) Nitrogen fixation by the cyanobacterial symbiont of the diatom genus *Hemiaulus*, *Mar. Ecol.– Progress Series* **76**, 201-204.

Villareal, T.A. (1994) Widespread occurrence of the *Hemiaulus* – cyanobacterial symbiosis in the Southwest North Atlantic Ocean, *Bull. Mar. Sci.* **54**, 1-7.

Wanntorp, L., Wanntorp, H.E., Oxelman, B. and Kallersjo, M. (2001) Phylogeny of *Gunnera*, *Plant Systematics Evol.* **226**, 85-107.

Wilkinson, C.R. (1992) Symbiotic interactions between marine sponges and algae, in W. Reisser (ed.), *Algae and Symbioses*, Biopress, Bristol, pp 111-152.

Wilkinson, C.R., Summons, R. and Evans, E. (1999) Nitrogen fixation in symbiotic marine sponges: ecological significance and difficulties in detection, *Memoirs Queensland Museum* **44**, 667-673.

Young, J.P.W. (1992) Phylogenetic classification of nitrogen-fixing organisms, in G. Stacey, R.H. Burris and H.J. Evans (eds), *Biological Nitrogen Fixation*, Chapman and Hall, New York, pp 43-86.

Zheng, W.W., Nilsson, M., Bergman, B. and Rasmussen, U. (1999) Genetic diversity and classification of cyanobacteria in different. *Azolla* species by the use of PCR fingerprinting, *Theor. Appl. Genet.* **99**, 1187-1193.

Zerges, W. (2000) Translation in chloroplasts, *Biochimie* **82**, 583-601.

INDEX

A

B

C

H

I

L

M

N

O

P

R

S

T

U

W

X

Z

MIX
Papier aus verantwortungsvollen Quellen
Paper from responsible sources
FSC® C105338

If you have any concerns about our products,
you can contact us on
ProductSafety@springernature.com

In case Publisher is established outside the EU,
the EU authorized representative is:
Springer Nature Customer Service Center GmbH
Europaplatz 3, 69115 Heidelberg, Germany

Printed by Libri Plureos GmbH
in Hamburg, Germany